# FRAGMENTS OF FULLERENES AND CARBON NANOTUBES

# FRAGMENTS OF FULLERENES AND CARBON NANOTUBES
## Designed Synthesis, Unusual Reactions, and Coordination Chemistry

Edited by

**MARINA A. PETRUKHINA**
**LAWRENCE T. SCOTT**

**WILEY**

A JOHN WILEY & SONS, INC., PUBLICATION

Published by John Wiley & Sons, Inc., Hoboken, New Jersey
Published simultaneously in Canada

For general information on our other products and services or for technical support, please contact
our Customer Care Department within the United States at (800) 762-2974, outside the United States
at (317) 572-3993 or fax (317) 572-4002.

Wiley also publishes its books in a variety of electronic formats. Some content that appears in print
may not be available in electronic formats. For more information about Wiley products, visit our
web site at www.wiley.com.

*Library of Congress Cataloging-in-Publication Data:*

Fragments of fullerenes and carbon nanotubes : designed synthesis, unusual reactions, and
coordination chemistry / edited by Marina A. Petrukhina and Lawrence T. Scott.
    p. cm.
  Includes bibliographical references and index.
  ISBN 978-0-470-56908-5 (cloth)
1. Fullerenes. 2. Fragmentation reactions. I. Petrukhina, Marina A. II. Scott, Lawrence T.
  QD181.C1F73 2011
  546'.681–dc22                                                    2010041015

Printed in the United States of America

oBook ISBN: 978-1-118-01126-3
ePDF ISBN:  978-1-118-01122-5
ePub ISBN:  978-1-118-01125-6

10  9  8  7  6  5  4  3  2  1

# CONTENTS

# PREFACE

On January 20, 1966, the *Journal of the American Chemical Society* published a communication by Richard Lawton, a young assistant professor from the University of Michigan, and one of his first graduate students, Wayne Barth, that marked the birth of what has become a vibrant branch of modern chemistry. The bowl-shaped, $C_{20}H_{10}$, polycyclic aromatic hydrocarbon reported for the first time in that communication was given the name "corannulene" by the authors and represented a new class of strained, nonplanar aromatic compounds. Other families of bent benzenoid hydrocarbons were already well known at the time (e.g., cyclophanes and helicenes); however, corannulene was the first "geodesic polyarene," a polycyclic aromatic compound characterized by earthlike curvature at the molecular level, having both convex and concave $\pi$ surfaces.

The minute quantities of corannulene available by Barth and Lawton's 17-step synthesis severely limited the opportunities to explore the novel properties of this marvelous new compound. An X-ray crystal structure confirmed the geodesic shape of the molecule, and preliminary electrochemical and photophysical studies were reported, but the initial flurry of activity quickly subsided as the original supply of material dwindled. A long dormant period followed.

Then suddenly, in 1985, a second geodesic polyarene, buckminsterfullerene ($C_{60}$), was discovered by accident. What novel properties would this new allotrope of carbon have? The concave surface is completely inaccessible to external reagents, but the convex surface is completely exposed. Traditional aromatic substitution reactions are impossible, because there are no hydrogen atoms to replace. Conditions were optimized for generating $C_{60}$ in the gas phase, and many beautiful experiments were conducted under these conditions. Concurrently, scientists eager to study the "wet chemistry" of $C_{60}$ and the higher fullerenes worked feverishly to develop methods for producing isolable quantities of these unique materials; however, success did not come easily. A year passed, then another, and another, and another, but nobody could figure out a way to prepare isolable quantities of $C_{60}$ by the vaporization of graphite.

These tantalizing but frustrating activities in the late 1980s turned inevitably to the question of whether one might be able to synthesize $C_{60}$ and higher fullerenes in the laboratory by rational chemical methods. At the time, the toolbox for synthetic organic chemistry had zero methods for bending planar polycyclic aromatic hydrocarbons into geodesic shapes, but efforts to address these deficiencies were begun.

In 1991, a new synthesis of corannulene that required only six steps (later shortened to just three) was reported by the Scott laboratory. A flood of studies on this archetypal geodesic polyarene soon followed, but, more importantly, the new strategy for synthesizing this curved network of trigonal carbon atoms provided the key to accessing dozens of larger fragments of fullerenes, or "buckybowls," as they have sometimes been called. In these syntheses, flash vacuum pyrolysis (FVP) at $\geq 1000°C$ was employed to distort the synthetic intermediates temporarily away from their preferred planar conformations so that remote portions of the molecules could become joined by $C-C$ bond formation, thereby locking the polyarene in a geodesic shape. The recognition that FVP could create high-temperature conditions in the gas phase that somewhat mimic the conditions under which fullerenes are produced certainly served as inspiration for the examination of FVP as a potential new tool for synthesizing geodesic polyarenes.

In 2002, the Scott laboratory extended this strategy to a 12-step chemical synthesis of $C_{60}$ that relied on FVP in the final step. The synthetic $C_{60}$ was obtained in isolable quantities, uncontaminated by other fullerenes, but the yield was quite low. There is clearly still a long way to go before the routine synthesis of isomerically pure fullerenes "from the ground up" can be considered a solved problem. In the meantime, fullerenes have become available in bulk by the arc vaporization of graphite, by chemical vapor deposition methods, and even by combustion processes. Commercial plants have now been built that can produce metric tons of $C_{60}$ (!), and the literature on fullerenes and materials derived therefrom has grown explosively.

Elsewhere, also in the 1990s, Siegel and his students worked out the first solution-phase method for imposing geodesic curvature on polyarenes. Others soon followed. These methods offer the distinct advantage over FVP that they can be scaled up. Continued refinements and improvements in this area have recently made corannulene available on a kilogram (kg) scale.

The first five chapters of this book describe a range of research in the United States, Israel, and Japan on the synthesis, chemistry, properties, and potential applications of various fullerene fragments and their derivatives. Our intent has been to provide not a comprehensive compilation of everything known in this rapidly expanding field but, rather, to give the reader a broad sampling of work in progress worldwide.

In the fullerene arena, inorganic chemists raced to prepare the first metal complexes of $C_{60}$ as soon as it became available in usable quantities. The first structurally characterized example, $[Pt(PPh_3)_2(\eta^2\text{-}C_{60})]$, was reported by Fagan et al. in 1991.

This was followed by preparation of a great number of *exo* and *endo* transition metal complexes of fullerenes over the two succeeding decades. As corannulene and other fullerene fragments became increasingly available in the 1990s, it was only natural that their transition metal complexes also became targets for synthesis. The Siegel laboratory achieved the first success in 1997, when they isolated and spectroscopically characterized a corannulene complex, $[(C_5Me_5)Ru(\eta^6\text{-}C_{20}H_{10})]^+$. However, it was not until 2004 that Angelici and coworkers finally succeeded in characterizing $\eta^6$ complexes of corannulene by X-ray crystallography for the first time. From their structural work, the dramatic impact that transition metals can have on bowl-shaped ligands was brought to light. In 1998, an unusual aryl–aryl bond breaking at the peripheral five-membered ring of the hemifullerene, a $C_3$-symmetric half of $C_{60}$, was observed in the Rabideau group when they attempted to make a $\pi$ complex by the method that worked on $C_{60}$. This accidental discovery of oxidative addition to $C_{30}H_{12}$ clearly illustrated the different reactivity of buckybowls compared to that of all closed buckyballs. By now, $\sigma$ complexes of nonplanar and strained polyarenes are beginning to attract increased attention, especially due to efforts of the Sharp laboratory.

The first $\pi$ complexes of corannulene were structurally characterized by X-ray crystallography in the Petrukhina laboratory in 2003. Their preparation was based on an original gas-phase deposition approach that proved to be very successful for isolation of crystalline $\eta^2$ complexes of buckybowls. This technique allowed synthesis and structural characterization of the first $\pi$ complexes of several larger bowls, including dibenzo[*a,g*]corannulene, indenocorannulene, and hemifullerene, accomplished in the Petrukhina group.

These complexation studies demonstrated the general preference of the convex face of corannulene for metal coordination. In this regard, the first selective concave coordination reported by Hirao and coworkers in 2007 for sumanene has been a breakthrough. It illustrated that the inside concave carbon faces of buckybowls can also be engaged in metal binding. The reader will find a detailed overview of the more recent progress in and the challenges facing coordination and organometallic chemistry of bowl-shaped polyarenes in Chapters 6–8.

Not long after the carbon arc method for preparing fullerenes was introduced, carbon nanotubes were discovered as unexpected byproducts. Intentional methods for their production were quickly developed, and an appreciation for their phenomenal properties soon spawned countless ideas about their potential uses in nanotechnology. Unfortunately, all of the methods currently known for preparing carbon nanotubes give inseparable mixtures of products that contain nanotubes of different diameters, different lengths, and, most importantly, different orientations of the six-membered rings along the shaft (i.e., different "chiralities"). For most of the proposed applications of carbon nanotubes in molecule-scale electronics, homogeneous supplies of single-chirality nanotubes will be required. To address this need, synthetic organic chemists have begun working toward the development of methods for the rational chemical synthesis of single-chirality, uniform-diameter carbon nanotubes. The final six chapters of this book describe research underway on this topic in laboratories from the United States, Canada, Germany, Switzerland, and Japan.

It is our hope that this book will inspire more chemists to become actively involved with the designed synthesis, unusual reactions, and coordination chemistry of fullerene fragments and carbon nanotubes.

MARINA A. PETRUKHINA

*The University at Albany*

LAWRENCE T. SCOTT

*Boston College*
*July 2010*

# FOREWORD

Before the discovery in 1985 of $C_{60}$, Buckminsterfullerene, the truncated icosahedral cage molecule consisting of sixty carbon atoms arranged in a closed cage structure with the pattern of a modern soccer ball, some prescient scientists, in particular David Jones in the UK and Eiji Osawa in Japan, had thought of the possibility of closed graphene cages, and at least one imaginative and intrepid chemist, Orville Chapman at UCLA, had tried to devise a synthesis. However, when it was discovered, some did not believe it, but other chemists suddenly became aware that a new vista of synthetic chemistry had suddenly opened up. One interesting study, which with hindsight foretold of such a possibility, was the study by Barth and Lawton in 1966, which indicated that the bowl-shaped polyarene corannulene, consisting of a fully-unsaturated five-membered ring surrounded by five hexagonal ones, formed unexpectedly easily in the final step from a mostly saturated precursor. Not all chemists were able to break free from the stranglehold of received misconceptions about how symmetry-based entropy factors and out-of-plane strain factors are involved. Larry Scott was one of the small number of chemists who immediately foresaw the possibility of this new organic synthetic paradigm, and a key breakthrough was his development of a much shorter synthetic route to corannulene.

Thus, no one is more qualified than Larry Scott and his colleague Marina Petrukhina to edit the first detailed compendium of articles that survey the present state-of-the-art of this truly exciting area. The only way in which carbon-based Nanoscience and Nanotechnology can contribute the revolutionary advances promised is if absolutely perfect structure control of the creation of fullerenes and nanotubes can be achieved, and the chapters in this book represent the crucial first steps up this fascinating, exciting, yet demanding ladder.

Scott and Petrukhina have assembled articles from the researchers who have made the key contributions in the construction of "fullerene and nanotube fragments," detailing the current status and future potential of the field. The chapters reflect the intense and rapidly growing interest, specifically focusing on the synthesis, physical properties and reactivity in metal binding and redox reactions, that will lead to future applications. In this compendium we find details of how the "fragments" field has taken seed and grown up as a fundamental component of the fullerene and nanotube steam-roller. The geodesic polyarenes created during this new era have, in turn, spawned a whole new branch of coordination and organometallic chemistry, targeting the unusual reactivity of curved and strained carbon-rich polyarenes and revealing the new rules of structure and bonding between transition metals and non-planar $\pi$-systems.

The fullerenes have opened our eyes to the fundamental value of embedding fully-unsaturated five-membered rings in graphene lattices and opened up previously unrecognized and unexplored "undulating" fields of aromatic hydrocarbon chemistry. We now know that carbon nanotubes offer the possibility of super-strong, ultra-thin materials and indicate how synthetic chemistry could in principle create intrinsically nanostructured materials that approach the ultimate limits of tensile strength and electrical/magnetic behaviour. These advances, however, will only be realized if the daunting task of controlling exact structure-specific synthesis of all types of fullerenes and nanotubes can be achieved. The development of chemical methods for the synthesis of single-chirality, uniform-diameter carbon nanotubes is but one vital first step, and the chapters collected here augur well for this revolution in materials science. The book brings together chemists from all over the world to assemble an in-depth expert perspective on the first steps along the arduous and demanding road to achieve the mouth-watering advances that our present knowledge about the unique behaviour that carbonaceous nanoscale structures offer. The authors tell their intriguing stories of their pioneering efforts in this burgeoning branch of science.

HARRY KROTO

*Florida State University*
*February 2011*

# CONTRIBUTORS

Toru Amaya, Department of Applied Chemistry, Graduate School of Engineering, Osaka University, Japan

Praveen Bachawala, Department of Chemistry, University of Cincinnati, Cincinnati, OH

Anthony P. Belanger, Merkert Chemistry Center, Boston College, Chestnut Hill, MA

Graham J. Bodwell, Department of Chemistry, Memorial University, St. John's, Newfoundland, Canada

Willard E. Collier, Department of Chemistry, Mississippi State University, Mississippi State, MS

Hu Cui, C. Eugene Bennett Department of Chemistry, University of West Virginia, Morgantown, WV

David Eisenberg, Institute of Chemistry and the Lise Meitner-Minerva Center for Computational Quantum Chemistry, The Hebrew University of Jerusalem, Israel

Alexander S. Filatov, Department of Chemistry, University at Albany, State University of New York, Albany, NY

Rainer Herges, Department of Chemistry, Christian-Albrechts University of Kiel, Germany

Toshikazu Hirao, Department of Applied Chemistry, Graduate School of Engineering, Osaka University, Japan

Masahiko Iyoda, Department of Chemistry, Graduate School of Science and Engineering, Tokyo Metropolitan University, Japan

Ramesh Jasti, Department of Chemistry, Division of Materials Science and Engineering, and the Center for Nanoscience and Nanobiotechnology, Boston University, Boston, MA

DEREK R. JONES, Department of Chemistry, University of Cincinnati, Cincinnati, OH

YOSHIYUKI KUWATANI, Department of Chemistry, Graduate School of Science and Engineering, Tokyo Metropolitan University, Japan

JAMES MACK, Department of Chemistry, University of Cincinnati, Cincinnati, OH and Merkert Chemistry Center, Boston College, Chestnut Hill, MA

KATHARINE A. MIRICA, Merkert Chemistry Center, Boston College, Chestnut Hill, MA

YASUSHI MORITA, Department of Chemistry, Graduate School of Science, Osaka University, Japan

TOHRU NISHINAGA, Department of Chemistry, Graduate School of Science and Engineering, Tokyo Metropolitan University, Japan

TOMOHIKO NISHIUCHI, Department of Chemistry, Graduate School of Science and Engineering, Tokyo Metropolitan University, Japan

MARINA A. PETRUKHINA, Department of Chemistry, University at Albany, State University of New York, Albany, NY

MORDECAI RABINOVITZ, Institute of Chemistry and the Lise Meitner-Minerva Center for Computational Quantum Chemistry, The Hebrew University of Jerusalem, Israel

UNIKELA KIRAN SAGAR, Department of Chemistry, Memorial University, St. John's, Newfoundland, Canada

GASTON R. SCHALLER, Department of Chemistry, Christian-Albrechts University of Kiel, Germany

A. DIETER SCHLÜTER, Institute of Polymers, Department of Materials, ETH Zurich, Switzerland

LAWRENCE T. SCOTT, Merkert Chemistry Center, Boston College, Chestnut Hill, MA

PAUL R. SHARP, Department of Chemistry, University of Missouri, Columbia, MO

ROY SHENHAR, Institute of Chemistry and the Lise Meitner-Minerva Center for Computational Quantum Chemistry, The Hebrew University of Jerusalem, Israel

MALTE STANDERA, Institute of Polymers, Department of Materials, ETH Zurich, Switzerland

ANDRZEJ SYGULA, Department of Chemistry, Mississippi State University, Mississippi State, MS

MASAYOSHI TAKASE, Department of Chemistry, Graduate School of Science and Engineering, Tokyo Metropolitan University, Japan

XIA TIAN, Department of Chemistry, Division of Materials Science and Engineering, and the Center for Nanoscience and Nanobiotechnology, Boston University, Boston, MA

AKIRA UEDA, Department of Chemistry, Graduate School of Science, Osaka University, Japan

GANDIKOTA VENKATARAMANA, Department of Chemistry, Memorial University, St. John's, Newfoundland, Canada

KUNG K. WANG, C. Eugene Bennett Department of Chemistry, University of West Virginia, Morgantown, WV

BO WEN, C. Eugene Bennett Department of Chemistry, University of West Virginia, Morgantown, WV

# ACRONYMS*

| | |
|---|---|
| AM1 | Austin method 1 semiempirical molecular orbital calculations |
| ASE | aromatic stabilization energy |
| BP-LED | bipolar longitudinal eddy-current delay |
| BSSE | basis set superposition error |
| CCSD(T) | coupled cluster (method including) singles and doubles (triples) |
| CID | collision-induced dissociation |
| CIP/FIP/SSIP | contact/free/solvent-separated ion pair |
| CNT | carbon nanotube (DWCNT/MWCNT/SWCNT—double-walled/ multiwalled/single-walled CNT) |
| CV | cyclic voltammetry |
| DA | Diels–Alder (RDA—retro-DA) |
| DFT | density functional theory |
| DMT | dynamic molecular tweezers |
| DOSY | diffusion-ordered spectroscopy |
| EA/EI | electron affinity/electron ionization |
| ENDOR | electron–nuclear double resonance |
| EPS | electrostatic potential surface |
| ESI | electrospray ionization |
| ESTN | electronic spin transient nutation |
| EXSY | exchange (NMR) spectroscopy |
| FAB | fast-atom bombardment |
| FMO | frontier molecular orbital |
| FVP | flash vacuum pyrolysis |
| GIAO | gauge-independent atomic orbital |
| GPC | gel permeation chromatography |
| HFCC | hyperfine coupling constant |
| HOMA | harmonic oscillator model of aromaticity |
| ILD | interlayer distance |
| IPR | isolated pentagon rule |
| MALDI-TOF | matrix-assisted laser desorption/ionization–time of flight |

---

* Partial list only; common (LUMO, NMR, UV, etc.) and chem. compd. (COT, DBU, TCNB, etc.) acronyms omitted here.

| | |
|---|---|
| MD/MM | molecular dynamics/mechanics |
| MM2 | molecular mechanics method 2 |
| MP2 | second order Møller–Plesset theory |
| NB | nonbonding |
| NBO | natural bond orbital(s) |
| NICS | nucleus-independent chemical shift |
| NOE | nuclear Overhauser effect (NOESY—NOE spectroscopy) |
| NOON | natural orbital occupation number |
| OFET | organic field effect transistor |
| OLED | organic light-emitting diode |
| PAC | polycyclic aromatic compound |
| PAH | polyaromatic hydrocarbon |
| PES | photoelectron spectroscopy |
| PGSE | pulsed-gradient spin echo (PGStE—pulsed-gradient stimulated echo) |
| POAV | $\pi$-orbital axis vector (analysis) |
| QCISD(T) | quadratic configuration interaction (calculation including) singles and doubles (triples) |
| RCM/ROM | ring-closing/opening metathesis (REM-ring enlargement metathesis) |
| SCF | self-consistent field |
| SIFT | selected-ion flowtube |
| TPA | two-photon absorption |
| VID | valence isomerization and dehydrogenation |
| VT | variable temperature |
| ZINDO | zero intermediate neglect of differential overlap |
| ZTC | zeolite-templated carbon |

# CHAPTER 1

# MOLECULAR CLIPS AND TWEEZERS WITH CORANNULENE PINCERS

ANDRZEJ SYGULA and WILLARD E. COLLIER

## 1-1. INTRODUCTION

Molecular receptors have long intrigued researchers, and many structural motifs have been synthesized. Chen and Whitlock, in 1978, described molecular tweezers designed to selectively bind planar π-conjugated molecules (Figure 1-1).[1] They listed three characteristics molecular tweezers needed in order to enhance complexation: (1) a rigid bridge or spacer preventing self-association of the arms or pincers, (2) a pincer-to-pincer distance sufficient for insertion of the "tweezed" molecule, and (3) a *syn* geometry of the pincers. Chen and Whitlock called attention to the rigid tweezerlike structure of echinomycin, a potent antineoplastic and antibiotic that intercalates into DNA, decades before the design of molecular clips for DNA.

Since 1978, numerous molecular tweezers have been synthesized[2] along with more recent structures termed *molecular clips*.[3] Logically, tweezers require a force to close onto the target molecule and will release the target when the applied force is released. In contrast, clips require a force to open and then, when the force is removed, the clip will close onto the target; thus clips likewise require a force to open and release the target. Several groups are working on dynamic molecular tweezers,[4] but as Harmata has noted, there is currently little difference between molecular clips and tweezers.[5] The distinction between clips and tweezers may soon be realized as more sophisticated structures are synthesized, but for now Harmata's observation remains valid.

Molecular tweezers and clips have the basic design illustrated above: a rigid spacer separating two pincers. Until relatively recently, all the molecular tweezers and clips designed to complex π-conjugated systems have incorporated planar pincers. But are all π-conjugated molecules planar? Why not design molecular clips and tweezers with curved pincers? That might seem a simple task. In reality, molecular clips and

*Fragments of Fullerenes and Carbon Nanotubes: Designed Synthesis, Unusual Reactions, and Coordination Chemistry*, First Edition. Edited by Marina A. Petrukhina and Lawrence T. Scott.

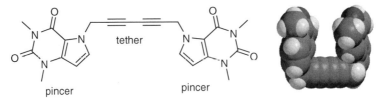

**Figure 1-1.** Chen and Whitlock's molecular tweezers. One of the possible conformations is shown.

tweezers with bowl-shaped pincers resulted from the convergence of a chance discovery that earned a Noble Prize and a 50-year quest to improve synthetic methodologies for a simple molecule. This account chronicles one of the author's contributions to the latter and their synthesis of the first buckycatcher.

### 1-1-1. PAHs: To Curve or Not to Curve?

A planar or near-planar geometry of carbon-based molecular networks is often cited as necessary for the efficient conjugation of π-electronic systems. However, it turns out that since the energy required for modest bending or twisting of polycyclic aromatic hydrocarbons (PAHs) is quite low, the crystal structures of several PAHs show severe distortion from planarity. For example, numerous significantly twisted PAHs retaining their conjugation have been synthesized and characterized,[6] including "the most highly twisted PAH ever prepared" (**1**) with an end-to-end twist of 144° (Figure 1-2).[7] The nonplanarity of **1** and related PAHs is induced by the steric repulsion of substituents (including hydrogen atoms) on the rim of the conjugated core.

In 1966, Barth and Lawton reported the first synthesis of a $C_{20}H_{10}$ hydrocarbon that they named *corannulene* (**2**).[8] This highly nonplanar PAH consists of five benzene rings arranged around the central five-membered "hub" ring and introduces a novel structural motif with two nonequivalent faces of the carbon network: *concave* and *convex* (Figure 1-3). In contrast to the previously mentioned twisted PAHs, the curvature of corannulene is not caused by bulky substituents but instead is a

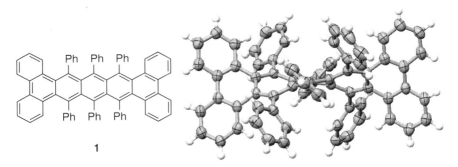

**Figure 1-2.** Pascal's " most highly twisted PAH ever prepared."

convex (exo)

concave (endo)

**2**          **3**

**Figure 1-3.** Corannulene and its structural relationship with fullerene $C_{60}$.

consequence of the incompatibility of abutting pentagons and hexagons in the construction of a planar sheet. Such a combination must result in formation of bowl-shaped surfaces and, in accord with Euler's theorem, if the number of pentagons reaches 12, a closed surface can be created.

In the mid-1980s, fullerenes, a new allotropic form of carbon, were discovered and earned Kroto, Curl, and Smalley a Noble Prize in 1996.[9] Icosahedral $C_{60}$ (**3**), commonly called *buckminsterfullerene* (after R. Buckminster Fuller, 1895−1983, American engineer), is the most studied member of the fullerene family. This nanoscale cage of $sp^2$-hybridized carbon atoms is constructed from six-membered rings along with 12 five-membered rings. Consequently, the corannulene carbon network is amply represented on the surfaces of fullerenes (Figure 1-3). This structural relation renewed interest in corannulene and related curved-surface PAHs, known as "buckybowls" or *fullerene fragments*.[10] There are numerous reasons for studying buckybowls. These bowl-shaped hydrocarbons may serve as models for fullerenes; they are potential precursors for rational syntheses of carbon cages, and, because of the complementary character of the carbon surfaces of buckybowls and fullerenes, there is a potential for host−guest supramolecular chemistry for these systems. An obvious but important observation is that buckybowls offer accessibility to both their convex and concave surfaces, in contrast to fullerenes, which possess an accessible convex surface only.

In this chapter, we will review more recent progress in the synthetic methodologies for buckybowls. These synthetic schemes have generated buckybowls in sufficient quantities to allow their incorporation into larger molecular architectures. We will then review novel synthetic methodologies for the construction of molecular clips and tweezers incorporating buckybowls as pincers. Finally, we will also discuss the potential of these structures to act as molecular clips and tweezers for the guest fullerene cages.

Since supramolecular tweezer−fullerene complexes depend on $\pi$−$\pi$ interactions for molecular recognition, we first briefly review the performance of computational models in description of $\pi$−$\pi$ interactions between conjugated carbon networks. Theoretical predictions of the binding potential of new molecular tweezers and clips will play an increasingly important role in identifying target molecules with desired complexation ability.[11] Therefore, we will examine the available experimental data from synthesized molecular clips and tweezers to assess the reliability of the theoretical models. Understanding the limitations and range of applicability of theoretical methods is imperative if these methods are to guide the synthesis of large molecular tweezers.

## 1-2. $\pi-\pi$ INTERACTIONS IN CURVED SURFACE SYSTEMS

### 1-2-1. $\pi-\pi$ Stacking: Is It Important?

Planar aromatic molecules have long intrigued both experimental and theoretical chemists. The $\pi-\pi$ interactions between various aromatic molecules are especially interesting. The benzene dimer has been scrutinized both experimentally and theoretically as the simplest model of $\pi-\pi$ interactions. Despite its small size, 12 carbon and 12 hydrogen atoms, the benzene dimer is computationally challenging when investigators attempt to use the highest levels of theory {e.g., coupled cluster or quadratic configuration interaction method using single, double and perturbative triple excitations [CCSD(T) or QCISD(T)] with large basis sets} to predict chemically accurate properties.[12] Experimentalists and theoreticians are still not in complete agreement on the structure of the most stable complexes formed by the archetypal aromatic molecule, benzene.[13]

Once corannulene was synthesized and the concept of a curved aromatic surface became accepted, it was natural to ask questions about the strengths and types of interactions between bowl-shaped PAHs. The common packing motifs for planar PAHs are "herringbone" (characterized by edge-to-face interactions) and $\pi-\pi$ stacking. The X-ray crystallography data for corannulene revealed no long-range stacking order.[14] In contrast, the crystal structure of cyclopentacorannulene, the second buckybowl ever synthesized, exhibited a stacked geometry.[15] Since then several other buckybowls have been synthesized, and the X-ray crystal data have revealed that $\pi-\pi$ stacking is a common but not a general feature of the solid state of buckybowls.

The discovery of fullerenes prompted the vision of buckybowl–fullerene complexes since the concave buckybowl face nicely fits the convex fullerene surface. Although mass spectral data revealed the existence of corannulene–fullerene radical cation complexes in the gas phase,[16] no evidence exists for the presence of the neutral complexes in the condensed phases. Some pentakis- and decakis-substituted corannulenes with different arylthioether arms were synthesized and screened by NMR titration experiments for the formation of the substituted corannulene–fullerene complex in solution.[17] Moderate association constants with $C_{60}$ in the range of $60-1400\,M^{-1}$ were reported, confirming some complexation occurring in toluene solutions. While intriguing, these experiments did not prove the existence of strong $\pi-\pi$ interactions between corannulenes and fullerenes. The reported association constants were relatively low and strongly substituent-dependent, so it can be argued that the rim substituent–fullerene interactions account for most of the binding, not $\pi-\pi$ interactions between the bowl and the fullerene. The early investigations indicating that the interactions between corannulene and fullerene were not as strong as expected led to the conclusion that "the attractive force of the concave–convex $\pi-\pi$ interaction is not so significant, if at all."[18]

The first definitive evidence for large $\pi-\pi$ interactions between corannulene and $C_{60}$ was published in 2007, when we described the synthesis of a molecular tweezers with corannulene pincers, dubbed the "buckycatcher."[19] The buckycatcher is shown to have strong $\pi-\pi$ interactions with the fullerene guest both in solution and in the solid state. X-ray crystallographic data show the fullerene molecule located in the center of a doubly concave cleft with most of the corannulene pincer carbon atoms in van der Waals contact with the fullerene cage with the shortest distance of 3.128 Å. An NMR titration experiment gave an association ($K_a$) constant value of $8600 \pm 500\,\text{M}^{-1}$.

A more recent variable-temperature scanning tunneling microscopy (STM) study has provided further evidence for significant $\pi-\pi$ interactions between corannulene and $C_{60}$ by imaging corannulene-$C_{60}$ complexes on a Cu(110) surface.[20] Xiao et al. first vapor-deposited corannulene onto the Cu(110) surface and then vapor deposited the fullerene $C_{60}$. Topographic STM images revealed corannulene-$C_{60}$ complexes, and attempts to relocate the $C_{60}$ away from the corannulene using the STM tip indicated a strongly bound corannulene-$C_{60}$ complex.

## 1-2-2. Theoretical Models of $\pi-\pi$ Interactions

The complexity of the synthesis of molecular tweezers and clips mandates the utilization of computational models to design and identify synthetic targets with the desired $\pi-\pi$ interactions. Unfortunately, the large size of these molecules and their complexes currently precludes using methods such as CCSD(T) or QCISD(T) that approach chemical accuracy. Therefore, it is critical to understand the limitations and advantages of the lower-level methods, such as molecular mechanics, Hartree–Fock (HF), and density functional theory (DFT), in their application to fullerene receptor design.

Since dispersion interactions arise from electron correlation, low-level computational methods cannot be expected to accurately predict binding energies of weakly bonded molecular complexes. Hartree–Fock molecular orbital (MO) methods (both semiempirical and *ab initio*) do not see dispersion interactions at all and, as a result, underestimate van der Waals interactions.

Density functional theory (DFT) is cost-effective, and calculations on large molecular systems can be run in a reasonable time. It has long been known that although *exact* DFT would include all correlation effects (dispersion interactions), the most popular functionals do not.[21] Until 2006, commonly used density functional methods still did not include long-range interactions, which are the basis of dispersion forces.[22] Since 2000 there has been a surge in research aimed at including dispersion interactions in DFT in order to couple the cost efficiency of DFT with the increased chemical accuracy achieved by accounting for dispersion interactions.[23] Numerous methods are now available that allow DFT to account for dispersion interactions.[24] Faced with these competing methods and new ones being added yearly, researchers encounter the dilemma of which method to employ and how accurate it will be. This can be resolved only by understanding each method's range of application and limitations.

We will discuss two methods commonly used to include dispersion interactions. While there may be better methods available, we have experience using these methods and have found them reliable after comparing our theoretical results to our experimental data.

One method of including dispersion interactions in DFT adds an empirically derived dispersion term to the DFT total energy and is known as DFT-D.[25] We have used one of these DFT-D methods, B97-D developed by Grimme,[25a] in our research. The details of our calculations are discussed later. The B97-D method has proved to be in close agreement with the CCSD(T) potential energy curves of several weakly bound dimer systems, including the benzene dimer.[26]

The second avenue, taken by Truhlar and others, is to reparameterize a functional to include dispersion interactions.[27] Truhlar's latest suite of functionals, M06 class, supercede his earlier M05 class. The M06-2X, M05-2X, M06-HF, M06, and M06-L are all recommended for the study of noncovalent interactions. One test compared the binding energy curves for the benzene–methane dimer calculated using 12 density functionals [each with the 6-311+G(2df,2p) basis set] to both MP2/CBS and CCSD (T)/CBS results. The M05-2X, M06-2X, and M06-HF functionals compared well with the CCSD(T)-calculated binding curve.[27a]

A few years ago there were no options for calculating the noncovalent interactions between large molecules. Now there are several, and more are added yearly. It can be confusing and as Truhlar points out, some have started to question the need for the frequent debut of new functionals.[27b] The reality is that while these methods are improving, none give chemically accurate results. The good news is that, until computer power and the computational codes are able to handle large molecules with the highest level of theory, we now have viable and reasonably accurate methods to calculate noncovalent interactions.

Considering the computational limitations, it is not surprising that the influence of buckybowl curvature on buckybowl $\pi$–$\pi$ stacking interactions has not been exhaustively studied by theoretical methods. After all, the model system, the corannulene dimer, possesses 40 carbon atoms and 20 hydrogen atoms, a significant size increase as compared to the benzene dimer. The early study by Tsuzuki considered the convex–convex corannulene dimer as a model for the convex–convex interaction of two $C_{60}$ cages (Figure 1-4). At the MP2/6-311G(2d) level, the binding energy was calculated as 13.4 kcal/mol with the minimum-energy separation of the central five-membered rings of 3.2 Å.[28] More recently, the binding energy obtained with the

| convex-convex | convex-concave | "planar" |

**Figure 1-4.** Conformations of the $\pi$–$\pi$ stacked corannulene dimers.

reparameterized B97-D method of 6.2 kcal/mol at the separation distance of 3.55 Å was mentioned by Peverati and Baldridge without any further discussion of the result.[29] A more detailed study of the stacking interactions of the convex–concave dimer and its "planarized" version (Figure 1-4) was reported by Sygula and Saebo.[30] The best estimate of the binding was 17–18 kcal/mol with the equilibrium distance of 3.64 Å at the SCS-MP2 level of theory with augmented cc-pVTZ basis set. This study also demonstrated that the π–π stacking of curved surfaces is of comparable magnitude as for planar systems of the same size, since the best estimation of the binding in the "planar" dimer (i.e., the dimer in which both corannulene subunits were forced to adopt planar conformations) was 18–20 kcal/mol. However, it has to be pointed out that ~15% of the binding energy in the concave–convex dimer can be attributed to the electrostatic dipole–dipole attraction, which is absent in the "planar" dimer.

Since the dispersion-corrected DFT functionals became available, more computational studies have been reported for even larger systems, including geometry optimization and binding energy calculations for the $C_{60}$@buckycatcher supramolecular complex. These results will be discussed in greater detail later.

While computation[al chemistry can predict properties, eventually the molecule must be synthesized. We start with the half-century pursuit of a simple bowl.

## 1-3. SYNTHESIS OF BUCKYBOWLS

### 1-3-1. Corannulene: From Solution to the Gas Phase and Back

The presence of a pentagonal hub in corannulene induces its bowl shape but also introduces extra strain as compared to planar benzenoid compounds. It is therefore not surprising that the standard synthetic procedures leading to the formation of additional benzene rings in planar PAHs are not successful when applied to the synthesis of buckybowls from the strainless precursors. The original synthesis of corannulene was achieved by a 17-step synthesis starting from acenaphthene with an overall yield of 0.4% (Scheme 1-1).[8] The remaining three six-membered rings were built one at a time around the core five-membered ring to produce a highly hydrogenated intermediate **4**, which was subsequently reduced to corannulene (**2**). Apparently, the main motif of the synthesis was to build up the strain stepwise and as late as possible.

The novelty of **2** attracted considerable attention following the Barth–Lawton communication, but the chemistry and properties of corannulene were not studied

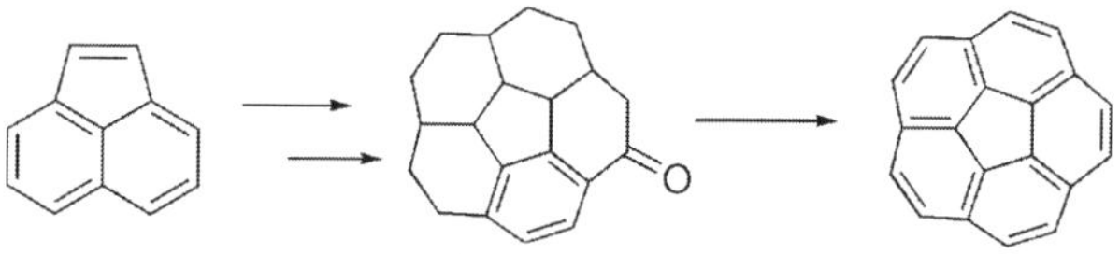

**4**                          **2**

**Scheme 1-1.** Barth–Lawton synthesis of corannulene.

thoroughly because of the very limited supply of this compound. The original synthetic route was simply too lengthy and laborious. A better synthetic route had to be found.

It was recognized early that fluoranthene (**5**) derivatives represent very attractive synthetic precursors to corannulenes, since they possess four of the six rings of **2** already placed in the right locations. Unfortunately, several Friedel–Crafts cyclization attempts using fluoranthene-7,10-diacetic acid (**6**) or the corresponding acid chloride failed to produce any of the expected corannulene derivatives.[31]

**5:** R = H

**6:** R = COOH, COCl

Obviously, the strain associated with the formation of the corannulene core makes the ring closure processes too slow to be competitive with the intermolecular side reactions leading to oligo- and polymeric products.

A major breakthrough was Scott's 1991 flash vacuum pyrolysis (FVP) methodology for the ring-closing step to produce corannulene from 7,10-diethynylfluoranthene **7** (Scheme 1-2).[32] FVP was performed at high temperatures (~1100°C), providing enough thermal energy to overcome the high activation barriers for the intramolecular ring closures. In addition, the double-cyclization step occurs in the gas phase minimizing unwanted intermolecular side reactions. Subsequently, the FVP-based preparation of corannulene was simplified to only three steps from commercially available starting materials with an overall yield of 26%.[33] In this case the more robust bischlorovinyl (**8**) served as an immediate precursor for corannulene. Other groups have applied FVP methodology for the synthesis of **2** from various precursors, but in most cases no significant improvement in overall yield has been achieved.

Numerous buckybowls have been synthesized by the FVP method,[10b] most notably cyclopentacorannulene (**9**, $C_{22}H_{12}$),[34] the first buckybowl larger than corannulene; two isomeric semibuckminsterfullerenes (**10** and **11**),[35] and circumtriindene (**12**, $C_{36}H_{12}$).[36] Ultimately, formation of buckminsterfullerene $C_{60}$ was achieved by FVP of a $C_{60}H_{27}Cl_3$ polyarene precursor.[37]

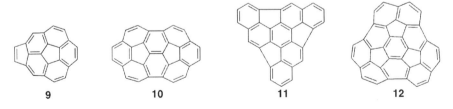

9  10  11  12

Despite FVP's undisputable success in preparation of buckybowls, it has several drawbacks. These include technical difficulties with scaling it up; basically no

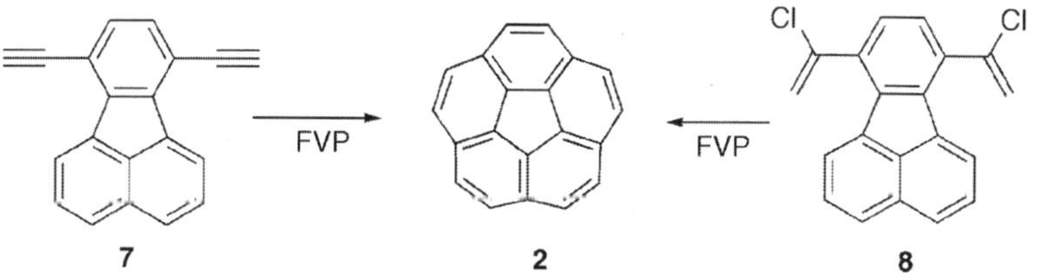

**Scheme 1-2.** Scott's FVP synthesis of corannulene.

functional group tolerance; and dramatically lower yields for larger, less volatile precursors. It soon became clear that practical solution-phase synthetic methodologies needed to be developed to allow for more thorough studies of buckybowls.

In 1996, Siegel reported a solution-phase synthesis of 2,5-dimethylcorannulene (**13**) from tetrabromide (**14**) by low-valence titanium coupling followed by DDQ dehydrogenation with a combined 18% yield for the two steps.[38] Siegel's approach represented just the second successful "wet" synthesis of the corannulene network, 30 years after the original Barth–Lawton work. The success of the formation of the strained corannulene framework was originally attributed to the intermediacy of high-energy organotitanium compounds. Indeed, low-valence titanium and vanadium intramolecular coupling of carbonyl (McMurry coupling)[39] or bromomethyl groups[40] has been known for some time as an effective methodology for the formation of strained ring systems. However, as we will discuss later in greater detail, the major reason for the success of the corannulene formation by Siegel's approach is that the precursor **14** (Scheme 1-3) suffers from severe steric strain caused by the proximity of the bromomethyl and bromoethyl groups in the "bay" regions of the fluoranthene core. The X-ray crystal structures of **14** and its parent hydrocarbon reveal a significant twist of the fluoranthene unit clearly caused by the steric hindrance of the alkyl or bromoalkyl substituents at C1, C6, C7, and C10.[38b]

Following Siegel's success, we tried to apply the same methodology to the synthesis of semibuckminsterfullerene (**10**), which we prepared previously using FVP but in low yield.[35a] Octamethylindenofluoranthene (**15**) was synthesized, and its X-ray crystal structure confirmed its highly nonplanar nature, indicating the presence of significant steric strain caused by the methyl groups located on the rim of the aromatic network.[41] However, numerous attempts using low-valence titanium or vanadium treatment of the octabromide **16** failed to produce any detectable amounts of the expected intramolecular coupling product (Scheme 1-4). In contrast, a respectable 20% yield of **10** was

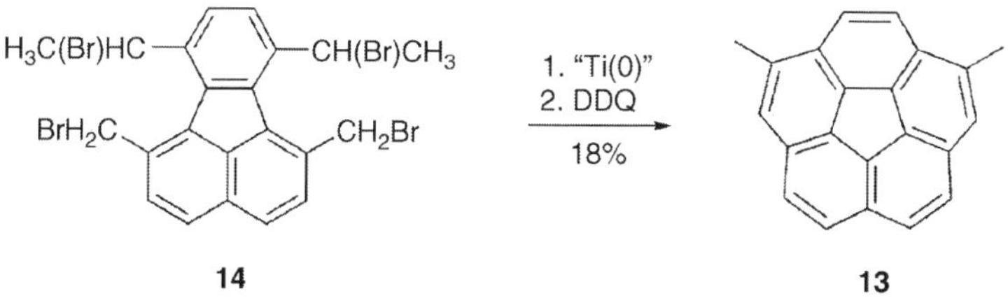

**Scheme 1-3.** Siegel's "wet" synthesis of 2,5-dimethylcorannulene.

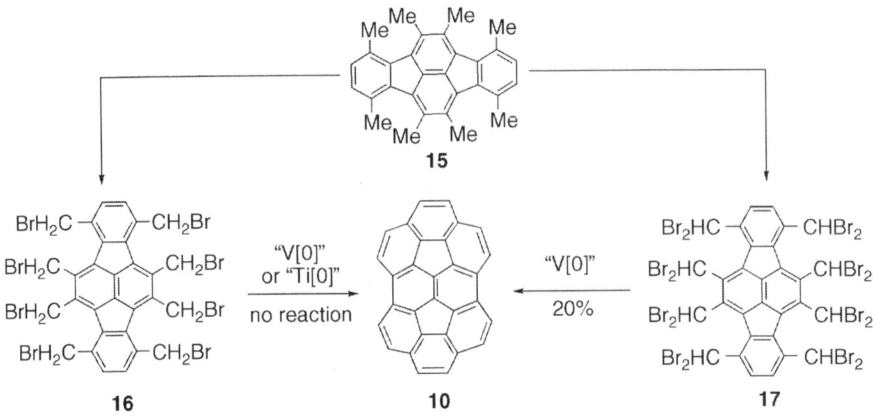

**Scheme 1-4.** Nonpyrolitic synthesis of semibuckminsterfullerene **10**.

achieved by low-valence vanadium treatment of dodecabromide (**17**). We were delighted to find that in contrast to the coupling of bromomethyl groups in **14**, which leads to the formation of a tetrahydrocorannulene core, the coupling of bromomethyl with dibromomethylene groups produces the aromatized product **10** presumably by HBr elimination following the initial ring forming coupling.[41]

The synthesis described above achieved the first solution-phase preparation of a buckybowl larger than corannulene. In addition, it raised the question of whether dibromomethyl groups are more effective precursors than bromomethyl analogs for the coupling leading to formation of the strained ring systems. This was tested by us and independently, by Siegel, on octabromide (**18**), which, on treatment with low-valence vanadium[42] or titanium[43] under high-dilution conditions, produces corannulene in an impressive 70–80% yield (Scheme 1-5).

In an attempt to produce corannulene tetraaldehyde by basic hydrolysis of the dibromomethyl groups in **18**, we serendipitously discovered a novel, low-cost procedure for formation of the corannulene core.[44] The classic hydrolysis method of refluxing **18** in aqueous acetone with $Na_2CO_3$ failed to produce any detectable amount of the expected aldehyde but instead resulted in a carbenoid double ring closure leading to a mixture of brominated corannulenes $C_{20}H_{10-n}Br_n$ ($n = 1–4$) in high yield! This mixture can subsequently be debrominated to produce corannulene in 50–55% yield for the two steps combined. Optimization of the reaction conditions resulted in a simple procedure consisting of a brief 15-min reflux of **18** in aqueous

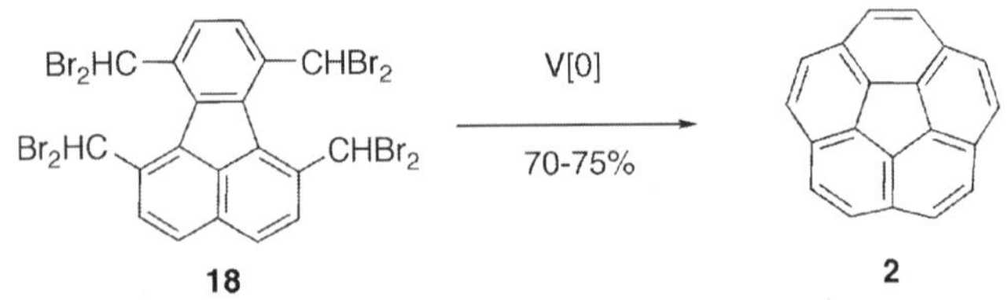

**Scheme 1-5.** High-yielding low-valence vanadium coupling of dibromomethyl groups.

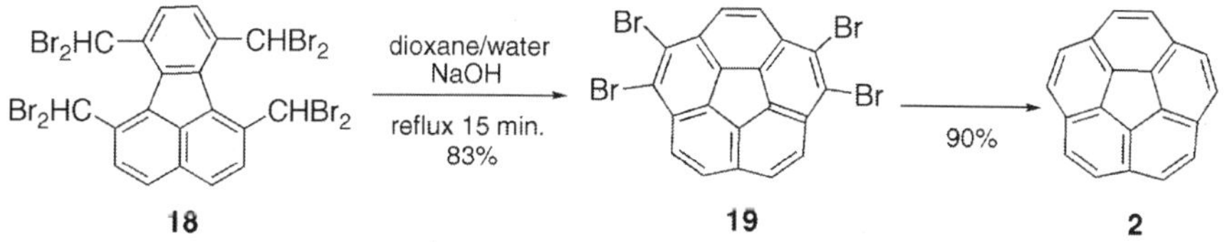

**Scheme 1-6.** Synthesis of the corannulene core by carbenoid coupling of octabromide (**18**).

dioxane with NaOH that produced 1,2,5,6-tetrabromocorannulene (**19**) in 83% isolated yield (Scheme 1-6).

This procedure has several advantages compared to the previous low-valence titanium/vanadium method, including cost, time, and simplicity of the preparation. The previous preparations required high-dilution applications (i.e., a slow, usually overnight addition of the solution of **18** by a syringe pump) and strict anhydrous/oxygen-free conditions (inert-gas atmosphere, dry degassed DME or THF solvents). In contrast, the present preparation can be successfully performed by a first-year undergraduate student assembling a round-bottomed flask with a reflux condenser. The reaction is completed in 15 min, and the separation of the product is trivial since **19** is quite insoluble in common organic solvents, so it is simply filtered from the reaction mixture, washed with water, and dried. The resulting 1,2,5,6-tetrabromocorannulene can be reduced to corannulene in high yield by palladium-catalyzed debromination[45] or used as a starting material for the synthesis of the rim-derivatized corannulenes.[46]

We discovered later that the ring closure of polybrominated systems such as **18** to form the corannulene core can be achieved under milder conditions with metal powders. For example, heating of hexabromide (**20**) with commercial nickel powder in DMF produced 1,2-dicarbmethoxycorannulene (**21**) in 60% yield (Scheme 1-7).[47]

The successful high-yield formation of the corannulene core from benzylbromides of 1,6,7,10-tetramethylfluoranthene under relatively mild conditions may look surprising considering the well-documented fruitless attempts to synthesize corannulene by Friedel–Crafts-type chemistry using 7,10-disubstituted fluoranthenes (e.g., **6**) as precursors.[31] The main reason for the success of the former approach is that a significant amount of strain is built into the precursors (1,6,7,10-tetrasubstituted fluoranthenes), which renders the final intramolecular ring closure favorable, due to the release of the strain. Crystal structure determination proves that both the hydrocarbons (e.g., tetramethylfluoranthene and **15**) as well as their

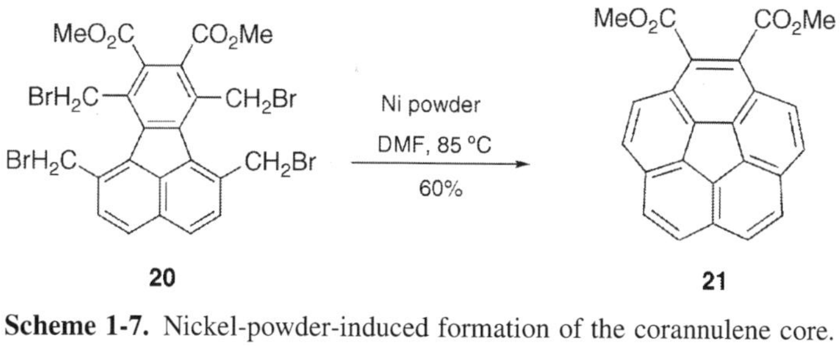

**Scheme 1-7.** Nickel-powder-induced formation of the corannulene core.

brominated derivatives suffer severe steric hindrance of their rim methyl, bromomethyl, and/or dibromomethyl substituents, resulting in a significant twisting of the PAH cores.[38b,48] In contrast, 7,10-disubstituted fluoranthenes can avoid the hindrance by a rotation of the rim substituents away from the hydrogen atoms at C1 and C6. Simple molecular mechanics method 2 (MM2) calculations confirm the explanation that the steric repulsion introduced by the bulky rim substituents raises the energy of the molecule enough to render product formation energetically favorable.[46a]

Availability of corannulene on a gram scale allowed for more systematic studies of its chemistry and for preparation of several derivatives using **2** as a starting material. More recent review articles summarize the progress in the field.[10a,c] Now we will detail the synthetic methodologies for construction of large molecular architectures embedding corannulene units with a potential for the formation of the supramolecular assemblies with various guest molecules, including fullerenes.

### 1-3-2. Corannulene-Based Synthons: 1,2-Didehydrocorannulene and Isocorannulenofuran

The Diels–Alder cycloaddition reaction is often used to construct large conjugated systems from the proper diene and dienophile pair of building blocks. In our pursuit of a practical synthesis of polycyclic aromatic hydrocarbons possessing corannulene subunits, we focused our attention on the previously unknown 1,2-didehydrocorannulene (**22**, a reactive dienophile), and the complementing diene, isocorannulenofuran (**23**). These building blocks promised an easy access to Diels–Alder adducts with corannulene subunits.

|       22       |       23       |

Originally we attempted the preparation of corannulyne by the well-known anthranilic acid route starting from the available dimethylcorannulenodicarboxylate (**21**). However, despite numerous attempts, we failed to find a useful procedure for the conversion of imide (**24**) to antranilic acid (**25**) [Scheme 1-8(a)]. We therefore turned our attention to alternative methods of benzyne generation and discovered that the *ortho*-deprotonation of easily accessible bromocorannulene (**26**) with sodium amide/potassium tertbutoxide in THF resulted in the formation of corannulyne adducts with the trapping dienes and amines [Scheme 1-8(b)].[49] These adducts were isolated with good yields (70–80%) demonstrating the synthetic usefulness of the approach.

It must be noted that the very harsh conditions required for corannulyne generation by the *ortho*-deprotonation procedure seriously limits the scope of this approach to base-resistant trapping agents. We therefore developed an alternative method for the mild generation of **22** from 2-TMS-corannulennyl triflate (**28**).[50]

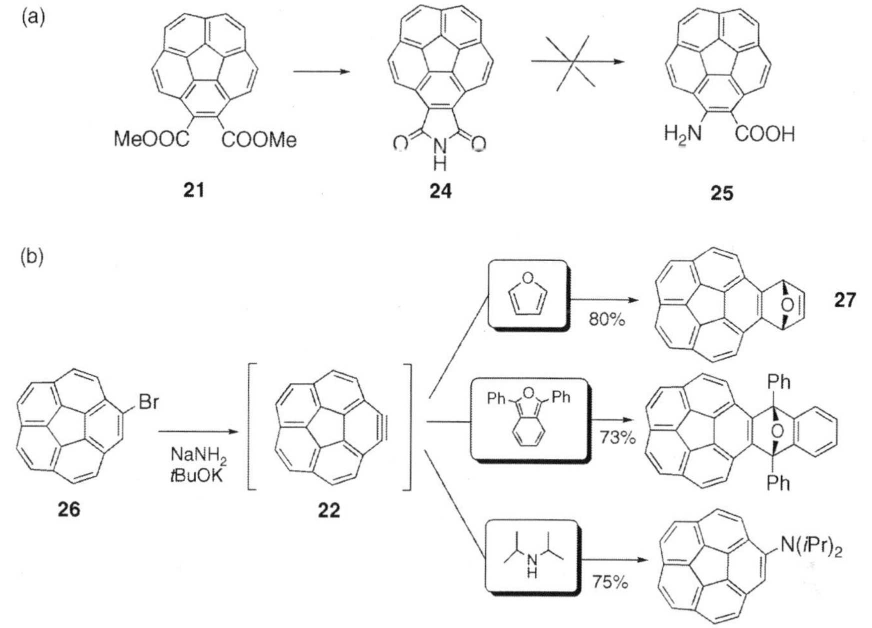

(a)

MeOOC    COOMe

**21**

**24**

H₂N    COOH

**25**

(b)

80%    **27**

73%    Ph ... Ph

75%    N(iPr)₂

**26**    NaNH₂ tBuOK    **22**

**Scheme 1-8.** (a) Unsuccessful attempts to corannulyne by the anthranilic acid route; (b) generation of corannulyne from bromocorannulene.

o-Trimethylsilylaryltriflates have been used as benzyne precursors because the elimination of the TMA and OTf groups can be achieved under very mild conditions by treatment with fluoride anions. Appropriate phenols are typically used as starting materials in the synthesis of these precursors.[51] Unfortunately, hydroxycorannulene was found to be quite unstable and could not be used as a starting material for the synthesis of **28**. We therefore developed an alternative route that started with bromocorannulene (**26**) (Scheme 1-9). Copper-catalyzed etheration of **26** with sodium methoxide led to methoxycorannulene, which was then *ortho*-brominated to produce **29** in good yield. The conversion of **29** to the corresponding o-bromophenol

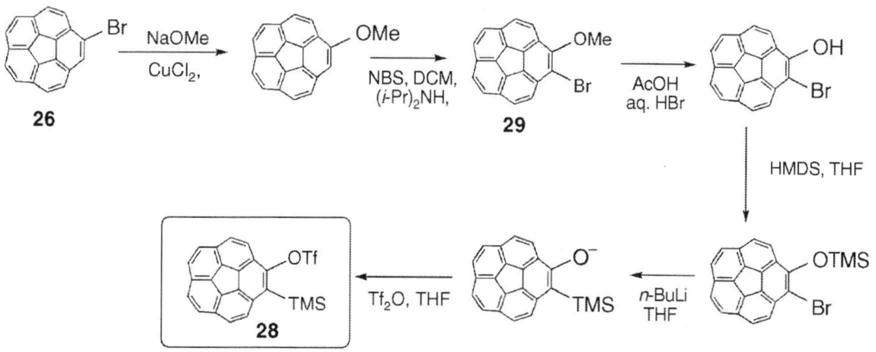

**26**    NaOMe CuCl₂,    OMe    NBS, DCM, (i-Pr)₂NH,    OMe Br    **29**    AcOH aq. HBr    OH Br

HMDS, THF

**28**    OTf TMS    Tf₂O, THF    O⁻ TMS    n-BuLi THF    OTMS Br

**Scheme 1-9.** Synthesis of o-TMS triflate (**28**).

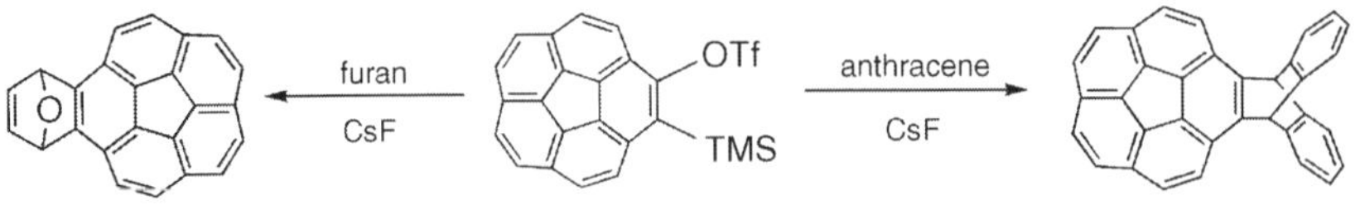

**Scheme 1-10.** Mild generation of corannulyne from **28**.

was achieved by brief reflux in AcOH with aqueous HBr, and the resulting phenol was quickly converted to the *o*-bromotrimethylsilyl ether, which, on metallation with *n*-BuLi at low temperature followed by quenching with triflic anhydride, produced **28** in modest yield of ca. 40% (from **29**). Both *o*-bromophenol and its TMS ether are rather unstable, so they were not purified but quickly converted to the final product **28**. In contrast, *o*-TMS-corannulenyl triflate is stable enough to be stored for weeks with no any evidence of decomposition. Its potential as a precursor for corannulene was tested by trapping experiments with dienes (e.g., furan and anthracene), which produced high yields (78–85%) of the respective adducts. The best results were achieved when the mixture of **26** with the trapping diene in acetonitrile was sonicated at slightly elevated temperatures with two equivalents (2 eq) of CsF.[50]

While trapping corannulyne generated from **28** with furan produces adduct **27** in yields quite similar to that of our original base-induced *o*-deprotonation method, the high-yield formation of anthracene adduct **30**[50] represents an example of the process that we failed to complete by the previous method.[49] This and other examples that we will discuss later prove the versatility of *o*-TMS triflate (**28**) for the synthetically useful generation of 1,2-didehydrocorannulene (Scheme 1-10).

The availability of endoxide (**27**) allows the generation of isocorannulenofuran (**23**) by the well-established *s-tetrazine approach*. Indeed, brief heating of **27** with the commercially available 2,6-bis-2-pyridyl-1,2,4,5-tetrazine provides **23** in an excellent 94% yield (Scheme 1-11).[52]

Isobenzofurans, known for their versatility as reactive dienes in Diels–Alder cycloadditions, are usually too unstable to be isolated, so they are often generated *in situ*. We were pleased to find that **23** is stable enough to be isolated, fully characterized by spectroscopic methods, and stored (as a solid) for several weeks under inert atmosphere with no sign of decomposition. On the other hand, solutions of **23** undergo slow decomposition when exposed to air and/or light, indicating its reactivity. The versatility of **23** in Diels–Alder cycloaddition reactions with benzynes was proved by the formation of endoxides **31–33** with good yields (Scheme 1-11). Deoxygenation of these and related endoxides provides a synthetically useful route to large PAHs with embedded corannulene subunits.[52]

In summary, the synthetic strategy for preparation of large PAHs with corannulene subunits developed in our laboratory is based on employment of three methods:

*Method A*. Introduction of 1,6,7,10-tetramethylfluoranthene unit to the appropriate molecular scaffold followed by a radical (di)bromination of the dangling methyl groups and a subsequent ring closure induced either by treatment with NaOH in aqueous dioxane or by nickel powder in DMF to generate the corannulene core.

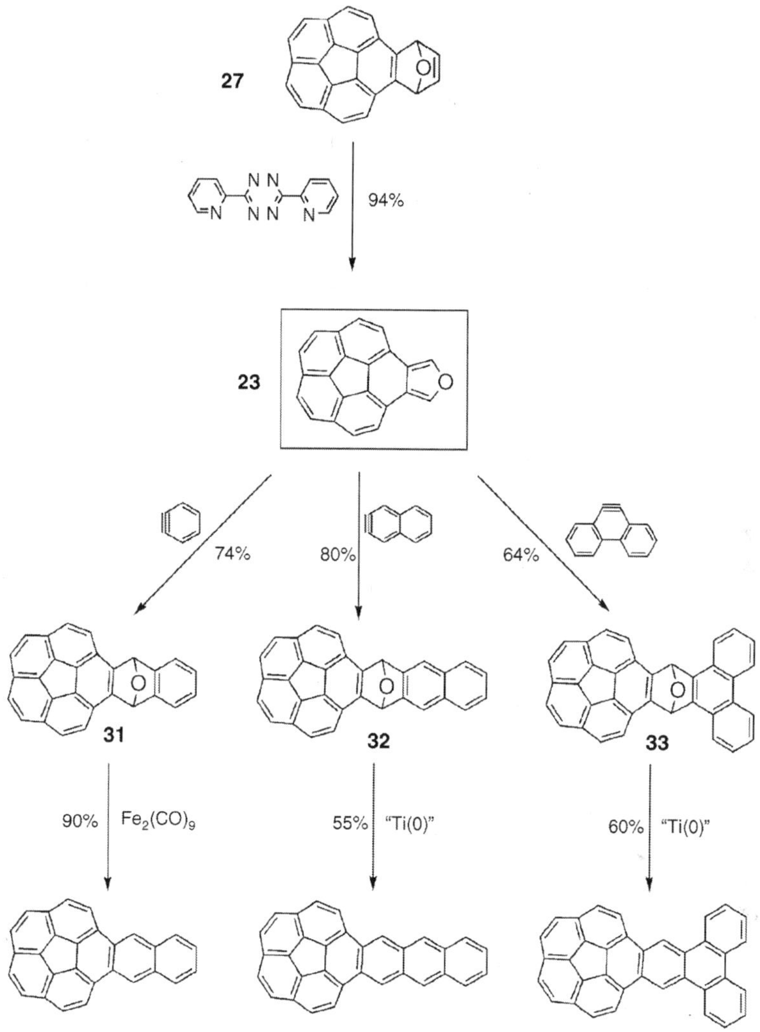

**Scheme 1-11.** Synthesis and Diels–Alder cycloaddition reactions of isocorannulenofuran (**23**).

*Method B.*  Introduction of the corannulene subunit by Diels–Alder cycloaddition
reaction of corannulyne (**22**) [generated from either bromocorannulene (**26**) or
from *o*-TMS triflate (**28**)] with an appropriate diene.

*Method C.*  Introduction of the corannulene subunit by Diels–Alder cycloaddition
reaction of isocorannulenefuran (**23**) with an appropriate dienophile.

The following section describes our successes as well as failures encountered when
applying this strategy to construct highly nonplanar molecular constructs with
corannulene subunits that have the potential to act as molecular receptors for guest
molecules with a special focus on fullerene cages.

## 1-4. TOWARD MOLECULAR CLIPS AND TWEEZERS WITH CORANNULENE PINCERS

### 1-4-1. Twin Corannulene: The First Attempt

The lack of experimental evidence for the formation of concave–convex stacked supramolecular assemblies of corannulenes with fullerenes in solution indicates that the attractive dispersion interaction of a single corannulene bowl with the convex surface of a fullerene cage is insufficient to overcome the expected entropy and solvation penalties associated with the dimeric complex formation. We therefore turned our attention to the design and synthesis of more complex molecular architectures with at least two corannulene pincers attached to a tether in a preorganized way to form a cavity large enough to accommodate $C_{60}$ as a host molecule.

Our first attempt led to the synthesis of "twin corannulene" (**34**), namely, dicorannulenobarrelene dicarboxylate.[53] The synthesis started with cyclopentadienone (**35**) (an intermediate in the corannulene synthesis), which undergoes a double Diels–Alder cycloaddition to 1,2-benzoquinone (Scheme 1-12). This one-pot process afforded target anthraquinone (**36**) in good yield. Quinone (**36**) was subsequently reduced to anthracene (**37**) by aluminum cyclohexide in cyclohexanol. Diels–Alder

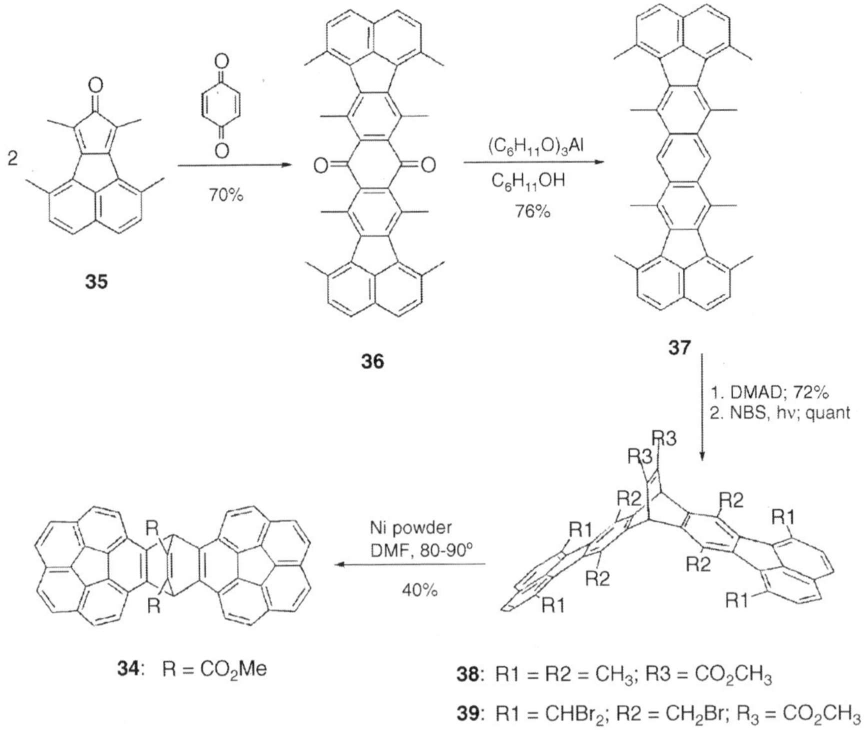

**Scheme 1-12.** Synthesis of "twin corannulene" (**34**).

cycloadditon of dimethylacetylenedicarboxylate to **37** provided difluoranthene barrelene (**38**), which was subsequently brominated under forcing conditions to produce dodecabromo derivative **39**. The bromide underwent fourfold ring closure on treatment with nickel powder to form dicorannulenebarrelene (**34**) in a respectable 40% yield.[53]

Twin corannulene (**34**) represents the first example of a potential molecular clip with two corannulene pincers and a barrelene tether. Three possible conformers of **34** should exist because of the inversion of the corannulene bowls. While the room temperature (rt) $^1$H NMR spectrum of **34** is quite simple, exhibiting only five different proton signals due to the time-averaged $C_{2v}$ symmetry of the molecule, at low temperatures the presence of three different species with approximate populations of 0.82, 0.13, and 0.05 is revealed by the splitting of the carbomethoxymethyl protons (a singlet at 3.80 ppm at rt) into three signals at 4.01, 3.71, and 3.40 ppm at −78°C (Figure 1-5). In

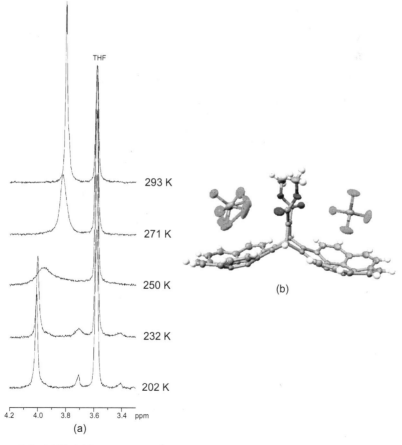

**Figure 1-5.** (a) Variable-temperature $^1$H NMR spectra of the carbmethoxymethyl protons of **34** in THF-$d_8$; (b) X-ray crystal structure of **34** with two CCl$_4$ solvating molecules (one CCl$_4$ molecule is disordered).

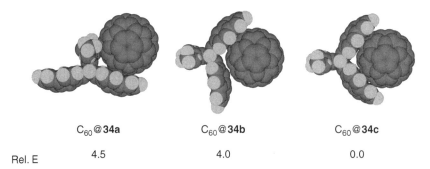

$C_{60}$@**34a**          $C_{60}$@**34b**          $C_{60}$@**34c**

Rel. E          4.5                    4.0                    0.0

**Figure 1-6.** MM2 geometries and relative stabilities of $C_{60}$@**34** assemblies.

line with the experiment, B3LYP/6-311G$^{**}$//B3LYP/3-21G calculations reveal that all three conformers of **34** have quite similar energies with the *exo–exo* conformer (**34a**) predicted to be most stable and the *endo–endo* conformer (**34c**) to be least stable. X-ray analysis shows that **34a** is also preferred in the solid state. Compound **34** incorporates two $CCl_4$ molecules into the crystal lattice, occupying the concave surfaces of the corannulene pincers (Figure 1-5).[53]

The presence of a significant cavity formed by the corannulene pincers permits the intriguing possibility that **34** might act as a host molecule in supramolecular assemblies. The *endo–endo* variant (**34c**) appears to be the best candidate to host a fullerene cage inside the cavity defined by the two concave surfaces of the pincers. Its slightly higher energy, as compared to the other two conformers of **34**, does not eliminate **34c** from the competition since MM2 calculations predict the $C_{60}$@**34c** complex to be more stable than the analogous complexes of $C_{60}$ with either **34a** or **34b** by ∼4 kcal/mol (Figure 1-6). Obviously, the *endo–endo* conformation of **34** allows for more efficient π–π stacking of both concave surfaces of the pincers with the convex surface of $C_{60}$. Because of the relatively low inversion barrier of corannulenes, equilibration of various conformers of **34** is quickly achieved at ambient temperatures. In addition, our MM2 calculations estimate a significant binding energy for the $C_{60}$@**34** complex formation (19.5 kcal/mol). Taking these results into account, we concluded that **34** would be a legitimate candidate for a molecular clip capable of hosting $C_{60}$ in a supramolecular assembly.

Unfortunately, we were not able to detect the complexation of $C_{60}$ by **34** experimentally. Several attempts to cocrystallize the two compounds from various solvents failed to produce any mixed crystals. Also, nuclear magnetic resonance (NMR) experiments did not show any significant changes in the chemical shift of the protons of **34** on titration with $C_{60}$ solutions.

### 1-4-2. A Buckycatcher Enters the Game

Searching for alternative tethers to incorporate into biscorannulene molecular clips, we turned our attention to dibenzocyclooctadiyne (**40**), which is an attractive candidate for the synthesis of highly nonplanar molecular networks. Although this

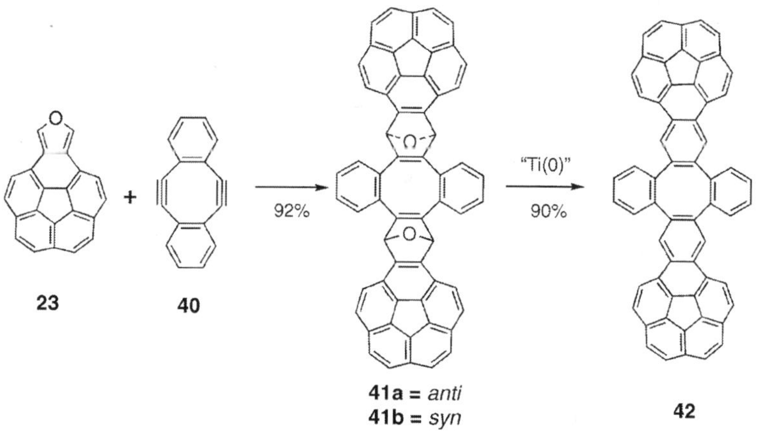

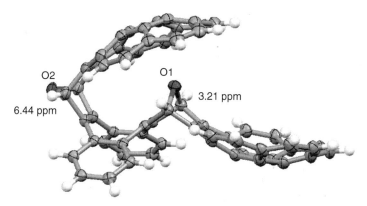

**Scheme 1-13.** Synthesis of buckycatcher $C_{60}H_{28}$.

dienophile is quite reactive toward dienes, it is stable enough to be stored for several weeks. Preparation of **40** is quite straightforward and, in addition, a double-Diels–Alder cycloaddition of an appropriate diene will generate the tetrabenzocyclooctatetraene scaffold, which exhibits some steric flexibility but, at the same time, has a significant inversion barrier keeping the newly generated pincers on the same side of the core.[54] We used isocorannulenofuran **23** as the diene that contributes corannulene subunits to the Diels–Alder adducts.[19] Thus, 2 eq of **23** reacted with dibenzocyclooctadiyne (**40**) to produce a mixture of two isomeric adducts, **41a** and **41b,** with a combined yield of 92% (Scheme 1-13). As judged by their NMR spectra, the major adduct exhibits $C_s$ symmetry, while the minor product has $C_{2v}$ symmetry. The X-ray crystal structure of **41a** confirmed that the major product resulted from the *anti*-bis addition of **23** to diyne (**40**) (Figure 1-7).

**Figure 1-7.** Crystal structure of the major adduct **41a**. A solvating toluene molecule is omitted for clarity. [1]H NMR chemical shifts of the bridgehead hydrogen atoms "inside" and "outside" the upper corannulene unit are shown.

Adduct **41a** exhibits a high degree of nonplanarity; as a result of the anti addition, one of the bridging oxygen atoms (denoted O1 in Figure 1-7) as well as the two bridgehead hydrogen atoms connected to the O1-neighboring carbon atoms are located "under the umbrella" of one of the neighboring corannulene subunits. These protons are strongly magnetically shielded by the corannulene fragment, resulting in an upfield shift (3.21 ppm), while the resonance of the other two bridgehead protons (close to O2) is observed at 6.44 ppm.[19]

The minor adduct **41b** has not been characterized by X-ray crystallography; however, on the basis of its NMR spectra, which indicate $C_{2v}$ symmetry and the typical chemical shifts of the four symmetry equivalent bridgehead hydrogen atoms (6.33 ppm), we conclude that it results from *syn* bis addition of **23** to diyne (**40**).

Deoxygenation of both isomers of **41** was achieved by the low-valent titanium method to produce the target hydrocarbon $C_{60}H_{28}$ (**42**).[19] In analogy to the case of dicorannulenobarrelene (**34**), we anticipated three possible conformations of **42a–42c**, quickly interconverting through bowl-to-bowl inversions of the corannulene subunits. Quite recently we realized that the tether's flexibility makes possible an unrecognized conformer (**42d**), exhibiting internal $\pi-\pi$ stacking of the corannulene pincers.[55] Our recent computational study demonstrates that the $\pi-\pi$ binding energy of the concave–convex corannulene dimer is quite substantial,[30] which could compensate for a significant bending penalty in the deformed TBCOT tether in **42d**. At the MM2 level of theory, **42d** represents a global minimum, more stable by $\sim$2 kcal/mol than the "open" conformers **42a–42c**. The geometries of the four conformers of **42** along with their relative energies calculated at B97-D/TZVP and B97-D/cc-pVQZ levels of theory are shown in Figure 1-8.[55]

The TZVP and cc-pVQZ basis sets are large enough to minimize the basis set superposition error (BSSE), a very important factor for the accurate calculation of binding energies of $\pi-\pi$ stacked systems. As shown previously by Grimme, BSSE accounts for approximately 10% of the dispersion interaction energy at B97-D/TZVP.[56] We have demonstrated that BSSE is almost completely eliminated when the very large cc-pVQZ basis set is used (thus reaching the *complete basis set* limit), so the numbers shown in Figure 1-8 represent the accuracy limit of the B97-D functional.

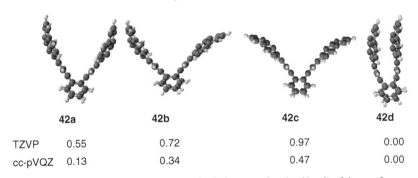

|          | 42a  | 42b  | 42c  | 42d  |
|----------|------|------|------|------|
| TZVP     | 0.55 | 0.72 | 0.97 | 0.00 |
| cc-pVQZ  | 0.13 | 0.34 | 0.47 | 0.00 |

**Figure 1-8.** B97-D calculated structures and relative energies (kcal/mol) of the conformers of buckycatcher **42**.

The stabilities of all four conformations of **42** lie within 0.5 kcal/mol of each other at the highest level of theory applied. Considering the accuracy limitations of the computational theory, it is difficult to declare with certainty which of the isomers is the most stable in the gas phase. However, well-defined but thermally accessible energy barriers separating the conformers bring interesting possibilities of a conformational diversity of **42** in condensed phases since in the presence of a proper complexing solvent or a cosolute, the open conformations may prevail, while in the gas phase or in poorly solvating media, the $\pi-\pi$ stacked **42d** could be favored. Generally, solvation effects will favor the open conformations because of their larger solvent-accesible surfaces as compared to **42d**.

The "closed" **42d** represents an unprecedented conformational motif of the terabenzocyclooctatetraene system induced by an intramolecular $\pi-\pi$ stacking of the corannulene pincers.[55] Thus, **42d** and the open *exo–endo* **42b** equilibrate through a transition state **42TS** (Figure 1-9). At the B97-D/cc-pVQZ level, **42TS** is higher in energy by 4.5 or 4.2 kcal/mol than **42d** or **42b**, respectively. In order to better understand the origins of the unusual "double-dip" bending potential of **42b–42d**, we further analyzed the contributions of the attractive $\pi-\pi$ stacking of the corannulene pincers and the deformation penalties of the dibenzocyclooctatetraene tether to the total potential energy.[55] The strong $\pi-\pi$ stacking of the corannulene pincers in the "closed" conformation **42d** overcomes the energy penalty caused by a significant distortion of the tether. On the other hand, in the open **42b** conformation the relaxation of the tether overcomes the loss of van der Waals attraction of the pincers and the total potential energy reaches another minimum. The origin of a transition state in the "opening" of the tether when going from **42d** to **42b** can then be explained by the loss of the dispersion attraction of the corannulene pincers, which is faster than the relaxation of the TBCOT fragment.

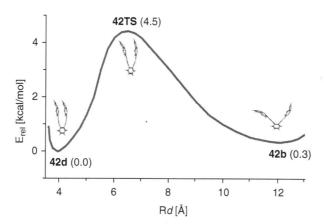

**Figure 1-9.** B97-D/cc-pVQZ "double-dip" bending potential of the *exo–endo* conformations of **42**. $R_d$ is the distance between the two most separated carbon atoms of the central five-membered rings of both corannulene subunits. Relative energies (kcal/mol) of the critical points are shown in parentheses.

### 1-4-3. Buckycatcher Complexes with $C_{60}$ and $C_{70}$: A Good Catch

Three of the four conformations of **42** exhibit large cavities defined by the cor-annulene pincers. *Endo–endo* **42a** appears to be the best candidate to accommodate a molecule of buckminsterfullerene with the two concave surfaces of the pincers available for $\pi-\pi$ stacking with the convex surface of $C_{60}$. The distances between symmetry-related carbon atoms in the hub five-membered rings of the two corannu-lene subunits of the geometry optimized **42a** range from 9.8 to 11.9 Å.[19,55] Since the effective van der Waals radius of $C_{60}$ in its pristine state and in several of its inclusion complexes falls in the range of 9.8–10.2 Å,[57] it is evident that **42a**'s cavity size is compatible with the size of buckminsterfullerene. MM2 calculations predict a binding energy of 24.1 kcal/mol for $C_{60}$@**42a**, significantly higher than the 19.5 kcal/mol calculated for the previously synthesized dicorannulenobarrelene clip **34**. These findings indicated that **42a** should be able to form stable inclusion complexes with fullerenes, in contrast to **34**.

Slow evaporation of an approximately 1 : 1 solution of **42** and $C_{60}$ in toluene produces dark-red crystals.[19] The X-ray crystal structure reveales these crystals to be the inclusion complex $C_{60}$@**42a** with two solvating toluene molecules. There is some disorder in the crystal, which involves the solvating toluene and $C_{60}$. The $C_{60}$ molecule lies on a crystallographic mirror plane, but its molecular mirror plane is rotated out of the crystallographic mirror by approximately $15°$ about a vector joining opposite six-membered ring centroids. Thus, its 60 carbon sites were assigned half-population. Also, the toluene molecule was modeled with two orientations having populations of 0.54 and 0.46. Buckycatcher **42** also lies on the crystallographic mirror plane that relates the corannulene subunits. Both **42** and the solvating toluene molecules wrap around the "equatorial" region of buckminsterfullerene and allow for the columnar arrangement of $C_{60}$ cages along the crystallographic *a* axis with close contacts of the cages in their "polar" regions [Figure 1-10(a)].[19] The buckminster-

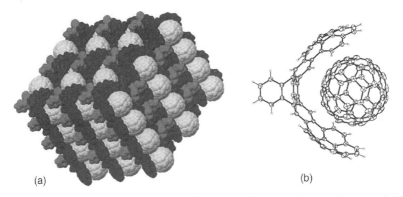

(a)                                        (b)

**Figure 1-10.** (a) Crystal packing pattern of **42** (blue) with $C_{60}$ (gold) and toluene (red). Both disordered orientations of $C_{60}$ and toluene are shown. (b) Ortep representation of the arrangement between **42** and $C_{60}$ in the crystal. Only one of the two disordered orientations of $C_{60}$ is shown.

fullerene separation in the crystal defined by the centroid-to-centroid distance of 9.7730(15) Å is at the lower end of the usual 9.80–10.2 Å range observed for other inclusion complexes of $C_{60}$.

The fullerene molecule is placed in the center of a doubly concave cavity formed by the corannulene pincers of **42a** [Figure 1-10(b)]. Examination of the nonbonding carbon–carbon distances between the corannulene pincers and $C_{60}$ indicates close contacts of the two (with the shortest distance of 3.128 Å), confirming that **42**, indeed, acts as a molecular receptor for $C_{60}$. Any detailed discussion of the host–guest interactions in $C_{60}$@**42a** based on the carbon–carbon distances from the X-ray crystal structure is complicated by the disorder of $C_{60}$. Therefore, we analyzed the geometry relations in greater detail by considering distances between the centroid of the fullerene cage and the carbon atoms of the corannulene pincers.[55] This approach allows for a better understanding of π–π host–guest interactions and renders the observed disorder of $C_{60}$ in the crystal state unimportant. Table 1.1 presents the fullerene centroid–buckycatcher carbon distances, which are averaged over the highest possible $C_{2v}$ symmetry of the corannulene pincers of **42a**, regardless of the supramolecular complex symmetry. The experimental[19] distances are compared to the distances calculated at B97-D/TZVP[55] and by M06-L/MIDI![58] levels. Generally, both computational models reproduce the experimental distances quite well. The calculated intramolecular distances are consistently shorter than the experimental data, with an average deviation of 0.08 Å (B97-D/TZVP) or 0.16 Å (M06-L/MIDI!). On average, B97-D/TZVP reproduces the experimental geometry of the complex slightly better than do the M06-L/MIDI! calculations.

Considering the radius of $C_{60}$ in the complex (3.543 ± 0.035 as found in the crystal) and the van der Waals radius of the $sp^2$-hybridized carbon atom (~1.8 Å), essentially all the carbon atoms of both corannulene pincers are in van der Waals contact with the convex surface of $C_{60}$. The shortest carbon–carbon distance between

**TABLE 1-1.** $C_{2v}$ **Symmetry Averaged Distances (Å) between the Centroid of the Fullerene Cage and the Pincer Carbon Atoms in $C_{60}$@42a**

| Method | C1 | C2a | C3 | C4 | C4a | C5 | C6 | C6a | C10b | C10d | C10e | Ref. |
|--------|------|------|------|------|------|------|------|------|------|------|------|------|
| X-ray | 6.77 | 6.71 | 6.73 | 6.76 | 6.77 | 6.83 | 6.85 | 6.80 | 6.85 | 6.88 | 6.90 | 19 |
| B97-D | 6.68 | 6.63 | 6.65 | 6.68 | 6.69 | 6.73 | 6.74 | 6.71 | 6.80 | 6.82 | 6.84 | 55 |
| M06-L | 6.54 | 6.51 | 6.58 | 6.63 | 6.62 | 6.72 | 6.74 | 6.67 | 6.65 | 6.69 | 6.72 | 58 |

the host $C_{60}$ and the corannulene pincers in the crystal is 3.13 Å, closely reproduced by both computational models (3.09 Å and 3.10 Å by B-97-D[55] and M06-L,[58] respectively). The distance between the two most widely separated carbon atoms of the central five-membered rings of both corannulene subunits ($R_d$, C10c–C10c'), which describes the size of a cleft in between the corannulene pincers in $C_{60}$@**42** complex, is 12.81 Å in the crystal. This crucial distance is also very well reproduced by both computational methods [12.80 (B97-D) and 12.70 Å (M06-L)].

Nuclear magnetic resonance titration experiments prove that the complexation of **42** with buckminsterfullerene also takes place in toluene solution.[19] Systematic downfield changes in the chemical shifts of the four symmetry-independent protons in the corannulene subunits (i.e., at C3, C4, C5 and C6; see Table 1.1) of **42** are observed on addition of $C_{60}$. The maximum changes in their chemical shifts, defined as $\Delta\delta_{max} = \delta_{bound} - \delta_{free}$, are 0.118, 0.134, 0.102, and 0.099 ppm, while the $\Delta\delta_{max}$ of the remaining three symmetry-independent protons of the TBCOT tether are only $-0.06$, $-0.01$, and $-0.01$ ppm. Assuming the existence of a fast (on the NMR timescale) equilibrium between the supramolecular dimer and the monomeric buckycatcher and $C_{60}$, we can write the following equation:

$$K_a = \frac{[C_{60}@\mathbf{42}]}{[C_{60}] * [\mathbf{42}]} \qquad \text{and} \qquad \delta_H = x * \delta_{bound} + (1-x) * \delta_{free}$$

where $x$ describes the mole fraction of the buckycatcher bound in the supramolecular assembly, that is, $x = [C_{60}@\mathbf{42}]/[\mathbf{42}]_{total}$. These relations lead to the following equation:

$$\Delta\delta = \frac{L(1 + K_a * X + K_a * Y) - (L^2 * (1 + K_a * X + K_a * Y)^2 - 4K_a^2 XYL^2)^{1/2}}{2K_a Y}$$

Here, $X = [C_{60}]_{total}$, $Y = [\mathbf{42}]_{total}$, and $L = \Delta\delta_{max}$.

Nonlinear curve fitting of the experimental changes of the chemical shifts of the clip protons as a function of $[C_{60}]_{total}$ and $[\mathbf{42}]_{total}$ gives an estimation of $K_a$ and $\Delta\delta_{max}$, which are treated as parameters. Figure 1-11 illustrates the experimentally determined chemical shift changes of one of the buckycatcher protons as a function of the concentration of $C_{60}$ along with the curve-fitting results. The curve fitting provided four independent evaluations of $K_a$ values in the range of 8000–9300 $M^{-1}$ with the averaged value of $K_a = 8600 \pm 500 \, M^{-1}$.[19]

As we discussed earlier, unsubstituted corannulene does not show any evidence of complexation with $C_{60}$. The same is true for biscorannulenobarrelene **34**, our first potential molecular tweezers with two corannulene pincers. The high value of the association constants for $C_{60}$@**42** complex formation proves, therefore, that the TBCOT scaffold aligns the corannulene pincers properly for the accomodation of fullerene cages in solutions. Not only do the previously reported complexations of *sym*-pentakis(arylthio)corannulenes with $C_{60}$ in toluene ($K_a$ in the range of 60–450 $M^{-1}$)[17a] and decakis(arylthio)corannulene in carbon disulfide ($K_a = 1400 \, M^{-1}$)[17b] have much lower association constants; the majority of the binding can be attributed to the rim substituents interacting with $C_{60}$ in these systems. In the case of

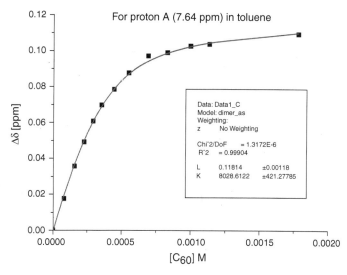

**Figure 1-11.** Chemical shift changes of one of the pincer hydrogen atoms in **42** on $C_{60}$ titration with the nonlinear curve-fitting results.

buckycatcher **42**, the strength of the binding comes from pure dispersion interactions of the convex surface of the fullerene cage with the properly oriented concave surfaces of the corannulene pincers, since no heteroatoms or polar substituents contribute to the binding.

Formation of the supramolecular assemblies of the buckycatcher with fullerenes is strongly influenced by the entropy and solvation penalties. These severe penalties have to be overcome by a very significant binding enthalpy of the supramolecule for a favorable free energy ($\Delta G$) of association. As we discussed previously, accurate assessment of the dispersion energies by the commonly used theoretical models is quite challenging, especially for larger systems. Our report describing the characterization of the $C_{60}$@buckycatcher complex inspired some theoretical studies of the system. The first assessment of the binding energy in $C_{60}$@**42** was reported by Wong, who published the results of a comparative study of the performance of nine popular density functionals on the complex.[59] It is not surprising that seven of the nine functionals failed to predict any binding between $C_{60}$ and the buckycatcher. Only two of the tested functionals—MPWB1K and M05-2X—predict some binding for the complex by 10.2 and 20.7 kcal/mol, respectively. Later, Zhao and Truhlar calculated the binding energy in $C_{60}$@**42** as 26.4 kcal/mol at the M06-2X/DIDZ//M06-L/MIDI! level.[58] Our more recent B94-D/TZVP calculations gave a dramatically higher estimate for the binding energy at ~43 kcal/mol.[55] It has to be pointed out that both M05-2X and M06-2X calculated binding energies were BSSE-corrected, and the corrections were quite substantial, due to the relatively low-quality basis sets employed. In contrast, B97-D/TZVP binding energies were not corrected for BSSE since the error is expected to be lower than 10% of the binding energies at this level of calculation (for a more detailed discussion, see Ref. 56).

The experimental $\Delta G_{293}$ of association of $C_{60}@\mathbf{42}$ in toluene is $-5.3$ kcal/mol, as calculated from $K_a = 8600 \text{ M}^{-1}$, determined by NMR titration experiment.[19] Because of the expected influence of the entropy and solvation on the equilibrium, it is not possible to assess the quality of the computed gas-phase binding energies using the experimentally determined $\Delta G$ value in solution. Zhao and Truhlar estimated the gas-phase free energies of $C_{60}@\mathbf{42}$ formation by the harmonic oscillator–rigid rotator approximation as approximately $-7$ kcal/mol, while their BSSE-corrected M06-2X/DIDZ//M06-L/MIDI! gas-phase energy of association was $-26.4$ kcal/mol.[58] The dramatic difference emphasizes the magnitude of the entropy penalty for the supramolecular $C_{60}@\mathbf{42}$ formation. In addition, the association of buckycatcher with $C_{60}$ in solution results in a considerable loss of the solvent-accessible surfaces on both host and guest molecules. Employment of a simple solvation model predicts solvent-accesible surfaces of 429, 776, and 973 $\text{Å}^2$ for $C_{60}$, buckycatcher, and $C_{60}@\mathbf{42}$ complex, respectively.[58] Clearly, considerable solvent accesible surface is lost in formation of the supramolecular assembly, which further decreases the exergonicity of the association. The combination of the entropy and solvation penalties imposes a requirement for quite high binding energies of the supramolecular assemblies to enable the hosts to act as molecular receptors for the fullerene cages.

On the basis of previously reported studies of similar and even larger systems, we believe that the binding energy of $C_{60}@\mathbf{42}$ calculated using the B97-D functional represents a more accurate estimate than do the previously reported M05-2X or M06-2X results. The M05-2X functional underestimates the binding in the benzene dimer and, while M06-2X performs better, it still significantly underestimates the dispersion energies at intermolecular distances greater than the equilibrium distance.[60] Since several noncovalent carbon–carbon distances in $C_{60}@\mathbf{42}$ are longer than the sum of the van der Waals atomic radii, it is understandable that M06-2X (or other asymptotically incorrect functionals) predict smaller binding energies than do DFT-D methods.

Two related systems exhibiting $\pi$–$\pi$ stacking of bowl-shaped conjugated carbon networks have been studied: the corannulene dimer $(\mathbf{43})$[29,30] and the $C_{60}@$corannulene complex $(\mathbf{44})$.[20] The *exo–endo*-corannulene dimer may be considered as a model system for larger systems exhibiting the convex–concave stacking of curved conjugated carbon surfaces. At the reliable spin-component scaled second order Moller-Plesset (SCS-MP2) level of theory with aug-cc-pVTZ basis set, the best computational estimate of the binding energy in $\mathbf{43}$ is $\sim 17$–$18$ kcal/mol, with the minimum-energy separation of the corannulene monomers of 3.64 Å.[30] B97-D/TZVP calculations gave very similar results (17.0 kcal/mol binding with 3.63 Å separation) confirming the reliability of this computational model.[55]

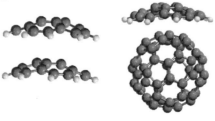

**43**               **44**

Calculations using the MP2 method predicted binding energies for the $C_{60}$@corannulene supramolecule (**44**) of 16.1, 17.5, and 24.7 kcal/mol with 6-31G, 6-31G**, and cc-pVDZ basis sets, respectively.[20] The MP2 method is known to seriously overestimate the dispersion forces at the basis set limit, but it was also observed some time ago that with limited basis sets this method can give satisfactory results, due to the cancellation of the two errors, namely, *overestimation* of the attraction by the MP2 approach and its *underestimation* due to the incomplete basis set.[61] We therefore believe that the reported range of 16–24 kcal/mol represents a reasonable estimate of the binding energy in **44**. B97-D/TZVP predicts the binding energy in **44** as 19.5 kcal/mol, in line with the MP2 results. Again, a significantly lower gas-phase binding energy of 12.4 kcal/mol was calculated for **44** at the M06-2X/DIDZ//M06-L/MIDI! level.[58]

Zhao and Truhlar provided a theoretical explanation for the lack of experimental evidence for the supramolecular assembly between $C_{60}$ and corannulene in solution.[58] Calculations of the entropic contributions provided a positive (+5.4 kcal/mol) gas-phase $\Delta G_{293}$ for the formation of **44**. In solution there is an additional solvation penalty caused by the solvent-accesible surface loss on the complex formation, which will surely increase $\Delta G_{293}$ by a few kcal/mol. In contrast to the $C_{60}$@buckycatcher complex, entropy and solvation penalties in $C_{60}$@corannulene override the significant host–guest dispersion attraction, and these results explain why this supramolecular complex has not been experimentally detected in solution.

### $C_{70}$ *Complexation by Buckycatcher.*
The ultimate goal in the design and synthesis of molecular clips is to achieve their specificity in binding of guest molecules (molecular recognition). Such specificity may lead to numerous applications of the molecular clips in (among others) separation sciences, catalysis, and material sciences. We tested the potential of buckycatcher **42** for molecular recognition of fullerene cages of diffrent sizes and shapes by a detailed computational study of the energetics of its supramolecular assembly with $C_{70}$ (**45**).[55] We considered three orientations of the oblong $C_{70}$ cage inside the cleft of the buckycatcher (Figure 1-12). Thus, in **45a** the $C_{70}$ axis is aligned within the mirror plane relating the two corannulene bowls of the buckycatcher, but somewhat away from the $C_2$ axis of the clip to better fit the concave shape of the pincers. In **45b**, the $C_{70}$ axis is perpendicular to both the $C_2$ axis and $C_s$ planes of symmetry halving the corannulene pincers. Finally, in **45c** the fullerene axis is perpendicular to the $C_2$ axis but is aligned with the aforementioned plane of symmetry of the clip. B97-D/TZVP gas-phase binding energies and $R_d$ distances, defined as distances between the two most separated carbon atoms of the central five-membered rings of the two corannulene subunits (C10e–C10e′), are presented in Figure 1-12. $C_{60}$@**42** complex data are also included in Figure 1-12 for comparison. The $R_d$ value for the unperturbed *endo–endo* conformer **42a** is 11.15 Å, and any change in this distance within the supramolecular assemblies **45a–45c** describes the degree of buckycatcher deformation caused by inclusion of the fullerene cage. Obviously the corannulene tweezers in **42** have to move apart in order to adopt the guest $C_{70}$ or $C_{60}$. However, as pointed out by Zhao and Truhlar,[58] the buckycatcher is quite flexible for modest changes in its tether geometry,

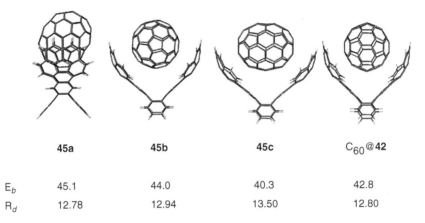

|         | **45a** | **45b** | **45c** | **C$_{60}$@42** |
|---------|---------|---------|---------|----------------|
| $E_b$   | 45.1    | 44.0    | 40.3    | 42.8           |
| $R_d$   | 12.78   | 12.94   | 13.50   | 12.80          |

**Figure 1-12.** B97-D/TZVP minimum-energy geometries of three C$_{70}$@**42** and C$_{60}$@**42** compounds with their gas-phase binding energies and $R_d$ values.

so the clip deformation penalties in **45a**, **45b**, and **45c** are also quite small (1.6, 2.1, and 2.5 kcal/mol, respectively).[55] Apparently **45a** represents the lowest-energy arrangement of C$_{70}$@**42** complex exhibiting the smallest deformation of the clip.

The most important result from the study of C$_{70}$@**42** is that its binding energy is quite similar to that of the C$_{60}$@**42** complex. This is, however, not entirely unexpected since the buckycatcher interacts with roughly a half of both C$_{60}$ and C$_{70}$ surfaces in the inclusion complexes, so $\pi$–$\pi$ stacking interactions should be similar in both cases. The calculated geometries of **42** in both **45a** and C$_{60}$@**42** are very similar, and their degrees of deformation are almost identical, as demonstrated by their calculated $R_d$ distances of 12.78 and 12.80 Å, respectively.[55]

To gain a deeper insight into the origin of the supramolecular binding in the complexes of the buckycatcher, an energy decomposition analysis was performed on C$_{60}$@**42** and C$_{70}$@**42**.[55] This approach separates the contributions to the binding energy of a supramolecule into exchange repulsion ($E_{ER}$), electrostatic ($E_{ES}$), and charge transfer–orbital interaction ($E_{OC}$) energies, which constitute the standard DFT interaction energy ($E_{DFT}$). Addition of the empirical dispersion correction ($E_{disp}$) yields the total interaction energy $E_{tot}$, which differs from the calculated binding energy $\Delta E_b$ by the deformation penalties of both the host and guest molecules. Table 1.2 presents the binding energy decomposition analysis of the buckycatcher complexes with C$_{60}$ and C$_{70}$ performed at the B97-D/TZVP level.[55]

**TABLE 1-2. B97-D/TZVP Binding Energy Decomposition Analysis of C$_{60}$@42 and C$_{60}$@42$^a$**

| Compound | $\Delta E_{ER}$ | $\Delta E_{ES}$ | $\Delta E_{OC}$ | $\Delta E_{DFT}$ | $\Delta E_{disp}$ | $\Delta E_{tot}$ |
|----------|-----------------|-----------------|-----------------|------------------|-------------------|------------------|
| C$_{60}$@**42** | 103.7 | −50.7 | −21.6 | 31.5 | −75.9 | −44.5 |
| C$_{70}$@**42** | 108.9 | −55.4 | −20.6 | 32.9 | −79.7 | −46.8 |

$^a$All contributions are given in kilocalories per mole (kcal/mol).

A comparison of the two complexes reveals no difference in the nature of the binding. In both cases the standard DFT contributions are repulsive but are overridden by the strongly attractive dispersion term. The dispersion contribution to binding is slightly larger in $C_{70}$@**42** as a result of a higher number of van der Waals contacts between the carbon atoms of the carbon cage and the corannulene pincers. In line with our original prediction of a "pure" $\pi$–$\pi$ dispersion character of binding in $C_{70}$@**42**, the energy decomposition analysis classifies both fullerene and buckycatcher supramolecules as typical van der Waals complexes with no specific electrostatic, orbital, or charge transfer interactions.[55] The same conclusion was reached by other authors.[58,59]

***C<sub>70</sub>@42: Experiment.*** Association of the buckycatcher with $C_{70}$ in solution was studied by NMR titration of **42** with $C_{70}$ in toluene-$d_6$ at 293 K.[55] Addition of $C_{70}$ to a solution of **42** causes measurable changes of the chemical shifts of three out of four symmetry-independent protons in the corannulene pincers. Following the procedure described earlier, an association constant of $6800 \pm 400\,M^{-1}$ was estimated. The $K_a$ for $C_{70}$@**42**, although quite high, is lower than the analogous value for $C_{60}$@**42** ($8600 \pm 500\,M^{-1}$). However, both association constants yield very similar free enthalpies of $-5.1$ and $-5.3$ kcal/mol, respectively, proving that the buckycatcher does not exhibit significant specificity in molecular recognition of the two fullerene cages, at least in toluene solutions.

Several batches $C_{70}$@**42** crystals were grown from various solvents, including toluene, $CS_2$, and chlorobenzene. Unfortunately, X-ray crystal structure analysis failed to yield acceptable results, due to crystal defects and/or multiple disorder problems. However, in all cases the presence of 1 : 1 inclusion complexes of $C_{70}$ with buckycatcher was confirmed, along with molecules of the solvents used. Also, in some cases the X-ray data were sufficient to obtain a basic structure of the inclusion complex.[62] As shown in Figure 1-13, the solid-state arrangement of $C_{70}$ inside the

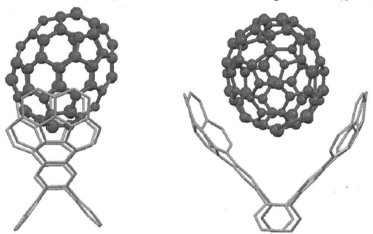

**Figure 1-13.** Two orthogonal views of $C_{70}$@**42**$^*$2PhCl in the crystal. The solvating chlorobenzene molecules are omitted for clarity.

cleft of the buckycatcher closely resembles the lowest-energy structure **45a** calculated at the B97-D/TZVP level.

In conclusion, the buckycatcher proved to be an effective molecular receptor for both $C_{60}$ and $C_{70}$ fullerenes in both solution and the solid state. Both inclusion complexes can be described as typical van der Waals complexes bound primarily by London dispersion forces. The existence of the strongly bonded $\pi$–$\pi$ stacked assemblies such as $C_{60}@\mathbf{42}$ and $C_{70}@\mathbf{42}$ proves the importance of $\pi$–$\pi$ stacking of the curved conjugated carbon networks in supramolecular assemblies.

We are currently studying inclusion complexes of the buckycatcher with other guests, including small polar molecules such as trinitrobenzene or DDQ. Several inclusion complexes have been characterized by X-ray crystallography with both *endo–endo* (**42a**) and *endo–exo* (**42b**) conformations of buckycatcher present. The results of these studies will be published in due course.

## 1-4-4. Potential Clips with Dibenzocyclooctadiene or Dibenzocyclooctatetraene Tethers

A series of highly nonplanar systems with clefts of various sizes and shapes constructed by combining corannulene pincers and cyclooctadiene (COD) or cyclooctatetraene (COT) tethers was synthesized and characterized by X-ray crystal structure determination in our laboratory.[63] The tetrabenzocycloctatetraene tether present in buckycatcher **42** exhibits some flexibility around its minimum-energy tub-shaped conformation but possesses a high barrier for the tub-to-tub inversion and rather limited conformational space. In contrast, the conformational flexibility of COD or COT tethers allows for an adoption of conformations unavailable for the buckycatcher.

Again, the syntheses utilized isocorannulenofuran (**23**) and 2-trimethylsilylcorannulenyltrifluoromethanesulfonate (**28**, a precursor for 1,2-didehydrocorannulene) as synthons (Scheme 1-14).

Synthesis of **46**, a molecular clip with a COT tether and corannulene and benzene pincers, was achieved by a Diels–Alder reaction of **23** with benzocyclobutadiene, generated *in situ* from dibromide **47**, which leads to the formation of two isomeric adducts, *endo*-**48** and *exo*-**48** (~9 : 1). The major adduct was characterized by X-ray crystallography (Figure 1-14), showing a highly nonplanar "closed clam" structure, with the benzene ring of the benzocyclobutane moiety located over the convex side of the corannulene unit. Not surprisingly, a significant shielding of the four hydrogen atoms of the AA′BB′ system of the benzene ring was

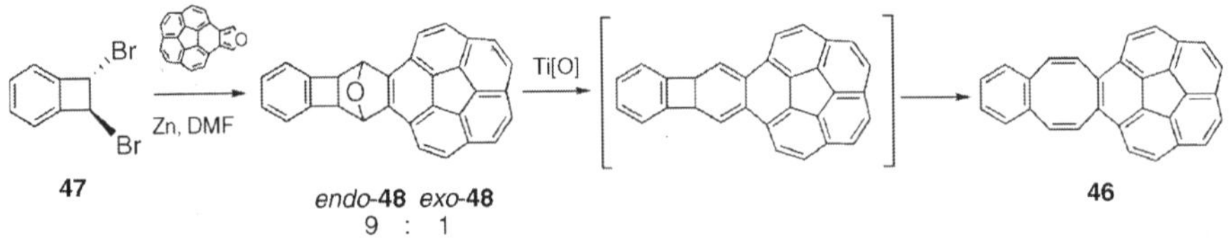

47        *endo*-48 *exo*-48        46
             9  :  1

**Scheme 1-14.** Synthesis of **46**.

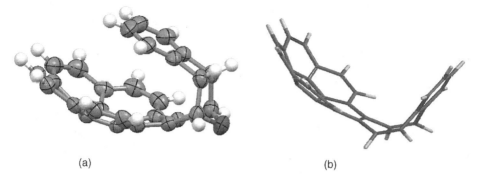

(a)                                                                                          (b)

**Figure 1-14.** (a) Crystal structure of *endo*-**48**; (b) MM2-optimized structure of **46**.

observed with their chemical shifts appearing in the $^1$H NMR spectrum as broad multiplets centered at 5.4 and 5.9 ppm, significantly upfield from the usual chemical shift of aromatic protons. In contrast, the analogous protons in *exo*-**48** appear as sharp AA'BB' multiplets centered in the usual range (7.2 and 7.3 ppm).

*Exo*-**48** and *endo*-**48** were deoxygenated with low-valence titanium to form $C_{30}H_{16}$ hydrocarbon **46**, presumably through the dihydrobiphenylene intermediate (Scheme 1-14). The target **46** was not characterized by X-ray crystallography but is expected to adopt an "open-clam" conformation of a clip, with corannulene and benzene pincers located on the tub-shaped COT tether (Figure 1-14).[63]

Molecular clips with more flexible 1,5-COD subunits were also synthesized (Scheme 1-15).[63]

Again, isofuran (**23**) reacted with an excess of the reactive alkyne generated by dehydrobromination of a mixture of dibromocyclooctadienes, producing a mixture of two major adducts, **49** and **50**. X-ray crystal structure determination of **49** shows an interesting and rather unusual arrangement of the cyclooctadiene ring in which six carbon atoms are almost coplanar and lie in a plane approximately perpendicular

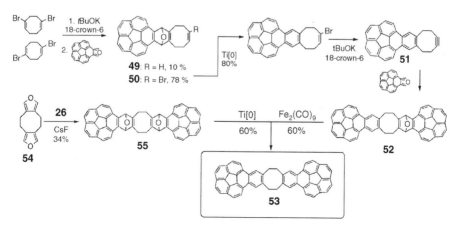

**Scheme 1-15.** Two alternative synthetic routes to **53**.

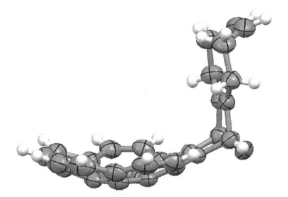

**Figure 1-15.** Crystal structure of the adduct **49**.

to the average plane of the corannulene subunit in an open-clam arrangement (Figure 1-15).

Endoxide (**50**) was deoxygenated and later dehydrobrominated to the reactive dienophile **51**. Diels–Alder cycloaddition of **51** to isofuran **23** lead to endoxide **52**, which, in turn, was deoxygenated with diiron nonacarbonyl to bis(benzocorannulene) cyclooctadiene (**53**), a $C_{52}H_{28}$ molecular clip with a COD tether and two benzocorannulene pincers.[63]

An alternative synthetic route to **53** utilizes the 1,2-didehydrocorannulene precursor **28** and difuran **54**, synthesized in one step from commercially available dipropargyl ether.[64] A double-Diels–Alder cycloaddition yields a mixture of two isomeric products, presumably *syn-* and *anti-***55**, which were reduced with low-valence titanium to produce **53** in a moderate yield of ∼60%.[63]

The $C_{52}H_{28}$ hydrocarbon (**53**) represents a more flexible version of buckycatcher with a COD tether and two benzocorannulene pincers. We have studied its conformational preferences by both MM2 and DFT calculations in comparison with the simpler analog of **53**, 5,6,11,12-tetrahydrodibenzo[*a,e*]cyclooctene (**56**). The conformational space of **56** was previously thoroughly studied by both computational and experimental methods.[65] Three conformers of **56** were located: twist–boat (**56TB**), chair (**56C**) and twist (**56T**) (Figure 1-16). Low-temperature NMR studies reveal the

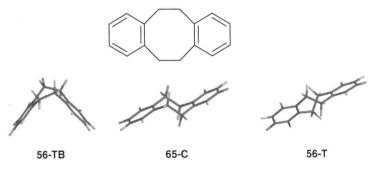

| 56-TB | 65-C | 56-T |

**Figure 1-16.** Three conformations of 5,6,11,12-tetrahydrodibenzo[*a,e*]cyclooctene.

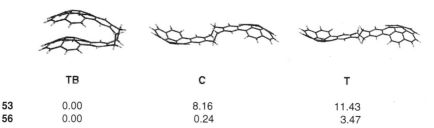

|    | TB | C | T |
|----|------|------|-------|
| **53** | 0.00 | 8.16 | 11.43 |
| **56** | 0.00 | 0.24 | 3.47 |

**Figure 1-17.** B97-D/TZVP calculated minimum-energy conformations of **53** and the relative energies (kcal/mol) of the twist–boat (**TB**), chair (**C**), and twist (**T**) conformers of **53** and **55**.

existence of two conformers, namely, **56TB** and **56C,** in comparable amounts. Accordingly, theoretical studies predicted the energies of the two conformers to be quite similar, with the twist **56T** conformer having significantly higher energy, which precludes its detection by NMR experiments. By analogy to the model system, three conformers of **53** were also considered and optimized at the B97-D/TZVP level. Their geometries and relative energies are presented in Figure 1-17. Relative energies of the three corresponding conformers of the dibenzo analog **56** calculated at the same level of theory are also shown in Figure 1-17 for comparison.[63]

In contrast to **56**, twist–boat conformer of **53** is calculated to be very strongly favored over the remaining **53C** and **53T** by over 8 kcal/mol. The strong preference for **53TB** is also predicted by MM calculations. The presence of the large benzocorannulene pincers located on a flexible cyclooctadiene tether causes a strongly stabilizing intramolecular π–π stacking interaction in the **TB** conformation. The two pincers in **53TB** are clearly arranged to maximize van der Waals attraction with the shortest nonbonding carbon–carbon distances between 3.44 and 3.94 Å.

It has been observed that **53** has a very limited solubility in organic solvents, in contrast to other corannulene-based molecular clips.[63] Its solubility is significantly lower than that of the larger and less flexible buckycatcher **42**. The strong gas-phase preference for the internally π–π stacked **TB** conformation may explain the low solubility of **53** since the solvent-accessible surface of this conformation is dramatically reduced in comparison to the usual "open" conformations of other buckybowls. There is no doubt that **53TB** should prevail in the gas phase or in weakly solvating solvents. Unfortunately, because of the aforementioned low solubility of **53,** low-temperature NMR studies were not attempted. On the other hand, in the presence of strongly solvating solvents the open conformer **53C** may become preferred. Indeed, X-ray crystal structure determination of a 1:1 solvate of **53** with nitrobenzene revels that the chair conformation prevails in the solid state of the solvate (Figure 1-18).[63] The disordered nitrobenzene molecule is encapsulated by two neighboring molecules of **53** and is in van der Waals contacts with the concave sides of the benzocorannulene pincers.

## 1-4-5. Tweezers in a Haystack: Molecular Mechanics as a Screening Tool

Future research will seek to synthesize molecular clips and tweezers that will selectively complex fullerenes and other guest molecules. The classic approach,

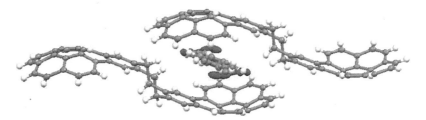

**Figure 1-18.** Crystal structure of **53**·PhNO$_2$ solvate. Several overlapping orientations of the disordered molecules of PhNO$_2$ are shown. The second molecule of **53** is added to demonstrate the encapsulation of nitrobenzene.

which includes (1) selecting tethers and pincers that appear likely to maximize binding of the target guests (fullerenes) with the host, (2) synthesizing numerous molecular clips and tweezers, and then (3) testing their molecular recognition abilities in solution (by spectroscopic titration experiments) and in the solid state (by X-ray crystallography), is a lengthy, tedious, and expensive venture. The rational design of molecular clips and tweezers with desired properties hinges on computational methods capable of calculating supramolecular assembly binding energies with reasonable accuracy.

As we have noted earlier, the new DFT methods with dispersion interactions are capable of reasonable accuracy. Unfortunately, these calculations for large supramolecular assemblies are still quite time-consuming. The even more time-demanding MP2, coupled-cluster or related highly electron-correlated methods, are prohibitively expensive for the large supramolecular systems we plan to synthesize.

We therefore have explored molecular mechanics (MM) methods for a crude but fast screening of the ability to form complexes between the pool of potential molecular clips and tweezers and guest molecules. MM methods suffer from several limitations and drawbacks, but at least they recognize van der Waals nonbonding attractions. Our strategy is to compare the MM results for a series of similar host−guest assemblies to provide useful information about trends rather than supply exact binding energies. To illustrate this approach, we show below the MM2-calculated binding energies of the supramolecular complexes of C$_{60}$ with some molecular clips equipped with corannulene pincers (Figure 1-19).

The MM2-calculated binding energies of the supramolecular dimers studied vary from 19.4 to 25.7 kcal/mol. We can assess these numbers by comparing the experimentally determined association behavior of some of the molecular clips described earlier. Thus, **34** and its simpler analog **57**, both synthesized in our laboratory, failed to exhibit any significant association with C$_{60}$ in solutions, while buckycatcher **42** forms a strong inclusion complex in both solid state and toluene solutions. The difference in their complexing abilities is reflected in the calculated binding energies, predicted to be stronger for C$_{60}$@**42** by ∼5 kcal/mol than in C$_{60}$@**34**. According to these data, **59** is not a strong candidate for an effective receptor for buckminsterfullerene, whereas **58** may show promise. Interestingly, **60**, an unknown analog of **34** that is augmented with two additional benzene rings, may

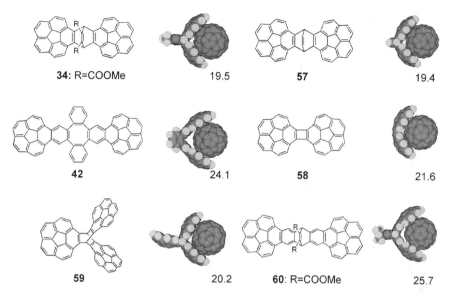

**Figure 1-19.** MM2 structures and binding energies (kcal/mol) of the inclusion complexes of $C_{60}$ with various molecular clips.

be an even better acceptor of $C_{60}$ than buckycatcher **42**. Its binding to $C_{60}$ is predicted to be 1.6 kcal/mol stronger than **42**, and if the MM2 prediction is correct, **60** may exhibit an association constant for complex formation an order of magnitude higher than **34**.

The physical basis for relying on trends in the binding energies is simple. The binding energy is calculated as the energy difference between product and substrate and relates to $\Delta H°$ of the complex formation in the gas phase. As we discussed earlier, two other factors also affect the free energy of association ($\Delta G°$) in solution: the entropy and the solvation effects. Both these factors work against the supramolecular assembly formation but should be similar in a series of related assemblies. Therefore, the larger the binding energy of a supramolecular assembly, the more likely the complex will be observed in condensed phases.

The focus on trends and not individual binding energies opens the door to freedom from only one MM method. Although the binding energies of the supramolecular complexes of $C_{60}$ with our molecular clips calculated with other force fields are quite different from those calculated using MM2, the trends are very similar. For example, Amber 96 calculations give binding energies of 32.1, 37.3, and 39.5 kcal/mol for $C_{60}@\textbf{34}$, $C_{60}@\textbf{42}$, and $C_{60}@\textbf{60}$, respectively, which are significantly higher than the MM2 results. However, the *relative* binding energies of these three supramolecular complexes are very similar with both force fields. Therefore, we conclude that MM should be a useful tool for rapid screening for the best synthetic targets if trends, not the individual binding energies, are examined and compared with available experimental data.

## 1-5. IF YOU BUILD IT, WHAT'S THE USE?

We now possess the synthetic methodology to construct novel molecular clips and tweezers with bowl-shaped pincers. The progress we have detailed in the practical production of these compounds allows for investigation of their properties and for consideration of their possible applications. So what possible uses could buckycatchers have?

Predicting the future uses of any new technology is often an exercise in futility and embarrassment. However, two major uses of buckycatchers that we envision are as stationary phases in liquid chromatography for the separation of fullerenes and as buckycatcher–fullerene complexes in photovoltaic devices.

Fullerenes can now be produced in multiton quantities and are being commercialized in numerous medicinal and industrial applications. There is no agreement on fullerenes as an occupational hazard or ultimately as a consumer or environmental hazard, but more recent studies prove that fullerenes and their derivatives pose potential health risks.[66] The need for chromatographic separation of fullerenes and detection at low concentration will assuredly grow in the coming years.

We have proved that molecular clips and tweezers with buckybowl pincers interact with fullerenes through $\pi$–$\pi$ convex–concave stacking of the curved carbon networks. These clips can be attached to stationary chromatographic phases and will modify the absorption of fullerenes onto the surface and change their elution times. We envision buckycatchers, engineered for specificity, immobilized on silica gel or other porous stationary phases for the separation of fullerene mixtures.

The second major use of buckycatchers is in molecular electronics, specifically photovoltaic devices. The fullerene-hosting abilities of our molecular clips should enable deposition of the guest (fullerene) cages onto a surface if a monolayer of the host buckycatchers can be generated on a solid support. The molecular clips with polar anchors such as **60** (or its dicarboxy analog) are good candidates for successful deposition onto the surface of a conducting support. Once the host monolayer is formed, it will be exposed to a solution of fullerene and, after solvent removal, a double host–guest layer will be formed on the solid support surface (Figure 1-20).

These bilayers are potential light acceptors for photovoltaic devices. Fullerenes and their derivatives have been widely employed in photovoltaic devices since they absorb light very efficiently and, in addition, they are good electron acceptors. Not

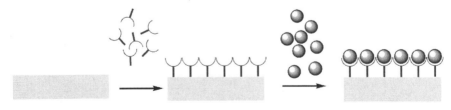

**Figure 1-20.** A cartoon representation of a bilayer formation on a solid support.

surprisingly, our $C_{60}@\mathbf{42}$ has already been studied theoretically for possible charge transfer mediation.[67] This study predicts that the hole transfer from fullerene to another fullerene mediated by **42** is three orders of magnitude faster than the analogous excess electron transfer. Another theoretical study predicts a possibility of a significant modulation of the two-photon absorption (TPA) process in $C_{60}@\mathbf{42}$, thus making it possible to apply these inclusion complexes as TPA active materials.[68]

Other possible uses of buckycatchers include hydrogen storage materials and novel transition metal ligands. Likely, the most important uses of buckycatchers have yet to be imagined. While the future may be uncertain, the long journey to buckycatchers illustrates the sometimes frustrating but ultimately satisfying pursuit of novel organic molecules.

## ACKNOWLEDGMENTS

Support to AS from the Chemical Sciences, Geosciences and Biosciences Division, Office of Basic Energy Sciences, Office of Science, US Department of Energy through Grant DE-FG02-04ER15514 is greatly appreciated. We thank all the students and research associates for their contributions to this project. AS acknowledges a long and fruitful collaboration with Professor Peter W. Rabideau in the area of novel PAHs.

## REFERENCES

1. C.-W. Chen, H. W. Whitlock, Jr., *J. Am. Chem. Soc.* **1978**, *100*, 4921–4922.

2. (a) M. Havlik, V. Král, R. Kaplánek, B. Dolenský, *Org. Lett.* **2008**, *10*, 4767–4769; (b) S. C. Zimmerman, *Top. Curr. Chem.* **1993**, *165*, 71–102.

3. (a) B. Branchi, V. Balzani, P. Ceroni, M. C. Kuchenbrandt, F.-G. Klärner, D. Bläser, R. Boese, *J. Org. Chem.* **2008**, *73*, 5839–5851; (b) P. Polavarapu, H. Melander, V. Langer, A. Gogoll, H. Grennberg, *New J. Chem.* **2008**, *32*, 643–651; (c) R. P. Sijbesma, R. J. M. Nolte, *Top. Curr. Chem.* **1995** *175*, 25–56.

4. (a) M. Skibiński, R. Gómez, E. Lork, V. A. Azov, *Tetrahedron* **2009**, *65*, 10348–10354; (b) T. Nishiuchi, Y. Kuwatani, T. Nishinaga, M. Iyoda, M. *Chem. Eur. J.* **2009**, *15*, 6838–6847.

5. M. Harmata, *Acc. Chem. Res.* **2004**, *37*, 862–873.

6. R. A. Pascal, Jr. *Chem. Rev.* **2006**, *106*, 4809–4819.

7. J. Lu, D. M. Ho, N. J. Vogelaar, C. M. Kraml, R. A. Pascal, Jr., *J. Am. Chem. Soc.* **2004**, *126*, 11168–11169.

8. (a) W. E. Barth, R. G. Lawton, *J. Am. Chem. Soc.* **1966**, *88*, 380–381; (b) W. E. Barth, R. G. Lawton, R. G. *J. Am. Chem. Soc.* **1971**, *93*, 1730–1745.

9. (a) H. W. Kroto, J. R. Heath, S. C. O'Brien, R. F. Curl, R. E. Smalley, *Nature* **1985**, *318*, 162–163; (b) K. M. Kadish, R. S. Rouff (eds.), *Fullerenes: Chemistry, Physics and Technology*, Wiley, New York, 2000.

10. (a) Y.-T. Wu, J. S. Siegel, *Chem. Rev.* **2006**, *106*, 4843–4867; (b) V. M. Tsefrikas, L. T. Scott, *Chem. Rev.* **2006**, *106*, 4868–4884; (c) A. Sygula, P. W. Rabideau, *Carbon Rich*

*Compounds*, M. M. Haley, R. R. Tykwinski, (eds.), Wiley-VCH, Weinheim, 2006, pp. 529–565.

11. J. M. Hermida-Ramón, M. Mandado, M. Sánchez-Lozano, C. M Estévez, *Phys. Chem. Chem. Phys.* **2010**, *12*, 164–169.

12. (a) M. O. Sinnokrot, C. D. Sherrill, *J. Phys. Chem. A* **2006**, *110*, 10656–10668; (b) E. C. Lee, D. Kim, P. Jurecka, P. Tarakeshwar, P. Hobza, K. S. Kim, *J. Phys. Chem. A* **2007**, *111*, 3446–3457; (c) T. Janowski, P. Pulay, *Chem. Phys. Lett.* **2007**, *447*, 27–32.

13. T. F. Headen, C. A. Howard, N. T. Skipper, M. A. Wilkinson, D. T. Bowron, A. K. Soper, *J. Am. Chem. Soc.* **2010**, *132*, 5735–5742.

14. (a) J. C. Hanson, C. E. Nordman, *Acta Cryst. B* **1976**, *B32*, 1147–1153; (b) M. A. Petrukhina, K. W. Andreini, J. Mack, L. T. Scott, *J. Org. Chem.* **2005**, *70*, 5713–5716.

15. A. Sygula, H. E. Folsom, R. Sygula, A. H. Abdourazak, Z. Marcinow, F. R. Fronczek, P. W. Rabideau, *Chem. Commun.* **1994**, 2571–2572.

16. H. Becker, G. Javahery, S. Petrie, P.-C. Chemg, H. Schwarz, L. T. Scott, D. K. Bohme, *J. Am. Chem. Soc.* **1993**, *115*, 11636–11637.

17. (a) S. Mizyed, P. E. Georghiou, M. Bancu, B. Cuadra, A. K. Rai, P. Cheng, L. T. Scott, *J. Am. Chem. Soc.* **2001**, *123*, 12770–12774; (b) P. E. Georghiou, A. H. Tran, S. Mizyed, M. Bancu, L. T. Scott, *J. Org. Chem.* **2005**, *70*, 6158–6163.

18. T. Kawase, H. Kurata, *Chem. Rev.* **2006**, *106*, 5250–5273.

19. A. Sygula, F. R. Fronczek, R. Sygula, P. W. Rabideau, M. M. Olmstead, *J. Am. Chem. Soc.* **2007**, *129*, 3842–3843.

20. W. Xiao, D. Passerone, P. Ruffieux, K. Aït-Mansour, O. Gröning, E. Tosatti, J. S. Siegel, R. Fasel, *J. Am. Chem. Soc.* **2008**, *130*, 4767–4771.

21. S. Kristyán, P. Pulay, *J. Chem. Phys. Lett.* **1994**, *229*, 175–180.

22. E. R. Johnson, G. A. DiLabio, *J. Chem. Phys. Lett.* **2006**, *419*, 333–339.

23. V. R. Cooper, L. Kong, D. C. Langreth, *Phys. Procedia* **2010**, *3*, 1417–1430.

24. E. R. Johnson, I. D. Mackie, G. A. DiLabio, *J. Phys. Org. Chem.* **2009**, *22*, 1127–1135.

25. (a) S. Grimme, *J. Comput. Chem.* **2006**, *27*, 1787–1799; (b) S. Grimme, J. Antony, T. Schwabe, C. Mück-Lichtenfeld, *Org. Biomol. Chem.* **2007**, *5*, 741–758.

26. Á. Vázquez-Mayagoitia, C. D. Sherrill, E. Aprà, B. G. Sumpter, *J. Chem. Theory Comput.* **2010**, *6*, 727–734.

27. (a) Y. Zhao, D. G. Truhlar, *Acc. Chem. Res.* **2008**, *41*, 157–167; (b) Y. Zhao, D. G. Truhlar, *Theor. Chem. Acc.* **2008**, *120*, 215–241.

28. S. Tsuzuki, T. Uchimaru, K. Tanabe, *J. Phys. Chem. A* **1998**, *102*, 740–743.

29. R. Peverati, K. K. Baldridge, *J. Chem. Theory Comput.* **2008**, *4*, 2030–2048.

30. A. Sygula, S. Saebø, *J. Int. Quant. Chem.* **2009**, *109*, 65–72.

31. (a) J. T. Craig, M. D. W. Robins, Aust. *J. Chem.* **1968**, *21*, 2237–2245; (b) R. H. Jacobson, PhD dissertation, Univ. California, Los Angeles, **1986**; (c) J. R. Davy, M. N. Iskander, J. A. Reiss, *Tetrahedron Lett.* **1978**, 4085–4088.

32. L. T. Scott, M. M. Hashemi, D. T. Meyer, H. B. Warren, *J. Am. Chem. Soc.* **1991**, *113*, 7082–7084.

33. L. T. Scott, P.-C. Cheng, M. M. Hashemi, M. S. Bratcher, D. T. Meyer, H. B. Warren, *J. Am. Chem. Soc.* **1997**, *119*, 10963–10968.

34. (a) A. H. Abdourazak, A. Sygula, P. W. Rabideau, *J. Am. Chem. Soc.* **1993**, *115*, 3010–3011; (b) A. Sygula, A. H. Abdourazak, P. W. Rabideau, *J. Am. Chem. Soc.* **1996**, *118*, 339–343.

35. (a) P. W. Rabideau, A. H. Abdourazak, H. E. Folsom, Z. Marcinow, A. Sygula, R. Sygula, *J. Am. Chem. Soc.* **1994**, *116*, 7891 7892; (b) A. H. Abdourazak, A. Z. Marcinow, A. Sygula, R. Sygula, P. W. Rabideau, *J. Am. Chem. Soc.* **1995**, *117*, 6410–6411.

36. (a) L. T. Scott, *Angew. Chem. Int. Ed.* **2004**, *43*, 4994–5007; (b) L. T. Scott, M. S. Bratcher, S. Hagen, *J. Am. Chem. Soc.* **1996**, *118*, 8743–8744.

37. L. T. Scott, M. M. Boorum, B. J. McMahon, S. Hagen, J. Mack, J. Blank, H. Wegner, A. de Meijere, *Science* **2002**, *295*, 1500–1503.

38. (a) T. J. Seiders, K. K. Baldridge, J. S. Siegel, *J. Am. Chem. Soc.* **1996**, *118*, 2754–2755; (b) A. Borchardt, K. Hardcastle, P. Gantzel, J. S. Siegel, *Tetrahedron Lett.* **1993**, *34*, 273–276.

39. J. E. McMurry, *Chem. Rev.* **1989**, *89*, 1513–1524 and references cited therein.

40. G. A. Olah, G. K. S. Prakash, *Synthesis* **1976**, 607–609.

41. A. Sygula, P. W. Rabideau, *J. Am. Chem. Soc.* **1998**, *120*, 12666–12667.

42. A. Sygula, P. W. Rabideau, *J. Am. Chem. Soc.* **1999**, *121*, 7800–7803.

43. (a) T. J. Seiders, K. K. Baldridge, E. L. Elliott, G. H. Grube, J. S. Siegel, *J. Am. Chem. Soc.* **1999**, *121*, 7439–7440; (b) T. J. Seiders, E. L. Elliott, G. H. Grube, J. S. Siegel, *J. Am. Chem. Soc.* **1999**, *121*, 7804–7813.

44. A. Sygula, P. W. Rabideau, *J. Am. Chem. Soc.* **2000**, *122*, 6323–6324.

45. P. Bachavala, M. S. Thesis, Mississippi State Univ., Mississippi State, **2006**.

46. (a) A. Sygula, G. Xu, Z. Marcinow, P. W. Rabideau, *Tetrahedron* **2001**, *57*, 3637–3644; (b) G. Xu, A. Sygula, Z. Marcinow, P. W. Rabideau, *Tetrahedron Lett.* **2000**, *41*, 9931–9934.

47. A. Sygula, S. D. Karlen, R. Sygula, P. W. Rabideau, *Org. Lett.* **2002**, *4*, 3135–3137.

48. A. Sygula, F. R. Fronczek, P. W. Rabideau, *Tetrahedron Lett.* **1997**, *38*, 5095–5098.

49. A. Sygula, R. Sygula, P. W. Rabideau, *Org. Lett.* **2005**, *7*, 4999–5001.

50. A. Sygula, R. Sygula, L. Kobryn, *Org. Lett.* **2008**, *10*, 3927–3930.

51. (a) Y. Himeshima, T. Sonoda, H. Kobayashi, *Chem. Lett.* **1983**, 1211–1214; (b) D. Peña, A. Cobas, D. Pérez, E. Guitián, *Synthesis* **2002**, 1454–1458.

52. A. Sygula, R. Sygula, P. W. Rabideau, *Org. Lett.* **2006**, *8*, 5909–5911.

53. A. Sygula, R. Sygula, A. Ellern, P. W. Rabideau, *Org. Lett.* **2003**, *5*, 2595–2597.

54. (a) S. M. Bacharach, *J. Org. Chem.* **2009**, *74*, 3609–3611 and references cited therein; (b) H. Huang, T. Stewart, M. Gutmann, T. Ohhara, N. Niimura, Y.-X. Li, J.-F. Wen, R. Bau, H. N. C. Wong, *J. Org. Chem.* **2009**, *74*, 359–369 and references cited therein.

55. C. Mück-Lichtenfeld, S. Grimme, L. Kobryn, A. Sygula, *Phys. Chem. Chem. Phys.* **2010**, *12*, 7091–7097.

56. S. Grimme, C. Mück-Lichtenfeld, J. Antony, *J. Phys. Chem. C* **2007**, *111*, 11199–11207.

57. M. Makha, A. Purich, C. L. Raston, A. N. Sobolev, *Eur. J. Inorg. Chem.* **2006**, 507–517.

58. Y. Zhao, D. G. Truhlar, *Phys. Chem. Chem. Phys.* **2008**, *10*, 2813–2818.

59. B. M. Wong, *J. Comput. Chem.* **2009**, *30*, 51–56.

60. (a) C. D. Sherrill, T. Takatani, E. G. Hohenstein, *J. Phys. Chem. A* **2009**, *113*, 10146–10159; (b) K. E. Riley, M. Pitonak, J. Cerny, P. Hobza, *J. Chem. Theory Comput.* **2010**, *6*, 66–80.

61. (a) P. Hobza, H. L. Selzle, E. W. Schlag, *J. Phys. Chem.* **1996**, *100*, 18790–18794; (b) V. Špirko, O. Enkvist, P. Soldan, H. L. Selzle, E. W. Schlag, P. Hobza, *J. Chem. Phys.* **1999**, *111*, 572–582.

62. F. R. Fronczek, A. Sygula, unpublished results.

63. L. Kobryn, W. P. Henry, F. R. Fronczek, R. Sygula, A. Sygula, *Tetrahedron Lett.* **2009**, *50*, 7124–7127.

64. H. Detert, B. Rose, W. Mayer, H. Meier, *Chem. Ber.* **1994**, *127*, 1529–1532.

65. M. L. Jimeno, I. Alkorta, J. Elguero, J. E. Anderson, R. M. Claramunt, R. M. Lavandera, *New J. Chem.* **1998**, *22*, 1079–1083 and references therein.

66. (a) J. Gao, H. L. Wang, A. Shreve, R. Iyer, *Toxicol. Appl. Pharmacol.* **2010**, *244*, 130–143; (b) E.-J. Park, H. Kim, Y. Kim, J. Yi, K. Choi, K. Park, *Toxicol. Appl. Pharmacol.* **2010**, *244*, 226–233.

67. A. A. Voityuk, M. Duran, *J. Phys. Chem. C* **2008**, *112*, 1672–1678.

68. S. Chakrabarti, K. Ruud, *J. Phys. Chem. A* **2009**, *113*, 5485–5488.

# CHAPTER 2

# SYNTHESIS OF BOWL-SHAPED AND BASKET-SHAPED FULLERENE FRAGMENTS VIA BENZANNULATED ENYNE–ALLENES

KUNG K. WANG, HU CUI, and BO WEN

## 2-1. INTRODUCTION

Benzannulated enyne–allenes bearing an aryl substituent at the alkynyl terminus as shown in **1** undergo the Schmittel cyclization reactions under mild thermal conditions to form the corresponding benzofulvenyl biradicals represented by **2** (Scheme 2-1).[1,2] In the cases where an aryl substituent is also present at the allenic terminus, an intramolecular radical–radical coupling reaction through a carbon–carbon double bond of the aryl substituent at the allenic terminus could occur, leading to the corresponding Diels–Alder adducts as depicted in **3**. While the transformation from **1** to **3** could also be regarded as a concerted Diels–Alder reaction, mechanistic studies suggest that the reaction proceeds through a stepwise process involving a benzofulvenyl biradical.[3–8] A subsequent prototropic rearrangement to regain aromaticity then produces a 5-aryl-11$H$-benzo[$b$]fluorenyl derivative as shown in **4**, which represents a small fragment of the $C_{60}$ and other fullerene surfaces. The mildness of the reaction conditions and the availability of several synthetic methods for benzannulated enyne–allenes allow the design of new synthetic pathways leading to bowl-shaped and basket-shaped polycyclic aromatic hydrocarbons bearing a significant curvature.[9–12]

One of the synthetic methods for benzannulated enyne–allenes involves an initial condensation between pivalophenone (**5**) and lithium acetylide (**6**), derived

*Fragments of Fullerenes and Carbon Nanotubes: Designed Synthesis, Unusual Reactions, and Coordination Chemistry*, First Edition. Edited by Marina A. Petrukhina and Lawrence T. Scott.
© 2012 John Wiley & Sons, Inc. Published 2012 by John Wiley & Sons, Inc.

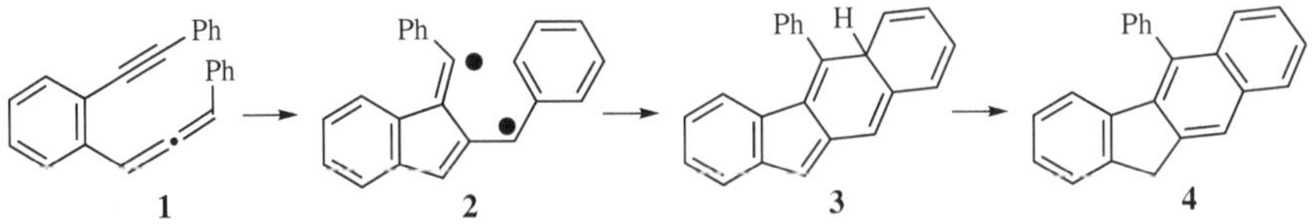

**Scheme 2-1.** Schmittel cyclization reaction of a benzannulated enyne–allene.

from 1-ethynyl-2-(2-phenylethynyl)benzene, to form propargylic alcohol (**7**) (Scheme 2-2).[13] Reduction of **7** with triethylsilane in the presence of trifluoroacetic acid then produces benzannulated enediyne (**8**). Treatment of **8** with potassium *tert*-butoxide in refluxing toluene at 110°C promotes a 1,3-prototropic rearrangement to produce benzannulated enyne–allene (**9**), *in situ*, which then undergoes the cascade cyclization sequence outlined in Scheme 2-1 to give benzofluorene (**10**) in excellent yield.

The chlorinated enyne–allene systems are also readily accessible from the corresponding propargylic alcohols. Specifically, treatment of propargylic alcohol (**12**), derived from 9-fluorenone (**11**) and lithium acetylide (**6**), with thionyl chloride at ambient temperature, promotes an $S_Ni'$ reaction to form chlorinated enyne–allene (**14**), which, in turn, promptly undergoes the cascade cyclization sequence to form the chlorinated benzofluorenyl derivative (**15**) (Scheme 2-3).[13] Rapid hydrolysis of **15** then furnishes alcohol **16** in 74% overall yield from **12**. A minor amount of **17**, derived from the intramolecular [2 + 2] cycloaddition reaction of **14**, is also produced in 12% yield.

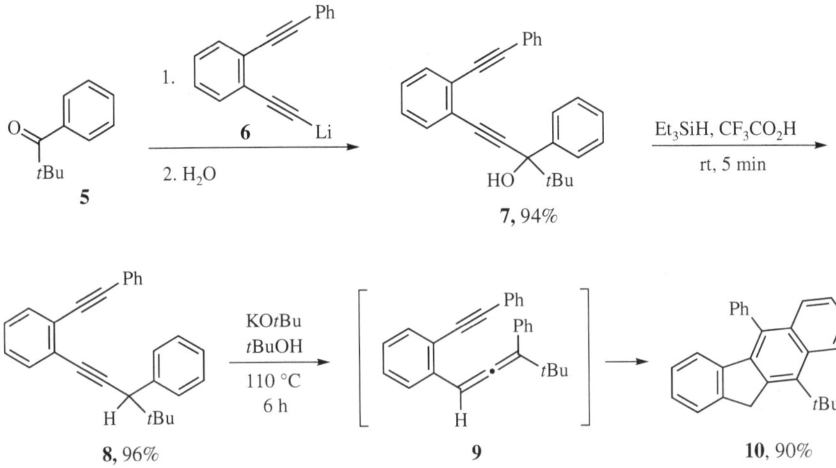

**Scheme 2-2.** Synthesis of 10-(1,1-dimethylethyl)-5-phenyl-11*H*-benzo[*b*]fluorene via a benzannulated enyne–allene.

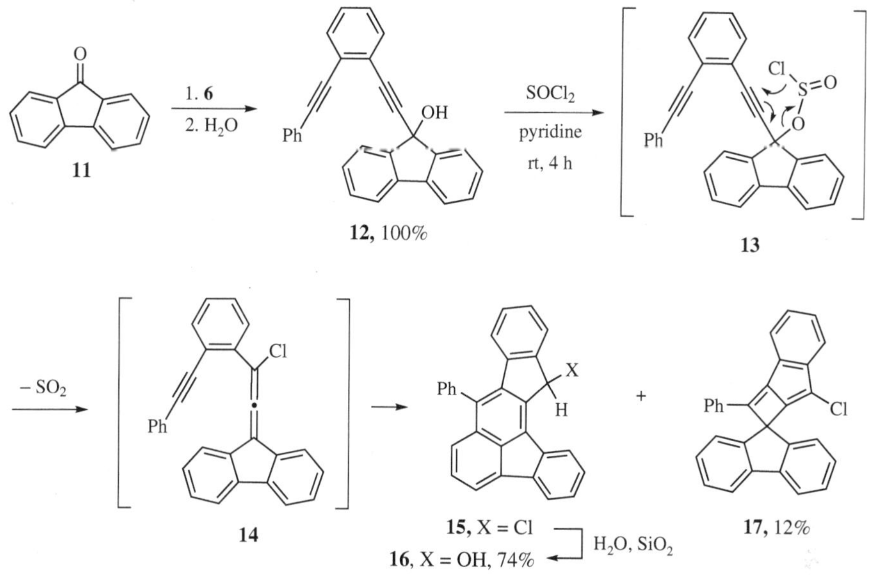

**Scheme 2-3.** Synthesis and cyclization of a chlorinated enyne–allene.

## 2-2. BOWL-SHAPED AND BASKET-SHAPED HYDROCARBONS

### 2-2-1. Bowl-Shaped Polycyclic Aromatic Hydrocarbons via Palladium-Catalyzed Intramolecular Arylation Reactions

The synthetic sequence outlined in Scheme 2-2 was adopted for the synthesis of benzofluorene (**24**) bearing a 2,6-dibromophenyl substituent at the C5 position (Scheme 2-4).[14] The Sonogashira reaction between (2,6-dibromophenyl)ethyne (**18**) and [(2-iodophenyl)ethynyl]trimethylsilane (**19**) produced **20**, which was desilylated to afford benzannulated enediyne (**21**). Condensation between **21** and pivalophenone (**5**) followed by reduction of the resulting propargylic alcohol then gave the requisite benzannulated enediyne (**23**) for subsequent cascade cyclization reactions. Treatment of **23** with potassium *tert*-butoxide in refluxing toluene at 110°C produced benzofluorene (**24**).

The presence of two properly situated bromo substituents in **24** allowed the application of the palladium-catalyzed intramolecular arylation reactions for additional carbon–carbon bond formation. The palladium-catalyzed intramolecular arylation reactions had found useful applications in the synthesis of bowl-shaped and basket-shaped polycyclic aromatic hydrocarbons, including dibenzo[*a,g*]corannulene (**25**)[15] and *as*-indaceno[3,2,1,8,7,6-*pqrstuv*]picenes (**26a,b**)[16] (Scheme 2-5). More recently, this method also found success in the preparation of several indenofused corannulenes, including pentaindenocorannulene (**27**).[17,18] On subjecting **24** to the reaction conditions reported by Scott et al.,[15] Kim et al. converted benzofluorene (**24**) to cyclopentaindenotriphenylene (**28**) in 37% yield (Scheme 2-6).[14] Similarly,

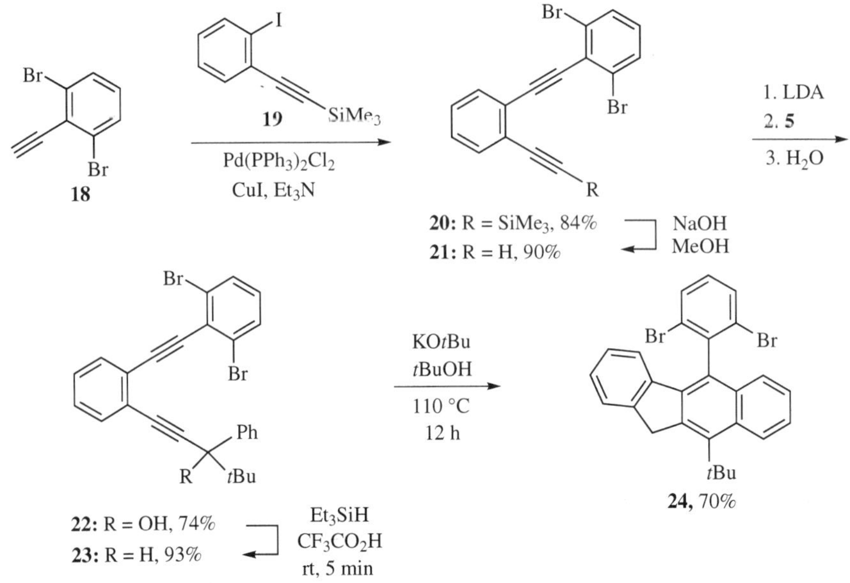

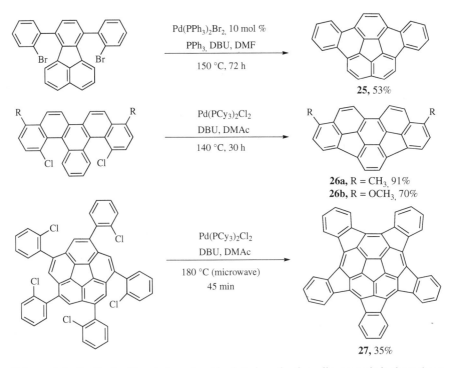

**Scheme 2-4.** Synthesis of 5-(2,6-dibromophenyl)-10-(1,1-dimethylethyl)-11*H*-benzo[*b*] fluorene.

**Scheme 2-5.** Synthesis of bowl-shaped and basket-shaped polycyclic aromatic hydrocarbons via palladium-catalyzed intramolecular arylation reactions.

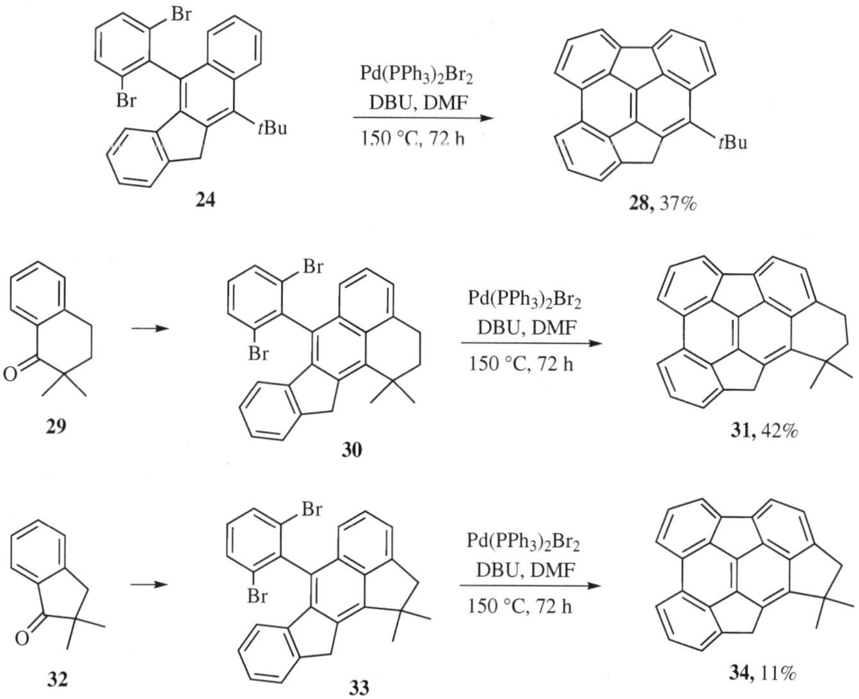

**Scheme 2-6.** Synthesis of bowl-shaped cyclopentaindenotriphenylenes.

dibromides (**30**) and (**33**), readily prepared from aryl ketones (**29**) and (**32**) by condensation with **21**, were transformed to indenochrysenes (**31**) and (**34**), respectively.

The X-ray structures of **28**, **31**, and **34** indicate the presence of significant curvatures. Compared to diindenochrysene (**35**),[19] the pyramidalization angles, defined as $\Theta_{\sigma\pi} - 90$ using the π-orbital axis vector analysis (POAV),[20–22] of the central ethylene carbon atoms of **28** and **31** are less pronounced (Figure 2-1). However, for **34**, the POAV angle of one of the central ethylene carbon atoms shows a greater extent of pyramidalization at 10.3°, corresponding to 88% of that found for $C_{60}$ with all carbon atoms having a POAV angle of 11.64°. Apparently, the presence of an additional five-membered ring in **34** causes its structure to be more strained and creates a more pronounced curvature. As a result, the transformation from **33** to **34** is also less efficient.

It was reported that, unlike corannulene, dihydrocyclopentacorannulene (**36**) (Figure 2-2) with an additional five-membered ring "locked" the bowl configuration, preventing a rapid bowl-to-bowl inversion process.[23] However, the dynamic ¹H NMR studies of **34** indicate rapid bowl-to-bowl inversion on the NMR timescale with the observation of only a singlet for each of the diastereotopic methylene hydrogens and a singlet for the diastereotopic methyl groups at 25°C. At −33°C, the methylene signals, recorded on a 600-MHz NMR spectrometer, became two broad humps, and the methyl signal also appeared as a broad peak,

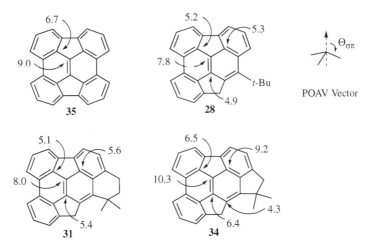

**Figure 2-1.** POAV angles of bowl-shaped polycyclic aromatic hydrocarbons.

**36**

**Figure 2-2.** Structure of dihydrocyclopentacorannulene (**36**).

suggesting rapid bowl-to-bowl inversion on the NMR timescale even at such a low temperature.

### 2-2-2. A $C_{44}H_{26}$ Hydrocarbon Bearing a 44-Carbon Framework of $C_{60}$

The presence of two keto groups in acenaphthenequinone (**37**) provides an opportunity for condensation with two equivalents (2 eq) of lithium acetylide (**6**) to form the corresponding propargylic diol. Unfortunately, several such attempts were unsuccessful.[24] However, mono protection of one of the two keto groups was successful in producing monoketal **38** in 90% yield (Scheme 2-7).[25] Condensation of **38** with lithium acetylide (**6**) followed by hydrolytic workup furnished propargylic alcohol (**39**). Treatment of **39** with thionyl chloride then promoted a sequence of cyclization reactions, as outlined in Scheme 2-3, to furnish the chlorinated benzo-fluorene (**40**). Removal of the chloro substituent by reduction with sodium borohydride produced **41** in 51% overall yield from **39**. Hydrolysis of the ketal group in **41** provided ketone (**42**). Initial attempts to prepare propargylic alcohol (**44**) by condensation of **42** with lithium acetylide (**6**) were unsuccessful, presumably due

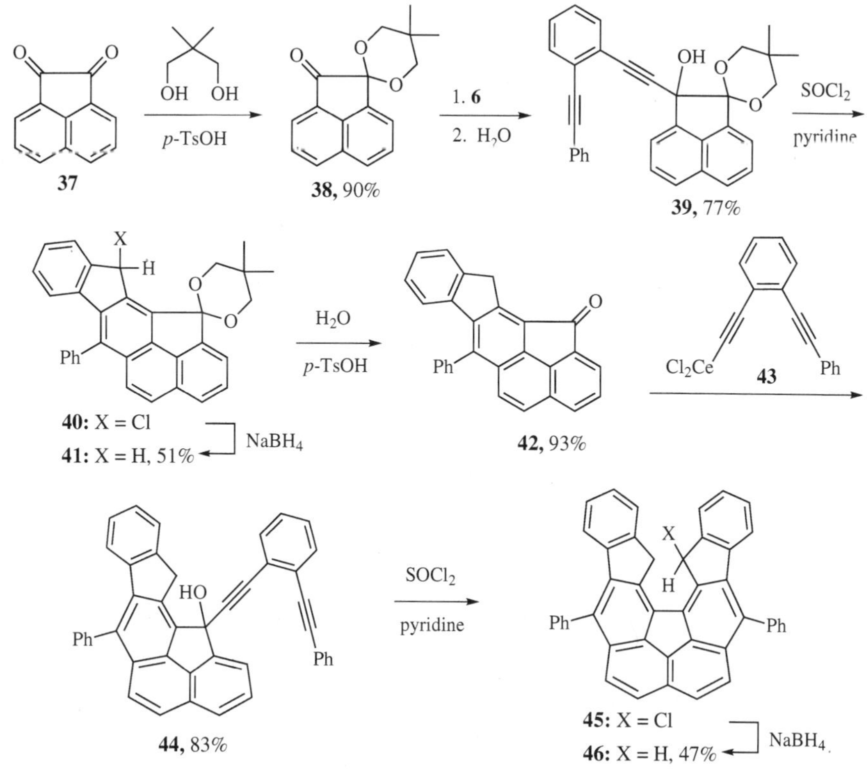

**Scheme 2-7.** Synthesis of the $C_{44}H_{26}$ benzodiindenofluoranthene (**46**).

to the presence of acidic methylene hydrogens on the benzofluorenyl substructure. Fortunately, by first converting **6** to the corresponding cerium derivative (**43**),[26] the condensation reaction occurred to afford **44** in 83% yield. Treatment of **44** with thionyl chloride followed by reduction with sodium borohydride then produced the $C_{44}H_{26}$ benzodiindenofluoranthene (**46**). The transformation from **37** to **46** involved eight steps with a 12.8% overall yield.

The structure of **46** bears a 44-carbon framework represented on the surface of $C_{60}$. The MM2-optimized structure indicates an essentially planar geometry of the benzodiindenofluoranthene moiety. Further connection of the methylene carbons to form a carbon–carbon double bond could yield **47** containing a diindenocorannulene unit (Scheme 2-8). Two additional intramolecular carbon–carbon bond formations could then lead to tetraindenocorannulene (**48**), which was synthesized from a corannulene derivative.[17,18] However, a variety of attempts to try to convert **46** to **47** were unsuccessful. Specifically, transformation of **46** to diketone **49** followed by the McMurry coupling produced only the reduced products **46** and **50** (Scheme 2-9).[24] Bromination of **46** with NBS to form dibromide **51** followed by treatment with potassium *tert*-butoxide also failed to give **47**.[24] Apparently, the essentially planar

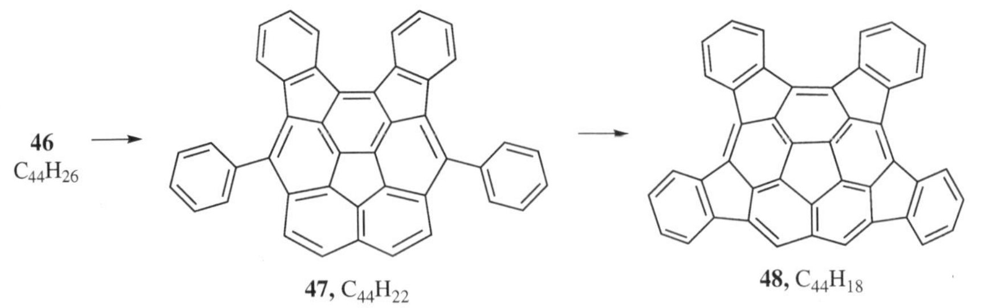

**46**
$C_{44}H_{26}$

**47, $C_{44}H_{22}$**

**48, $C_{44}H_{18}$**

**Scheme 2-8.** Indenocorannulenes from the $C_{44}H_{26}$ hydrocarbon **46**.

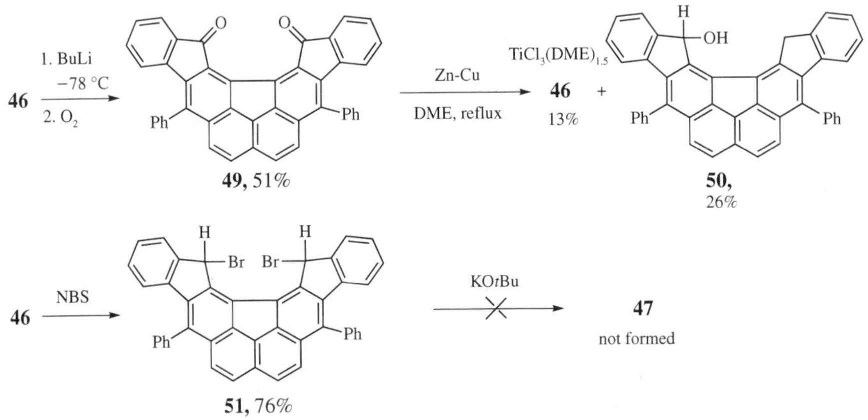

**Scheme 2-9.** Derivatives of the $C_{44}H_{26}$ hydrocarbon **46** for potential intramolecular carbon–carbon bond-forming reactions.

geometry of **46** having a long distance of $\sim 3.58$ Å between the methylene carbons prevents the connection of these two carbon atoms.

## 2-2-3. A Basket-Shaped $C_{56}H_{40}$ Hydrocarbon Bearing a 54-Carbon Framework of $C_{60}$

The failure to connect methylene carbons in **46** to form **47** prompted a change of strategy for the construction of curved carbon frameworks that can be mapped onto the surface of $C_{60}$. In the first synthesis of corannulene (**54**), the strategy of constructing **52** containing multiple tetrahedral $sp^3$-hybridized carbons as the precursor of **53** prior to the final transformations to the fully aromatized **54** was employed by Barth and Lawton (Scheme 2-10).[27,28] More recently, this approach was adopted in the synthesis of sumanene (**57**), a symmetric $C_{21}H_{12}$ subunit of $C_{60}$ (Scheme 2-11).[29] The presence of multiple tetrahedral $sp^3$-hybridized carbons in **52** and the ring-opening product of **55a** causes several parts of the molecules to be in close proximity of one another, greatly facilitating the intramolecular carbon–carbon bond

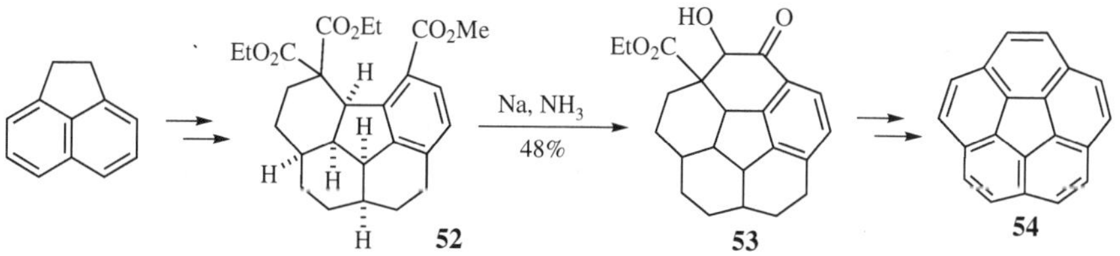

**Scheme 2-10.** Barth and Lawton's[27,28] synthesis of corannulene.

formations leading to **53** and **56**, respectively. This strategy also found success in the construction of a basket-shaped $C_{56}H_{40}$ hydrocarbon bearing a 54-carbon framework of $C_{60}$.[30]

The first stage of the synthetic sequence involved the preparation of diketone (**65**) as a key intermediate leading toward the $C_{56}H_{40}$ hydrocarbon (Scheme 2-12).[30] Condensation between dimethyl acetonedicarboxylate (**58**) and benzil (**59**) followed by dehydration produced cyclopentadienone (**60**) as reported previously.[31] As a reactive diene, **60** underwent the Diels–Alder reaction with 2,5-dihydrofuran (**61**) smoothly to give the cycloaddition adduct **62** as a mixture of the *endo* and *exo* (14 : 1) isomers. Treatment of **62** with water to induce decarbonylation then led to diester (**63**). The decarbonylation reaction presumably proceeded through the addition of water to the keto carbonyl to form the corresponding hydrate, which in turn underwent a retro-Claisen condensation followed by decarboxylation. Saponification of **63** to produce the corresponding diacid followed by transformation to the diacid chloride with thionyl chloride for the $AlCl_3$-catalyzed intramolecular Friedel–Crafts acylation reactions then furnished diketone (**64**). Dimethylation from the less hindered side opposite to the tetrahydrofuran ring then furnished diketone (**65**).

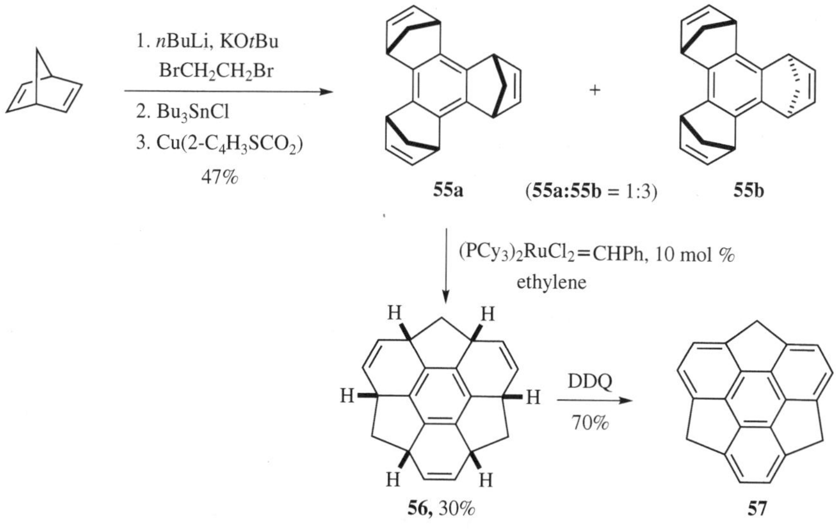

**Scheme 2-11.** Sakurai, Daiko, and Hirao's[29] synthesis of sumanene.

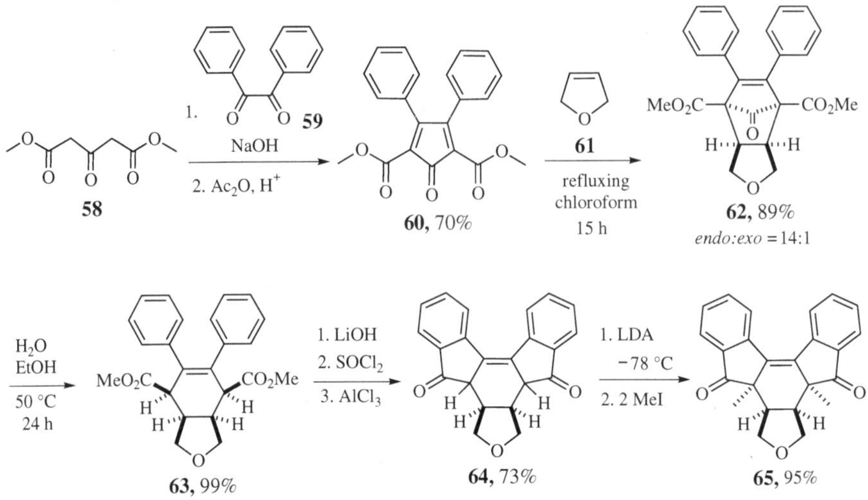

**Scheme 2-12.** Synthesis of diketone **65** as a precursor toward the basket-shaped $C_{56}H_{40}$ hydrocarbon **77**.

Condensation of **65** with one equivalent (1 eq) of lithium acetylide (**6**) produced propargylic alcohol (**66**) (Scheme 2-13). Unlike acenaphthenequinone (**37**), condensation of **65** with 2 eq of **6** was successful in producing propargylic diol (**68**).[32] Treatment of **66** and **68** with thionyl chloride followed by air oxidation then furnished diketones (**67**)[30] and (**69**),[32] respectively.

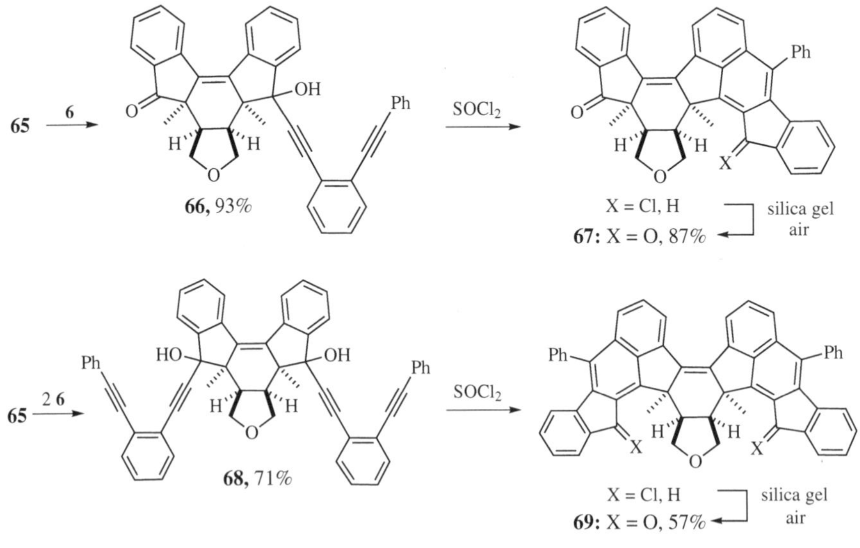

**Scheme 2-13.** Thionyl chloride-induced cascade cyclizations of benzannulated enediynyl alcohols.

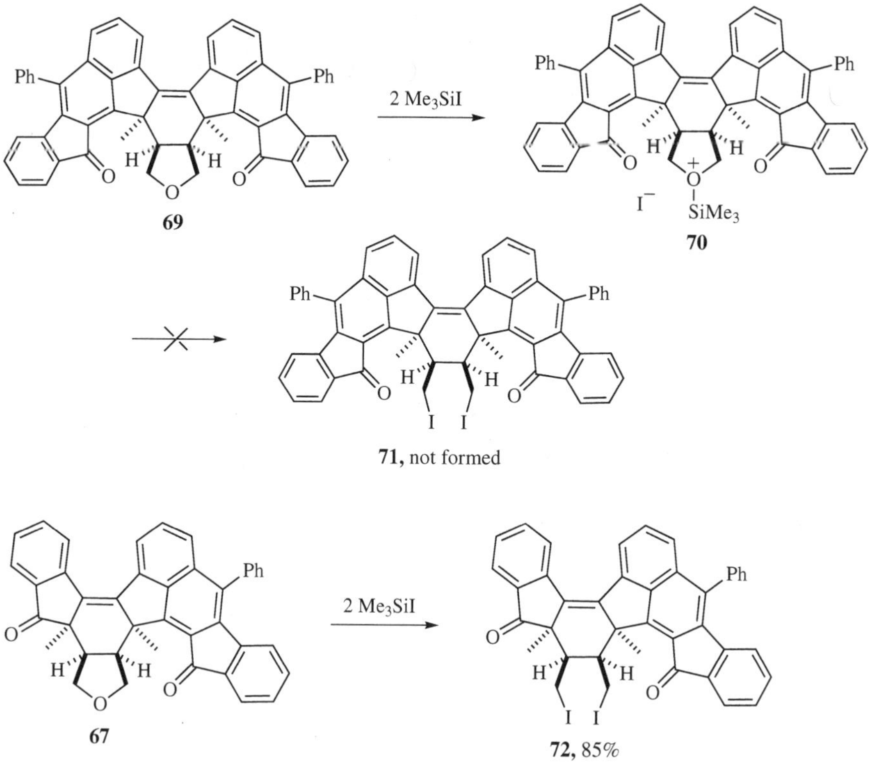

**Scheme 2-14.** Trimethylsilyl iodide-promoted tetrahydrofuran ring opening.

The tetrahydrofuran ring in **67** and **69** represents a latent 1,4-diiodide if the ring could be cleaved by trimethylsilyl iodide as reported previously.[33] However, repeated attempts to try to cleave the tetrahydrofuran ring in **69** with trimethylsilyl iodide were unsuccessful (Scheme 2-14).[32] Treatment of **69** with diiodosilane, a reagent also reported to cleave tetrahydrofuran,[34] resulted in the reduction of the two keto groups to the corresponding methylene groups. The reason for the failure to cleave the tetrahydrofuran ring became apparent after examination of the X-ray structure of **69**.[32] The X-ray structure of **69** shows that the tetrahydrofuran ring is folded inside the two benzofluorenyl moieties, blocking the backsides of the α carbons of the resulting silyloxonium ion (**70**) for an $S_N2$ attack by an iodide ion. One the other hand, diketone (**67**), wherein one side of the tetrahydrofuran ring remained relatively unhindered, was readily cleaved by trimethylsilyl iodide to form diiodide (**72**) in 85% yield.[30]

Condensation of **72** with 1 eq of lithium acetylide (**6**) resulted in selective attack of the less hindered keto group to give propargylic alcohol (**73**), identical to **71** depicted in Scheme 2-14, (Scheme 2-15). On exposure to $SOCl_2$, followed by oxidation with $MnO_2$, diketone (**74**) along with the [2 + 2] cycloaddition adduct (**75**) were isolated. However, attempts to promote the intramolecular Barbier reactions of **74** with $SmI_2$[35]

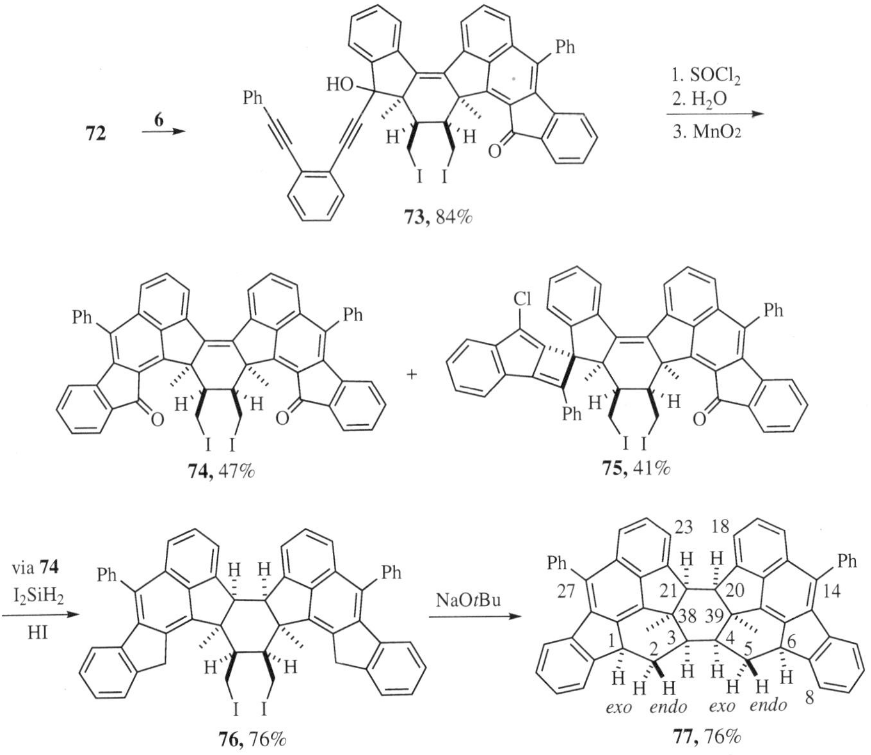

**Scheme 2-15.** Synthesis of the basket-shaped $C_{56}H_{40}$ hydrocarbon **77**.

to connect the carbons bearing an iodo substituent with the neighboring carbonyl carbon were unsuccessful, presumably because the keto groups in **74** are conjugated with aromatic systems. However, an alternative pathway involving treatment of **74** with diiodosilane in the presence of hydrogen iodide to reduce the keto groups was successful with concomitant reduction of the central carbon–carbon double bond to give **76**. The newly formed C–H bonds on the central cyclohexyl ring of **76** are *cis* to the methyl groups. The all-*cis* relationship among the methine hydrogens and the methyl groups on the central cyclohexyl ring along with the presence of multiple $sp^3$-hybridized carbons in **76** make it possible to move the iodo-bearing carbons close to the neighboring methylene carbons on the five-membered rings, setting the stage for the intramolecular alkylation reactions. The methylene hydrogens on the five-membered rings of the benzofluorenyl substructures are relatively acidic, allowing the corresponding carbanions to be readily accessible. Treatment of **76** with sodium *tert*-butoxide, prepared from sodium hydride in mineral oil and *tert*-butyl alcohol, was successful in promoting the intramolecular alkylation reactions, giving rise to the basket-shaped $C_{56}H_{40}$ hydrocarbon **77**.

The $^1H$ NMR spectrum of **77** indicates that it is unsymmetric with a skewed conformation and the rate of racemization is relatively slow on the NMR timescale.

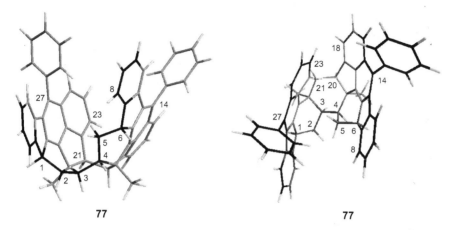

77                                                     77

**Figure 2-3.** MM2-optimized structure of the $C_{56}H_{40}$ hydrocarbon **77** viewed from two different perspectives. (Reprinted with the permission of the American Chemical Society.)

In addition, several aromatic and aliphatic hydrogen signals also show unusual upfield shifts with the aromatic hydrogen on C23 appearing at $\delta$ 3.98, and the *endo* and *exo* hydrogens on C5 appearing at $\delta$ −3.32 and 0.29, respectively. The MM2-optimized structure (Figure 2-3) suggests that the central cyclohexyl ring adopts a twist−boat conformation with C4 and C21 holding the flagpole C−C bonds. The twist motion involves moving C38 and C39 up and C3 and C20 down, causing the two benzo-fluorenyl substructures to approach each other and providing significant magnetic shieldings for the upfield-shifted hydrogens.

The two phenyl substituents are orientated roughly perpendicular to the benzo-fluorenyl substructures to avoid nonbonded steric interactions, and the rate of rotation of the phenyl substituent on C27 is relatively slow on the NMR timescale as observed previously in related structures.[13] The upfield-shifted signal at $\delta$ 5.45 is attributed to the *ortho* hydrogen of the phenyl substituent on C27 pointing inward toward the endohedral side of the molecule. A significant nuclear Overhauser effect (NOE) between this *ortho* hydrogen and the aromatic hydrogen on C8 indicates that they are in close proximity to each other, although they are located on opposite sides of the structure. This observation further supports the twist−boat conformation of the central cyclohexyl ring, which could bring these two aromatic hydrogens in close proximity to each other.

The structure of **77** has a 54-carbon framework that can be mapped onto the surface of $C_{60}$, missing only three 2-carbon fragments. Diketone **65** provides 22 carbons of the 54-carbon framework, while the remaining 32 carbons come from two molecules of lithium acetylide (**6**). The interior 30-carbon core of **77** can be regarded as a partially hydrogenated [5,5]circulene (**78**) (Figure 2-4), a $C_{30}H_{12}$ semibuckmin-sterfullerene,[36,37] with only the carbon−carbon bond between C18 and C23 remaining unconnected. Further connection of C18 with C23 and formations of the carbon−carbon bonds between the four *ortho* carbons of the phenyl substituents

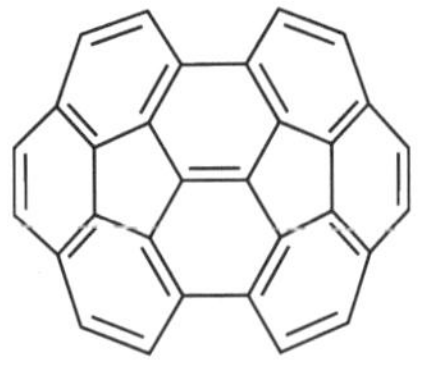

**78**

**Figure 2-4.** Structure of [5,5]circulene.

with the benzofluorenyl substructures could lead to the basket-shaped $C_{56}H_{30}$ hydrocarbon **79** (Scheme 2-16). Two more carbon–carbon bond formations could then lead to a $C_{56}H_{26}$ buckyball bearing three holes due to the absence of three 2-carbon fragments.

## 2-2-4. A Basket-Shaped $C_{56}H_{38}$ Hydrocarbon as a Precursor Toward an Endcap Template for (6,6) Carbon Nanotubes

The synthetic pathway leading to **69** having a 54-carbon framework of $C_{60}$ by the condensation reactions of 2 eq of lithium acetylide (**6**) with diketone (**65**) outlined in Scheme 2-13 is very attractive in assembling large fullerene fragments for subsequent transformations to basket-shaped polycyclic aromatic hydrocarbons. The design and synthesis of a tetraketone for condensation with 2 eq of lithium acetylide to take full advantage of this efficient pathway for the construction of a basket-shaped $C_{56}H_{38}$ hydrocarbon have been achieved.[38]

The synthetic sequence outlined in Scheme 2-17 illustrates the preparation of tetraketone (**88**) as a key intermediate leading to the basket-shaped $C_{56}H_{38}$ hydrocarbon **99**. The silyl enol ether (**81**) was produced in quantitative yield from treatment of the commercially available 4-bromo-1-indanone (**80**) with triisopropylsilyl trifluoromethanesulfonate.[39] The methylene hydrogens of the indene moiety in **81** are relatively acidic, making it possible to produce the corresponding carbanion by lithiation with lithium diisopropylamide (LDA). The Cu(II) chloride–promoted coupling[40,41] of the resulting carbanion followed by desilylation with tetrabutylammonium fluoride (TBAF) then produced an essentially 1 : 1 mixture of the *rac*-**82** and

$77, C_{56}H_{40}$ →

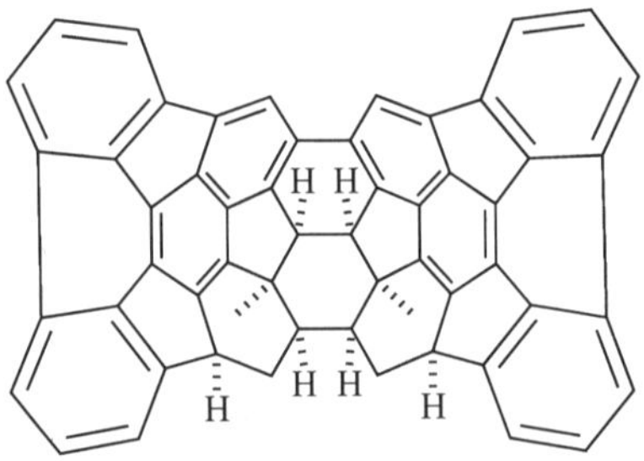

$79, C_{56}H_{30}$

**Scheme 2-16.** Structure of the basket-shaped $C_{56}H_{30}$ hydrocarbon **79**.

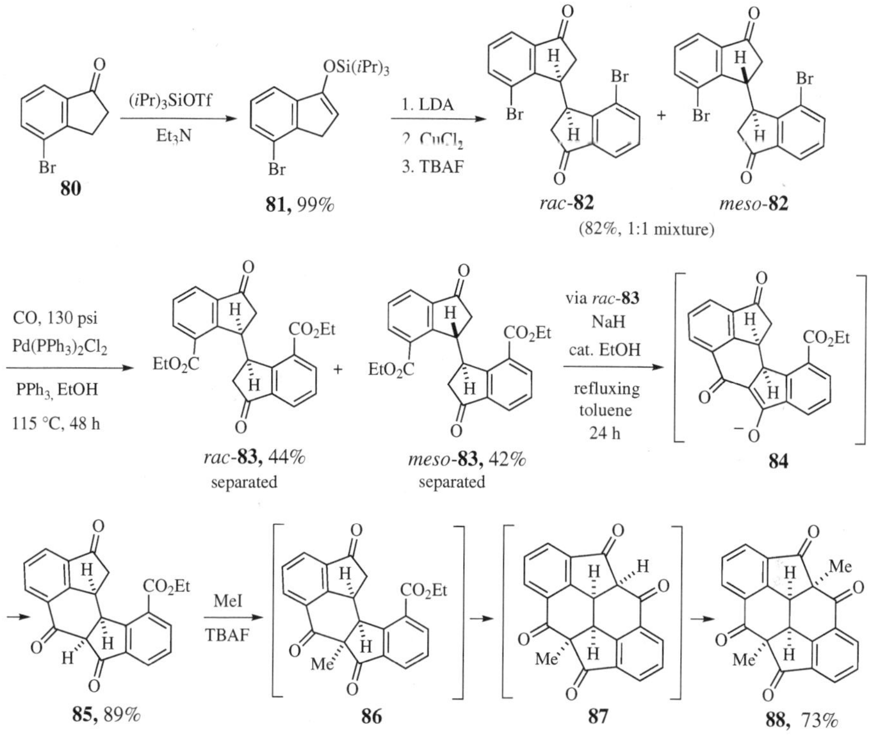

**Scheme 2-17.** Synthesis of tetraketone **88** as a precursor toward the basket-shaped $C_{56}H_{38}$ hydrocarbon **99**.

*meso*-**82** isomers in 82% combined yield. After the palladium-catalyzed carboethoxylation reactions,[42,43] the resulting diketodiesters *rac*-**83** and *meso*-**83** were separated by silica gel column chromatography.

Treatment of *rac*-**83** with sodium hydride in the presence of ethanol promoted an intramolecular Claisen-type condensation to form triketone (**85**). The NOE measurements indicate that the hydrogen between the two keto carbonyls is *cis* to the two central methine hydrogens. However, the second intramolecular Claisen-type condensation did not occur to form the corresponding tetraketone. The lack of a second intramolecular Claisen-type condensation was unexpected because triketone (**85**) appeared to be even better suited for the second intramolecular condensation. An analysis of the reaction pathway revealed that after the first intramolecular Claisen-type condensation to form **85**, deprotonation of the more acidic α-hydrogen between the two keto carbonyls must have occurred to form enolate (**84**) under the reaction condition. The formation of the enolate resulted in a more planar structure for **84**, preventing the positioning of the ester carbonyl in a parallel orientation on top of the π electrons of the second enolate for condensation. Interestingly, in an attempt to replace the acidic hydrogen between the two keto carbonyls with a methyl group by methylation with methyl iodide in the presence of TBAF at room temperature,[44] the second Claisen-type condensation also occurred along with a subsequent methylation to give tetraketone (**88**) directly. Apparently, after an initial methylation to form **86**, the second condensation reaction occurred readily even under such a mild reaction condition because the ester carbonyl could now be placed in a parallel orientation on top of the $\pi$ electrons of the enolate or the corresponding enol for condensation to form **87**, which, in turn, was methylated to give **88**.

The structure of tetraketone **88** was established by X-ray structure analysis, showing the presence of a $C_2$ symmetry. The methyl groups and the methine hydrogens in **88** are all *cis* to one another, indicating that the two methylation reactions occurred from the less hindered convex side as observed in the case of **65**. The all-*cis* relationship causes **88** to have a bent structure with the two benzene rings in essentially perpendicular orientation with each other.

The presence of four keto groups in **88** provided opportunities for condensation with lithium acetylides as observed in the formation of propargylic diol (**68**) from diketone (**65**). However, unlike the case for **65**, the issue of chemoselectivity of lithium acetylides toward **88** for condensation needed to be investigated first. Treatment of **88** with an excess of lithium acetylide–ethylenediamine complex (**89**) produced propargylic diol (**90**) as the major product (**90** : **91** = 5 : 1) (Scheme 2-18). The structure of **90** was established by X-ray structure analysis, showing that the keto carbonyls on the six-membered rings of **88** were attacked preferentially from the convex side. Interestingly, even in the presence of a large excess of **89**, only symmetric diols (**90** and **91**) were produced without the formation of the unsymmetrical diols, triols, and tetraols.

Alternatively, treatment of **88** with an excess of lithium acetylide (**6**) also produced a symmetric diol from attacking the keto groups on the six-membered rings preferentially as observed for **90**.[45] However, attempts to induce the Schmittel cyclization reactions with thionyl chloride appeared to give predominantly the corresponding [2 + 2] cycloaddition adduct[45] as observed in the formation of **75**. Attempts to reduce the diol with triethylsilane in the presence of trifluoroacetic acid were unsuccessful.

Treatment of the mixture of **90** and **91** (5 : 1) with thionyl bromide then produced allenic dibromide (**92**) as the major product and its isomers as minor products. The structure of **92** was established by X-ray structure analysis, showing that the two bromo substituents pointed toward the concave side of the molecule. The palladium-catalyzed coupling reactions[46] between **92** and arylzinc chloride (**93**) produced, *in situ*, the benzannulated enyne–allene **94**. It is worth noting that the coupling of **92** with **93** represents a new synthetic pathway for the benzannulated enyne–allene systems. After 12 h at room temperature followed by heating at 50°C for one hour, a complete transformation of **94** to benzofluorenyl diketone (**95**) was observed. The appearance of the characteristic AB splitting pattern in the $^1$H NMR spectrum with a large geminal coupling constant of 23.0 Hz attributable to the presence of two diastereotopic methylene hydrogens on the five-membered rings of the benzofluorenyl substructures as observed previously[47] indicated the successful formation to **95**.

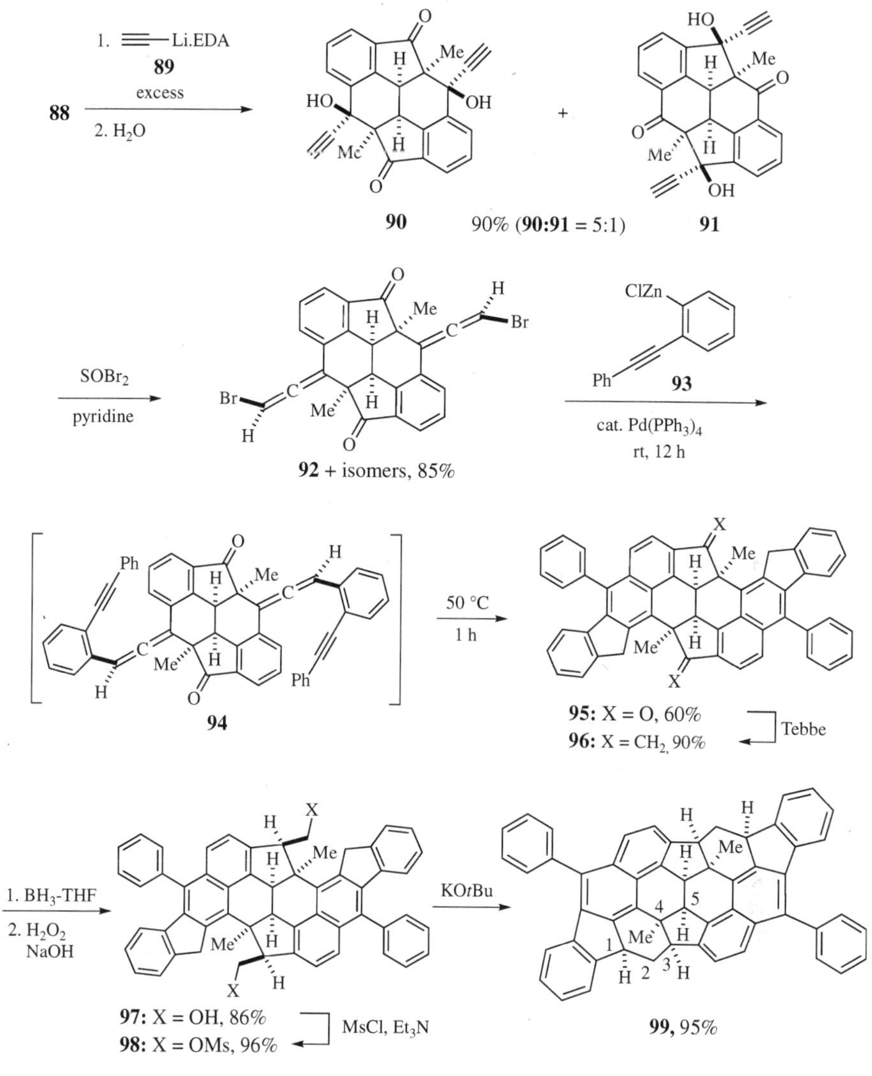

**Scheme 2-18.** Synthesis of the basket-shaped $C_{56}H_{38}$ hydrocarbon **99**.

The presence of two keto groups in **95** allowed methylenation with the Tebbe reagent[48,49] to produce diene **96**, which on treatment with $BH_3$–THF followed by oxidation then provided diol **97**. The hydroboration reactions occurred from the less hindered convex side. As a result, the two hydroxymethyl groups were oriented toward the endohedral (concave) side of **97**. The endohedral orientations of the two hydroxymethyl groups were of crucial importance for the subsequent intramolecular carbon–carbon bond-forming reactions. Diol (**97**) was converted to the corresponding methanesulfonate (**98**) with methanesulfonyl chloride in the presence of triethylamine. With the methanesulfonate ion serving as a good leaving group, the stage

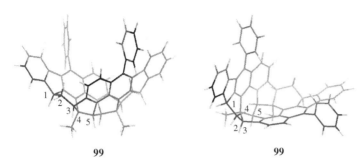

99                              99

**Figure 2-5.** MM2-optimized structure of the $C_{56}H_{38}$ hydrocarbon **99** viewed from two different perspectives. (Reprinted with the permission of the American Chemical Society.)

again was set for the intramolecular alkylation reactions as observed for **76**. Treatment of **98** with potassium *tert*-butoxide produced **99** in excellent yield. The high acidity of the methylene hydrogens on the benzofluorenyl units along with their close proximities to the neighboring carbons bearing a methanesulfonate leaving group again contribute to the high efficiencies of the intramolecular alkylation reactions. The ability to employ the same type of reaction twice for most of the 12-step synthetic sequence starting from 4-bromo-1-indanone (**80**) greatly enhances the overall efficiency in producing **99**.

The methine hydrogens and the methyl groups in **99** are *cis* to one another, causing **99** to have a basket-shaped structure. The molecule is chiral, possessing only $C_2$ symmetry. The MM2-optimized structure of **99** (Figure 2-5) indicates that the six-membered ring containing C1 to C4 carbons would adopt a boatlike conformation with H1 and the methyl group on C4 assuming the flagpole positions. Such a conformation is supported by the NMR measurements of vicinal coupling constants of aliphatic hydrogens. The two phenyl substituents are oriented essentially perpendicular to the neighboring benzofluorenyl group, as observed previously.[13]

The structure of **99** contains a 30-carbon difluorenonaphthacenyl core that can be mapped onto the surface of $C_{78}$.[10,50] Compared to the basket-shaped $C_{56}H_{40}$ hydrocarbon **77**, the 30-carbon core in **99** is fully connected. Hydrocarbon **99** or its derivatives could serve as potential precursors leading to the $C_{68}H_{26}$ hydrocarbon (**100**) and the fully aromatized $C_{66}H_{12}$ polycyclic aromatic hydrocarbon (**101**) as endcap templates[51,52] for (6,6) carbon nanotubes (Figure 2-6). Compared to **99**, there are two additional benzene units in **100** at the periphery. The structure of **100** retains the 30-carbon interior core but is fused at the rim with a [6] cycloparaphenylene, which represents a nanohoop segment of (6,6) carbon nanotubes.[53–55]

The synthetic sequence outlined in Schemes 2-17 and 2-18 could be adopted for the introduction of two additional benzene units at the periphery for further construction of a [6]cycloparaphenylene rim. The two keto groups in **95** could provide the needed handles for the introduction of two functionalized benzene units by condensation

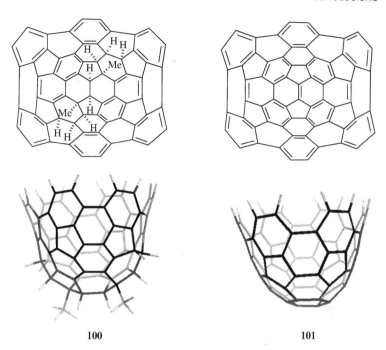

**100**                                        **101**

**Figure 2-6.** Structures of the basket-shaped $C_{68}H_{26}$**100** and $C_{66}H_{12}$**101**. (Reprinted with the permission of the American Chemical Society.)

with 2 eq of 2,6-dichlorobenzylmagnesium bromide followed by dehydration. The presence of 10 $sp^3$-hybridized carbons in the interior core places the phenyl groups at the periphery in close proximity to one another, making it potentially feasible to connect them by palladium-catalyzed intramolecular arylation reactions to form a paraphenylene rim.

## 2-3. CONCLUSIONS

The propargylic alcohols, readily prepared from condensation between ketones and lithium acetylides, are excellent precursors of the benzannulated enyne–allenes. These unsaturated molecules undergo cascade cyclization reactions under mild thermal conditions to produce benzofluorenyl derivatives bearing carbon frameworks of fullerene fragments. The reactions can tolerate the presence of keto and halogen functionalities, making it possible for subsequent intramolecular carbon–carbon bond formations leading to bowl-shaped and basket-shaped hydrocarbons possessing significant curvatures. While this overall synthetic strategy has found success in producing several curved hydrocarbons, the field is still in its infancy, requiring further exploration to realize its full potential for the construction of larger fullerene fragments and carbon nanotube endcaps.

## REFERENCES

1. M. Schmittel, M. Strittmatter, S. Kiau, *Tetrahedron Lett.* **1995**, *36*, 4975–4978.

2. K. K. Wang, in *Modern Allene Chemistry*, N. Krause, A. S. K. Hashmi (eds.), Wiley-VCII, Weinheim, **2004**, Vol. 2, pp. 1091–1126.

3. M. Schmittel, M. Strittmatter, S. Kiau, *Angew. Chem. Int. Ed. Engl.* **1996**, *35*, 1843–1845.

4. M. Schmittel, M. Strittmatter, K. Vollmann, S. Kiau, *Tetrahedron Lett.* **1996**, *37*, 999–1002.

5. M. Schmittel, S. Kiau, *Liebigs Ann./Recueil* **1997**, 733–736.

6. M. Schmittel, M. Keller, S. Kiau, M. Strittmatter, *Chem. Eur. J.* **1997**, *3*, 807–816.

7. M. Schmittel, M. Maywald, *Chem. Commun.* **2001**, 155–156.

8. M. Schmittel, M. Maywald, M. Strittmatter, *Synlett* **1997**, 165–166.

9. Y.-T. Wu, J. S. Siegel, *Chem. Rev.* **2006**, *106*, 4843–4867.

10. V. M. Tsefrikas, L. T. Scott, *Chem. Rev.* **2006**, *106*, 4868–4884.

11. G. Mehta, H. S. P. Rao, *Tetrahedron* **1998**, *54*, 13325–13370.

12. P. W. Rabideau, A. Sygula, *Acc. Chem. Res.* **1996**, *29*, 235–242.

13. H. Li, H.-R. Zhang, J. L. Petersen, K. K. Wang, *J. Org. Chem.* **2001**, *66*, 6662–6668.

14. D. Kim, J. L. Petersen, K. K. Wang, *Org. Lett.* **2006**, *8*, 2313–2316.

15. H. A. Reisch, M. S. Bratcher, L. T. Scott, *Org. Lett.* **2000**, *2*, 1427–1430.

16. L. Wang, P. B. Shevlin, *Org. Lett.* **2000**, *2*, 3703–3705.

17. B. D. Steinberg, E. A. Jackson, A. S. Filatov, A. Wakamiya, M. A. Petrukhina, L. T. Scott, *J. Am. Chem. Soc.* **2009**, *131*, 10537–10545.

18. E. A. Jackson, B. D. Steinberg, M. Bancu, A. Wakamiya, L. T. Scott, *J. Am. Chem. Soc.* **2007**, *129*, 484–485.

19. H. E. Bronstein, N. Choi, L. T. Scott, *J. Am. Chem. Soc.* **2002**, *124*, 8870–8875.

20. R. C. Haddon, L. T. Scott, *Pure Appl. Chem.* **1986**, *58*, 137–142.

21. R. C. Haddon, *J. Am. Chem. Soc.* **1987**, *109*, 1676–1685.

22. R. C. Haddon, *Science* **1993**, *261*, 1545–1550.

23. A. H. Abdourazak, A. Sygula, P. W. Rabideau, *J. Am. Chem. Soc.* **1993**, *115*, 3010–3011.

24. Hai-Ren Zhang, PhD dissertation, C. Eugene Bennett Dept. Chemistry, West Virginia Univ., Morgantown, WV, **May 2000**.

25. H.-R. Zhang, K. K. Wang, *J. Org. Chem.* **1999**, *64*, 7996–7999.

26. T. Imamoto, Y. Sugiura, N. Takiyama, *Tetrahedron Lett.* **1984**, *25*, 4233–4236.

27. W. E. Barth, R. G. Lawton, *J. Am. Chem. Soc.* **1966**, *88*, 380–381.

28. W. E. Barth, R. G. Lawton, *J. Am. Chem. Soc.* **1971**, *93*, 1730–1745.

29. H. Sakurai, T. Daiko, T. Hirao, *Science* **2003**, *301*, 1878.

30. K. K. Wang, Y.-H. Wang, H. Yang, N. G. Akhmedov, J. L. Petersen, *Org. Lett.* **2009**, *11*, 2527–2530.

31. D. M. White, *J. Org. Chem.* **1974**, *39*, 1951–1952.

32. H. Yang, PhD dissertation, C. Eugene Bennett Dept. Chemistry, West Virginia Univ., Morgantown, WV, **May 2006**.

33. M. E. Jung, M. A. Lyster, *J. Org. Chem.* **1977**, *42*, 3761–3764.

34. E. Keinan, D. Perez, *J. Org. Chem.* **1987**, *52*, 4846–4851.

35. A. Krief, A.-M. Laval, *Chem. Rev.* **1999**, *99*, 745–777.

36. A. Sygula, P. W. Rabideau, *J. Am. Chem. Soc.* **1999**, *121*, 7800–7803.

37. P. W. Rabideau, A. H. Abdourazak, H. E. Folsom, Z. Marcinow, A. Sygula, R. Sygula, *J. Am. Chem. Soc.* **1994**, *116*, 7891–7892.

38. H. Cui, N. G. Akhmedov, J. L. Petersen, K. K. Wang, *J. Org. Chem.* **2010**, *75*, 2050–2056.

39. J.-Q. Yu, H.-C. Wu, E. J. Corey, *Org. Lett.* **2005**, *7*, 1415–1417.

40. N. E. Heimer, M. Hojjatie, C. A. Panetta, *J. Org. Chem.* **1982**, *47*, 2593–2598.

41. P. Nicolet, J.-Y. Sanchez, A. Benaboura, M. J. M. Abadie, *Synthesis* **1987**, 202–203.

42. A. El-ghayoury, R. Ziessel, *J. Org. Chem.* **2000**, *65*, 7757–7763.

43. L. J. Charbonnière, N. Weibel, R. F. Ziessel, *Synthesis* **2002**, 1101–1109.

44. P. C. B. Page, A. S. Hamzah, D. C. Leach, S. M. Allin, D. M. Andrews, G. A. Rassias, *Org. Lett.* **2003**, *5*, 353–355.

45. H. Cui, C. Eugene Bennett Dept. Chemistry, West Virginia Univ., Morgantown, WV, unpublished results.

46. K. Ruitenberg, H. Kleijn, C. J. Elsevier, J. Meijer, P. Vermeer, *Tetrahedron Lett.* **1981**, *22*, 1451–1452.

47. H. Li, J. L. Petersen, K. K. Wang, *J. Org. Chem.* **2001**, *66*, 7804–7810.

48. S. H. Pine, G. S. Shen, H. Hoang, *Synthesis* **1991**, 165–167.

49. F. N. Tebbe, G. W. Parshall, G. S. Reddy, *J. Am. Chem. Soc.* **1978**, *100*, 3611–3613.

50. S. Hagen, M. S. Bratcher, M. S. Erickson, G. Zimmermann, L. T. Scott, *Angew. Chem. Int. Ed.* **1997**, *36*, 406–408.

51. T. J. Hill, R. K. Hughes, L. T. Scott, *Tetrahedron* **2008**, *64*, 11360–11369.

52. E. H. Fort, P. M. Donovan, L. T. Scott, *J. Am. Chem. Soc.* **2009**, *131*, 16006–16007.

53. R. Jasti, J. Bhattacharjee, J. B. Neaton, C. R. Bertozzi, *J. Am. Chem. Soc.* **2008**, *130*, 17646–17647.

54. B. D. Steinberg, L. T. Scott, *Angew. Chem. Int. Ed.* **2009**, *48*, 5400–5402.

55. B. L. Merner, L. N. Dawe, G. J. Bodwell, *Angew. Chem. Int. Ed.* **2009**, *48*, 5487–5491.

# CHAPTER 3

# ANIONS OF BUCKYBOWLS

DAVID EISENBERG, ROY SHENHAR, and MORDECAI RABINOVITZ

## 3-1. INTRODUCTION

Research on anions of polycyclic aromatic hydrocarbons (PAHs)[1,2] began in 1867, with the potassium reduction of naphthalene by Berthelot.[3] Later, Schlenk and coworkers studied metal reductions of anthracene and other polyarenes, with emphasis on structure and synthetic applications.[4] Ever since, chemists have been fascinated by carbanions of PAHs for their role as important reactive intermediates and for their special electronic properties.

Within the family of PAH carbanions, there exists a fascinating subgroup: anions of *curved* PAHs (Figure 3-1). Hopf classified curved PAHs according to the curvature-inducing element:[5] (1) angularly annelated benzene rings (e.g., helicenes), (2) bridged systems (e.g., cyclophanes), and systems with non-six-membered rings (e.g., buckybowls). The interplay of strain and aromaticity in curved PAHs can shed light on the concept of aromaticity. Furthermore, the strain–aromaticity balance shifts unpredictably in the anions of curved PAHs, since the addition of electrons enables various transformations, electronic or structural, that serve to stabilize the reduced compounds.

Buckybowls are curved PAHs that can be mapped onto the surfaces of fullerenes. They are especially intriguing because of their structural similarity to fullerenes and carbon nanotubes (CNT).[2] Buckybowls serve as models for these novel carbon allotropes, highlighting the interplay of strain and conjugation, magnetic and electronic behavior, and other properties of such compounds. Because of their capacity for supercharging (similar to other symmetric PAHs), buckybowl anions can be viewed as charge storages; however, as they are bowl-shaped, they also store structural strain. Under the right circumstances, the combined strain might be harvested in synthesis, such as in the gradual construction of fullerenes by closing

*Fragments of Fullerenes and Carbon Nanotubes: Designed Synthesis, Unusual Reactions, and Coordination Chemistry*, First Edition. Edited by Marina A. Petrukhina and Lawrence T. Scott.
© 2012 John Wiley & Sons, Inc. Published 2012 by John Wiley & Sons, Inc.

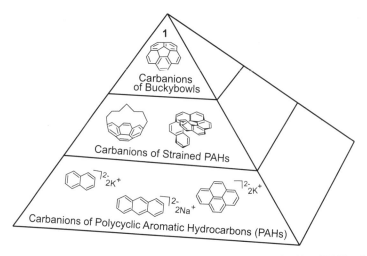

**Figure 3-1.** A family of buckybowl carbanions as part of the larger family of PAH carbanions. Corannulene (**1**) is the archetypal buckybowl.

bay regions. Thus, the reactivity of buckybowl carbanions might lead to applications in the rational synthesis of fullerenes and CNTs, a long-term goal in chemistry.

In this chapter, we attempt to review the field of buckybowl anions, which have been at the center of research in our lab and in others since the 1980s. Corannulene (**1**) is the archetypal buckybowl, and the reduction of corannulene is a key reaction in the study of buckybowl anions.

### 3-1-1. Preparation and Stabilization of Buckybowl Anions

Most buckybowl anions studied so far were prepared by contact with alkali metals. The metals used were Li (wire), Na, K, Rb, or Cs (mirror). The reduction was carried out in dry aprotic solvents under inert atmosphere (vacuum or argon gas). Electron transfer was often accomplished in ethereal solvents, such as deuterated tetrahydrofuran (THF-$d_8$), where the oxygen atom of the solvent stabilizes the alkali metal cations.

Buckybowls can accept several charges (up to six electrons), alternating their magnetic behavior from paramagnetic in odd-number reduction states to diamagnetic with even numbers of electrons. The electron transfer proceeds in equilibrium, as depicted in Figure 3-2.[1]

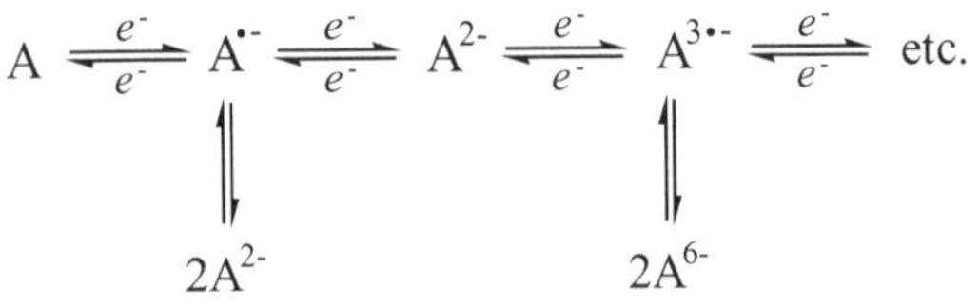

$$A \underset{e^-}{\overset{e^-}{\rightleftharpoons}} A^{\bullet -} \underset{e^-}{\overset{e^-}{\rightleftharpoons}} A^{2-} \underset{e^-}{\overset{e^-}{\rightleftharpoons}} A^{3 \bullet -} \underset{e^-}{\overset{e^-}{\rightleftharpoons}} \text{etc.}$$

$$2A^{2-} \qquad\qquad 2A^{6-}$$

**Figure 3-2.** Typical electron transfer sequence in buckybowl anions (A).

In many buckybowls, the lowest unoccupied molecular orbital (LUMO) energy levels are degenerate, due to their high symmetries. As a result, the electrons added on charging often thermally populate a low-lying triplet spin state at room temperature, rendering the compounds invisible to nuclear magnetic resonance (NMR). At low temperatures, a lower-energy singlet spin state can often be observed with NMR, existing as a result of Jahn–Teller distortion.[6] The thermal singlet–triplet equilibrium depends on the electronic structure, as well as on the counterion and solvent.[6]

Buckybowl anions can also be prepared electrochemically. However, this route is less common because the experimental reduction potentials that can be applied are not as low as those of alkali metals, and the higher reduction states are often impossible to reach by this method.[7] Another alternative for the formation of PAH carbanions is the deprotonation of precursors containing protons attached to $sp^3$-hybridized carbons at benzylic positions (such as sumanene; see Section 3-5).

The bulky buckybowl anions are stabilized by counterions, which are involved in ion–solvation equilibrium with the solvent molecules.[8] The limiting cases in this equilibrium are free ion pairs (FIPs) and contact ion pairs (CIPs), and in between there are several forms of solvent-separated ion pairs (SSIPs). FIPs are highly atypical for carbanions. The ion–solvation equilibrium of buckybowl anions is influenced by various factors.[8,9] CIP is preferred when charge is localized on the skeleton of the anion, when the temperature is increased (due to entropic considerations) and with soft cations (e.g., $K^+$). SSIP is preferred in ethereal solvents, due to their high dielectric constants, and with hard cations such as $Li^+$. However, lithium cations would form CIP inside a molecular cage or in between layers of anions in sandwich-type dimers.

Following reduction, PAHs undergo various stabilizing transformations. For example, charge segregation or electron redistribution on the skeleton might serve to decrease the antiaromatic character or increase the aromatic character in the molecule (e.g., Section 3-3-2). Other stabilizing phenomena might include the formation of supramolecular oligomers, in which alkali metal cations bind together anionic layers (e.g., Sections 3-2-5 and 3-2-8). In some cases, even more stabilization might be gained by bond formation or cleavage in the molecule (e.g., Section 3-3-1).

## 3-1-2. Characterization of Buckybowl Anions

Nuclear magnetic resonance (NMR) spectroscopy is the most informative method for the study of diamagnetic anions of buckybowls, and electron spin resonance (ESR) spectroscopy complements it for paramagnetic anions. UV–visible absorption spectroscopy assists in monitoring the progress of the reduction process, which is often accompanied by a series of color changes. Electrochemical studies are also conducted.

Full structural elucidation of the buckybowl anions is enabled by combinations of 1D and 2D NMR methods. By applying homonuclear and heteronuclear correlation methods for studying through-bond and through-space interactions, complete structural and conformational assignment of the molecule can be obtained.

The proton ($^1$H) and carbon ($^{13}$C) chemical shifts reveal the charge densities, mode of conjugation,[10] and magnetic anisotropy in the molecule.[6,11,12] Furthermore, the carbon atoms' hybridization can be identified by typical ranges of $^1J_{C,H}$ coupling constants.[13] The relationship between carbon chemical shifts and charge density has been studied both experimentally and theoretically.[8,14,15] Fraenkel and coworkers suggested a universal linear dependence,[15] formulated by the following empirical relationship

$$\Delta\delta = K\Delta q_\pi$$

where $\Delta\delta$ is the chemical shift difference relative to the neutral polycycle, $\Delta q_\pi$ is the charge density difference, and $K$ is a constant ($K_H = 10.7$ ppm/$e^-$, $K_C = 160$ ppm/$e^-$).[16] An improved relationship that takes into account the anisotropy of the anions was reported by Eliasson et al.[11] Their equation better evaluates the degree of charging and mode of charge delocalization in the molecules:

$$K_C = F_C + \frac{n_C}{Q_\pi}a\chi_H$$

Here, $F_C$ is the pure chemical shift/charge factor (estimated to be 134 ppm/$e^-$), $n_C$ is the total number of carbon atoms in the $\pi$ system, $Q_\pi$ is the total $\pi$-charge change (i.e., $-2$ for dianions), $a$ is a negative constant (estimated to be $-2.4$).[11] The proton anisotropy term ($\chi_H$) is calculated from the average shifts change of the $^1$H chemical shifts ($<\Delta\delta_H>$), and the average change in $\pi$ charge at the proton-bearing carbon atoms ($<\rho_\pi>$, obtained from quantum mechanical calculations):

$$\chi_H = <\Delta\delta_H> - K_H <\rho_\pi>$$

The $^7$Li NMR spectroscopy is commonly used to study the ion-pairing equilibria in lithium reductions. In addition, the lithium cations serve as aromaticity probes, reporting the strength and magnitude of the induced ring current in their vicinity (Figure 3-3).[17] In cases of supramolecular aggregation, lithium cations sandwiched between anionic layers have telltale absorptions from $-10$ ppm and higher. These absorptions assist in the unequivocal identification of such aggregation (e.g., Section 3-2-5).

The most commonly used techniques for the study of dynamic processes that occur on the timescale of the NMR experiments are variable-temperature (VT) NMR (usually for $^1$H or $^7$Li) and exchange NMR spectroscopy (EXSY).[18] In VT-NMR the

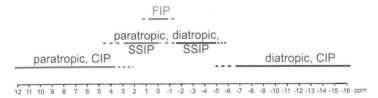

**Figure 3-3.** Ranges of $^7$Li NMR absorptions, indicating ion pairing and induced ring currents.[17]

coalescence temperature for two interchanging peaks is determined, and the energy barrier for the dynamic process can be calculated. EXSY is a 2D NMR technique that enables the study of exchange processes at a single temperature, by measuring the magnetization transfer due to chemical exchange.[19] The energy barriers are calculated from the volume of the 2D exchange peaks using the matrix method described in the literature.[20] These methods allow for the determination of energy barriers in the range of 5–25 kcal/mol, depending on the properties of the solvent.[18]

Diffusion NMR is a powerful technique for the study of aggregation processes. It has been applied in the study of aggregation in buckybowl anions. In diffusion NMR, the rate of self-diffusion of a compound along a pulsed-gradient magnetic field is measured, allowing the characterization of solvation and ion-pairing effects in monomeric anions, as well as determination of size in supramolecular aggregates.[21] Several pulse sequences are available, from the early pulsed-gradient spin echo (PGSE)[22] sequence to the advanced bipolar longitudinal eddy-current delay (BP-LED)[23] pulse sequence. The latter is commonly used in commercial spectrometers with the diffusion-ordered spectroscopy (DOSY)[24] technique, which allows for two-dimensional mapping of diffusion constants versus chemical shifts. The extraction of a hydrodynamic radius ($r_H$) from the diffusion constant ($D$) is usually based on the Stokes–Einstein equation

$$D = \frac{k_B T}{6\pi\eta r_H}$$

where $k_B$ is Boltzmann constant, $T$ is the temperature, and $\eta$ is the solution viscosity (e.g., Section 3-2-7).[21]

Quantum mechanical calculations[25] have been essential to the field of carbanions throughout its existence, with theoretical analysis explaining observed phenomena and guiding experimental efforts. Different levels of complexity were used to gain useful information: from simple Hückel and $\omega\beta$ calculations, through semiempirical methods to density functional theory (DFT), and *ab initio* level calculations. In buckybowl anions research, theoretical calculations were applied to predict minimum-energy geometries, relative energies of conformations, counterion effects, magnetically induced currents, and dynamic processes such as bowl-to-bowl inversion barriers. NMR chemical shifts are calculated with the gauge-independent atomic orbital (GIAO) method.[26] Charge distributions are often calculated with the natural population analysis method, based on the natural bond orbitals (NBO) theory.[27]

Calculations of the electronic structure of anions, such as the shape and energy of molecular orbitals, have been instrumental in rationalizing and predicting various structural changes that occur on charging. In the case of large buckybowls, the identification of various substructures on the skeleton affords important insights on charge distribution, NMR chemical shifts, and reactivity (e.g., see discussion of bicorannulenyl dianions in Section 3-2-8, and Section 3-4).

Electron spin resonance (ESR) spectroscopy is another auxiliary tool used in the study of buckybowl radical anions.[28] Although the signal of the uncoupled electron (ESR) is much stronger than that of the nuclei (NMR), the collected signals contain less structural information. The method has been applied for radical anions of buckybowls to study their symmetry, structural dynamics, and ion-pairing effects. Fourier transform (FT) ESR was used with laser flash photolysis in studies of the dynamics of electron photoejection, in which light irradiation induces an ejection of an electron from a diamagnetic anion, to give a transient paramagnetic radical observed in ESR (e.g., Section 3-2-3).

Electrochemical studies of the reduction of buckybowls have also been conducted (e.g., Section 3-2-1). Cyclic voltammetry (CV) was used to study the thermodynamics and kinetics of redox processes, as well as ion-pairing phenomena. The main drawback of this method is the available experimental reduction potentials, which are seldom negative enough to achieve high reduction states.

## 3-2. REDUCTION OF CORANNULENE AND SUBSTITUTED CORANNULENES

### 3-2-1. Full Spectroscopic Characterization of Corannulene Reduction

Corannulene is the archetypal buckybowl, and its radical anion was the first buckybowl anion to be investigated,[29] immediately after its first synthesis in 1966.[30] New synthetic routes toward corannulene were developed in the early 1990s,[31] opening the way to extensive research into the properties of this fascinating buckybowl and its anions.

Because of the high symmetry of corannulene ($C_{5v}$), its LUMO energy levels are doubly degenerate, allowing it to accept four electrons.[32] Each reduction state of corannulene has been detected and characterized experimentally and theoretically. The progress of the reduction was monitored visually by color changes; the monoanion radical is emerald green, the dianion is purple, and the tetraanion is brown.[33] The color of the trianion was not observed, since it exists only in steady-state concentrations due to disproportionation to the di- and tetraanions.[34]

The ultimate reduction state of corannulene depends on the counterion. Lithium metal reduces corannulene up to a tetraanion,[33] and it was initially reported that other alkali metals cannot reduce it past the dianion.[34] However, prolonged reaction times afford a trianion when charging with sodium metal, and a tetraanion for potassium, rubidium, and cesium.[35] The reversible electrochemical reduction up to the trianion of corannulene was also achieved, enabled by a rational choice of solvent/electrolyte system.[7,36,37]

The equilibrium electron transfer processes in corannulene are depicted in Figure 3-4.[34] Up to the triply reduced corannulene, each anion is in equilibrium with its predecessor along the reduction coordinate. However, the formation of the tetraanion was found to be irreversible in a lithium reduction, due to the emergence of a thermodynamically favorable supramolecular dimer (see Section 3-2-5).[38]

The view that corannulene consists of two concentric aromatic annulenes (Figure 3-5) was put forward in the seminal paper on synthesis by Barth and Lawton,[30] who also gave the compound its common name (*cor + annula*, meaning heart + ring in Latin)

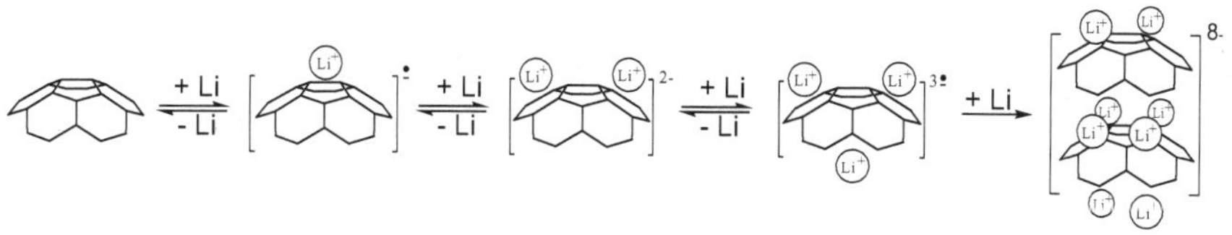

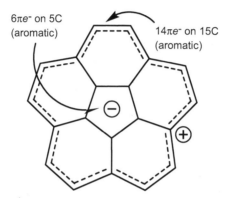

**Figure 3-4.** The reduction pathway of corannulene.[34]

6πe⁻ on 5C
(aromatic)

14πe⁻ on 15C
(aromatic)

**Figure 3-5.** Valence bond structure of neutral corannulene in the annulene-within-an-annulene model.[30]

based on this model. The effect of reduction on the magnetic properties of the compound can be rationalized in the annulene-within-an-annulene framework, at least as a first approximation. The $^1$H NMR spectrum of neutral corannulene consists of one singlet at a low field ($\delta = 7.93$ ppm in THF-$d_8$),[33] indicating a diatropic ring current about the perimeter. On charging to a dianion, the ring current becomes paratropic, indicated by a broad absorption at $\delta = -5.6$ ppm (186 K, Section 3-2-3).[34] Finally, the tetraanion ring current reverts to diatropic again, absorbing at $\delta = 6.95$ ppm (at 298 K).[33] Theoretical calculations at the DFT level suggested that the added charge goes to the perimeter,[7] causing the ring current to alternate between diatropic (neutral, 14πe⁻), paratropic (dianion, 16πe⁻), and diatropic (tetraanion, 18πe⁻). However, molecular orbital calculations, carried out at the semiempirical and *ab initio* levels,[7] have shown the structure of the anions to be more complicated than the simplistic representation of the annulene-within-an-annulene model.

## 3-2-2. Corannulene Monoanion

The radical monoanion of corannulene was initially prepared both electrochemically and with sodium and potassium metals, and the resulting anions were shown to be equivalent using ESR spectroscopy.[29] The monoanion was revisited in several studies, repeating the electrochemical and ESR characterizations.[7,34,36,37]

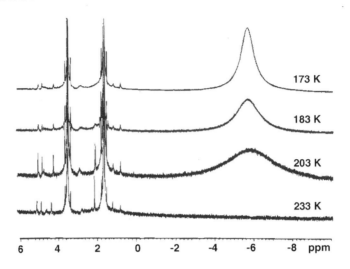

**Figure 3-6.** Variable-temperature $^{1}$H NMR of the corannulene dianion. (Reprinted with permission from Ref. 34. Copyright 1995 American Chemical Society.)

### 3-2-3. Corannulene Dianion

The dianion of corannulene was discovered several years after the tetraanion.[34] It appears in the $^{1}$H NMR as a high-field, broad absorption at very low temperatures (Figure 3-6). The dianion was observed in reductions with all alkali metals.[35]

The disappearance of the spectrum at higher temperatures was attributed to a thermal population of the triplet spin state.[6] Although corannulene dianion should have a triplet ground state, due to the doubly degenerate LUMO energy levels in the compound, a Jahn–Teller distortion leads to a close-lying singlet ground state.[34,39]

Theoretical semiempirical and *ab initio* calculations predict that corannulene flattens on charging (Figure 3-7).[7,32] This prediction was supported by experiments on 1,8-dicorannulenyloctane tetraanion (see Section 3-2-7).[35] The bowl inversion energy barrier of each dianionic charged bowl was determined by variable-temperature NMR, utilizing the methylene protons as diastereotopic probes. The barriers were found to be some 2 kcal/mol lower than in neutral corannulene:[40] 8.8 and 9.3 kcal/mol for the potassium- and cesium-derived dianions, respectively. These values were in good agreement with the early MP2/3-21G calculations (9.1–10.2 kcal/mol), and were later supported by DFT calculations (B3PW91/Midi). The flattening can be explained in two ways: (1) since the newly occupied orbitals are of $\pi^{*}$ nature with multicenter bonding, they force the involved carbons to lie on the same geometric plane, while also weakening and elongating the bond, thus allowing the flexibility required for flattening; and (2) Coulomb repulsion exists between the charges, especially as most of them are on the periphery, causing mutual repulsion and thus flattening. The flattening effect was predicted to be even more pronounced in the tetraanion, and the observation of a single, average $^{1}$H NMR signal at all temperatures

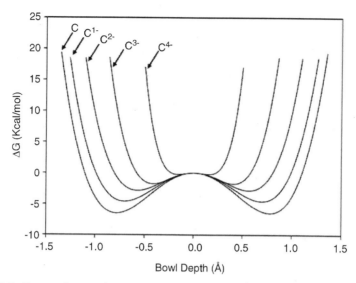

**Figure 3-7.** Energy diagram for the bowl-to-bowl inversion barrier of corannulene and its anions, calculated with B3PW91/Midi in vacuum. (Reprinted with permission from Ref. 7. Copyright 2007 American Chemical Society.)

for the corannulene tetraanion dimer (Section 3-2-5) provides an experimental support for this prediction.

The magnetically induced ring currents were computed at the *ab initio* level of theory for the corannulene di- and tetraanion.[41] The magnitude of the π-ring currents was found to be large relative to benzene, and strongly dependent on the oxidation state. Together with neutral corannulene and the dication, these four molecules constitute a very special set that spans all possible patterns of -hub–rim circulation: diatropic hub–diatropic rim (tetraanion), diatropic hub–paratropic rim (dianion), paratropic hub–diatropic rim (neutral), and paratropic hub–paratropic rim (dication). This analysis also suggests that the corannulene dianion, observed in [1]H NMR as a "diamagnetic" singlet only at low temperatures, is actually a paramagnetic closed-shell species.

The light-induced electron ejection from the corannulene dianion was studied by time-resolved laser flash photolysis and FT-ESR, demonstrating a facile, $Li^+$-mediated recombination pathway.[42] The initially formed intimate cage complex, consisting of excited corannulene and an electron, undergoes dissociation due to electrostatic repulsion, through a low-energy path enabled by the formation of a $[Li^+, e^-]$ ion pair.

## 3-2-4. Corannulene Trianion

The corannulene trianion was observed in a reduction with lithium metal, and an ESR study established that no dimerization occurs at this reduction state.[34] The first evidence was the lack of variation in the characteristic hyperfine pattern over the entire reduction stage from the di- to the tetraanion. This indicates that the number of coupled counterions does not change during the reduction, refuting a dimerization hypothesis. In addition, the ESR spectrum changed reversibly and with no decrease in intensity as the temperature was varied between 200 and 290 K, presumably as a result of tightening and loosening of the associated $Li^+$ cations. The reversibility of the temperature variation also excludes the possibility of a dimerization process for the trianion.

### 3-2-5. Corannulene Tetraanion

The corannulene tetraanion, initially prepared with lithium as the reducing metal, was quite distinctive as a rare example of a "supercharged" PAH.[33] It holds the world record with 0.2 additional electrons per carbon atom, more than $C_{60}$ hexaanion $(0.1\ e^-/C)$[43] and superseding even potassium-intercalated graphite $(C_8K, 0.125\ e^-/C)$.[44] The formation of the tetraanion with all other alkali metals except sodium was also demonstrated, however, after significantly longer reduction times.[35] The $^1H$ NMR spectrum of the lithium-reduced tetraanion consists of a single sharp line $(\delta = 6.95\ ppm,\ THF\text{-}d_8)$, while corannulene tetraanions prepared with the other metals affords spectra with broad absorptions that appeared only at low temperatures, indicating rapid equilibrium with the trianion radical or an energetically close triplet state. The degree of charge of the observed species was first established by quenching with water, which produced a tetrahydrocorannulene derivative as a major product.[33] In addition, the average change in the $^{13}C$ NMR shifts was found to equal 180.5 ppm/$e^-$, similar to the typical average value of $\sim 160$ ppm/$e^-$ for an assignment of a four-electron reduction.[16]

The most intriguing property of the corannulene tetraanion was discovered serendipitously, when *tert*-butylcorannulene was charged with lithium metal.[38] It was found that corannulenic bowls self-assemble into supramolecular dimers, where two anionic decks are bound with multiple lithium ions. Since a single substitution drastically lowers the symmetry of the corannulene moiety (from $C_{5v}$ to $C_1$), dimers of monosubstituted corannulene exist in two forms: a *meso* dimer and a *d,l* dimer (Figure 3-8). As a result, the $^1H$ NMR absorptions of *tert*-butylcorannulene tetraanion were split into two sets, revealing that the previously observed corannulene tetraanion was actually a highly symmetric supramolecular dimer. The formation of a "mixed

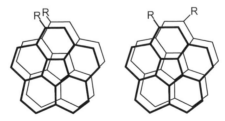

**Figure 3-8.** Supramolecular stereochemistry of monosubstituted corannulene dimers: *meso* and *d,l*.

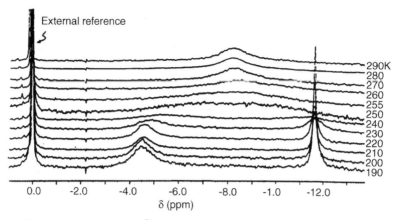

**Figure 3-9.** Variable-temperature $^7$Li NMR spectrum of the corannulene tetraanion dimer. Labels next to spectra denote temperature. The external reference is LiBr.[38]

sandwich" of corannulene and *tert*-butylcorannulene further corroborated the proposed self-assembly.

The dimeric nature of the tetraanion was supported by evidence from $^7$Li NMR (Figure 3-9). Absorptions at unusually high field ($\delta = -11.7$ ppm, THF-$d_8$) were assigned to very strongly shielded Li$^+$ ions, which are sandwiched between tetraanionic decks and thus experience both diatropic ring currents. These lithium cations exist in CIP with the corannulene, while the absorption at $-4.5$ ppm belongs to SSIP lithium cations, probably located at the exterior of the bowls. This nature of the ion pairing was demonstrated by addition of a common ion (LiBr) to the THF-$d_8$ solution: no peak for free lithium appeared at 0.0 ppm, and the peak at $-4.5$ ppm disappeared, while a new absorption of the interchanging FIP and SSIP lithium cations appeared at $-2.5$ ppm. The peak at $-11.7$ remained unaffected, attesting to a more intimate CIP bonding of the sandwiched lithium cations.

The interior and exterior lithium cations interchange on the NMR timescale, as evident from the coalescence of the $^7$Li absorptions at $T_c = 265$ K. The energy barrier ($\Delta G^{\ddagger}$) for this process was calculated to equal 13.2 kcal/mol at this temperature. The cation interchange occurs without dissociation of the dimer. Corannulene continues its rapid bowl inversion dynamics within the dimer, as suggested by the equivalence of protons from both bowls. No splitting is observed down to 175 K, corroborating the prediction that corannulene tetraanion is flatter than the neutral compound.

Diffusion NMR studies have confirmed the dimerization hypothesis.[45,46] Charging in itself suffices to slow down the diffusion of monomeric PAHs by some 20–30%, due to ion pair formation that leads to an increased solvation shell.[45] The diffusion constant of corannulene tetraanion, however, decreased by as much as 45%, from 15.7 to 8.4 ($\times 10^{-10}$ m$^2$/s), demonstrating a significant increase in size of the diffusing species, well beyond solvation effects.[45]

Electron photoejection was also observed in lithium-charged tetraanionic corannulene.[47] The emissive ESR spectra of the resulting corannulene trianion and photoelectron were measured, and various electron spin polarization mechanisms were discussed. A single model, the radical triplet pair mechanism, could explain the ESR results nicely. The appeal of the system is the medial nature of the binding, enabling the study of spin dynamics and photochemistry of ion-bound complexes that lie somewhere between freely diffusing ion pairs and fixed-distance donor–acceptor systems.

### 3-2-6. Penta-*t*-Butyl-Corannulene

A noteworthy aspect of the corannulene dimerization was revealed in the reduction of 1,3,5,7,9-penta-*t*-butylcorannulene (**2**, Figure 3-10) with lithium metal.[48] Monomeric corannulene tetraanion was observed for the first time, in addition to the two expected dimers (*d,l* and *meso*). The [1]H-NMR absorption of the monomer appeared at an even lower field ($\delta = 7.86$ ppm for the rim hydrogen atoms, THF-$d_8$) than the two peaks of the diastereomeric dimers ($\delta = 7.18$ and 6.98 ppm for the rim hydrogen atoms). The [7]Li-NMR absorptions were also assigned to CIP and SSIP lithium cations in the dimers and the monomer.

Interestingly, the monomer disappears slowly over time, and eventually only dimers remain. This demonstrates that bulky groups have little influence on the final outcome of the reduction, and there is only a kinetic hindrance to dimerization.

### 3-2-7. Dicorannulenyloctane

In 1,8-dicorannulenyloctane (**3**, Figure 3-11), two corannulene units are tethered with an octane chain.[35] The reduction of **3** with lithium (and with other alkali metals) proceeded similarly to that of corannulene (more about the dianion in Section 3-2-3). At the last stage of the reduction, when each bowl is charged to a tetraanion, high-field absorptions in the [7]Li NMR ($\delta = -11.5$ ppm) served again as evidence for a dimerization event.

Two modes of aggregation can be expected for its octaanion, *intra*- and *inter*-molecular, where the latter is expected to lead to supramolecular oligomers. To

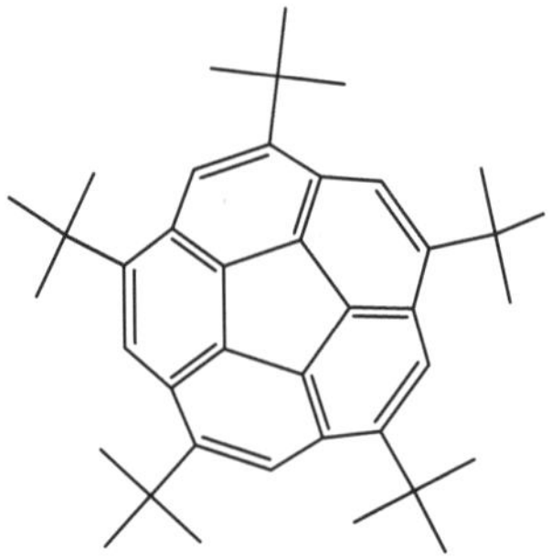

**2**

**Figure 3-10.** 1,3,5,7,9-penta-*t*-butyl-corannulene (**2**).

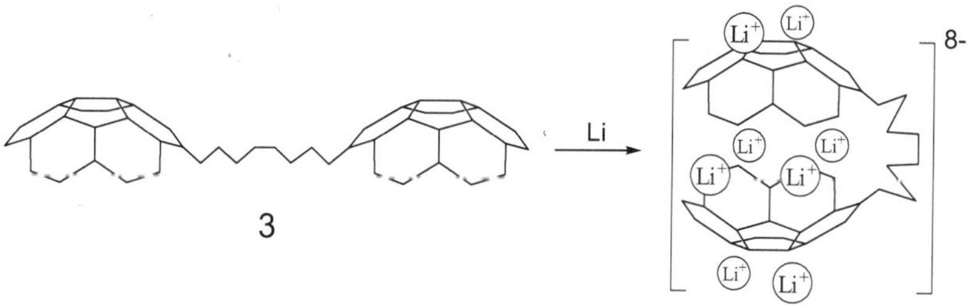

**Figure 3-11.** Reduction of 1,8-dicorannulenyloctane with lithium metal.

determine the type of stacking, an estimate for the size of the aggregate was sought. To that end, the diffusion rate of the compound was measured in a diffusion NMR experiment by the pulsed-gradient stimulated echo (PGSE) technique. The diffusion constant of $3^{8-}$ ($5.5 \times 10^{-10}$ m$^2$/s) is only slightly lower than that of corannulene tetraanion dimer $1^{4-}$ ($7.2 \times 10^{-10}$ m$^2$/s), a deceleration probably caused by the alkyl chain. This finding is consistent with intramolecular sandwich formation and less so with intermolecular stacking, which would have led to the formation of oligomers and a more significant decline in the diffusion rate. Intramolecular aggregation was further supported by reduction experiments of **3** with corannulene, where no "mixed sandwiches" were observed.

## 3-2-8. Bicorannulenyl

***Bicorannulenyl Dianion.*** Bicorannulenyl is a biaryl composed of two corannulene bowls (**4**, Figure 3-12).[49] The first diamagnetic anion afforded by a potassium reduction[50] was assigned a charge of two electrons, based on a comparison of calculated [13]C NMR shifts and the fully assigned experimental shifts. Two isomers were observed in the [1]H NMR, and the stereochemistry of the compound was analyzed with DFT calculations. An intriguing phenomenon was observed—the central bond was found to shorten, and the twist angles about it were flattened (Figure 3-12). In other words, **4** became an overcrowded ethylene on charging to a dianion.

The transformation of a biaryl into an overcrowded ethylene is quite surprising, as the opposite process is the more widely known.[51] In fact, overcrowded ethylenes respond to many kinds of stimuli (charging, heat, pressure, and light) and switch to a twisted conformation to minimize steric crowding.[52] In such transformations,

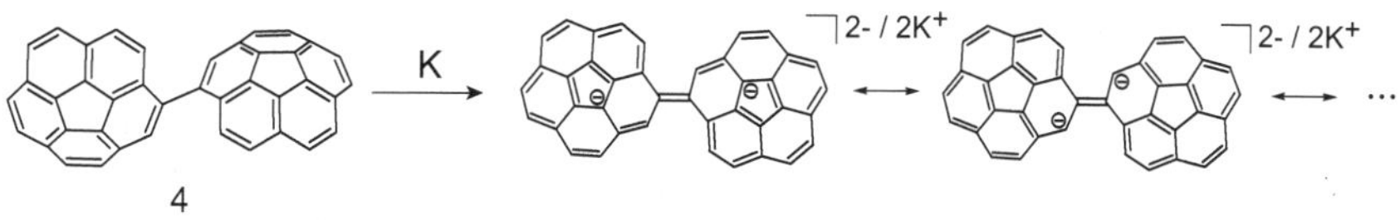

**Figure 3-12.** Reduction of bicorannulenyl (**4**) with potassium metal to the first reduction state.

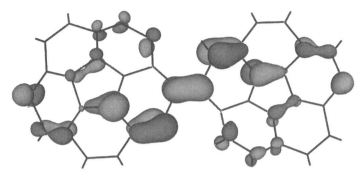

**Figure 3-13.** Highest occupied molecular orbital of the bicorannuenyl dianion (B3LYP/ 6-31G$^*$).

the central bond becomes longer and the twist angle becomes closer to 90°. However, in simple, noncrowded biaryls the central bond is known to become shorter on charging, such as in biphenyl and bipyridine radical ions.[53] The observation that the same occurs in **4**—in which a sterically crowded 1,1′-binaphthyl substructure can be identified—can be explained by the introduction of a curved element into the structure. The major isomer adopts, according to the calculations, a conformation with a 180° central angle, in which the bowls fit comfortably to minimize repulsion.

The calculated LUMO orbital of **4** demonstrates that the added electrons strengthen the bonding on the central bond (Figure 3-13). In the Lewis resonance structures that stem from a central double bond, there is significant bond length alternation along the periphery. Experimental proof for such alternation was obtained from the NMR $^1J_{H,H}$ couplings, which become significantly alternant in magnitude along the rim compared to the neutral **4** case.

**_Bicorannulenyl Octaanion._** The remarkable self-assembly of corannulene tetraanions suggests the possibility for a unique implementation—a novel binding motif for supramolecular polymers.[54] The bonding in supramolecular polymers is based on secondary interactions, such as hydrogen bonds, metal coordination, and π–π stacking.[55] Because of the reversibility of the bonding, the length of the polymer—which exists in equilibrium with its monomers—is dependent on monomer concentration, temperature, solvent, and so on. Thereby, new properties, such as self-healing, sensing, and improved processing conditions, may result.

It was demonstrated that charging monomer **4** to a tetraanion with lithium metal yields an assembly of reversibly bonded oligomers. This is a "proof of concept" for a new binding motif in the world of supramolecular polymers (Figure 3-14), termed charged polyarene stacking. Moreover, such a self-assembled anionic polymer is also a negative-charge reservoir, with eight electrons per monomer available for reversible transfer. In contrast with **3**, with its flexible octyl spacer, the rigidity of **4** does not allow for intramolecular aggregation.

The $^1$H and $^7$Li NMR spectra of the **4** octaanion are very complex, with over 35 absorptions in the proton spectrum at room temperature. The very complexity of the proton and lithium spectra, as compared to neutral **4** and the corannulene dimer

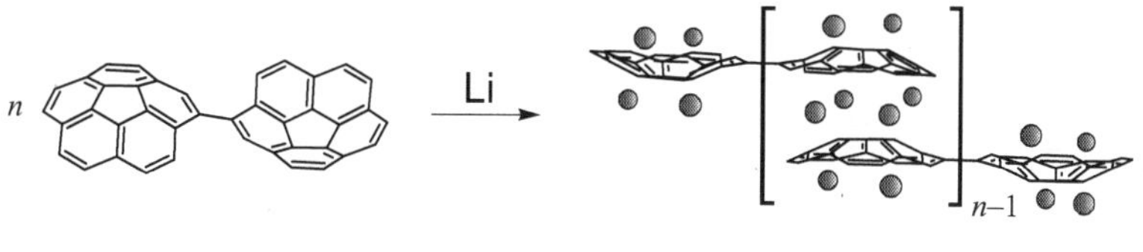

$4_n^{8-}/8n\,Li^+$

**Figure 3-14.** Reduction of bicorannulenyl (**4**) with lithium metal to the last reduction state.

spectra, respectively, is indicative of a complex aggregation phenomenon similar to the NMR spectra of polymers.

The extent of the aggregation was demonstrated with diffusion NMR experiments, demonstrating a significant decrease in the diffusion constants ($D$) on charging. From $D = 8.3 \pm 0.2 \times 10^{-10}\,m^2/s$ in neutral bicorannulenyl, the diffusion constant decreased to $D = 3.1 \pm 0.3 \times 10^{-10}\,m^2/s$ on average in the charged mixture. The degree of polymerization was determined with the Stokes–Einstein equation (Section 3-1-2) and appropriate polymer-folding models. According to the analysis, oligomers of four or five units (molecular weight = 2000–2500 g/mol), either linear or cyclic, exist in the mixture at equilibrium.

The reversibility of the binding was demonstrated with mixtures containing methylcorannulene, acting as a chain stopper. The addition of monofunctional monomers is a common tool used to probe and control the degree of polymerization in supramolecular polymers.[56] In this study, it was found that methylcorannulene homodimers diffuse faster when the chain stopper is present in the solution. This finding indicates that methylcorannulene tetraanions interchange, on the timescale of the NMR experiment, between the homodimer and the oligomers, proving the reversible nature of bonding in the oligomers.[54]

### 3-2-9. Ball-and-Socket Li$^+$-Mediated Stacking: Corannulene and Fullerene Anions

The anionic self-assembly motif of corannulene was further broadened, and applied to the organization of fullerenes. This was accomplished by the demonstration of a supramolecular heterodimer between with corannulene tetraanion and anions of *penta*-derivatized fullerenes (Figure 3-15).[57]

On reduction of pentamethylfullerene ($Me_5C_{60}H$) and pentaphenylfullerene ($Ph_5C_{60}H$) with lithium metal, the compounds undergo deprotonation, and a cyclopentadienyl moiety is formed at one pole of the fullerene. Further reduction introduces four more electrons into the fullerene derivatives, to give the pentaanions. A nonhomogeneous charge distribution serves to enhance electrostatic interactions, and thus promotes aggregation of the fullerene with the structurally complementary corannulene bowl.

At this stage, the formation of two heterodimer isomers was determined by various NMR techniques. New corannulene absorptions were evident in the $^1H$ and $^{13}C$ NMR

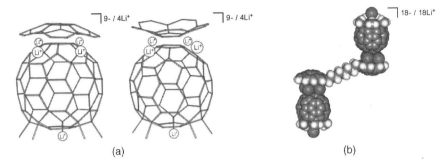

**Figure 3-15.** (a) The two isomers of the supramolecular heterodimer of corannulene tetraanion and the pentamethylfullerene pentaanion; (b) a possible structure of the pentamethylfullerene-1,8-dicorannulenyloctane bis(heterodimer). (Reprinted with permission from Ref. 57. Copyright 2005 American Chemical Society.)

spectra, as well as new high-field absorptions in the $^7$Li NMR. Diffusion NMR experiments demonstrated that the new corannulene peaks diffuse at the same rate as do the fullerene peaks.

The structure of the two heterodimer isomers has been elucidated [Figure 3-15 (a)].[57] It was suggested that the isomers stem from either concave or convex conformation of the bowl in the dimer, and that the bowl is coordinated to the nonderivatized pole of the fullerenes. This assignment was supported by symmetry considerations for the methyl groups of the fullerene, lack of through-space interactions between the methyl groups and corannulene, and experiments with substituted corannulenes. The utility of the self-assembly motif in the organization of fullerenes was demonstrated in principle using tethered corannulenes as host molecules, with the formation of a bis(heterodimer) [Figure 3-15(b)] between a derivatized fullerene and 1,8-dicorannulenyloctane.

## 3-3. ANIONS OF ANNELATED CORANNULENES

Annelation of corannulene changes the π topology of the molecule, which often leads to surprising phenomena on reduction. Various new moieties can be identified on the skeleton of such extended corannulenes, steering the molecules toward novel modes of reactivity. Two general types of annelated corannulene derivatives were studied in the past: *peri*-annelated with five-membered rings (5MRs) and annelated with six-membered rings (6MRs).

### 3-3-1. Anions of Indenocorannulenes

On reduction of indenocorannulene (**5**) with potassium metal, a four-step alternating reductive dimerization/bond cleavage process was demonstrated (Figure 3-16).[58] Covalent dimers were formed in the mono- and trianion reduction stages via electron coupling of the "corner" (C1) carbon atoms. The presence of a dibenzofulvene moiety

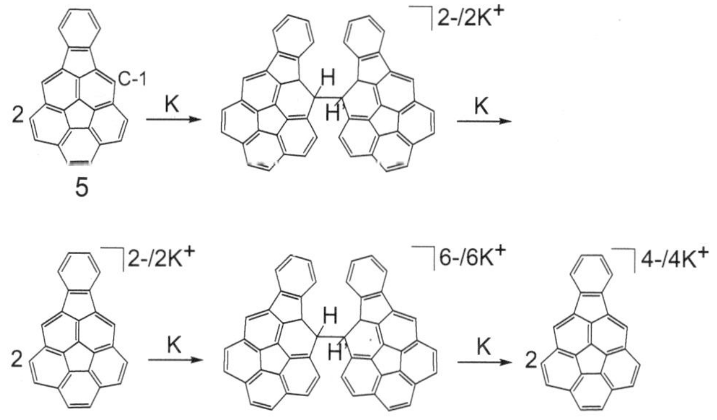

**Figure 3-16.** The stepwise reduction of indenocorannulene with potassium metal.

on the extended corannulene skeleton explains the dimerization behavior, as dibenzofulvene radical anion is known to undergo similar dimerization.[59] However, the reduction process in indenocorannulene is unique since it includes as many as four alternating dimerization/bond cleavage steps, while only two-step alternating processes of this sort were known previously.[59,60]

The dimerization reaction was revealed by NMR spectroscopy. The formation of a new σ bond was evident from the coupling constants of C1 carbon atoms ($^1J_{C,H}$ and $^3J_{H,H'}$), as measured in $^1H$ NMR and heteronuclear single-quantum coherence (HSQC) NMR experiments. Carbon atom C1 undergoes radical anion coupling due to its high charge density, which is evident from theoretical DFT calculations and the high-field $^{13}C$ NMR absorption (δ = 56.5 and 47.6 ppm for the dimer dianion and hexaanion, respectively).

When lithium served as the reducing metal, a single step of reductive dimerization was observed in the monoanionic stage, after which an unstable trianion radical was formed.[61] The trianion underwent irreversible side reactions (e.g., solvation, protonation), probably due to the decreased stabilization the smaller lithium cation offers to the charged C1 atoms.

The reduction process of diindeno[1,2,3-*bc*;1,2,3-*hi*]corannulene (**6**) proceeded altogether differently, demonstrating the drastic effect of π topology on the reactivity pattern.[61] The compound was reduced with potassium to the mono-, di-, and trianion, but not further. Moreover, no reductive dimerization was observed. Evidently, the addition of a second indeno group alters the charge distribution of the buckybowl anion relative to that of indenocorannulene **5**, leading to stabilization by charge delocalization and reduced reactivity toward dimerization.

## 3-3-2. Other Extended Corannulenes

The leading motif to follow in the reduction of extended corannulenes is changes in annulenic behavior that accompany annelation at various positions. For example, three extended corannulenes were charged to dianions with lithium or potassium metal (Figure 3-17): dibenzo[*a,g*]corannulene (**7**),[62] naphtho[2,3-*a*]corannulene

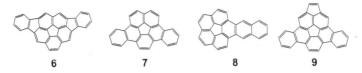

**Figure 3-17.** Diindeno[1,2,3-*bc*:1,2,3-*hi*]corannulene (**6**), dibenzo[*a,g*]corannulene (**7**), naphtho[2,3-*a*]corannulene (**8**), and dibenzo[*a,g*]cyclopenta[*kl*]corannulene (**9**).

(**8**),[61] and dibenzo[*a,g*]cyclopenta[*kl*]corannulene (**9**).[62] The first two dianions (**7**$^{2-}$ and **8**$^{2-}$) exhibited paratropic ring currents, as evident from their unusually low $K_C$ values ($-21.1$ and $-25.0$ ppm/$e^-$, respectively, for the lithium-charged dianions). However, they are still significantly less paratropic than corannulene dianion ($K_C = -183$ ppm/$e^-$). Furthermore, dianion **9**$^{2-}$ is even diatropic ($K_C = 89.0$ ppm/$e^-$).

These differences were rationalized relative to corannulene. The paratropicity of dianions **7**$^{2-}$ and **8**,$^{2-}$ similar to that of corannulene dianion but smaller in magnitude, is attributed to the small perturbation to the $\pi$ framework of corannulene. The annelated benzene rings partially quench the paratropic ring current of the corannulene rim, but the compound remains, essentially, a perturbed corannulene skeleton. However, in the case of a 5MR *peri*-annelation (**9**$^{2-}$), the $\pi$ topology is more severely altered. For example, a [5,5]fulvalene moiety can be identified on the skeleton, and it is not unreasonable to assume that the tendency to stabilize a dipolar fulvalene leads to significant distortion of the $\pi$ bonding in the compound relative to corannulene.

Compounds **7** and **8** could be charged up to trianion radicals with lithium metal, but not further. This demonstrates that these extended corannulenes do not self-assemble into lithium-stitched dimers as does corannulene. Reduction with potassium metal did afford tetraanions in the case of **7** and **9**, where the larger K$^+$ cations could stabilize the high degree of charging. Dimerization was not observed in **7**, **8**, or **9**.[61,62]

### 3-3-3. Cyclopenta[*bc*]corannulene Tetraanion: A New Type of Coordinative Dimer

Cyclopenta[*bc*]corannulene (**10**, Figure 3-18) was reduced stepwise up to a trianion with either lithium or potassium, in a manner similar to that observed for corannulene.[61] However, the potassium-charged trianion was very reactive and underwent irreversible reactions with the solvent. Further reduction with lithium metal afforded the tetraanion, and then a unique phenomenon was observed: coordinative dimerization (Figure 3-18).

The C1 carbon atom has become $sp^3$-hybridized, as indicated by its high-field shift and coupling constant ($^1J_{C,H} = 129.56$ Hz). High-field $^7$Li NMR absorptions ($\delta = 9.5$ and $-10.1$ ppm) were observed in the region typical for strongly shielded lithium cations sandwiched between aromatic anionic layers. However, no new C–C bonds were created, as demonstrated by the lack of new H–H couplings or new long-range proton–carbon correlations. In addition, self-diffusion experiments demonstrated that the new species diffuses at approximately the same rate as the corannulene tetraanion dimer.

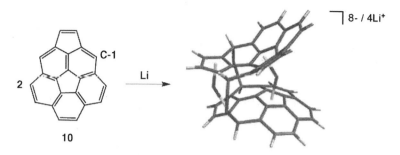

**Figure 3-18.** Reduction of cyclopentacorannulene (**10**) with lithium metal. (Reprinted with permission from Ref. 61. Copyright 2006 American Chemical Society.)

On the basis of this evidence, a coordinative dimer assignment was suggested, in which the concentration of high charge density on carbon atoms C1 induces the formation of $C-Li-C$ bonds, in addition to $\pi-Li-\pi$ bonds. Although the ability of lithium cations to interact with multiple carbanion centers has been studied in the past, this was the first observation of coordinative dimerization of a buckybowl. The nature of bonding in the coordinative dimer places it between the supramolecular dimer of corannulene tetraanion and the covalent indenocorannulene dimer.

Both **5** and **10** demonstrate the special reactivity of buckybowls, which arises also from their curved topology. The surface curvature contributes to the pyramidalization of carbon atoms on the skeleton, promoting charge localization on these carbons, and thus their bond-forming reactions.

Overall it can be said that *peri*-annelation with 5MRs alters the charge distribution much more drastically than does 6MR annelation, since 5MRs serve as strong electron-withdrawing groups. Another generalization can be made regarding "supercharging" of buckybowls; in the case of lithium reduction, it seems that dimerization is the actual driving force for the formation of a tetraanion, since lithium cations are too small for sufficient stabilization of high charges. Potassium, however, is capable of such stabilization without dimerization, as was demonstrated in a few cases.

### 3-3-4. Phenalene-Fused Corannulene Monoanion

Phenalene-fused corannulene (**11**, Figure 3-19) is a 6MR *peri*-annelated corannulene.[63] The anion of **11** was generated by deprotonation of a singly hydrogenated derivative with a base (*n*-BuLi).

A phenalene moiety can be identified on the skeleton of **11**. Both phenalene radical and phenalene anion feature an extensively spin-delocalized nature, and the phenalenyl substructure in **11** contributes strongly to its electronic behavior. DFT calculations reveal the phenalenyl anion-type substructure in the electrostatic potential surface of **11**⁻ [Figure 3-19(b)]. Further support for the prominence of this moiety in **11**⁻ is obtained from its HOMO image, prominent resonance structures, and nucleus-independent chemical shift (NICS)[64] aromaticity analysis.

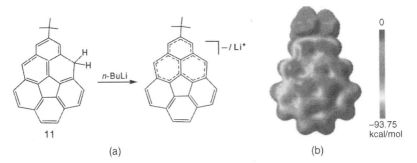

**Figure 3-19.** (a) Deprotonation of phenalenyl-fused corannulene (**11**) with a base; (b) electrostatic potential energy surface of **11**⁻ anion. (Reprinted with permission from Ref. 63. Copyright 2008 American Chemical Society.)

The bowl-to-bowl inversion energy barrier of neutral **11** was calculated to be 11.3 kcal/mol, larger than the calculated barrier for corannulene bowl inversion at the same calculation level (9.2 kcal/mol). This was attributed to the effect of *peri*-annelation.[73] Furthermore, the bowl inversion energy barriers were also calculated for neutral, di-, and trianionic **11**, affording values of 12.6, 8.1, and 5.4 kcal/mol, respectively. The continuous decrease of the bowl inversion barriers on charging is similar to that found for corannulene.

## 3-4. DIINDENOCHRYSENE ANIONS

One of the smallest symmetric buckybowls is diindeno[1,2,3,4-*defg*;1′,2′,3′,4′-*mnop*] chrysene (**12**, Figure 3-20), a concave $C_{26}H_{12}$ fullerene fragment possessing $C_{2v}$ symmetry.[65] The central double bond is exposed to nucleophilic attack, similar to the double bonds in fullerenes. Thus, whereas corannulene is a good model for the magnetic and aromatic character of fullerenes, **12** complements it by reflecting the surface reactivity of fullerenes. The reduction of **12** was explored using a combined experimental and theoretical approach.[66]

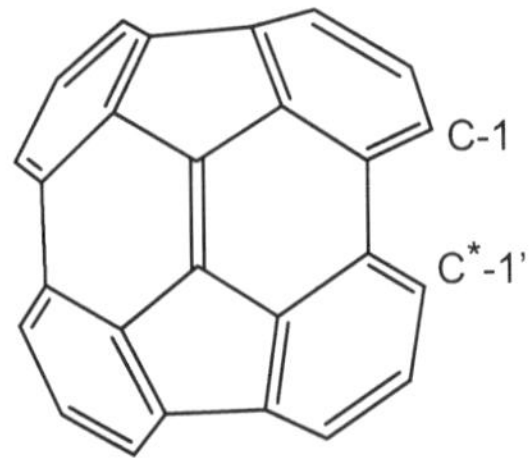

**12**

**Figure 3-20.** Diindeno[1,2,3,4-*defg*;1′,2′,3′,4′-*mnop*]chrysene (**12**).

Both **12** and a tetra-*tert*-butyl derivative of **12** were reduced with lithium metal to yield dianions and tetraanions. However, the full structural assignment of the buckybowl and its anions was not straightforward. Because of the high symmetry of the molecule, the standard 2D NMR methods fail to distinguish between the protons and carbons in the phenanthrene- and fluorene type bays since they are isochronous, such as the carbons C1 and C1'. A novel carbon-edited version of the nuclear Overhauser effect spectroscopy (NOESY) method was developed to assist in the assignment, based on measuring through-space interactions between bay protons (e.g., H1 and H1') in those molecules where one of these protons is attached to a $^{13}$C.[67] This method allowed unequivocal distinction between the more crowded phenanthrene bay and the less crowded fluorene bay even in the anion species, where charging did not allow reliance on crowding-induced chemical shift differences for the assignment. GIAO/DFT-calculated NMR shifts also assisted in completing the assignment of the NMR spectra of the anions.

Once full structural assignment was obtained, the charge distribution on the π framework could be calculated, based on the difference between the $^{13}$C NMR spectra of the anion and those of the neutral species. The experimental charge distribution compared well with the DFT calculations, both for NMR chemical shift values and when comparing the NBO charge distribution. However, the most interesting comparison is between the experimental charge distribution and the calculated electronic structure, as depicted by the atomic orbital (AO) coefficients of the relevant frontier orbitals (Table 3-1). It was found that the experimental *charge density* of an anionic species resembles the *electron density* of its LUMO orbital (Table 3-1, two leftmost columns). In other words, predictions about the *electronic structure* of the closest LUMO orbitals could be made *experimentally*.

Buckybowl anions, as well as most PAH anions, comply with the structural prerequisites for such an analysis: they are *rigid*, and therefore their orbitals do not change significantly on charging, and they are sufficiently *large* to allow for a reliable description by a single determinant, with no need for mixing of the LUMO orbitals.

The true electronic structure of **12** is revealed in the orbital analysis, which shows that it is composed of canonical benzene orbitals, with a double-bond character in the center (Table 3-1, two rightmost columns). The isolated nature of the central double bond gives rise to the reactivity pattern of **12**, which resembles that of fullerenes.

The bowl depths and bowl inversion barriers of **12** and its di- and tetraanions were also calculated, and were found to alternate along the reduction pathway. From an inversion barrier of 6.7 kcal/mol (B3LYP/6-311G$^{**}$) in the neutral **12**, the barrier rises to 11.0 kcal/mol in the dianion, and decreases to 6.5 kcal/mol for the tetraanion. This alternating behavior is in contrast to the monotonic decrease in bowl depth that is predicted and observed in the charging of corannulene.[7,32] In each case, it can be rationalized from the structure of the frontier orbital. The frontier orbital of the dianion (LUMO of neutral **12**) exhibits a strong antibonding interaction of the central double-bond fragment orbital, thus deepening the bowl by inducing a strong electronic repulsion between the two fluorene-type fragments of **12** at the hub of the bowl. As the compound is charged further, the frontier orbitals of the tetraanion (the LUMO+1 of neutral **12**) exhibit a node on the central double bond, so the charge

**TABLE 3-1. Experimental Charge Distributions and Calculated Frontier Orbitals of $12^{2-}$ and $12^{4-}$, Compared to the Canonical Orbitals of Benzene**

| Referenced Charge Distribution from $^{13}$C-NMR Spectra | Frontier Orbitals from DFT Calculations | Benzene Canonical Orbitals |
|---|---|---|
| Tetraanion | LUMO+1 | $\Psi_5$ |
| Dianion | LUMO | $\Psi_4$ |
| Neutral | HOMO | $\Psi_2$ |

*Source*: Reprinted with permission from Ref. 66. Copyright 2001 American Chemical Society.

of the last two electrons is distributed on the rim, thus flattening and widening the bowl.

In another study, **12** played a central role in a series of charged $C_{26}$ PAHs (Figure 3-21), in which an interesting correlation was demonstrated between the skeletal structure of the dianions and the order of their anisotropy relative to the neutral compounds. This correlation led to correct predictions regarding the order of reactivity of the neutral compounds toward reduction with alkali metals.[12]

Advanced charge distribution analyses were performed on the dianions of the $C_{26}$ PAHs. The proton anisotropy values ($\chi_H$, Table 3-2) revealed that the dianions are ordered on an *aromaticity scale*, from the most aromatic $13^{2-}$, to the most antiaro-

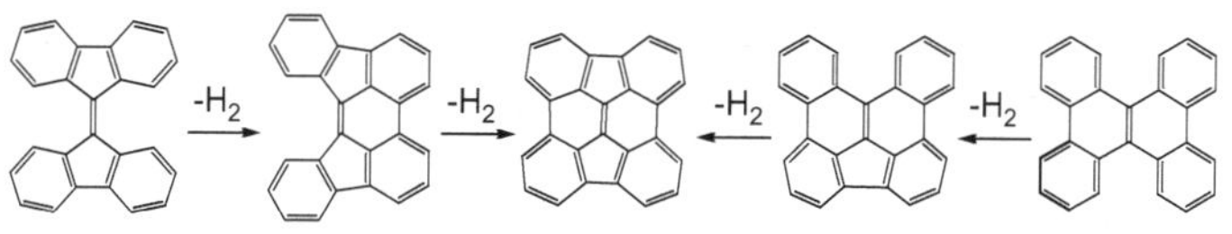

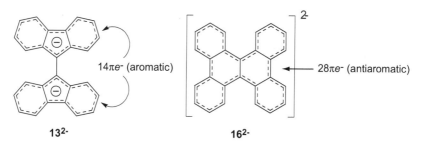

**Figure 3-21.** Five $C_{26}$ PAHs with similar bonding structures.[12]

**TABLE 3-2. Experimental and Theoretical Properties of the Dianions of Series Studied in Ref. 12**

| Compound | Anisotropy $\chi_H$ (ppm) | HOMO-LUMO Gap (kcal/mol) | Homodesmotic Stability Relative to **13** (kcal/mol) |
|---|---|---|---|
| **13**$^{2-}$ | +0.40 | 39.6 | — |
| **14**$^{2-}$ | −0.07 | 20.2 | −20.1 |
| **12**$^{2-}$ | −0.75 | 10.8 | −30.8 |
| **15**$^{2-}$ | −1.15 | 16.2 | −39.2 |
| **16**$^{2-}$ | −2.54 | 12.5 | −44.4 |

matic **16**$^{2-}$. The π-electron count of these two limiting cases (Figure 3-22) corroborates this trend. The increasing antiaromatic character along the series correlated to a decreasing HOMO-LUMO gap (Table 3-2) and suggested a decreasing relative stability. DFT calculations corroborated this prediction (Table 3-2, last column), showing that relative homodesmotic stability decreased as the antiaromatic character gathered strength. Furthermore, reducing the compounds with lithium metal in binary mixtures has provided unequivocal evidence for the predicted order of relative stabilities, showing that the PAHs with more aromatic character are reduced to anions before those with less aromatic (or more antiaromatic) character. Overall, the *anisotropy change* was suggested as a valid index for aromaticity in charged PAHs.

**Figure 3-22.** Lewis resonance structures for the most aromatic and most antiaromatic dianions of the series.[12]

## 3-5. SUMANENE TRIANION

Sumanene (**17**, Figure 3-23) is a special buckybowl, first prepared in 2003.[68] Buckybowl **17** represents a $C_{3v}$ symmetric fragment of a fullerene, with one crucial difference: the presence of three methylene carbons at benzylic positions. This disruption of the $\pi$ skeleton plays a major role in the charging of **17**. Thus, only the trianion is a true buckybowl anion, as it is composed of $sp^2$-hybridized carbons only. However, the charging process is interesting also in comparison with the corannulene reduction.

Sumanene was triply deprotonated in a stepwise fashion by treatment with equivalent amounts of $t$-BuLi in THF-$d_8$, to give the mono-, di-, and trianions (Figure 3-23).[69] Each subsequent anion exhibited $^1$H NMR absorptions at a higher field, and an increasing difference in chemical shifts between the *exo* and *endo* positions on the remaining benzylic carbons.

The charges are localized at the benzylic positions, as evident from the reaction of **17** with an electrophile.[69] On introduction of excess amounts of Me₃SiCl, the tris(trimethylsilyl) derivative **18** was afforded as the sole isomer. Perfect selectivity toward the *exo* position was exhibited, probably due to a steric effect. The strength of the diamagnetic ring current was increased along the charging of **17**, as seen in the mounting shielding effect on all of the protons of **17**$^{3-}$ in the $^1$H NMR.

Because of the high localization of charge, a reaction of sumanene anions with electrophiles allows for convenient and controlled functionalization at any desired benzylic positions.[69,70] Moreover, hexasubstituted sumanene could also be obtained from the neutral compound, through a nucleophilic substitution reaction.[70]

The reduction of **17** was studied electrochemically in $N,N$-dimethylformamide and acetonitrile.[71] Compound **17** was reduced to a monoanion in a partially chemically reversible reduction step. The limited chemical reversibility of the sumanene reduction, as compared to the fully chemically reversible formation of corannulene monoanion,[7,29,36] might stem from the decreased $\pi$ conjugation over the buckybowl skeleton, leading to a decreased ability to exchange electrons.

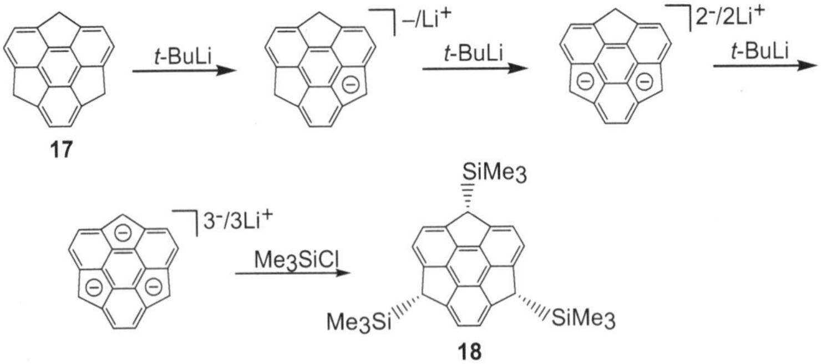

**Figure 3-23.** Deprotonation and quench of sumanene (**17**). (Reprinted with permission from Ref. 68. Copyright 2005 American Chemical Society.)

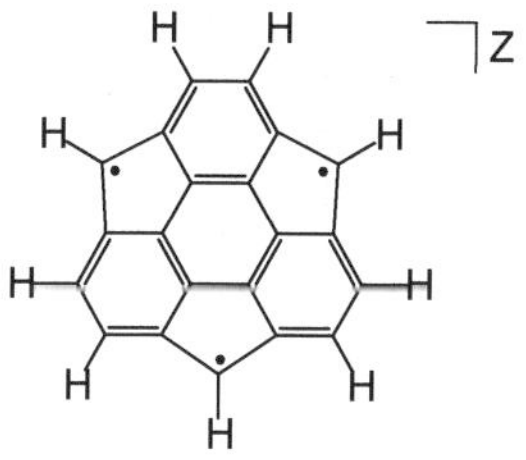

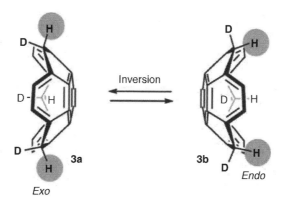

**19**

**Figure 3-24.** The sumanene structure studied by Sastry[71] ($Z = +3$ to $-3$).

The changes in the bowl-depth that occur on charging were predicted theoretically by Sastry et al. before sumanene was first synthesized.[72] They studied the hypothetical compound $C_{21}H_9^Z$ (**19**, $Z = +3$ to $-3$), which can be viewed as a $C_3$-symmetric fragment of $C_{60}$, or a pure $sp^2$ "sumanene," where a hydrogen is abstracted from each benzylic carbon (Figure 3-24).

Along the theoretical reduction pathway of **19**, from neutral to trianion, the bowl gradually flattens, similarly to corannulene. The bowl inversion barriers, calculated at the lowest-energy symmetries with B3LYP/6-31G*, decrease from 33.2 kcal/mol in the neutral $C_{21}H_9$ radical, to 29.6 in the monoanion, 20.1 in the radical dianion, and 12.2 kcal/mol in the trianion. The predicted flattening trend is analogous to that found in corannulene, arising similarly from increasing Coulombic repulsion as the degree of charging increases. However, the magnitudes of the bowl inversion barriers are larger than those of corannulene anions, as indeed sumanene is deeper than corannulene.[73]

The bowl inversion dynamics of sumanene and its anions were studied experimentally in the Hirao group, and their findings did not support the theoretically predicted trend. First, the bowl inversion energy barrier of neutral sumanene was determined using 2D EXSY NMR by following the chemical exchange between the *exo* and *endo* protons in deuteriosumanene (Figure 3-25). The barrier was calculated to be 20.3 kcal/mol in THF-$d_8$ at 318 K.

**Figure 3-25.** Bowl-to-bowl inversion in sumanene (**17**). (Reproduced from Ref. 70 by permission of The Royal Society of Chemistry.)

However, when the 2D EXSY NMR method was applied to the mono- and dianions of **17**, no crosspeak was observed even at elevated temperatures, suggesting that they are more rigid than the neutral molecule. The bowl inversion barriers could be determined by monitoring the equilibration of an *exo*-deuterated anion in ¹H NMR, and were, indeed, higher than that of neutral **17**: 21.8 and 21.5 kcal/mol for the mono- and dianions, respectively. The trianion could not be studied in a similar fashion, as it does not have different *endo/exo*-benzylic positions. No reevaluation of the theoretical predictions, which were not supported by the results of Hirao's experiments, has appeared yet.

## 3-6. HEMIFULLERENE

The $C_{30}H_{12}$ geodesic polyarene **20** is a $C_3$-symmetric hemifullerene isomer. Reduction of **20** with potassium metal produced the largest buckybowl anions so far.[74] Charging was observed with potassium metal but not with lithium, and only after weeks of contact with the metal at room temperature. The only diamagnetic species observed along the reduction route was the hexaanion. The degree of charging was determined using a quench with water, which cleanly yielded a hexahydro derivative ($C_{30}H_{18}$). The structure suggested for the new species was an exciting concave–concave tetrameric aggregate (Figure 3-26).

The tetramer structure assignment was based on experimental and theoretical evidence. Charge distribution analysis, calculated from ¹³C NMR shifts, revealed high charge localization on three carbon positions (Figure 3-27). These carbon atoms underwent pyramidalization, as evident from their $^1J_{C,H}$ values, typical of $sp^3$ carbons ($^1J_{C,H} = 124.7$ Hz).[13] DFT calculations on various alternative structures, such as a monomeric hexaanion with or without "covalently bonded" potassium cations, could

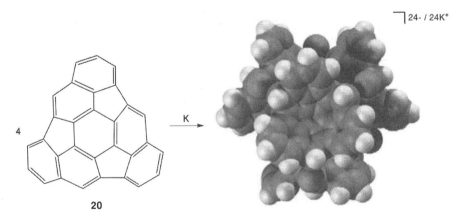

**Figure 3-26.** Reduction of hemifullerene (**20**) with potassium. Carbon atoms, hydrogen atoms, and potassium cations are colored blue, yellow, and red, respectively. (Adapted with permission from Ref. 74. Copyright Wiley-VCH Verlag GmbH & Co. KGaA.)

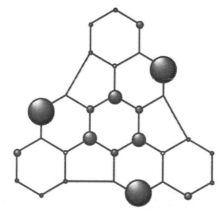

**Figure 3-27.** Calculated charge distribution on hemifullerene (**20**). (Adapted with permission from Ref. 74. Copyright Wiley-VCH Verlag GmbH & Co. KGaA.)

not explain the observed pyramidalization. The only fully consistent explanation was a tetrameric aggregate, in which the potassium cations bind hexaanionic units together at the pyramidalized carbons, with the entire supramolecular complex retaining the $C_3$ symmetry of the neutral molecule.

Supramolecular aggregation mediated by potassium cations also explains the strong counterion dependence of the reduction outcome; since C−Li distances are shorter, the charged buckybowls would repel each other both sterically and by Coulomb repulsion. Thus, the formation of a tetramer is disfavored and reduction with lithium does not proceed. The long reaction time that was required for the formation of this species can also be rationalized by the tetramer structure, as the self-assembly process proceeds with slow kinetics, and the system can reach the global thermodynamic minimum only after a sufficient period of equilibrating.

Interestingly, further reduction afforded additional sets of $^1$H NMR absorptions, possessing the same symmetry but with lower intensities. This suggests that even larger $C_3$-symmetric aggregates are present in the solution. Their irreversible formation indicates that they are even more stable than the smallest representative aggregate, the tetramer.

## 3-7. SUMMARY

Anions of buckybowls are important for the study of various topics in physical organic chemistry. Buckybowls can serve as "supercharge" reservoirs, and their anions serve as focal points for the complex interplay of strain and aromaticity. Buckybowl anions demonstrate different modes of charge distribution and localization, which depend, in part, on various moieties that can be identified on their skeletons, for example, dibenzofulvene in indenocorannulene (**5**) or [5,5]fulvalene in dibenzocyclopenta-corannulene (**9**). On charging with alkali metals, some of them form bonds of various types, from the supramolecular, $Li^+$-bound corannulene (**1**) tetraanion dimer, through the coordinative dimer of tetraanionic **9**, to the covalent dimers of **5** mono- and trianions.

In the greater family of PAH anions, the uniqueness of buckybowls lies in their curved topology. The relationship between degree of charge and bowl depth was investigated in several compounds, such as corannulene, diindenochrysene (**12**), and sumanene (**17**). The curvature of corannulene was employed as a recognition motif in the formation of a heterodimer between a corannulene tetraanion and a pentaanion of pentasubstituted fullerenes. In the case of bicorannulenyl (**4**), it was the introduction of a curved corannulene element into a biaryl structure that enabled its surprising transformation into an overcrowded ethylene, on charging to a dianion. Finally, it was shown that curvature directs reactivity in compounds such as **9**, **17**, and hemifullerene **20**, by inducing pyramidalization of certain carbon atoms in the skeleton and thus promoting charge localization and bond formation at these atoms.

Overall, anions of buckybowls constitute a vital ingredient in the charm of buckybowls, and add an exciting new chapter to the established science of carbanions.

## REFERENCES

1. For reviews on reduction of PAHs, see M. Szwarc, *Ions and Ion Pairs in Organic Reactions*, Wiley-Interscience, New York, **1974**; K. Müllen, *Chem. Rev.* **1984**, *84*, 603–646; M. Rabinovitz, *Top. Curr. Chem.* **1988**, *146*, 99–169; R. Benshafrut, E. Shabtai, M. Rabinovitz, L. T. Scott, *Eur. J. Org. Chem.* **2000**, 1091–1106.

2. I. Aprahamian, M. Rabinovitz, in *The Chemistry of Organolithium Compounds*, Z. Rappoport, I. Marek (eds.), Wiley, Chichester, UK, **2006**, Vol. 2, pp. 477–524; T. Sternfeld, M. Rabinovitz, in *Carbon Rich Compounds: From Molecules to Materials*, M. M. Haley, R. R. Tykwinski (eds.), Wiley-VCH Verlag GmbH, KGaA, Weinheim, **2006**, pp. 566–623.

3. M. Berthelot, *Ann. Chim.* **1867**, *12*, 155.

4. W. Schlenk, T. Weickel, *Chem. Ber.* **1911**, *44*, 1182–1189; W. Schlenk, A. Thal, *Chem. Ber.* **1913**, *46*, 2840–2854.

5. H. Hopf, *Classics in Hydrocarbon Chemistry*, Wiley-VCH, Weinheim, **2000**.

6. A. Minsky, A. Y. Meyer, R. Poupko, M. Rabinovitz, *J. Am. Chem. Soc.* **1983**, *105*, 2164–2172.

7. C. Bruno, R. Benassi, A. Passalacqua, F. Paolucci, C. Fontanesi, M. Marcaccio, E. A. Jackson, L. T. Scott, *J. Phys. Chem. B* **2009**, *113*, 1954–1962 and references cited therein.

8. D. H. O'Brien, in *Comprehensive Carbanion Chemistry*, Part A, E. Buncel, T. Durst (eds.), Elsevier, Amsterdam, **1980**, p. 271.

9. T. E. Hogen-Esch, J. Smid, *J. Am. Chem. Soc.* **1966**, *88*, 307–318; T. E. Hogen-Esch, E. J. Smid, *J. Am. Chem. Soc.* **1969**, *91*, 4580–4581.

10. F. Gerson, K. Müllen, E. Vogel, *Angew. Chem. Int. Ed.* **1971**, *10*, 920–921.

11. B. Eliasson, U. Edlund, K. Müllen, *J. Chem. Soc., Perkin Trans. 2* **1986**, 937–940.

12. R. Shenhar, R. Beust, S. Hagen, H. E. Bronstein, I. Willner, L. T. Scott, M. Rabinovitz, *J. Chem. Soc., Perkin Trans. 2* **2002**, 449–454.

13. H. O. Kalinowski, S. Berger, S. Braun, *Carbon-13 NMR Spectroscopy*, Wiley, Chichester, **1988**.

14. G. J. Martin, M. L. Martin, S. Odiot, *Org. Magn. Reson.* **1975**, *7*, 2–17; D. G. Farnum, V. Gold, in *Adv. Phys. Org. Chem.*, Academic Press, **1975**, Vol. 11, pp. 123–175; H. Baumann, H. Olsen, *Helv. Chim. Acta* **1980**, *63*, 2202–2211; S. Fliszár, G. Cardinal, M. T. Béraldin, *J. Am. Chem. Soc.* **1982**, *104*, 5287–5292.

15. G. Fraenkel, R. E. Carter, A. McLachlan, J. H. Richards, *J. Am. Chem. Soc.* **1960**, *82*, 5846–5850.

16. P. C. Lauterbur, *J. Am. Chem. Soc.* **1961**, *83*, 1838–1846; P. C. Lauterbur, *Tetrahedron Lett.* **1961**, *2*, 274–279; H. Spiesecke, W. G. Schneider, *Tetrahedron Lett.* **1961**, *2*, 468–472; T. Schaefer, W. G. Schneider, *Can. J. Chem.* **1963**, *41*, 966–982.

17. Based on A. Ayalon, PhD thesis, Hebrew Univ. Jerusalem, **1993**; D. Eisenberg, MSc thesis, Hebrew Univ. Jerusalem, **2006**.

18. M. Rabinovitz, A. Pines, *J. Am. Chem. Soc.* **1969**, *91*, 1585–1590; J. Sandström, in *Dynamic NMR Spectroscopy*, Academic Press, **1982**, pp. 93–123.

19. C. L. Perrin, T. J. Dwyer, *Chem. Rev.* **1990**, *90*, 935–967.

20. B. A. Johnson, J. A. Malikayil, I. M. Armitage, *J. Magn. Reson.* **1988**, *76*, 352–357.

21. Y. Cohen, L. Avram, L. Frish, *Angew. Chem. Int. Ed.* **2005**, *44*, 520–554 and references cited therein; A. Macchioni, G. Ciancaleoni, C. Zuccaccia, D. Zuccaccia, *Chem. Soc. Rev.* **2008**, *37*, 479–489 and references cited therein.

22. E. O. Stejskal, J. E. Tanner, *J. Chem. Phys.* **1965**, *42*, 288–292.

23. D. H. Wu, A. D. Chen, C. S. Johnson, *J. Magn. Reson. A* **1995**, *115*, 260–264.

24. C. S. Johnson, *Prog. Nucl. Magn. Reson. Spectrosc.* **1999**, *34*, 203–256 and references cited therein.

25. D. C. Young, *Computational Chemistry: A Practical Guide for Applying Techniques to Real World Problems*, Wiley-Interscience, New York, **2001**.

26. R. Ditchfield, *Mol. Phys.* **1974**, *27*, 789–807; K. Wolinski, J. F. Hinton, P. Pulay, *J. Am. Chem. Soc.* **1990**, *112*, 8251–8260.

27. J. P. Foster, F. Weinhold, *J. Am. Chem. Soc.* **1980**, *102*, 7211–7218; A. E. Reed, R. B. Weinstock, F. Weinhold, *J. Chem. Phys.* **1985**, *83*, 735–746.

28. F. Gerson, *High Resolution ESR Spectroscopy*, Wiley, New York, **1970**.

29. J. Janata, J. Gendell, C.-Y. Ling, W. Barth, L. Backes, H. B. Mark Jr., R. G. Lawton, *J. Am. Chem. Soc.* **1967**, *89*, 3056–3058.

30. W. E. Barth, R. G. Lawton, *J. Am. Chem. Soc.* **1966**, *88*, 380FF.

31. L. T. Scott, M. M. Hashemi, D. T. Meyer, H. B. Warren, *J. Am. Chem. Soc.* **1991**, *113*, 7082–7084; A. Borchardt, A. Fuchicello, K. V. Kilway, K. K. Baldridge, J. S. Siegel, *J. Am. Chem. Soc.* **1992**, *114*, 1921–1923; G. Zimmermann, U. Nuechter, S. Hagen, M. Nuechter, *Tetrahedron Lett.* **1994**, *35*, 4747–4750.

32. A. Sygula, P. W. Rabideau, *J. Mol. Struct. (THEOCHEM)* **1995**, *333*, 215–226.

33. A. Ayalon, M. Rabinovitz, P. C. Cheng, L. T. Scott, *Angew. Chem. Int. Ed.* **1992**, *31*, 1636–1637.

34. M. Baumgarten, L. Gherghel, M. Wagner, A. Weitz, M. Rabinovitz, P. C. Cheng, L. T. Scott, *J. Am. Chem. Soc.* **1995**, *117*, 6254–6257.

35. E. Shabtai, R. E. Hoffman, P.-C. Cheng, E. Bayrd, D. V. Preda, L. T. Scott, M. Rabinovitz, *J. Chem. Soc., Perkin Trans. 2* **2000**, 129–133.

36. R. J. Angelici, B. Zhu, S. Fedi, F. Laschi, P. Zanello, *Inorg. Chem.* **2007**, *46*, 10901–10906.

37. T. J. Seiders, K. K. Baldridge, J. S. Siegel, R. Gleiter, *Tetrahedron Lett.* **2000**, *41*, 4519–4522.

38. A. Ayalon, A. Sygula, P.-C. Cheng, M. Rabinovitz, P. W. Rabideau, L. T. Scott, *Science* **1994**, *265*, 1065–1067.

39. T. Yamabe, K. Yahara, T. Kato, K. Yoshizawa, *J. Phys. Chem. A* **2000**, *104*, 589–595; T. Sato, A. Yamamoto, T. Yamabe, *J. Phys. Chem. A* **2000**, *104*, 130–137.

40. L. T. Scott, M. M. Hashemi, M. S. Bratcher, *J. Am. Chem. Soc.* **1992**, *114*, 1920–1921.

41. G. Monaco, L. T. Scott, R. Zanasi, *J. Phys. Chem. A* **2008**, *112*, 8136–8147.

42. R. Shenhar, I. Willner, D. V. Preda, L. T. Scott, M. Rabinovitz, *J. Phys. Chem. A* **2000**, *104*, 10631–10636.

43. J. W. Bausch, G. K. S. Prakash, G. A. Olah, D. S. Tse, D. C. Lorents, Y. K. Bae, R. Malhotra, *J. Am. Chem. Soc.* **1991**, *113*, 3205–3206; Q. S. Xie, E. Perezcordero, L. Echegoyen, *J. Am. Chem. Soc.* **1992**, *114*, 3978–3980; R. C. Haddon, *Acc. Chem. Res.* **1992**, *25*, 127–133.

44. W. Rüdorf, E. Z. Schulze, *Anorg. Chem.* **1954**, *277*, 156.

45. Y. Cohen, A. Ayalon, *Angew. Chem. Int. Ed.* **1995**, *34*, 816–818.

46. R. E. Hoffman, E. Shabtai, M. Rabinovitz, V. S. Iyer, K. Müllen, A. K. Rai, E. Bayrd, L. T. Scott, *J. Chem. Soc., Perkin Trans. 2* **1998**, 1659–1664.

47. G. Zilber, V. Rozenshtein, P. C. Cheng, L. T. Scott, M. Rabinovitz, H. Levanon, *J. Am. Chem. Soc.* **1995**, *117*, 10720–10725.

48. A. Weitz, M. Rabinovitz, P. C. Cheng, L. T. Scott, *Synth. Met.* **1997**, *86*, 2159–2160.

49. D. Eisenberg, A. S. Filatov, E. A. Jackson, M. Rabinovitz, M. A. Petrukhina, L. T. Scott, R. Shenhar, *J. Org. Chem.* **2008**, *73*, 6073–6078.

50. D. Eisenberg, E. A. Jackson, J. M. Quimby, L. T. Scott, R. Shenhar, *Angew. Chem. Int. Ed.* **2010**, *49*, 7538–7542.

51. R. H. Cox, *J. Magn. Reson.* **1970**, *3*, 223–229; M. Walczak, G. D. Stucky, *J. Organomet. Chem.* **1975**, *97*, 313–323; E. Ahlberg, O. Hammerich, V. D. Parker, *J. Am. Chem. Soc.* **1981**, *103*, 844–849; P. Neta, D. H. Evans, *J. Am. Chem. Soc.* **1981**, *103*, 7041–7045; B. A. Olsen, D. H. Evans, *J. Am. Chem. Soc.* **1981**, *103*, 839–843; B. A. Olsen, D. H. Evans, I. Agranat, *J. Electroanal. Chem.* **1982**, *136*, 139–148; D. H. Evans, R. W. Busch, *J. Am. Chem. Soc.* **1982**, *104*, 5057–5062; D. H. Evans, N. Xie, *J. Am. Chem. Soc.* **1983**, *105*, 315–320; Y. Cohen, J. Klein, M. Rabinovitz, *J. Chem. Soc., Chem. Commun.* **1986**, 1071–1073; M. Jørgensen, K. Lerstrup, P. Frederiksen, T. Bjørnholm, P. Sommer-Larsen, K. Schaumburg, K. Brunfeldt, K. Bechgaard, *J. Org. Chem.* **1993**, *58*, 2785–2790.

52. For reviews on overcrowded ethylenes, see P. U. Biedermann, J. J. Stezowski, I. Agranat, in *Advances in Theoretically Interesting Molecules*, R. P. Thummel (ed.), JAI Press, Stamford, CT, **1998**, Vol. 4, pp. 245–322; B. L. Feringa, *Acc. Chem. Res.* **2001**, *34*, 504–513; B. L. Feringa, *J. Org. Chem.* **2007**, *72*, 6635–6652.

53. G. F. Pedulli, *Res. Chem. Intermed.* **1993**, *19*, 617–634 and references cited therein; M. F. M. Post, J. Langelaar, J. D. W. Vanvoorst, *Chem. Phys. Lett.* **1977**, *46*, 331–333; L. Ould-Moussa, O. Poizat, M. Castellá-Ventura, G. Buntinx, E. Kassab, *J. Phys. Chem.* **1996**, *100*, 2072–2082; M. Castellá-Ventura, E. Kassab, G. Buntinx, O. Poizat, *Phys. Chem. Chem. Phys.* **2000**, *2*, 4682–4689; M. S. Denning, M. Irwin, J. M. Goicoechea, *Inorg. Chem.* **2008**, *47*, 6118–6120; E. Gore-Randall, M. Irwin, M. S. Denning, J. M. Goicoechea, *Inorg. Chem.* **2009**, *48*, 8304–8316.

54. D. Eisenberg, E. A. Jackson, J. M. Quimby, L. T. Scott, R. Shenhar, *Chem. Commun.* **2010**, *46*, 9010–9012.

55. L. Brunsveld, B. J. B. Folmer, E. W. Meijer, R. P. Sijbesma, *Chem. Rev.* **2001**, *101*, 4071–4097; T. F. A. De Greef, M. M. J. Smulders, M. Wolffs, A. P. H. J. Schenning, R. P. Sijbesma, E. W. Meijer, *Chem. Rev.* **2009**, *109*, 5687–5754; J. D. Fox, S. J. Rowan, *Macromolecules* **2009**, *42*, 6823–6835.

56. P. M. Saville, E. M. Sevick, *Langmuir* **1998**, *14*, 3137–3139; F. Lortie, S. B. Boileau, L. Bouteiller, C. Chassenieux, F. Lauprêtre, *Macromolecules* **2005**, *38*, 5283–5287; W. Knoben, N. A. M. Besseling, M. A. C. Stuart, *Macromolecules* **2006**, *39*, 2643–2653.

57. I. Aprahamian, D. Eisenberg, R. E. Hoffman, T. Sternfeld, Y. Matsuo, E. A. Jackson, E. Nakamura, L. T. Scott, T. Sheradsky, M. Rabinovitz, *J. Am. Chem. Soc.* **2005**, *127*, 9581–9587.

58. I. Aprahamian, G. J. Bodwell, J. J. Fleming, G. P. Manning, M. R. Mannion, T. Sheradsky, R. J. Vermeij, M. Rabinovitz, *Angew. Chem. Int. Ed.* **2003**, *42*, 2547–2550.

59. M. Baumgarten, K. Müllen, *Top. Curr. Chem.* **1994**, *169*, 1–103; E. D. Bergmann, *Chem. Rev.* **1968**, *68*, 41–84 and references cited therein.

60. U. Edlund, B. Eliasson, *J. Chem. Soc., Chem. Commun.* **1982**, 950–952.

61. I. Aprahamian, D. V. Preda, M. Bancu, A. P. Belanger, T. Sheradsky, L. T. Scott, M. Rabinovitz, *J. Org. Chem.* **2006**, *71*, 290–298.

62. A. Weitz, E. Shabtai, M. Rabinovitz, M. S. Bratcher, C. C. McComas, M. D. Best, L. T. Scott, *Chem. Eur. J.* **1998**, *4*, 234–239.

63. S. Nishida, Y. Morita, A. Ueda, T. Kobayashi, K. Fukui, K. Ogasawara, K. Sato, T. Takui, K. Nakasuji, *J. Am. Chem. Soc.* **2008**, *130*, 14954–14955.

64. P. v. R. Schleyer, C. Maerker, A. Dransfel, H. Jiao, N. J. R. v. E. Hommes, *J. Am. Chem. Soc.* **1996**, *118*, 6317–6318.

65. S. Hagen, U. Nuechter, M. Nuechter, G. Zimmermann, *Tetrahedron Lett.* **1994**, *35*, 7013–7014; S. Hagen, U. Nuechter, M. Nuechter, G. Zimmermann, *Polycyclic Aromat. Compd.* **1995**, *4*, 209–217.

66. R. Shenhar, R. Beust, R. E. Hoffman, I. Willner, H. E. Bronstein, L. T. Scott, M. Rabinovitz, *J. Org. Chem.* **2001**, *66*, 6004–6013.

67. R. E. Hoffman, R. Shenhar, I. Willner, H. E. Bronstein, L. T. Scott, A. Rajca, M. Rabinovitz, *Magn. Reson. Chem.* **2000**, *38*, 311–314.

68. H. Sakurai, T. Daiko, T. Hirao, *Science* **2003** *301*, 1878.

69. H. Sakurai, T. Daiko, H. Sakane, T. Amaya, T. Hirao, *J. Am. Chem. Soc.* **2005**, *127*, 11580–11581.

70. T. Amaya, H. Sakane, T. Muneishi, T. Hirao, *Chem. Commun.* **2008**, 765–767.

71. P. Zanello, S. Fedi, F. F. de Biani, G. Giorgi, T. Amaya, H. Sakane, T. Hirao, *Dalton Trans.* **2009**, 9192–9197.

72. U. D. Priyakumar, G. N. Sastry, *J. Mol. Struct. (THEOCHEM)* **2004**, *674*, 69–75.

73. T. J. Seiders, K. K. Baldridge, G. H. Grube, J. S. Siegel, *J. Am. Chem. Soc.* **2001**, *123*, 517–525.

74. N. Treitel, T. Sheradsky, L. Q. Peng, L. T. Scott, M. Rabinovitz, *Angew. Chem. Int. Ed.* **2006**, *45*, 3273–3277.

# CHAPTER 4

# CURVED π-CONJUGATED STABLE OPEN-SHELL SYSTEMS POSSESSING THREE-DIMENSIONAL MOLECULAR/ELECTRONIC SPIN STRUCTURES

YASUSHI MORITA and AKIRA UEDA

## 4-1. INTRODUCTION

More than two decades have passed since the discovery of buckminsterfullerene $C_{60}$[1] and the development of its large-scale production method.[2] Inspired by its exotic ball-shaped fused polycyclic molecular/π-electronic structures, researchers have carried out a large number of fundamental and applied studies on this third carbon allotrope from both experimental and theoretical aspects in a wide range of scientific fields such as chemistry, physics, and materials science.[3] Consequently, the research field of new carbon materials possessing spherical or curved molecular and π-electronic structures has grown very rapidly. In particular, the development of superconductivity[4] and ferromagnetism[5] in the $C_{60}$-based molecular systems and the discovery of higher fullerenes[6] and carbon nanotubes[7] are widely known as the representative pioneering works. For further development of this intriguing and fascinating research field, the following issues are currently important: (1) design and synthesis of new spherical or curved molecular systems possessing a fused polycyclic carbon π-network and development of efficient synthetic method; (2) molecule-level elucidation of their molecular/electronic structures and physical properties; and (3) application to electronic devices, fuel cells, medicines, and areas. In this context, zeolite-templated carbon (ZTC), prepared by Kyotani and coworkers, attracts much attention as a new carbon nanomaterial,[8,9] as it consists of buckybowl-like nanographene assembled into a three-dimensionally regular network.[8f] Interestingly, this new carbon nanomaterial shows a remarkable potential for applications as materials for hydrogen storage and fuel cells.[9]

*Fragments of Fullerenes and Carbon Nanotubes: Designed Synthesis, Unusual Reactions, and Coordination Chemistry*, First Edition. Edited by Marina A. Petrukhina and Lawrence T. Scott.
© 2012 John Wiley & Sons, Inc. Published 2012 by John Wiley & Sons, Inc.

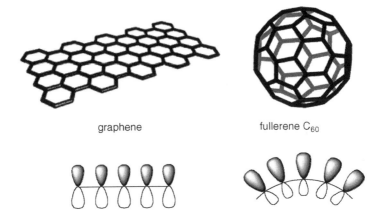

graphene                    fullerene $C_{60}$

**Figure 4-1.** Representative molecular structure and $p$ orbitals of graphene and fullerene $C_{60}$, respectively.

It is generally accepted that such unique physical properties and functionalities are intrinsically attributable to the intra- and intermolecular interactions within or between these spherical or curved molecules. Here, we focus on the difference in such interactions in spherical or curved molecules and planar molecules in terms of molecular shape. Graphene is a *planar,* sheetlike, gigantic π-conjugated electronic system comprising condensed six-membered carbon rings (Figure 4-1).[10] Thus the π electrons of graphene distribute over the two-dimensional sheet structure. In sharp contrast, the π electrons of $C_{60}$ spread over the curved molecular skeleton in a three-dimensional pattern (Figure 4-1), showing a remarkable difference in convex and concave faces. The reason could be attributed to (1) the unsymmetric nature of the $p$ orbitals of $C_{60}$ originating from a geometrically intermediate electronic structure between $sp^2$ and $sp^3$-hybrid orbitals and (2) different patterns and degree of overlap of the $p$ orbitals for each face due to the distortion of the π-electron system.[11] Therefore, intra- and intermolecular interactions of the curved $C_{60}$ system are intrinsically different from those of the *planar* graphene system in terms of molecular dimensionality. The comparison between their molecular structures in terms of the connecting mode of carbon atoms (topology of the π-electron network) provides another interesting aspect. As described above, graphene is composed only of six-membered rings of $sp^2$-hybridized carbon atoms. On the other hand, $C_{60}$ possesses six-membered and five-membered $sp^2$ carbon rings, giving rise to bending nature of the $sp^2$ carbon network of $C_{60}$ and thus the spherical molecular structure. Importantly, $C_{60}$ and graphene are categorized as nonalternant and alternant π-conjugated systems, respectively. Therefore, spherical or curved π-conjugated molecular systems such as fullerenes and carbon nanotubes are fundamentally different from planar ones seen in graphene not only in terms of molecular shape or dimensionality but also in the topology of the π-electron network.

By cutting out a curved fused polycyclic structure from fullerenes or a cap moiety of carbon nanotubes, we can obtain bowl-shaped π-conjugated molecules composed of five- and six-membered rings. Since the discovery of fullerenes and carbon nanotubes, bowl-shaped hydrocarbon molecules or buckybowls have been recognized as their fragments and attracted much attention in their syntheses, structures, and chemical and physical properties.[12] In addition, such hydrocarbons are expected to serve as possible synthetic intermediates for chemical syntheses of fullerenes and carbon nanotubes by a bottom–up method.[13] They have some characteristic features that are intrinsically different from those of fullerenes, such as (1) the presence of hydrogen atoms around the molecular skeleton, (2) the availability of both convex and concave faces,[14] and (3) dynamic bowl-to-bowl inversion behavior. Corannulene ($C_{20}H_{10}$, see discussion below)[15–17] and sumanene ($C_{21}H_{12}$)[18] are well-known buckybowls. The first synthesis of corannulene was achieved in 1966, about 20 years earlier than the discovery of $C_{60}$, by a 17-step synthetic method conducted by Barth and Lawton.[15] In 1976 single-crystal X-ray structure analysis experimentally revealed a bowl-shaped molecular structure of corannulene.[16] From the 1990s to the early 2000s, facile synthetic methods of corannulene were developed by Scott's group, Siegel's group, and Rabideau's group,[17] inducing today's developments of corannulene or buckybowl chemistry. Furthermore, in addition to the parent corannulene, various kinds of corannulene derivatives and their metal salts and complexes have been experimentally and theoretically investigated in terms of syntheses, molecular and electronic structures, bowl-to-bowl inversion barriers, aromaticity, and intra- and intermolecular interactions.[12,13,15–17]

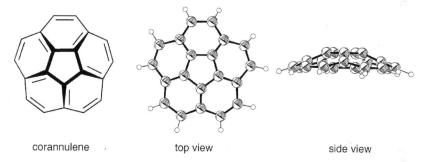

corannulene          top view          side view

Organic molecules possessing one or more unpaired electrons, organic open-shell molecules, or organic radicals, are generally known as *reaction intermediates* in organic synthetic reactions with shortlived unstable chemical species.[19] The studies on organic radicals have a 110-year history since the discovery of triphenylmethyl radical by Gomberg in 1900 that survives only under anaerobic condition.[20] However, appropriate chemical modifications are able to sufficiently stabilize organic radicals for handling in air atmosphere.[21] The successful stabilization method for open-shell molecules not only greatly contributes to fundamental studies for the elucidation of their molecular and electronic spin structures and physical properties but also plays very important roles as spin sources for a wide range of applied studies for the development of magnetic, conductive, and optical materials,[22–24] secondary batteries,[25] matter–spin qubit of quantum computers,[26] and spin probe and ESR imaging techniques.[27]

Figure 4-2 illustrates the schematic drawings of planar π radical, curved π radical, and tetrahedral σ radical. The studies on the stable radicals described above focus mainly on planar π radicals such as the phenalenyl system,[28,29] accumulating a variety of information about the structure–stability relationship and exotic physical

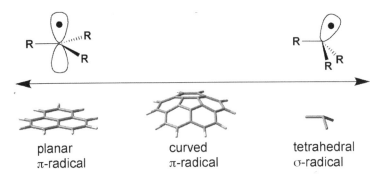

planar
π-radical

curved
π-radical

tetrahedral
σ-radical

**Figure 4-2.** Schematic representations of *planar* π-radical, *curved* π-radical, and *tetrahedral* σ-radical. R represents an arbitrary organic substituent.

properties. In a keen contrast, curved π radical is very rare, and molecule-level investigations have not been carried out because of low stability even in anaerobic conditions. As illustrated in Figure 4-2, curved π radical occupies an intermediate geometry between planar π radical and tetrahedral σ radical, and is unique in having robust three-dimensional molecular and electronic spin structures with π-electronic communication. Thus, investigation of the electronic spin distribution nature, intra- and intermolecular magnetic interactions, and local aromaticity of curved π-radical system is one of the most intriguing issues in the contemporary scientific research fields. In order to elucidate these properties at the molecular level, fundamental studies on the air-stable curved π-radical systems are crucially important, not only opening up a new research field of open-shell chemistry and molecular magnetism[21,22] but also developing functional materials based on 3D electronic spin structure and intra- and intermolecular interactions.

This chapter describes the curved π-conjugated open-shell molecules based on fullerenes and corannulene. After reviewing the open-shell chemistry of the fullerene, air-stable bowl-shaped neutral radical systems based on corannulene are discussed in detail in terms of design concepts, synthetic methods, molecular and electronic spin structures, and related physical properties.

## 4-2. OPEN-SHELL SYSTEMS BASED ON FULLERENES

As open-shell systems possessing curved π-electronic structures, fullerene-based radical anions and neutral radical species are extensively studied from both experimental and theoretical aspects.[30–32] Here we describe typical examples of these studies in view of the stability and molecular and electronic spin structures.

### 4-2-1. Radical Anion Species

Fullerene $C_{60}$ has a low-lying triply degenerated LUMO. Electrochemical measurements experimentally reveal that $C_{60}$ is able to accept up to six electrons.[33] Various kinds of radical anion salts and complexes of $C_{60}$ are prepared by chemical or electrochemical reduction, and their reactivity, molecular and electronic spin struc-tures, Jahn-teller effect, spin multiplicity, and magnetic properties are investigated in terms of X-ray crystal structure analysis, ESR, UV–visible–near infrared (NIR), and magnetic susceptibility measurements.[30]

***Monoradical Monoanion Species.*** In 1991, Wudl and coworkers synthesized a monoradical monoanion species $C_{60}^{\bullet-}$ as an air-stable $(Ph_4P)_2Cl$ salt and measured the solid-state ESR spectrum.[34a] As shown in Figure 4-3(a), the ESR spectrum gives a simple single line ($g = 1.9991$). Interestingly, this signal shows temperature-depen-dent linewidth changes. It is considered that this phenomenon is due to temperature-dependent electron spin–lattice relaxation.[34c] X-ray crystal analysis of the $(Ph_4P)_2Cl$ salt [Figure 4-3(b)] demonstrates that the $C_{60}^{\bullet-}$ molecule is surrounded by the countercations, and the intermolecular distance between two neighboring $C_{60}^{\bullet-}$ molecules is $> 6$ Å.[34b] Therefore, intermolecular magnetic interactions between the $C_{60}^{\bullet-}$ molecules are very weak (Curie temperature $\sim -1$ K).[34e] Because of the disordering of the crystal structure, a detailed discussion on molecular structure is not performed. In 1995, Broderick and coworkers reported the crystal structure of $[(C_5Me_5)_2Ni^+][C_{60}^{\bullet-}]$ as the first $C_{60}^{\bullet-}$ crystal structure without structure disorder-ing.[34d] As illustrated in Figure 4-3(c), the $C_{60}^{\bullet-}$ molecule is sandwiched between the two nickelocene cations $(C_5Me_5)_2Ni^+$. The five-membered rings of the $C_{60}^{\bullet-}$ and the nickelocene are oriented in a staggered manner, and the interplanar distance is about 3.2 Å. Analysis of the interatomic distance of the $C_{60}^{\bullet-}$ suggests that the $C_{60}^{\bullet-}$ skeleton is slightly distorted because of the sandwich structure. ESR and magnetic susceptibility measurements are not reported, and thus the electronic spin properties are not unveiled.

Interestingly, some $C_{60}^{\bullet-}$ derivatives form an intermolecular C–C single bond, giving a $\sigma$-bonded dimeric pair $(C_{60}^-)_2$.[35,36] The first $\sigma$-bonded dimer $(C_{60}^-)_2$ is observed in the metastable $M \cdot C_{60}$ (M = K, Rb, and Cs) phases.[35] Konarev, Saito, and coworkers prepared the charge transfer complexes of $C_{60}$ with chromocene or cobaltocene derivatives and experimentally revealed that the radical anion species of $C_{60}$ in the solid state exist as the closed-shell dimer $(C_{60}^-)_2$ at low temperature, while it exists as the open-shell monomer $C_{60}^{\bullet-}$ at room temperature (Figure 4-3(d,e)).[36] At 300 K, the $C_{60}^{\bullet-}$ molecule forms zigzag chains with equal intercage center-to-center distances (10.1 Å) [Figure 4-3(d)]. Such a zigzag chain arrangement is also observed at 100 K [Figure 4-3(e)]. However, the intercage center-to-center distances at 100 K are significantly different (9.28 and 9.91 Å), due to the formation of the $\sigma$-bonded dimer $(C_{60}^-)_2$. The intermolecular C–C distance in $(C_{60}^-)_2$ (1.60 Å) [Figure 4-3(e)] is longer than the usual C–C single-bond distance (1.54 Å). Reflecting this temperature-dependent structural change, the $C_{60}^{\bullet-}$ complexes are paramagnetic at room tempera-ture and diamagnetic at low temperature, as studied by magnetic susceptibility measurements. Furthermore, Konarev, Saito, and coworkers experimentally disclosed that $C_{70}^{\bullet-}$ also forms a $\sigma$-bonded closed-shell dimer $(C_{70}^-)_2$ at low temperature.[36b,37]

***Polyanion Species.*** Among the polyanion species of $C_{60}$, dianion and trianion species are extensively studied from both experimental and theoretical sides.[30,38,39]

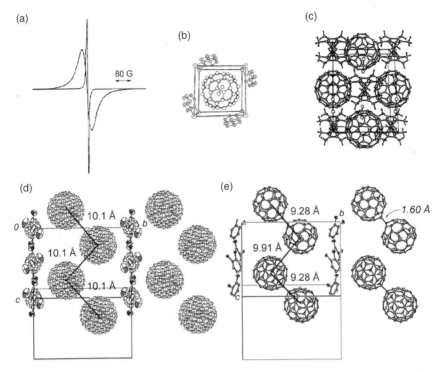

**Figure 4-3.** (a) Solid-state ESR spectra at 300 K (broad line) and at 80 K (narrow line). (Reprinted with permission from Ref. 34a. Copyright 1991 American Chemical Society.) (b) Packing structure of [(Ph$_4$P)$_2$Cl][C$_{60}$$^{\bullet-}$]. (Reprinted from U. Bilow, M. Jansen, *J. Chem. Soc. Chem. Commun.* **1994**, 403–404. Reproduced by permission of The Royal Society of Chemistry.) (c) Crystal structures of [(C$_5$Me$_5$)$_2$Ni$^+$][C$_{60}$$^{\bullet-}$]. (Reprinted with permission from Ref. 34d. Copyright 1995 American Chemical Society.) (d) Crystal structures of [(C$_5$Me$_5$)$_2$Cr$^+$][C$_{60}$$^{\bullet-}$] at 300 K. (Reprinted with permission from Ref. 36b. Copyright 2003 American Chemical Society.) (e) Crystal structures of [(C$_5$Me$_5$)$_2$Cr$^+$][C$_{60}$$^{\bullet-}$] at 100 K. (Reprinted with permission from Ref. 36b. Copyright 2003 American Chemical Society.) Values shown in nonitalic and italic (roman) type in (d) and (e) indicate the intercage center-to-center distance and the intercage C–C bond distance, respectively. Some phenyl groups and hydrogen and chlorine atoms in (b), hydrogen atoms in (c), and (C$_5$Me$_5$)$_2$Cr$^+$ cations in (d) and (e) are omitted for clarity.

According to Hund's rule, the dianion and trianion are expected to be ground-state triplet ($S = 1$) diradical C$_{60}$$^{2\bullet2-}$ and quartet ($S = \frac{3}{2}$) triradical C$_{60}$$^{3\bullet3-}$ species, respectively, due to the triply degenerated LUMO of C$_{60}$. However, if the degeneracy of the LUMO is broken, the dianion and trianion could give low-spin species such as closed-shell singlet ($S = 0$) C$_{60}$$^{2-}$ and doublet ($S = \frac{1}{2}$) monoradical C$_{60}$$^{\bullet3-}$ in the ground state, respectively. Many researchers are discussing their ground-state spin multiplicity in terms of ESR measurements in glassy solutions; however, this topic still remains controversial.

In 1995, Eaton and coworkers carried out variable-temperature ESR measurements of the electrochemically generated dianion species of C$_{60}$.[38d] The signal intensity attributable to the triplet ($S = 1$) diradical species $C_{60}^{2\bullet2-}$ decreases with temperature lowering, and thus the dianion is considered to be the ground-state singlet species ($S = 0$) with a thermally accessible triplet state ($S = 1$). The energy gap between the two states is estimated to be about $600 \, cm^{-1}$ (863 K). Furthermore, in 1996, Baumgarten and coworkers generated the dianion of $C_{60}$ by reduction with potassium metal and determined a zero-field splitting parameter $D$ in the triplet ($S = 1$) state to be 2.7 mT in terms of the fine-structure ESR spectrum in a frozen solution.[38e] On the basis of this experimentally determined $D$ value, they evaluate the average spin–spin distance $r$ to be 1 nm (10 Å) by the point dipole approximation method. This $r$ value is close to the mean atom-to-atom diameter of pristine $C_{60}$ in the crystal (7.1 Å),[40] demonstrating that the two spins locate separately on each $C_{60}$ skeleton. It is considered that the repulsion of electronic spins in the triplet state favors such a large spin–spin distance. In contrast to these observations, in 2002, Reed and coworkers claimed that the dianion is a diamagnetic ($S = 0$) species.[38g] This is because the dianion species generated by using a highly purified $C_{60}$ sample does not give any ESR signals in DMSO at 125–280 K. Thus, they conclude that previously reported triplet ESR signals of $C_{60}^{2\bullet2-}$ are attributable to the open-shell anionic species generated by the reduction of a $C_{120}O$ impurity [Figure 4-4(a)] in air-exposed $C_{60}$ samples.[38g]

In order to determine the spin multiplicity of the dianion in the ground state, magnetic susceptibility measurements of their salts and complexes in the solid state were also carried out. Reed's,[38a] Fässler's,[38f] and Saito's[38h] groups showed that the ground state of the dianion is a closed-shell singlet $C_{60}^{2-}$ ($S = 0$) in the solid samples of $[PPN^+]_2[C_{60}^{n\bullet2-}]$ ($n = 0$ or 2) ($PPN^+ = $ bis(triphenylphosphine)iminium), $[K^+([2.2.2]crypt)]_2[C_{60}^{n\bullet2-}]$ ($n = 0$ or 2) ([2.2.2]crypt = 4,7,13,16,21,24-hexaoxa-1,10-diazobicyclo-[8.8.8]-hexacosane), and $[(C_5Me_5)_2Co^+]_2[C_{60}^{n\bullet2-}]$ ($n = 0$ or 2), respectively. Notably, Saito and coworkers estimated that the energy gap between the ground singlet ($S = 0$) state $C_{60}^{2-}$ and the thermally accessible triplet ($S = 1$) state $C_{60}^{2\bullet2-}$ is $730 \pm 10 \, cm^{-1}$ (1050 K) by magnetic susceptibility and ESR measurements in the solid state.[38h] X-ray crystal structure analysis [Figure 4-4(b)] discloses that each $C_{60}^{n\bullet2-}$ ($n = 0$ or 2) is surrounded by $(C_5Me_5)_2Co^+$ cations and the center-to-center distances between the adjacent $C_{60}^{n\bullet2-}$ ($n = 0$ or 2) molecules are over 12 Å. Thus, the $C_{60}^{n\bullet2-}$ ($n = 0$ or 2) molecule is isolated in the crystal, and there are no effective intermolecular interactions.[38h]

The monoradical trianion species $C_{60}^{\bullet3-}$ ($S = \frac{1}{2}$) in a frozen solution gives a broad temperature-dependent ESR spectrum,[30,34c,38a,c,d,g] as it is similar to the monoradical monoanion species $C_{60}^{\bullet-}$. The $g$ value of $C_{60}^{\bullet3-}$ ($g = 2.0025$)[38c] is larger than that of $C_{60}^{\bullet-}$ ($g = 1.9963$). The quartet species of $C_{60}^{3\bullet3-}$ ($S = \frac{3}{2}$) was also detected by Shohoji, Takui, and coworkers by using continuous-wave (cw)-ESR and 2D electron spin transient nutation (2D ESTN) spectroscopy.[39b] By use of the spin dipolar approach, the average distance $r$ between electronic spins of the quartet state is estimated to be 7.8 Å, which is comparable with 7.1 Å for the mean atom-to-atom diameter of pristine $C_{60}$ in the crystal.[40] They also claim that the ground state is a quartet ($S = \frac{3}{2}$) in terms of temperature dependence of the ESR signal intensities. In a sharp contrast, Reed and coworkers claim that the ground state is a doublet ($S = \frac{1}{2}$).[38g,39c] Their 2D ESTN spectrum for the trianion generated by using the purified $C_{60}$ sample without any $C_{120}O$ impurity [Figure 4-4(a)] shows only a doublet

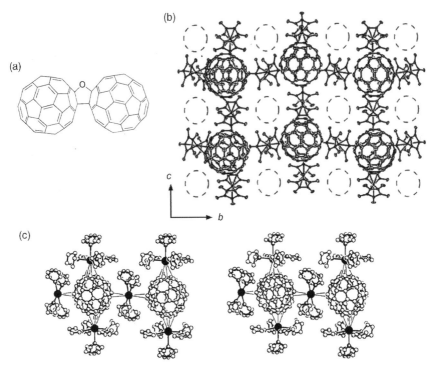

**Figure 4-4.** (a) Chemical structure of $C_{120}O$. (b) Crystal structure of $[(C_5Me_5)_2Co^+]_2[C_{60}^{n \cdot 2-}]$ ($n = 0$ or 2). (Reprinted with permission from Ref. 38h. Copyright 2003 American Chemical Society.). Dashed ellipses show the channels accommodating solvent molecules. (c) Stereoview of $[K^+_3(THF)_{14}][C_{60}^{n \cdot 3-}]$ ($n = 1$ or 3). (Reprinted from C. Janiak, S. Mühle, H. Hemling, K. Köhler, *Polyhedron* **1996**, *15*, 1559–1563 with permission from Elsevier.) Black circles represent the potassium ions. Hydrogen atoms are omitted for clarity.

signal at 4 K.[39c] Furthermore, they describe that the $C_{120}O$ radical anion species generated from the $C_{120}O$ impurity gives a quartet signal that is similar to the $C_{60}^{3 \cdot 3-}$ spectrum reported by Shohoji and Takui's group.[39b] More recently, Takui and coworkers experimentally demonstrated that molecular clusters composed of the dianion and trianion species in the ground state form triplet ($S = 1$), quintet ($S = 2$), and septet ($S = 3$) states by the intermolecular exchange interaction.[39d] This result suggests that these polyanion species have high-spin ground states such as triplet ($S = 1$) diradical $C_{60}^{2 \cdot 2-}$ and quartet ($S = \frac{3}{2}$) triradical $C_{60}^{3 \cdot 3-}$.

Magnetic susceptibility measurements for the polycrystalline sample of $[PPN^+]_3[C_{60}^{n \cdot 3-}]$ ($n = 1$ or 3) ($PPN^+$ = bis(triphenylphosphine)iminium) indicate that the trianion is a ground-state doublet ($S = \frac{1}{2}$) species $C_{60}^{\cdot 3-}$ in the solid state.[38a] While the crystal structure is not disclosed, it is considered that $C_{60}^{n \cdot 3-}$ ($n = 1$ or 3) molecules are well isolated in the solid state, due to the presence of three large $PPN^+$ countercations. This implies that no effective intermolecular magnetic interactions are operative, and the magnetic susceptibility obtained is ascribed to an intramolecular exchange interaction. In 1996, Janiak and coworkers reported the first X-ray crystal structure of a potassium salt of the trianion, $[K^+_3(THF)_{14}][C_{60}{}^{n\bullet 3-}]$ ($n = 1$ or 3) [Figure 4-4(c)].[39a] This is the only example of a crystal structure analysis of the trianion derivatives reported so far. As shown in Figure 4-4(c), the $C_{60}{}^{n\bullet 3-}$ ($n = 1$ or 3) molecules are surrounded by potassium cations. Therefore, intermolecular magnetic interactions of the $C_{60}{}^{n\bullet 3-}$ ($n = 1$ or 3) derivatives are expected to be negligible, while solid-state magnetic properties are not experimentally disclosed.

## 4-2-2. Neutral Radicals

***Neutral Radical Adducts.*** By the reaction of fullerenes with neutral radical derivatives with carbon-, oxygen-, sulfur-, phosphorus-, boron-, and silicon-centered functional groups, neutral radical adducts of $C_{60}$ and $C_{70}$ possessing an unpaired electron on the spherical $\pi$-electronic network are generated.[31,32,41–43] The detection and characterization of these species are carried out by ESR spectroscopy under anaerobic conditions.

In 1991 Krusic, Wasserman, and coworkers report that a fullerene-based neutral radical $R_nC_{60}{}^\bullet$ ($R = PhCH_2$) is formed by the reaction of $C_{60}$ with benzyl radical photochemically generated in the presence of di-*tert*-butyl peroxide in a toluene solution.[41] The neutral radicals with three benzyl moieties $R_3C_{60}{}^\bullet$ and five benzyl moieties $R_5C_{60}{}^\bullet$ [Figure 4-5(a,c)] are detected by ESR spectroscopy [Figure 4-5(b,d)], suggesting that each radical possesses the allyl- and cyclopentadienyl-type electronic spin structures, respectively. This means that the electronic spin resides mainly on the five-membered ring, and spin delocalization on the $C_{60}$ surface is not so large. These species are stable in a degassed solution, due to the steric hindrance of the benzyl moieties. However, isolation as the solid causes a decomposition, and thus the solid-state properties such as intermolecular magnetic interaction are not revealed. A pentaphenyl derivative $R_5C_{60}{}^\bullet$ ($R = Ph$) was investigated by Nakamura and coworkers in 1997.[44a] The cyclic voltammetry (CV) measurements of the anion $R_5C_{60}{}^-$ ($R = Ph$), which possesses the cyclopentadienyl anion-type electronic structure, give a reversible one-electron oxidation wave, suggesting generation of the corresponding cyclopentadienyl radical $R_5C_{60}{}^\bullet$ ($R = Ph$). Isolation and detection by ESR spectroscopy were not performed. Nakamura's group also prepared the radical dianion species $R_5C_{60}{}^{\bullet 2-}$ ($R = Ph$ and biphenyl) by reduction of $R_5C_{60}H$ ($R = Ph$ and biphenyl) and characterized their electronic structure in terms of ESR and UV–visible–NIR spectroscopy in a solution state.[44b] Interestingly, $R_5C_{60}{}^{\bullet 2-}$ ($R = biphenyl$) exists as a monomer in solution; however, in the solid state it forms a closed-shell dimer bearing an intermolecular C–C single bond (1.58 Å) as revealed by X-ray crystal structure analysis. The single bond is thermally stable even at 300 K in the crystalline state. This is in contrast to the intermolecular C–C bond in the dimer structure of the monoanion species $(C_{60}{}^-)_2$, which starts to break at 200–220 K.[36]

In the case of the reactions of $C_{60}$ with alkyl or fluoroalkyl neutral radicals, the corresponding monoadducts $RC_{60}{}^\bullet$ ($R = alkyl$, fluoroalkyl) are detected by ESR spectroscopy [Figure 4-5(e,f)].[42,43] An ESR spectrum of $RC_{60}{}^\bullet$ ($R = C(CD_3)_3$)

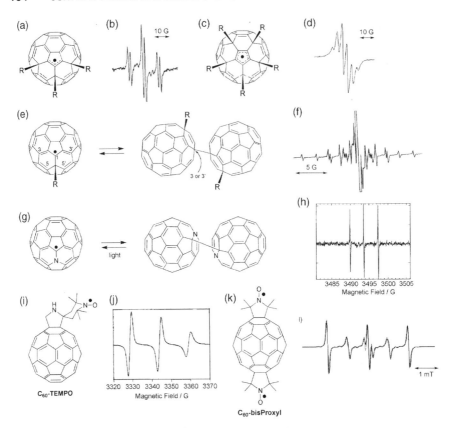

**Figure 4-5.** Chemical structures of (a) $R_3C_{60}^\bullet$ and (c) $R_5C_{60}^\bullet$ and solution-phase ESR spectra of (b) $R_3C_{60}^\bullet$ ($g = 2.0022$, $R = Ph^{13}CH_2$) and (d) $R_5C_{60}^\bullet$ ($g = 2.0022$, $R = Ph^{13}CH_2$) at 50 °C. (From P. J. Krusic, E. Wasserman, P. N. Keizer, J. R. Morton, K. F. Preston, *Science*, **1991**, *254*, 1183–1185. Reprinted with permission from The American Association for the Advancement of Science.) (e) Dimerization scheme of $RC_{60}^\bullet$ and (f) ESR spectrum of $RC_{60}^\bullet$ [$R = C(CD_3)_3$ in benzene at 300 K. (Reprinted from J. R. Morton, K. F. Preston, P. J. Krusic, E. Wasserman, *J. Chem. Soc. Perkin Trans.* **1992**, *2*, 1425–1429. Reproduced by permission of The Royal Society of Chemistry.) (g) Dimerization scheme of $C_{59}N^\bullet$ and (h) ESR spectrum of $C_{59}N^\bullet$ ($g = 2.0013$) in 1-chloronaphthalene at room temperature. (Reprinted with permission from Ref. 45h. Copyright 1997 American Chemical Society.) (i) Chemical structures of **$C_{60}$-TEMPO** and (k) **$C_{60}$-bisProxyl** and solution-phase ESR spectra of (j) **$C_{60}$-TEMPO** in toluene at 210 K. (Reprinted with permission from Ref. 46a. Copyright 1995 American Chemical Society.) (l) **$C_{60}$-bisProxyl** in toluene at 298 K. (From M. Mazzoni et al. *ChemPhysChem* **2002**, *6*, 527–531. Copyright Wiley-VCH Verlag GmbH & Co. KGaA. Reproduced with permission.)

[Figure 4-5(f)] indicates that the electronic spin is located mainly on the carbon atoms at the 1, 3, 3′, 5, and 5′ positions.[42c] In particular, the unpaired electron of this class of neutral radical system is relatively localized on the $C_{60}$ skeleton. Interestingly, these monoadducts $RC_{60}^\bullet$ (R = alkyl, fluoroalkyl) show a temperature-dependent equilibrium between the open-shell monomer $RC_{60}^\bullet$ and the closed-shell dimer

$RC_{60}-C_{60}R$ with an intermolecular $C-C$ single bond in solution [Figure 4-5 (e)].[42b-d,43a] Owing to the steric hindrance of the substituent R, the $C-C$ single bond is formed not at the 1 position with the largest spin density but at the 3 or $3'$ positions. Unfortunately, these species were also not isolated in the solid state.

Such a dimerization reaction is also reported for the neutral radical species of azafullerene $C_{59}N$ [Figure 4-5(g)].[45] This radical is generated by the photoirradiation of the $C-C$-bonded dimeric pair $(C_{59}N)_2$. Its ESR spectrum ($g = 2.0013$) [Figure 4-5 (h)] shows three-line hyperfine splittings that are attributable to the single nitrogen nucleus.[45h] When the photoirradiation stops, the ESR signal disappears immediately because of the formation of the closed-shell dimeric pair $(C_{59}N)_2$. This dimeric pair $(C_{59}N)_2$ is stable and is isolated in air. The structure of $(C_{59}N)_2$ is characterized by several measurements, such as MS,[45a] UV−visible,[45a] IR,[45a] and NMR[45a,f,i] spectroscopies, cyclic voltammetry,[45a] synchrotron X-ray powder diffraction,[45c] and theoretical calculation,[45d] indicating that the $C-C$ single bond is formed between the carbon atoms adjacent to the nitrogen atom [Figure 4-5(g)].

***Neutral Radical-Substituted Derivatives.*** In contrast to the abovementioned neutral radicals, which show low stability in air, some neutral radical-substituted $C_{60}$ derivatives are stable enough to be isolated in air.[32,46,47] In 1995 Corvaja, Maggini, and coworkers synthesized and isolated **$C_{60}$-TEMPO**, a $C_{60}$ derivative possessing a TEMPO (2,2,6,6-tetramethylpiperidine-1-oxyl) neutral radical moiety [Figure 4-5 (i)].[46a] **$C_{60}$-TEMPO** in a toluene solution gives a three-line ESR spectrum that is characteristic of TEMPO radical [Figure 4-5(j)], indicating that the electronic spin of the **$C_{60}$-TEMPO** is localized on the $N-O$ moiety. Thus, there are no sizable spin densities on the $C_{60}$ curved π surface. This is due to very small π conjugation between the radical moiety and the $C_{60}$ skeleton as well as the spin-localized nature of the TEMPO radical itself. For the same reasons, a series of the neutral radical-substituted $C_{60}$ monoradical derivatives prepared so far possess only low spin densities on the $C_{60}$ π-electronic network.[46] $C_{60}$-based neutral diradical systems possessing two nitroxide radicals such as TEMPO and proxyl(2,2,5,5-tetramethylpyrrolidine-1-oxyl) have also been prepared and investigated.[47] The ESR spectrum of **$C_{60}$-bisProxyl** in a solution state [Figure 4-5(k,l)] certainly demonstrates the presence of intramolecular exchange interaction between the two nitroxide radicals.[47b] The experimentally obtained intramolecular exchange interaction $J$ in this class of the diradical systems is, however, very weak ($|Jk_B^{-1}| < 0.1$ K).[47b] The origin of such a small $J$ value is due to the very long intramolecular spin−spin distance resulting from the small spin delocalization on the $C_{60}$ skeleton from the nitroxide radical substituents. Thus, for the quantitative discussion on intramolecular exchange interaction via a curved π-conjugated system, the design and synthesis of highly spin-delocalized curved π-diradical system with a strong intramolecular exchange interaction are crucially important.

## 4-3. OPEN-SHELL SYSTEMS BASED ON CORANNULENE

In sharp contrast to fullerenes and carbon nanotubes, corannulene possesses protons around the curved molecular skeleton. In general, much precise information relevant to molecular and electronic structures of organic molecules is obtained by the measurements of $^1$H NMR and ESR spectroscopy. Thus, an open-shell derivative based on corannulene provides a good model system for elucidation of the detailed molecular and electronic spin structures in a curved and nonalternant π-conjugated system. In particular, the electronic spin distribution nature on the convex and concave faces, 3D intra- and intermolecular magnetic interactions, and dynamic electronic spin properties attributable to the bowl-to-bowl inversion behavior are of great interest. So far, radical ion species generated by alkali metal reduction of corannulene derivatives (for further details, see Chapter 3) and corannulene-based neutral radical systems bearing neutral radical substituents have been experimentally studied. In the following sections, syntheses, structures, and physical properties of these kinds of corannulene-based open-shell molecules are discussed.

## 4-3-1. Monoradical Systems

***Radical Ion Species of Corannulene Derivatives.*** Corannulene has a low-lying doubly degenerate LUMO and is able to accommodate up to four excess electrons. Actually, the radical anion, dianion, radical trianion, and tetraanion species of corannulene are prepared by both chemical or electrochemical methods.[48,49] Among them, the tetraanion shows a very interesting self-aggregation behavior with a "sandwich" fashion.[48d,i] In contrast to the closed-shell dianion and tetraanion species studied by NMR spectroscopy, the radical anion[48a,e−h] and radical trianion[48e−g] are paramagnetic species possessing an unpaired electron and are characterized by solution-phase ESR measurements. In 1967, Lawton and coworkers reported an ESR spectrum of the radical anion species generated by electrochemical or chemical reduction [Figure 4-6(a)].[48a] This spectrum ($g = 2.0027$) shows one kind of hyperfine splitting (0.157 mT) due to 10 equivalent protons of the corannulene skeleton. This result implies high spin delocalization over the corannulene curved π-electronic system as suggested by DFT calculation [Figure 4-6(d)]. The Jahn–Teller effect of the radical anion has also been studied by variable-temperature ESR measurements.[48g,h] In addition to the radical anion, Lawton and coworkers attempted to generate the radical cation species by electrochemical oxidation of corannulene.[48a] However, the attempt was not successful because the oxidation product rapidly underwent a polymerization reaction. In 1995, Scott and coworkers detected the radical trianion species of corannulene by ESR spectroscopy.[48e] Its ESR spectrum in THF solution [Figure 4-6(b)] ($g = 2.0025$) gives well-resolved hyperfine splittings attributable to one kind of protons (0.162 mT) and counter-Li-cations (0.075–0.080 mT).[48e] The spectral simulation reproduces the experimental spectrum well [Figure 4-6(c)]. These results indicate the extensive electronic spin delocalized nature of the radical trianion species as well as the radical anion species [Figure 4-6(d,e)]. However, these radical anion species are studied only by spectroscopic measurements in degassed solution state, and an attempt to isolate these species in the solid state has failed. In addition, Scott, Rabinovitz, and coworkers reported the alkali metal reduction of several corannulene derivatives with annelated five- or six-membered rings,[49,50] such as dibenzo[*a,g*]corannulene,[50a] dibenzo[*a,g*]

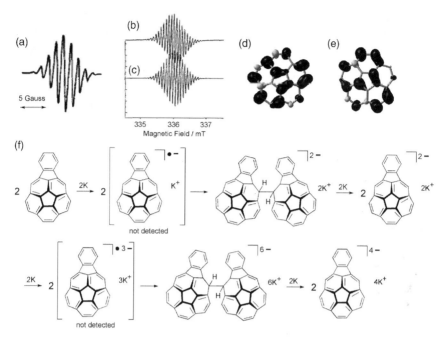

**Figure 4-6.** Experimental ESR spectra of (a) the radical anion species in THF at room temperature. (Reprinted with permission from Ref. 48a. Copyright 1967 American Chemical Society.) (b) Radical trianion species in THF at 200 K, and (c) the simulated one for (b). (Reprinted with permission from Ref. 48e. Copyright American Chemical Society.) Spin density distribution of (d) the radical anion and (e) radical trianion species calculated at the UB3LYP/6-31G(d,p) level. These calculated results show unbalanced spin density distributions while the proton HFCC for 10 proton nuclei of each species is equivalent in liquid-phase ESR measurements. This means that, in experimental conditions, the unbalanced spin density distribution is averaged. (f) Reduction process of indenocorannulene.

cyclopenta[*kl*]corannulene,[50a] indenocorannulene,[50b,c] diindeno[1,2,3-*bc*:1,2,3-*hi*] corannulene,[50c] naphtho[2,3-*a*]corannulene,[50c] and cyclopenta[*bc*]corannulene.[50c] Interestingly, it is proposed that indenocorannulene shows a dimerization/bond cleavage process on reduction [Figure 4-6(f)].[50b,c] Furthermore, reduction of cyclopenta[*bc*]corannulene gives a coordinative dimer that consists of its two tetraanion species, bound together in a convex–convex fashion via lithium cations.[50c] In these studies, the closed-shell anion species are experimentally characterized by NMR spectroscopy, contributing to understanding their reduction process. In contrast, the open-shell anion species such as radical monoanion and radical trianion are not detected by ESR spectroscopy, and their electronic spin structure is not revealed.

***Neutral Radical-Substituted Corannulene Derivatives.*** In order to evaluate the electronic spin structure on curved π-conjugated networks at the molecular level, synthesis and isolation of air-stable neutral radical derivatives with a sizable spin density on the curved π-conjugated system are crucially important. In this context, neutral radical-substituted corannulene derivatives have been investigated from both experimental and theoretical aspects.[51-53] This section concentrates on the syntheses, molecular and electronic-spin structures, intermolecular exchange interaction, and local aromaticity of the corannulene-based neutral monoradical derivatives with curved and nonalternant π-conjugated systems.[51,52]

In 2004, a verdazyl radical-substituted corannulene (**Cor-Ver**) was synthesized and isolated as the first air-stable bowl-shaped neutral radical.[51a] Since then, synthesis and isolation of corannulene-based air-stable neutral radicals with iminonitroxide, phenoxyl, and *tert*-butylnitroxide moieties **Cor-IN**,[51b]**Cor-PhO**,[51c] and **Cor-tBu-NO**[51d] have been achieved. The synthetic routes to these neutral radicals are shown in Scheme 4-1. All the neutral radical derivatives are prepared from the monobromo derivative of corannulene.[54]**Cor-Ver** and **Cor-IN** are synthesized in three steps from bromocorannulene, respectively. The formylated derivative is obtained by treatment of the monobromo derivative with *n*-BuLi followed by DMF. Condensation with 2,4-dimethylcarbonohydrazide or 2,3-bis(hydroxyamino)-2,3-dimehylbutane gives the radical precursor of **Cor-Ver** and **Cor-IN**, respectively. The neutral radicals **Cor-Ver** and **Cor-IN** are obtained as air-stable solids by oxidation reaction of the corresponding precursor with an excess amount of PbO$_2$. **Cor-Ver** is stable for a long period of time in a degassed solution but gradually decomposes in the presence of atmospheric oxygen. Interestingly, **Cor-IN** is very stable even in a solution state in air. The phenoxyl radical-substituted neutral radical **Cor-PhO** is synthesized in four steps from bromocorannulene. Suzuki coupling reaction of the boronic ester derivative of corannulene with methoxymethyl-protected (MOM-protected) di-*tert*-butylbromophenol and subsequent depotection by hydrochloric acid yields the radical precursor **Cor-PhOH** as colorless plates. Treatment with an excess of PbO$_2$ and recrystallization gives the neutral radical **Cor-PhO** as black plates. This neutral radical in the

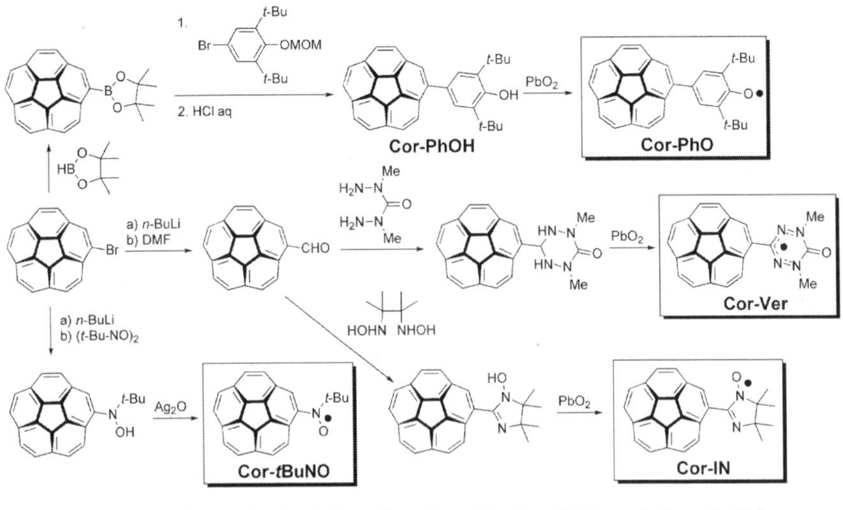

**Scheme 4-1.** Synthesis of **Cor-Ver, Cor-IN, Cor-PhO**, and **Cor-tBuNO**.

crystal is stable in air at −30°C for a few weeks, and is extremely stable in a degassed solution. The synthesis of **Cor-tBuNO** is achieved in only two steps from bromocorannulene. The hydroxylamine derivative, the radical precursor of **Cor-tBuNO**, is obtained by treatment with *n*-BuLi followed by 2-methyl-2-nitrosopropane dimer. Oxidation with Ag₂O and purification by silica gel column chromatography afford **Cor-tBuNO** as an orange solid, which survives in air at −30°C in the solid state and is also stable in a degassed solution.

For the evaluation of electronic spin structures of these corannulene-based curved π-radical systems, liquid-phase ESR, electron−nuclear double resonance (ENDOR), and electron-nuclear-nuclear triple resonance (TRIPLE) measurements, and DFT calculations were performed.[51]Figure 4-7 illustrates their hyperfine ESR spectra, experimentally and theoretically obtained hyperfine coupling constants (HFCCs), and spin density distributions. By use of ENDOR/TRIPLE spectroscopy in addition to the usual ESR spectroscopy, the magnitude and relative sign of HFCCs of these radicals are determined in a straightforward manner. All the neutral radicals give HFCCs ascribable to the $^1$H nuclei on the corannulene moieties as well as the $^1$H or $^{14}$N nuclei on the radical moieties (Figure 4-7, center). The experimentally obtained HFCCs of all neutral radicals show good agreement with the calculated ones. These results confirm that the electronic spin in these neutral radicals is certainly delocalized onto the π-conjugated system of the corannulene from the radical substituents as illustrated by the spin density distributions (Figure 4-7, right). UV−visible and cyclic voltammetry (CV) measurements also indicate that appreciable electronic communication arises between the corannulene and radical moiety in the ground state. Interestingly, the magnitude of spin delocalization onto the corannulene is significantly different in these neutral radical derivatives (Figure 4-7). The magnitude of the experimentally obtained absolute value of HFCC of $H^1$ and the ratio of spin densities on the corannulene skeleton of **Cor-PhO** (0.295 mT and 24%) and **Cor-tBuNO** (0.204 mT and 43%) are significantly larger than those of **Cor-Ver** (0.060 mT and 14%) and **Cor-IN** (0.037 mT and 12%) (Figure 4-7). This indicates that electronic spins of **Cor-PhO** and **Cor-tBuNO** are more widely delocalized onto the curved corannulene π surface than are those of **Cor-Ver** and **Cor-IN**. These differences in the spin-delocalized nature on the corannulene skeleton are attributable to the topological nature of spin density distribution of the employed radical substituents. In other words, the spin-delocalized nature is dependent on the sign of spin density at the C4 position, where the carbon atom of the radical moiety is attached to the corannulene skeleton (Figure 4-7, right). Actually, **Cor-PhO** and **Cor-tBuNO** possess positive spin density at C4 and show the more extensive spin delocalization, while **Cor-Ver** and **Cor-IN** possess a negative one and show smaller spin delocalization.

Although there are differences in the magnitude of spin delocalization depending on the neutral radical moieties employed, spin density distributions of all these neutral radicals show an "unbalanced spin delocalization nature," that is, an uneven spin distribution over spin-rich and spin-poor regions within the corannulene moiety (Figure 4-7). This feature is well illustrated in **Cor-PhO**[51c] and **Cor-tBuNO**,[51d] in which rings A and B possess significantly larger amounts of spin density than do rings D and E. In addition, the spin density on C8 is larger than that on C21 in all these

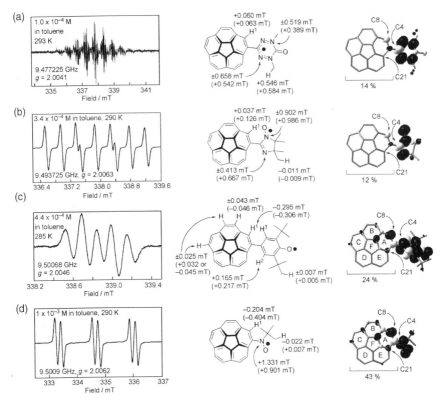

**Figure 4-7.** Hyperfine ESR spectra, experimentally and theoretically (in parentheses) obtained HFCCs, and calculated spin density distributions. (a) **Cor-Ver** (Reprinted with permission from Ref. 51a. Copyright 2004 American Chemical Society.) (b) **Cor-IN** (Reprinted from S. Nishida et al., *Polyhedron* **2005**, *24*, 2200–2204 with permission from Elsevier.) (c) **Cor-PhO** (From A. Ueda et al., *Angew. Chem. Int. Ed.*, **2010**, *49*, 1678–1682. Copyright Wiley-VCH Verlag GmbH & Co. KGaA. Reproduced with permission.) (d) **Cor-tBuNO**. The magnitude and relative sign of HFCCs are determined by ENDOR and TRIPLE measurements. Black and gray regions in the spin density distributions denote positive and negative spin densities, respectively. Values under the spin density distribution maps indicate the ratio of the calculated spin densities on the corannulene moiety to the total calculated spin densities for each derivative.

neutral radical systems. The origin of this unique nature in this class of bowl-shaped neutral radicals is interpreted as the topological effect due to the nonalternant π-conjugated system of corannulene (see Figures 4-8 and 4-9).

The first X-ray structural analysis of a neutral radical derivative of a curved π-conjugated system was carried out on a single crystal of **Cor-PhO** [Figure 4-8 (a,b)].[51c] The structural features of **Cor-PhO** are revealed by comparison with the molecular structure of its precursor, **Cor-PhOH**. While the bowl depth and POAV (π-orbital axis vector) angles[55] of **Cor-PhO** (0.91 Å and 8.5°, respectively) are similar to those of **Cor-PhOH** (0.92 Å and 8.4°, respectively), significant changes in bond

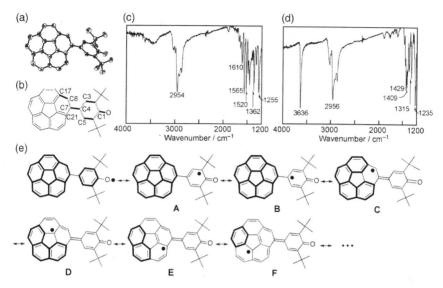

**Figure 4-8.** (a) Molecular structure of **Cor-PhO**. Hydrogen atoms are omitted for clarity. (b) Major changes of bond lengths in **Cor-PhO** from **Cor-PhOH**. Bold and dashed bonds represent shorter and longer bonds, respectively, in **Cor-PhO** as compared with corresponding bonds in **Cor-PhOH**. IR spectra of (c) **Cor-PhO** and (d) **Cor-PhOH** in KBr pellet at room temperature. (From A. Ueda et al., *Angew. Chem. Int. Ed.*, **2010**, *49*, 1678–1682. Copyright Wiley-VCH Verlag GmbH & Co. KGaA. Reproduced with permission.). (e) Possible canonical resonance structures of **Cor-PhO**. Six-membered rings represented by bold lines indicate 6π-electron (aromatic) system.

lengths of the phenoxyl moiety and C7−C8 and C8−C17 bonds of the corannulene moiety are observed [Figure 4-8(b)]. In particular, the O−C1 bond of **Cor-PhO** [1.250(2) Å] is closer to the C=O bond length of *p*-benzoquinone (1.222 Å)[56a] and *p*-terphenoquinone (1.231 Å),[56b] indicating that the C−O bond of **Cor-PhO** has substantial double-bond character. IR measurements of **Cor-PhO** and **Cor-PhOH** in both the solid and solution states also demonstrate the double-bond character of the O−C1 bond of **Cor-PhO**.[51c] In the radical **Cor-PhO**, O−H stretching found in **Cor-PhOH**—3636 cm$^{-1}$ in the KBr [Figure 4-8(d)], 3631 cm$^{-1}$ in CH$_2$Cl$_2$ solution—disappeared, and a new sharp absorption appeared at 1565 cm$^{-1}$ in the solid [Figure 4-8(c)] and at 1567 cm$^{-1}$ in solution. These new absorptions are similar to the C=O vibration frequency of *p*-terphenoquinone (1575 cm$^{-1}$ in the solid state).[56b] In addition, the dihedral angles between corannulene and the phenoxyl moiety of **Cor-PhO** [C3−C4−C7−C8 (35.9(2)°], C5−C4−C7−C21 [41.5(2)°] decreases slightly in comparison with those of **Cor-PhOH** (C3−C4−C7−C8 [38.6(5)°], C5−C4−C7−C21 [43.7(5)°]). These changes are reasonably interpreted by considering quinoidal structural contribution such as the structures **C**, **D**, **E**, and **F** in the classical canonical resonance structures of **Cor-PhO** [Figure 4-8(e)]. Furthermore, the resonance structures suggest that the contributions of structures **C**

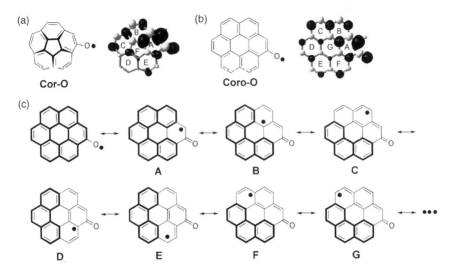

**Figure 4-9.** Molecular structure and calculated spin density distribution of (a) **Cor-O** and **Coro-O**. Black and gray regions in the spin density distributions denote positive and negative spin densities, respectively. Calculations are carried out at the UB3LYP/6-31G(d)//UB3LYP/6-31G level. (Reprinted from K. Fukui et al., *Polyhedron* **2005**, *24*, 2326–2329 with permission from Elsevier.) (c) Possible canonical resonance structures of **Coro-O**. Six-membered rings represented by bold lines indicate 6π-electron (aromatic) system.

and **D** are larger than those of structures **E** and **F** because structures **C** and **D** have more 6π-electron (aromatic) systems than do structures **E** and **F**. This topological feature of the nonalternant π-conjugated system of corannulene with a five-membered ring system gives rise to the unbalanced delocalization of spin on the corannulene π surface.

The topological effect on spin delocalization in the nonalternant corannulene system has also been investigated by comparison of spin density distribution of an oxyl radical of corannulene **Cor-O** with that of coronene **Coro-O** (Figure 4-9).[52] In contrast to corannulene, coronene has a planar and alternant condensed polycyclic π-conjugated system. Most of the spin density of **Cor-O** (85%) is delocalized onto the corannulene skeleton, where the unbalanced spin-delocalized nature is realized: spin-rich (rings A, B, C, E, and F) and spin-poor (ring D) regions [Figure 4-9(a)]. On the other hand, the spin densities of **Coro-O** are spread over the whole molecular skeleton rather evenly [Figure 4-9(b)]. This difference in the spin density distribution between **Cor-O** and **Coro-O** has been studied for both geometric and topological effects.[52] In order to examine the former effect associated with molecular shape, the spin density distribution of **Cor-O** with the planar geometry at the transition state in the bowl-to-bowl inversion phenomenon is calculated. It turns out that the unbalanced spin density distribution of **Cor-O** is maintained without a dependence on geometry. Furthermore, the resonance structures of **Coro-O** [Figure 4-9(c)] suggest that the origin of the evenly delocalized electronic spin on **Coro-O** is due mainly to the alternant nature of the π-conjugated system. Resonance structures **D** and **E** possess the same number of 6π-electron systems as do resonance structures **C**, **F**, and **G** [Figure 4-9(c)], occurring in the even spin distribution of **Coro-O** with the alternant π-conjugated system.

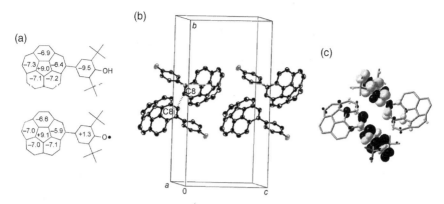

**Figure 4-10.** (a) NICS(0) values (ppm) of **Cor-PhOH** (upper) and **Cor-PhO** (lower) calculated at the UB3LYP/6-31G(d,p)//UB3LYP/6-31G level using the crystal structures as initial structures. (b) Packing structure of **Cor-PhO**. The dashed line represents the intermolecular short contact (3.791 Å). Hydrogen atoms and the *tert*-butyl groups are omitted for clarity. The thermal ellipsoids are shown at the 50% probability level. (From A. Ueda et al., *Angew. Chem. Int. Ed.*, **2010**, *49*, 1678–1682. Copyright Wiley-VCH Verlag GmbH & Co. KGaA. Reproduced with permission.) (c) Spin density distribution of dimeric pair. The calculation is carried out at the UB3LYP/6-31G(d,p) level on the basis of crystal structure. (From A. Ueda et al., *Angew. Chem. Int. Ed.*, **2010**, *49*, 1678–1682. Copyright Wiley-VCH Verlag GmbH & Co. KGaA. Reproduced with permission.)

The curved aromaticity[57] and the solid-state packing structure of **Cor-PhO** are also discussed with respect to the unique spin delocalization nature on the corannulene moiety.[51c] Figure 4-10(a) shows the nucleus-independent chemical shift (NICS)[58] for **Cor-PhO** and **Cor-PhOH** calculated by the DFT method using their crystal structures. The NICS method is known as a facile and efficient probe for evaluating aromaticity even for open-shell molecules.[59] Negative NICS values indicate the presence of induced diatropic ring currents and "local aromaticity," whereas positive values denote paratropic ring currents and "local antiaromaticity." In **Cor-PhOH**, negative NICS(0) values are obtained in all six-membered rings [Figure 4-10(a), upper structure]. In a sharp contrast, in **Cor-PhO**, more positive values are obtained in all ring systems on the corannulene moiety, especially for rings A and B [Figure 4-10(a), lower structure], which is in agreement with the positions having a large amount of spin density [Figure 4-7(c)]. These findings demonstrate that the local aromaticity of the ring system having a sizable spin density decreases significantly in the curved π conjugation of corannulene, which is consistent with the case of a planar odd–alternant π-radical system.[59b] This is the first elucidation of curved aromaticity of open-shell molecules with a bowl-shaped π-conjugated system.

The **Cor-PhO** forms a dimeric pair with an intermolecular separation of 3.791 Å between carbon atoms (C8...C8*i*) of the corannulene moieties (symmetry operation *i*: $-x, -y+1, -z+2$) in the crystal [Figure 4-10(b)].[51c] Because of the relatively large amounts of spin density on these carbon atoms [Figure 4-7(c)], a sizable intermolecular exchange interaction between the corannulene moieties is expected. To evaluate this intermolecular interaction and characterize the bulk magnetic properties of the crystalline state of **Cor-PhO**, the magnetic susceptibility of a polycrystalline sample is measured in the range from 1.9 to 300 K in a static magnetic field of 0.1 T. The result shows a ground-state spin-singlet formation with an antiferromagnetic intermolecular interaction ($Jk_B^{-1} = -22.5 \pm 0.2$ K) owing to the intermolecular exchange interaction between 3D delocalized electronic spins on the corannulene in the crystal. The experimental value is in good agreement with the calculated one ($Jk_B^{-1} = -27.0$ K) [Figure 4-10(c)].

## 4-3-2. Diradical System

Intra- and intermolecular interactions of the curved π-conjugated molecules are intrinsically three-dimensional (3D), in contrast to those of planar π-conjugated molecules, underlying their unique chemical and physical properties and functionalities, such as the superconductivity and ferromagnetism found in the $C_{60}$ system.[3-5] Among them, *intra*molecular magnetic interaction between the electronic spins on the curved π surface is extensively studied for *ionic* species of $C_{60}$ such as $C_{60}^{n\bullet 2-}$ ($n = 0$ or 2) and $C_{60}^{n\bullet 3-}$ ($n = 1$ or 3) (see Section 4-2-1, discussion of polyanion species). Their electronic structures are greatly influenced by not only the dynamic spin polarization of electrons but also the negative charges on the spherical π-conjugated system and the countercation species.

*Neutral* diradical systems are known as the most useful probes for studying *intra*molecular magnetic interactions in organic molecules.[60] While many neutral diradical derivatives relevant to planar π-conjugated systems have been investigated, in depth, study of a curved π-conjugated neutral diradical system is limited to only the $C_{60}$ fullerene derivative linked with two nitroxide radicals.[47] As discussed in Section 4-2-2 under "Neutral Radical-Substituted Derivatives," their *intra*molecular exchange interaction (*J*) through the $C_{60}$ skeleton is very weak ($|Jk_B^{-1}| < 0.1$ K) because of the small spin delocalization onto the $C_{60}$ π-electronic network from the nitroxide radicals due to the *spin-localized* nature on the N–O moieties. Therefore, in order to evaluate an *intra*molecular exchange interaction in the curved π-conjugated system in a *quantitative manner*, synthesis and isolation of a stable neutral diradical derivative with an extensively *spin-delocalized* nature on the curved π surface have been the focus of current attention in molecular magnetism and open-shell chemistry.[21,22,32]

As described in Section 4-3-1 under "Neutral Radical-Substituted Corannulene Derivatives," corannulene-based stable neutral *monoradical* systems such as the phenoxyl radical derivative **Cor-PhO** show highly spin-delocalized nature on the intrinsically 3D bowl-shaped and nonalternant π-conjugated network.[51] Thus, a corannulene-based neutral *diradical* derivative with two phenoxyl radical moieties **Cor-(PhO)₂** can be a good target molecular system for this purpose [Figure 4-11(a)].[53] Because of the highly spin-delocalized nature of **Cor-PhO**, strong *intra*molecular exchange interaction through the 3D corannulene π-electron network of **Cor-(PhO)₂** is expected. Interestingly, **Cor-(PhO)₂** can be represented by a closed Kekulé structure δ as well as diradical structures α–γ [Figure 4-11(b)]. In this section, the synthesis, structure, and 3D *intra*molecular

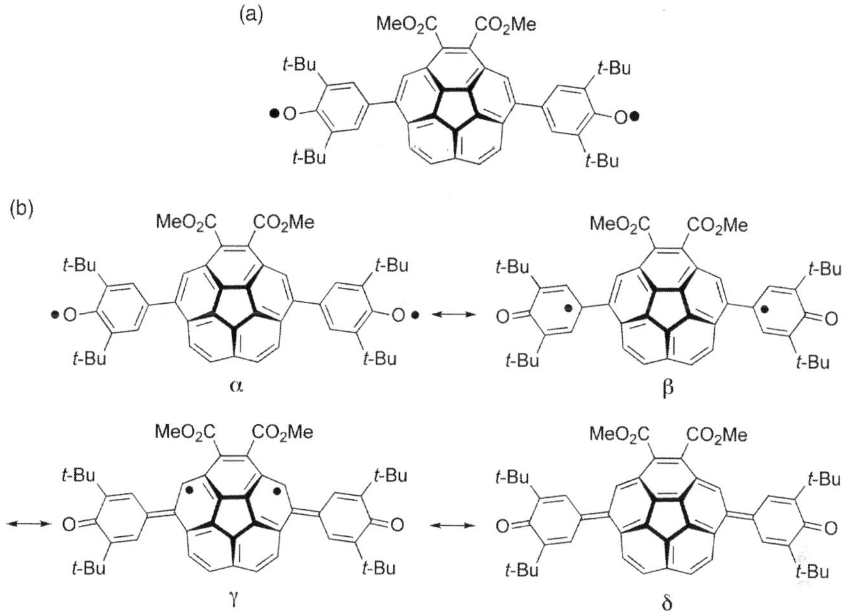

**Figure 4-11.** (a) Chemical structure and (b) canonical resonance structures of **Cor-(PhO)$_2$**.

exchange interaction of **Cor-(PhO)$_2$** with curved and nonalternant $\pi$-conjugated system are discussed.[53]

A synthetic route to **Cor-(PhO)$_2$** is depicted in Scheme 4-2.[53] The diradical precursor, bisphenol derivative **Cor-(PhOH)$_2$**, is obtained as yellow blocks by Suzuki coupling reaction of dibromo derivative[61] with the appropriate boronic acid, followed by deprotection of the methoxymethyl (MOM) groups. Oxidation of **Cor-(PhOH)$_2$** with an excess amount of PbO$_2$ and subsequent recrystallization give the single crystal of **Cor-(PhO)$_2$** suitable for X-ray crystal structure analysis. In the crystalline state, most of **Cor-(PhO)$_2$** survives in air at $-30°$C for a few weeks, and it is also stable in a degassed solution.

The X-ray crystal structure analysis of **Cor-(PhO)$_2$** gives the first example of the structural elucidation of neutral *diradical* derivatives with a curved $\pi$-conjugated network [Figure 4-12(a,c)] and plays a crucial role in revealing the physical properties.[53] In the crystal the molecules form 1D columnar structure in a convex–concave fashion, where three crystallographically independent molecules **A**, **B**, and **C** exist in

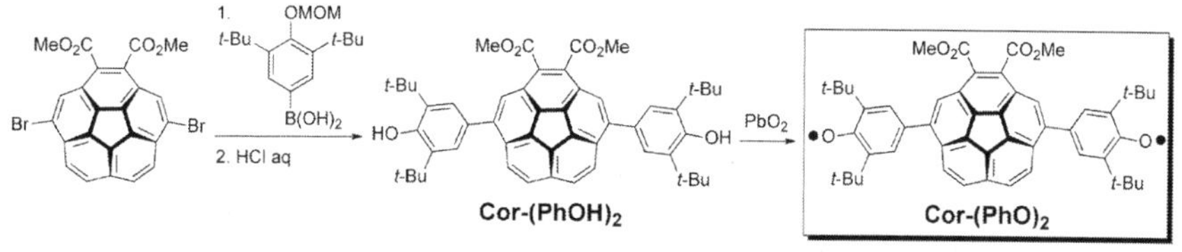

**Scheme 4-2.** Synthesis of **Cor-(PhO)$_2$**.

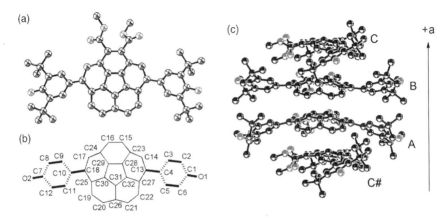

**Figure 4-12.** (a) Molecular structure of the molecule **A** and (c) packing structure of **Cor-(PhO)₂** along the *a* axis. # denotes the symmetry operation of $x + 1, y, z$. The thermal ellipsoids are scaled to the 50% probability level. Hydrogen atoms are omitted for clarity. (From Y. Morita et al., *Angew. Chem. Int. Ed.*, **2008**, *47*, 2035–2038. Copyright Wiley-VCH Verlag GmbH & Co. KGaA. Reproduced with permission.) (b) Major changes of bond lengths in **Cor-(PhO)₂** (the average values of **A–C**) from **Cor-(PhOH)₂**. Bold and dashed bonds represent shorter and longer bonds in comparison with the corresponding bonds in **Cor-(PhOH)₂**, respectively.

the unit cell [Figure 4-12(c)]. Interestingly, molecules **A–C** have significantly different curvature. Judging from bowl depths and π-orbital axis vector (POAV) angles of the corannulene skeleton (**A**—0.84 Å, 7.8°; **B**—0.80 Å, 7.7°; **C**—0.85 Å, 8.1°, respectively), the curvature decreases in the order of **C**, **A**, and **B**. As shown in Figure 4-12(b), the bond length of O1−C1 (or O2−C7) of **Cor-(PhO)₂** (1.282 Å) is much shorter than that of **Cor-(PhOH)₂** (1.385 Å), and is close to the corresponding C−O bond length of the monoradical **Cor-PhO** (1.250 Å)[51c] and the C=O bond length of *p*-terphenoquinone (1.231 Å).[56b] This indicates that the O1−C1 and O2−C7 bonds of **Cor-(PhO)₂** have a certain degree of C=O double-bond character. In addition, significant changes of bond lengths arise in the two six-membered rings (C1−C6 and C7−C12) and C4−C13 and C10−C18 bonds. On the other hand, the C23−C28, C24−C29, and C28−C29 bond lengths in the corannulene skeleton of **Cor-(PhO)₂** remain almost unchanged from those of **Cor-(PhOH)₂** [Figure 4-12(b)]. Thus, **Cor-(PhO)₂** possesses much larger contributions of the diradical structures with quinoidal character such as β and γ than does the closed Kekulé structure δ [Figure 4-11(b)]. Furthermore, the dihedral angles between the corannulene skeleton and the phenoxyl or phenol moieties decrease significantly (11°–19°) in **Cor-(PhO)₂** in comparison with **Cor-(PhOH)₂**. These changes can be interpreted as a result of the shortened C4−C13 and C10−C18 bond lengths in **Cor-(PhO)₂** and their increased double-bond character. These bond length and dihedral angle analyses experimentally illustrate that two electronic spins of **Cor-(PhO)₂** delocalize onto the corannulene skeleton with retention of diradical character in the crystal.

The packing pattern of the convex−concave nature is unique even in the crystal structure composed of the curved π-conjugated closed-shell molecules including sumanene.[18] The *inter*molecular short contacts between the corannulene skeletons are found between C31 in **A** and C26 in **B** (3.41 Å), between C18 in **B** and C21 in **C** (3.38 Å), and between C32 in **C#** and C19 in **A** (3.32 Å). Because these carbon atoms possess relatively smaller spin densities by one to two order(s) of magnitude than do C2, C4, C6, C8, C10, and C12 on the phenoxyl moieties and C23 and C24 on the corannulene [Figure 4-13(d)], there is almost no effective *inter*molecular magnetic interaction. However, these distances are close to the sum of the van der Waals radii of two carbon atoms (3.40 Å), indicating an occurrence of $\pi$-$\pi$ convex–concave interactions between the corannulene skeletons.[14]

The *intra*molecular magnetic interaction via the corannulene $\pi$-conjugated network is studied by temperature-dependent magnetic susceptibility measurements for a polycrystalline sample of **Cor-(PhO)$_2$** in the range of 1.9–298 K.[53] While the $\chi_p T$ values are temperature-independent [~0.0 emu (electromagnetic unit) mol$^{-1}$ K] below 140 K, they gradually increase above 140 K. The $\chi_p T$ value at 298 K (0.178 emu mol$^{-1}$ K) is much lower than the expected $\chi_p T$ value for noninteracting two $S = \frac{1}{2}$ spins (0.75 emu mol$^{-1}$ K), indicating the presence of a very strong antiferromagnetic interaction. Because of the lack of effective *inter*molecular magnetic contacts [Figure 4-12(c)], it is concluded that the very strong *intra*molecular antiferromagnetic interaction ($J k_B^{-1} = -405 \pm 2$ K) occurs through the curved and nonalternant 3D $\pi$ conjugation of corannulene in the crystalline state. Thus, the diradical **Cor-(PhO)$_2$** has a singlet ($S = 0$) ground state with a thermally accessible excited triplet ($S = 1$) state. The singlet–triplet energy gap $2 J k_B^{-1}$ is estimated to be $-810$ K.

Furthermore, the electronic spin structure of the triplet species is elucidated by liquid- and glass-phase ESR spectroscopy.[53] The experimentally determined HFCCs [Figure 4-13(a)] show good agreement with the calculated ones. Thus, the unpaired electronic spin in the triplet state is certainly delocalized onto the corannulene skeleton from the phenoxyl moieties as suggested by the calculated spin density distribution [Figure 4-13(d)]. By comparison of the experimentally obtained HFCCs of the diradical **Cor-(PhO)$_2$** with the corresponding HFCCs of the monoradical **Cor-PhO**,[51c] the HFCCs of H1 and H6 ($-0.151$ mT) and H7, H8, H9, and H10 ($+0.086$ mT) of **Cor-(PhO)$_2$** [Figure 4-13(a,d)] are approximately half of the HFCCs of H1 ($-0.295$ mT) and H2 and H3 ($+0.165$ mT) of **Cor-PhO** [Figure 4-7(c)], respectively. This obviously means that the *intra*molecular exchange interaction ($J$) of **Cor-(PhO)$_2$** is much larger than the hyperfine interaction ($A$), suggesting that in the solution state the delocalized unpaired electrons of **Cor-(PhO)$_2$** interact significantly through the curved $\pi$-conjugated network of corannulene. Glass-phase ESR measurements [Figure 4-13(b)] give the characteristic fine structure ($\Delta M_s = \pm 1$) and forbidden transition ($\Delta M_s = \pm 2$). The zero-field splitting parameters ($D$, $E$) and principal $g$ values are determined as $|D/hc| = 5.18 \times 10^{-3}$ cm$^{-1}$, $|E/hc| = 0.39 \times 10^{-3}$ cm$^{-1}$, $g_{xx} = 2.0033$, $g_{yy} = 2.0033$, and $g_{zz} = 2.0063$ by spectral simulation. Furthermore, the observed $E$ value is nonvanishing and $|E/D| = 0.0753$. The distance $r$ between the two radical centers is estimated as 7.95 Å in terms of a pointdipole approximation using the obtained $D$ value, where $D < 0$ is assumed. This value $r$ corresponds to the distance when each radical center locates on the C4–C13

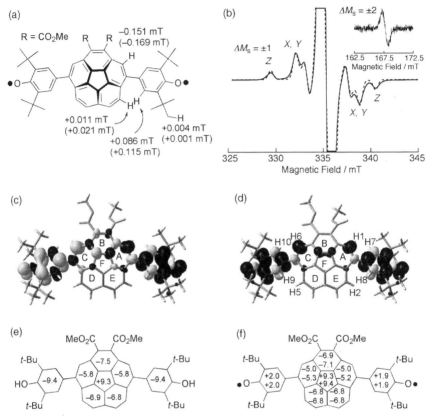

**Figure 4-13.** (a) Experimentally and theoretically (shown in parentheses) obtained HFCCs of **Cor-(PhO)₂**. The magnitude and relative sign of HFCCs are unequivocally determined by ENDOR and TRIPLE measurements in a degassed toluene solution. (b) Fine-structure ESR spectra of **Cor-(PhO)₂** ($\Delta M_s = \pm 1$, microwave frequency 9.40792 GHz) in a degassed frozen toluene glass ($1.3 \times 10^{-3}$ M) at 158 K. The solid and dashed lines indicate the observed and simulated spectra, respectively. Inset shows the forbidden transition ($\Delta M_s = \pm 2$, microwave frequency 9.40799 GHz) of **Cor-(PhO)₂** in the frozen toluene glass at 158 K. Calculated spin density distributions of molecule **A** in (c) broken-symmetry (BS) singlet state and (d) triplet state. Black and gray regions denote positive and negative spin densities, respectively. Calculations are performed at the UB3LYP/6-31G(d,p) level using the X-ray crystal structures. (From Y. Morita et al., *Angew. Chem. Int. Ed.*, **2008**, *47*, 2035–2038. Copyright Wiley-VCH Verlag GmbH & Co. KGaA. Reproduced with permission.) NICS(0) values (ppm) of (e) **Cor-(PhOH)₂** and (f) **Cor-(PhO)₂** (upper—BS singlet state; lower—triplet state). Calculations are performed at the RB3LYP/6-31G(d,p)//RB3LYP/6-31G(d,p) level for **Cor-(PhOH)₂** and the UB3LYP/6-31G(d,p)//UB3LYP/6-31G(d,p) level for **Cor-(PhO)₂** using the X-ray crystal structures as initial structures.

and C10−C18 bonds since the average interatomic distances between C4 and C10 atoms and C13 and C18 atoms in the crystal are 9.5 and 6.6 Å, respectively [Figure 4-12(a,b)]. The arithmetical mean of these two distances is 8.05 Å, which is almost equal to the value of the estimated distance of 7.95 Å. This agreement implies that there are considerable contributions of resonance structures β and γ in the triplet state [Figure 4-11(b)]. All these experimental results demonstrate that the extensive spin delocalization onto the corannulene moiety gives significant electronic spin communication via the curved and nonalternant π surface of the corannulene.

The diradical **Cor-(PhO)$_2$** has its own place as an excellent intermediate case between thus far reported planar π-conjugated systems and completely 3D π-conjugated species such as neutral C$_{60}$ in the triplet state.[39b,d] A neutral C$_{60}$ fullerene in its triplet excited state or the dianion in the ground state as completely 3D π-conjugated systems illustrates complete breakdown of the point dipole approximation and spin dipolar orbit model. The latter model has frequently been used for the successful interpretation of electronic and molecular structures of planar π-conjugated molecular systems. In view of the value of 0.0753 for |$E/D$| of **Cor-(PhO)$_2$**, the present curved diradical **Cor-(PhO)$_2$** gives an excellent testing ground for advanced and sophisticated quantum chemistry for theoretical zero-field splitting tensors, where spin–orbit contributions are taken into account because the significant contributions of the particular canonical resonance structures to the singlet-state diradical occurs.

The calculated spin density distributions suggest the extensive spin-delocalized nature of **Cor-(PhO)$_2$** in both broken-symmetry (BS) singlet and triplet states [Figure 4-13(c,d)].[53] In particular, in the BS singlet state [Figure 4-13(c)], rings A, B, and C on the corannulene have a large amount of spin density, while rings D and E possess a relatively small amount. Thus, the unbalanced spin-delocalized nature is also generated in the diradical system **Cor-(PhO)$_2$** as well as the monoradical system [Figures 4-7 and 4-9(a)]:[51c,52] spin-rich (rings A, B, and C) and spin-poor regions (rings D and E). The origin of this unique nature is the topological effect arising from the nonalternant π conjugation of the corannulene system, which is illustrated by canonical resonance structure studies.[53] In addition, the delocalized spin densities are distributed almost symmetrically on the right and left sides of the molecular skeleton with antiparallel signs, in agreement with the putative electronic spin structure for the singlet state. On the other hand, in the triplet state [Figure 4-13(d)], the spin densities of both sides reside with the parallel signs. Furthermore, the spin densities of rings A, B, and C in the triplet state are less than those in the BS singlet state.

These differences in the spin density distributions between the two spin states greatly affect their curved local aromaticity, as studied by the NICS method for **Cor-(PhO)$_2$** and **Cor-(PhOH)$_2$** [Figure 4-13(e,f)].[53] In **Cor-(PhOH)$_2$**, negative NICS(0) values are obtained in all six-membered rings [Figure 4-13(e)]. In a sharp contrast, in **Cor-(PhO)$_2$** for both spin states, more positive values are obtained in all ring systems of the corannulene skeleton as well as the phenoxyl moieties [Figure 4-13(f)]. In particular, rings A, B, and C, which have a sizable amount of spin densities [Figure 4-13(c)], show larger positive shifts than do rings D, E, and F. This result clearly indicates that the delocalized spin densities of **Cor-(PhO)$_2$** decrease the curved local aromaticity, as they are similar to the case of **Cor-PhO** [Figure 4-10(a)]. The BS singlet state shows slightly more positive values in of rings A, B, and C than does the triplet state. This trend is coincident with the spin density distribution of each spin state.

The intriguing relationship between the curvature and *intra*molecular exchange interactions $J$ is discussed by Ueda et al.[53] The three crystallographically independent

**TABLE 4-1. Calculated Intramolecular Exchange Interaction Parameters ($Jk_B^{-1}$) and Singlet Diradical Characters and Experimentally Obtained Bowl Depths and POAV Angles of Molecules A–C$^a$**

| Compound | A | B | C |
|---|---|---|---|
| $Jk_B^{-1}$ (K) | −425.0 | −525.4 | −339.5 |
| Singlet diradical character (%) | 83.7 | 80.1 | 84.6 |
| Bowl depth (Å) | 0.84 | 0.80 | 0.85 |
| POAV angles (°) | 7.8 | 7.7 | 8.1 |

$^a$Calculations are carried out by the UB3LYP/6-31G(d,p) level using the X-ray crystal structures. Singlet diradical characters are estimated by calculating the NOON of LUMO.

diradical molecules **A–C** show different curvature and different calculated large negative $J$ values (Table 4-1). Their average value ($Jk_B^{-1} = -430.0$ K) reproduces well the experimentally obtained one ($Jk_B^{-1} = -405 \pm 2$ K). In addition, the origin of these sizable differences between the calculated $J$ values of **A–C** are studied in detail with respect to their structural differences in (1) the dihedral angle between the corannulene skeleton and the phenoxyl moieties and (2) the curvature of the corannulene skeleton. As a result, it is concluded that the magnitude of the $J$ value increases with the decrease in curvature of the corannulene π-conjugated system. It is notable that such small differences of the curvature of **A–C** (0.05 Å in bowl depth) lead to very large differences in the $J$ values (186 K). This is due to the enhancement of the π conjugation between the corannulene skeleton and the radical moiety. In addition to the bond length analysis described above [Figure 4-12(b)], natural orbital occupation number (NOON) analysis (Table 4-1)[62] and comparison between the experimental and calculated bond lengths (Figure 4-14) strongly support the significant diradical contributions in the singlet ground state of **Cor-(PhO)₂**. Table 4-1 indicates that the contributions of diradical structures and closed Kekulé structure are about 83% and 17%, respectively. Furthermore, all experimental bond lengths obtained by X-ray structure analysis are significantly closer to the calculated bond lengths of the singlet diradical structure than to those of the Kekulé structure (Figure 4-14). The high diradical character of this system is probably attributed to the aromatic stabilization of the corannulene π system, which prevents **Cor-(PhO)₂** from forming the Kekulé structure δ with fewer 6π benzene structure contributions than β and γ [Figure 4-11(b)]. Importantly, the difference of the curvature also correlates with the magnitude of the diradical character of **Cor-(PhO)₂** as well as the *intra*molecular exchange interaction $J$ (Table 4-1). These results suggest that the singlet diradical character decreases with increasing magnitude of the antiferromagnetic *intra*molecular exchange interaction $J$. In this context, a theoretical analysis of the zero-field splitting tensor for **Cor-(PhO)₂** in the thermally accessible triplet state gives a clue to understanding the 3D electronic spin structure in a straightforward manner.

## 4-3-3. Phenalenyl-Fused Corannulene

Phenalenyl radical **PLY•** (see text below) is a planar polycyclic π radical with an extensive spin-delocalization. Because of the nonbonding molecular orbital (NBMO)

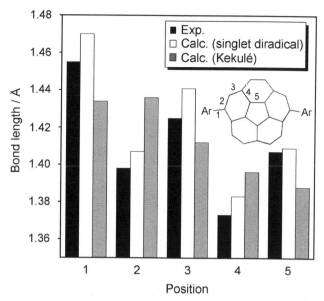

**Figure 4-14.** Comparison of experimentally obtained bond lengths with calculated ones (bonds 1−5 in inset) of **Cor-(PhO)₂**. The black, vacant, and gray regions denote experimental, calculated (singlet diradical structure), and calculated (Kekulé structure) bond lengths, respectively. The experimental values are the average bond lengths of molecules **A−C**. The calculated bond lengths of **Cor-(PhO)₂** are obtained by the DFT method [the UB3LYP/6-31G (d,p) level for the singlet diradical structure and the RB3LYP/6-31G(d,p) level for Kekulé structure] using the X-ray crystal structure as initial structures. (From Y. Morita et al., *Angew. Chem. Int. Ed.*, **2008**, *47*, 2035–2038. Copyright Wiley-VCH Verlag GmbH & Co. KGaA. Reproduced with permission.)

nature of the SOMO, the phenalenyl system has a high redox ability to form three redox species, cation, neutral radical, and anion.[63] More recent progress in phenalenyl chemistry was triggered by the first isolation of tri-*tert*-butylated phenalenyl radical in the crystalline state in 1999.[64] Since then, several kinds of air-stable phenalenyl systems have been synthesized and isolated by introduction of heteroatoms or extension of the π-conjugated system.[28,29] Such phenalenyl systems exhibit exotic and unprecedented physical properties and functionalities attributable to their unique electronic spin structures and redox abilities.[28,29] However, it should be noted that all of the phenalenyl derivatives investigated so far possess rigid planar molecular skeletons. In other words, the half-century-old phenalenyl chemistry has been limited to the chemistry of 2D molecular and electronic systems.

This situation greatly changed in 2008, when phenalenyl chemistry was expanded to intrinsically 3D molecular and electronic systems.[65] Inspired by the studies on the corannulene-based stable neutral radicals,[51–53] researchers designed and synthesized a phenalenyl-fused corannulene anion **Cor-PLY⁻** (see discussion below), creating a new field termed *curve-structured phenalenyl chemistry*.[65] Interestingly, the **Cor-PLY** system with a six-membered *peri*-annelated structure can be viewed as a substructure of fullerene $C_{70}$. The study gives an intriguing discussion on the electronic structure originating in the geometric effect, especially on the electronic communication between the phenalenyl odd–alternant system and the nonalternant system of corannulene having a five-membered ring.

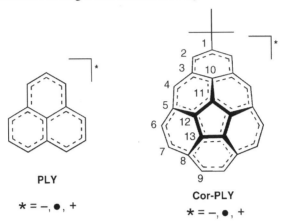

**PLY**

**★ = −, ●, +**

**Cor-PLY**

**★ = −, ●, +**

The anion $Li^{+} \cdot \textbf{Cor-PLY}^{-}$ was synthesized in five steps from corannulene (Scheme 4-3).[65] Friedel–Crafts acylation of corannulene gives a ketone derivative in quantitative yield. Taking advantage of the neighboring participation effect, a formyl group is introduced regioselectively into the *peri*-position of the alkylketo substituent. Aldol reaction with NaOEt yields a phenalenone derivative as colorless plates, and the subsequent reduction of the carbonyl group with $LiAlH_4$ and $AlCl_3$ affords a phenalene derivative. Finally, treatment of the phenalene with an equimolar amount of *n*-BuLi in degassed THF-$d_8$ gives $Li^{+} \cdot \textbf{Cor-PLY}^{-}$ as a red-purple solution. The anion is highly stable in a degassed THF solution.

Figure 4-15(a) summarizes the observed $^{1}$H and $^{13}$C chemical shifts of $Li^{+} \cdot \textbf{Cor-PLY}^{-}$.[65] Assignments of all chemical shifts are based on HMQC, HMBC, and NOESY NMR spectra. Importantly, high-field shifts of the protons (7.14 and 6.68 ppm) and the carbons (106.8 and 105.0 ppm) at positions 2 and 4 compared to the other ones indicate the occurrence of large negative-charge densities at these positions. Furthermore, calculated HOMO pictures and electrostatic potential surfaces of both $\textbf{Cor-PLY}^{-}$ and $\textbf{PLY}^{-}$ demonstrate that $\textbf{Cor-PLY}^{-}$ has the phenalenylanion-type electronic structure

**Scheme 4-3.** Synthesis of $Li^{+} \cdot \textbf{Cor-PLY}^{-}$.

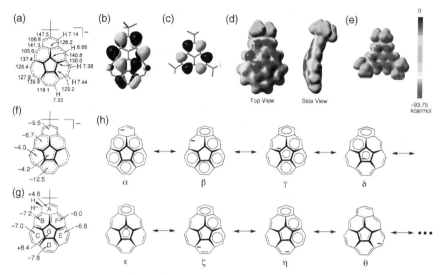

**Figure 4-15.** (a) Observed $^1$H and $^{13}$C chemical shifts (ppm) of Li$^+$·**Cor-PLY**$^-$; (b,c) HOMO structures; (d,e) electrostatic potential surfaces of **Cor-PLY**$^-$ and **PLY**$^-$, respectively. (Reprinted with permission from Ref. 65. Copyright 2008 American Chemical Society.) (f,g) NICS (0) values of **Cor-PLY**$^-$ and the phenalene derivative, respectively. Calculations for (b)–(g) are carried out at the RB3LYP/6-311G(d, p)//RB3LYP/6-311G(d, p) level. (h) Representative resonance structures of **Cor-PLY**$^-$. Circles in the structures denote 6π-electron (aromatic) systems.

with large negative-charge densities at positions 2, 4, and 12 [Figure 4-15(b–e)]. Additionally, an appreciable amount of the HOMO coefficients resides on the central five-membered rings, which implies the contribution of the electronic structure of cyclopentadienyl anion as well as that of the phenalenyl anion. The contribution of the phenalenylanion-type electronic structure is also evaluated by the NICS method [Figure 4-15(f,g)] and resonance structures [Figure 4-15(h)]. By the transformation of the phenalene to **Cor-PLY**$^-$ by deprotonation, NICS(0) values of rings A and G become negative, indicating that these ring systems have local aromaticity [Figure 4-15(f,g)]. Furthermore, the value of the ring D of **Cor-PLY**$^-$ is more negative than that of the phenalene, showing an increase in the local aromaticity of the ring D. Figure 4-15(h) illustrates that structures α and β have the phenalenylanion-type structure, δ and ε have the cyclopentadienylanion-type structures, and γ has both structures. By considering number of 6π systems in these resonance structures and NICS(0) values [Figure 4-15(f)], the phenalenylanion-type structure has the most important contribution to the whole electronic structure of **Cor-PLY**$^-$.

The phenalenyl-fused corannulene system **Cor-PLY** also gives valuable information on the relationship between the structure and electronic features and bowl-to-bowl inversion energy (Figure 4-16).[65] A bowl-inversion barrier closely relevant to the bowl depth is one of the crucial research issues of bowl-shaped molecules. Siegel and coworkers report that the bowl inversion barrier of corannulene derivatives

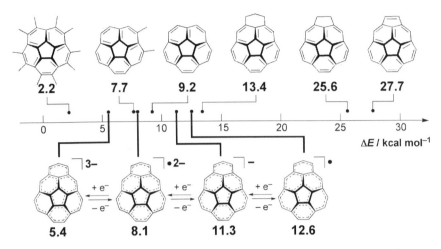

**Figure 4-16.** Calculated bowl inversion barriers ($\Delta E$) of the redox species of the **Cor-PLY** system (nonsubstituted) by using optimized structures obtained at the B3LYP/cc-pVDZ level (lower), and the corannulene derivatives calculated by Siegel and coworkers (upper). (Reprinted with permission from Ref. 66. Copyright 2008 American Chemical Society.)

varies significantly as a consequence of differences in their molecular structures [Figure 4-16, upper].[66] An alkyl substituent acts as a sterically repulsive force and flattens the bowl structure, causing a decrease in the bowl inversion barrier $\Delta E$. On the contrary, annelation across the *peri*-positions causes a deepening of the bowl structure, leading an increase in $\Delta E$. The calculated bowl inversion barrier (the B3LYP/cc-pVDZ level of theory) of the **Cor-PLY** neutral radical (12.6 kcal/mol) is significantly larger than that of corannulene (9.2 kcal/mol), indicating an increase in bowl depth by the annelation across the *peri*-positions of the corannulene skeleton. Importantly, the relationship between the redox state and the inversion energy is also calculated, showing a stepwise decrease in the energy on one-electron reduction: neutral radical (12.6 kcal/mol), anion (11.3 kcal/mol), radical dianion (8.1 kcal/mol), and trianion (5.4 kcal/mol). These studies are feasible by utilizing the phenalenyl system introduced in the corannulene skeleton, and the results suggest that the curvature is flattened stepwise with increasing negative charge.[67] Thus, the bowl depth of the **Cor-PLY** system can be interpreted in terms of the balance between the *peri*-annelation and negative charge effects.

## 4-4. CONCLUSIONS AND OUTLOOK

Continuous attention has been paid to the curved π-conjugated molecular systems from both fundamental and applied aspects. Since the pioneering work on the neutral radical derivatives of $C_{60}$ reported by Wasserman and coworkers in 1991,[41] chemists have devoted tremendous efforts to synthesize and isolate novel open-shell molecules with curved π conjugation. However, the number of curved π-conjugated open-shell systems investigated so far is overwhelmingly scarce in comparison with their closed-shell counterparts. This is probably due to the instability of the open-shell species such as the radical anion systems of $C_{60}$ and corannulene. The $C_{60}$ derivatives linked with nitroxide radical moieties are air-stable, while they have little spin density on the curved $\pi$ surface. Accordingly, synthesis and isolation of air-stable bowl-shaped neutral radicals with significant spin delocalization on the curved $\pi$-electron network are a topical issue, achieving a significant breakthrough in curved $\pi$-conjugated chemistry. Furthermore, the phenalenyl-fused corannulene system gives rise to *curve-structured phenalenyl chemistry*, after a half-century history of the phenalenyl system.

The chemistry of stable neutral curved or nonplanar $\pi$ radicals has just begun. The abovementioned studies mark the beginning of this attractive new research field. For further development of this field, it is crucially important to design and synthesize novel stable open-shell organic molecules with exotic nonplanar or strained $\pi$-conjugated molecular systems.[68] In addition to bowl-shaped $\pi$-conjugated open-shell molecules,[69] belt- or hoop-shaped,[70] helicenic,[71] and Möbius[72] molecules are fascinating synthetic targets. Their stability, spin-delocalized and spin-polarized natures, intra- and intermolecular magnetic interactions, and local aromaticity will be significantly influenced by their exotic 3D $\pi$-conjugated molecular and electronic structures, giving new insight into curved $\pi$-electronic chemistry and materials. In particular, formation of highly arranged molecular assemblies based on 3D intra- and intermolecular interactions of nonplanar open-shell systems is of great interest.

As one possible approach to forming such aggregated structures, we propose a hybrid *open-shell* system based on nonplanar $\pi$-conjugated molecules. Metal salts or complexes composed of nonplanar $\pi$-conjugated open-shell molecule and metal ion are certainly intriguing candidates.[69] In addition, nonplanar $\pi$-conjugated *closed-shell* molecule can also give such a hybrid open-shell system by construction of metal complexes with paramagnetic metal ions and charge transfer (CT) salts or complexes with electron donor or acceptor molecules. As pioneering examples of such hybrid systems, the *molecular sandwich*, a supramolecular aggregate composed of corannulene polyanion species and lithium ions,[48d,i] *buckyferrocene*, a hybrid system of ferrocene and fullerene,[73] $\pi$-*bowl complex*, a metal complex composed of bowl-shaped hydrocarbon such as corannulene or sumanene derivatives and metal ions,[12e,18c,f,h,74] and *supramolecular donor–acceptor $\pi$ complex*, an aggregate structure of fullerene $C_{60}$ with a bowl-shaped TTF (tetrathiafulvalene)-type electron donor molecule,[75] have been reported. While most of the reported hybrid systems are intrinsically *closed-shell*, they suggest promising strategies for the construction of hybrid *open-shell* systems. There is great interest in the electronic spin structure and magnetic properties of the hybrid open-shell systems in terms of 3D intra- and intermolecular magnetic interactions based on structural and electronic features of nonplanar $\pi$-conjugated molecules.

Actually, the synthesis of novel nonplanar open-shell systems is very challenging for various researchers, especially synthetic and materials scientists. Therefore, we again emphasize the importance of the continuous efforts to design and synthesize air-stable open-shell systems. We believe that unexpected and unprecedented phenomena and physical properties attributable to the intrinsically 3D molecular and electronic spin structures will lead to exploration of new multifunctional materials relevant to magnetic, electric, and optical properties such as molecular spin batteries,[25c–f] chiral magnets,[24,76] paramagnetic liquid crystals,[24] and molecular spin quantum computing.[26]

## ACKNOWLEDGMENT

We would like to thank Prof. Takeji Takui (Osaka City University), Prof. Kazunobu Sato (Osaka City University), Prof. Kazuhiro Nakasuji (Fukui University of Technology), and Dr. Shinsuke Nishida (Osaka City University) for helpful and encouraging discussions.

## REFERENCES AND NOTES

1. H. W. Kroto, J. R. Heath, S. C. O'Brien, R. F. Curl, R. E. Smalley, *Nature* **1985**, *318*, 162–163.

2. W. Krätschmer, L. D. Lamb, K. Fostiropoulos, D. R. Huffman, *Nature* **1990**, *347*, 354–358.

3. (a) K. M. Kadish, R. S. Ruoff (eds.), *Fullerenes: Chemistry, Physics and Technology*, Wiley, New York, **2000**; (b) W. Andreoni (ed.), *The Physics of Fullerene-Based and Fullerene-Related Materials*, Kluwer Academic Publishers, Netherlands, **2000**; (c) M. S. Dresselhaus, G. Dresselhaus, P. Avouris (eds.), *Carbon Nanotubes: Synthesis, Structure, Properties, and Applications*, Springer, Heidelberg, **2001**; (d) A. Hirsch, M. Brettreich, *Fullerenes: Chemistry and Reactions*, Wiley-VCH, New York, **2005**; (e) A. Jorio, M. S. Dresselhaus, G. Dresselhaus (eds.), *Carbon Nanotubes: Advanced Topics in the Synthesis, Structure, Properties, and Applications*, Springer, Heidelberg, **2008**.

4. (a) A. F. Hebard, M. J. Rosseinsky, R. C. Haddon, D. W. Murphy, S. H. Glarum, T. T. M. Palstra, A. P. Ramirez, A. R. Kortan, *Nature* **1991**, *350*, 600–601; (b) K. Tanigaki, T. W. Ebbesen, S. Saito, J. Mizuki, J. S. Tsai, Y. Kubo, S. Kuroshima, *Nature* **1991**, *352*, 222–223; (c) A. R. Kortan, N. Kopylov, S. Glarum, E. M. Gyorgy, A. P. Ramirez, R. M. Fleming, F. A. Thiel, R. C. Haddon, *Nature* **1992**, *355*, 529–532; (d) K. Tanigaki, I. Hirosawa, T. W. Ebbesen, J. Mizuki, Y. Shimakawa, Y. Kubo, J. S. Tsai, S. Kuroshima, *Nature* **1992**, *356*, 419–421; (e) E. Özdas, A. R. Kortan, N. Kopylov, A. P. Ramirez, T. Siegrist, K. M. Rabe, H. E. Bair, S. Schuppler, P. H. Citrin, *Nature* **1995**, *375*, 126–129.

5. (a) P.-M. Allemand, K. C. Khemani, A. Koch, F. Wudl, K. Holczer, S. Donovan, G. Grüner, J. D. Thompson, *Science* **1991**, *253*, 301–303; (b) B. Narymbetov, A. Omerzu, V. V. Kabanov, M. Tokumoto, H. Kobayashi, D. Mihailovic, *Nature* **2000**, *407*, 883–885.

6. (a) F. Diederich, R. Ettl, Y. Rubin, R. L. Whetten, R. Beck, M. Alvarez, S. Anz, D. Sensharma, F. Wudl, K. C. Khemani, A. Koch, *Science* **1991**, *252*, 548–551; (b) R. Ettl, I. Chao, F. Diederich, R. L. Whetten, *Nature* **1991**, *353*, 149–153; (c) C. Thilgen, A. Herrmann, F. Diederich, *Angew. Chem. Int. Ed. Engl.* **1997**, *36*, 2268–2280.

7. S. Iijima, *Nature* **1991**, *354*, 56–58.

8. (a) Z. Ma, T. Kyotani, A. Tomita, *Chem. Commun.* **2000**, 2365–2366; (b) Z. Ma, T. Kyotani, Z. Liu, O. Terasaki, A. Tomita, *Chem. Mater.* **2001**, *13*, 4413–4415; (c) Z. Ma, T. Kyotani, A. Tomita, *Carbon* **2002**, *40*, 2374–2377; (d) T. Kyotani, Z. Ma, A. Tomita, *Carbon* **2003**, *41*, 1451–1459; (e) K. Matsuoka, Y. Yamagishi, T. Yamazaki, N. Setoyama,

A. Tomita, T. Kyotani, *Carbon* **2005**, *43*, 876–879; (f) H. Nishihara, Q.-H. Yang, P.-X. Hou, M. Unno, S. Yamauchi, R. Saito, J. I. Paredes, A. Matínez-Alonso, J. M. D. Tascón, Y. Sato, M. Terauchi, T. Kyotani, *Carbon* **2009**, *47*, 1220–1230.

9. (a) F. Su, J. Zeng, Y. Yu, L. Lv, J. Y. Lee, X. S. Zhao, *Carbon* **2005**, *43*, 2366–2373; (b) Z. Yang, Y. Xia, R. Mokaya, *J. Am. Chem. Soc.* **2007**, *129*, 1673–1679; (c) H. Nishihara, P.-X. Hou, L.-X. Li, M. Ito, M. Uchiyama, T. Kaburagi, A. Ikura, J. Katamura, T. Kawarada, K. Mizuuchi, T. Kyotani, *J. Phys. Chem. C* **2009**, *113*, 3189–3196.

10. (a) K. S. Novoselov, A. K. Geim, S. V. Morozov, D. Jiang, Y. Zhang, S. V. Dubonos, I. V. Grigorieva, A. A. Firsov, *Science* **2004**, *306*, 666–669; (b) A. K. Geim, K. S. Novoselov, *Nat. Mater.* **2007**, *6*, 183–191; see also a recent review on graphene: (c) M. J. Allen, V. C. Tung, R. B. Kaner, *Chem. Rev.* **2010**, *110*, 132–145.

11. (a) R. C. Haddon, *Acc. Chem. Res.* **1992**, *25*, 127–133; (b) R. C. Haddon, *Science* **1993**, *261*, 1545–1550; (c) R. C. Haddon, *J. Am. Chem. Soc.* **1997**, *119*, 1797–1798; (d) B. R. Weedon, R. C. Haddon, H. P. Spielmann, M. S. Meier, *J. Am. Chem. Soc.* **1999**, *121*, 335–340; (e) T. Kawase, H. Kurata, *Chem. Rev.* **2006**, *106*, 5250–5273.

12. For overviews of bowl-shaped hydrocarbon and related molecules, see (a) L. T. Scott, *Pure Appl. Chem.* **1996**, *68*, 291–300; (b) P. W. Rabideau, A. Sygula, *Acc. Chem. Res.* **1996**, *29*, 235–242; (c) L. T. Scott, H. E. Bronstein, D. V. Preda, R. B. M. Ansems, M. S. Bratcher, S. Hagan, *Pure Appl. Chem.* **1999**, *71*, 209–219; (d) A. Sygula, P. W. Rabideau, Synthesis and chemistry of polycyclic aromatic hydrocarbons with curved surfaces: Buckybowls, in *Carbon-Rich Compounds*, M. M. Haley, R. R. Tykwinski (eds.), Wiley-VCH, Weinheim, **2006**, pp 529–565; (e) M. A. Petrukhina, L. T. Scott, *Dalton Trans.* **2005**, 2969–2975; (f) Y.-T. Wu, J. S. Siegel, *Chem. Rev.* **2006**, *106*, 4843–4867; (g) V. M. Tsefrikas, L. T. Scott, *Chem. Rev.* **2006**, *106*, 4868–4884.

13. (a) R. B. M. Ansems, L. T. Scott, *J. Am. Chem. Soc.* **2000**, *122*, 2719–2724; (b) L. T. Scott, *Angew. Chem. Int. Ed.* **2004**, *43*, 4994–5007; (c) E. A. Jackson, B. D. Steinberg, M. Bancu, A. Wakamiya, L. T. Scott, *J. Am. Chem. Soc.* **2007**, *129*, 484–485; (d) E. H. Fort, P. M. Donovan, L. T. Scott, *J. Am. Chem. Soc.* **2009**, *131*, 16006–16007.

14. E. M. Pérez, N. Martín, *Chem. Soc. Rev.* **2008**, *37*, 1512–1519; see also Ref. 11e.

15. (a) W. E. Barth, R. G. Lawton, *J. Am. Chem. Soc.* **1966**, *88*, 380–381; (b) W. E. Barth, R. G. Lawton, *J. Am. Chem. Soc.* **1971**, *93*, 1730–1745.

16. J. C. Hanson, C. E. Nordman, *Acta Crystallogr.* **1976**, *B32*, 1147–1153.

17. (a) L. T. Scott, M. M. Hashemi, D. T. Meyer, H. B. Warren, *J. Am. Chem. Soc.* **1991**, *113*, 7082–7084; (b) L. T. Scott, P.-C. Cheng, M. M. Hashemi, M. S. Bratcher, D. T. Meyer, H. B. Warren, *J. Am. Chem. Soc.* **1997**, *119*, 10963–10968; (c) T. J. Seiders, E. L. Elliott, G. H. Grube, J. S. Siegel, *J. Am. Chem. Soc.* **1999**, *121*, 7804–7813; (d) A. Sygula, P. W. Rabideau, *J. Am. Chem. Soc.* **2000**, *122*, 6323–6324; (e) A. Sygula, G. Xu, Z. Marcinow, P. W. Rabideau, *Tetrahedron* **2001**, *57*, 3637–3644.

18. Since the first synthesis in 2003, sumanene chemistry has rapidly grown in terms of synthesis, bowl-to-bowl inversion dynamics, chirality, metal complexation, and application to electrical materials. For typical examples, see (a) H. Sakurai, T. Daiko, T. Hirao, *Science* **2003** *301*, 1878; (b) H. Sakurai, T. Daiko, H. Sakane, T. Amaya, T. Hirao, *J. Am. Chem. Soc.* **2005**, *127*, 11580–11581; (c) T. Amaya, H. Sakane, T. Hirao, *Angew. Chem. Int. Ed.* **2007**, *46*, 8376–8379; (d) S. Higashibayashi, H. Sakurai, *J. Am. Chem. Soc.* **2008**, *130*, 8592–8593; (e) T. Amaya, S. Seki, T. Moriuchi, K. Nakamoto, T. Nakata, H. Sakane, A. Saeki, S. Tagawa, T. Hirao, *J. Am. Chem. Soc.* **2009**, *131*, 408–409; (f) H. Sakane, T.

Amaya, T. Moriuchi, T. Hirao, *Angew. Chem. Int. Ed.* **2009**, *48*, 1640–1643; (g) T. Amaya, T. Nakata, T. Hirao, *J. Am. Chem. Soc.* **2009**, *131*, 10810–10811; (h) T. Amaya, W.-Z. Wang, H. Sakane, T. Moriuchi, T. Hirao, *Angew. Chem. Int. Ed.* **2010**, *49*, 403–406.

19. (a) S. Z. Zard, *Radical Reactions in Organic Synthesis*, Oxford Univ. Press, New York, **2003**; (b) H. Togo, *Advanced Free Radical Reactions for Organic Synthesis*, Elsevier, Oxford, **2004**.

20. (a) M. Gomberg, *J. Am. Chem. Soc.* **1900**, *22*, 757–771; (b) M. Gomberg, *J. Am. Chem. Soc.* **1901**, *23*, 496–502; (c) M. Gomberg, *Chem. Rev.* **1924**, *1*, 91–141.

21. For overviews of stable radicals, see (a) A. R. Forrester, J. M. Hay, R. H. Thomson, *Organic Chemistry of Stable Free Radicals*, Academic Press, London and New York, **1968**; (b) R. G. Hicks, *Org. Biomol. Chem.* **2007**, *5*, 1321–1338; (c) R. G. Hicks (ed.), *Stable Radicals: Fundamental and Applied Aspects of Odd-Electron Compounds*, Wiley, Chichester, **2010**.

22. For overviews of molecule-based magnetic materials on the base of organic stable radicals, see (a) P. M. Lahti (ed.), *Magnetic Properties of Organic Materials*, Marcel Dekker, New York, **1999**; (b) P. Day, A. E. Underhill (eds.), *Metal–Organic and Organic Molecular Magnets*, The Royal Society of Chemistry, Cambridge, **1999**; (c) K. Itoh, M. Kinoshita (eds.), *Molecular Magnetism*, Kodansha, and Gordon & Breach Science Publishers, Tokyo, **2000**; (d) J. S. Miller, M. Drillon (eds.), *Magnetism: Molecules to Materials*, Wiley-VCH, Weinheim, **2001–2005**; Vols. *I–V*; (e) W. Linert, M. Verdaguer (eds.), *Molecular Magnets: Recent Highlights*, Springer-Verlag, Wien, **2003**; (f) T. L. Makarova, F. Palacio (eds.), *Carbon-Based Magnetism*, Elsevier, Amsterdam, **2006**; (g) M. Fourmigué, L. Ouahab (eds.), *Conducting and Magnetic Organometallic Molecular Materials*, Springer-Verlag, Berlin, **2009**.

23. J. Veciana, C. Rovira, D. B. Amabilino (eds.), *Supramolecular Engineering of Synthetic Metallic Materials*, Kluwer Academic Publishers, Dordrecht, **1999**.

24. R. Tamura, Organic functional materials containing chiral nitroxide radical units, in *Nitroxides: Applications in Chemistry, Biomedicine, and Materials Science*, Wiley-VCH, Weinheim, **2008**, pp. 303–329.

25. For information on organic radical batteries based on stable nitroxide radical polymers, see (a) K. Nakahara, S. Iwasa, M. Satoh, Y. Morioka, J. Iriyama, M. Suguro, E. Hasegawa, *Chem. Phys. Lett.* **2002**, *359*, 351–354; (b) H. Nishide, K. Oyaizu, *Science* **2008**, *319*, 737–738. In 2002, the possibility of realizing a high-performance secondary battery by utilizing stable neutral radical with multistage redox ability is also claimed; see (c) Y. Morita, T. Aoki, K. Fukui, S. Nakazawa, K. Tamaki, S. Suzuki, A. Fuyuhiro, K. Yamamoto, K. Sato, D. Shiomi, A. Naito, T. Takui, K. Nakasuji, *Angew. Chem. Int. Ed.* **2002**, *41*, 1793–1796; (d) Y. Morita, S. Nishida, J. Kawai, K. Fukui, S. Nakazawa, K. Sato, D. Shiomi, T. Takui, K. Nakasuji, *Org. Lett.* **2002**, *4*, 1985–1988. On the basis of the ideas reported in Ref. 25c,d, a secondary battery based on a 6OPO neutral radical with multistage redox ability is fabricated, and its charge–discharge properties are disclosed in detail; see (e) Y. Morita, T. Okafuji, M. Satoh, *Jpn. Kokai Tokkyo Koho JP* 2007227186 A20070906, **2007**; (f) S. Nishida, Y. Morita, M. Moriguchi, K. Fukui, D. Shiomi, K. Sato, M. Satoh, K. Nakasuji, T. Takui, Molecular crystalline secondary batteries: Properties of phenalenyl-based multi-stage redox systems as cathode active organic materials, Paper presented at 1st Russian-Japanese Workshop on Open Shell Compounds and Molecular Spin Devices, Novosibirsk, Russia, June 30–July 2, **2007**.

26. (a) R. Rahimi, K. Sato, D. Shiomi, T. Takui, Quantum information processing as studied by molecule-based pulsed ENDOR spectroscopy, in *Modern Magnetic Resonance*, G. A. Webb (ed.), Springer-Verlag, Dordrecht, **2007**, pp. 643–650; (b) K. Sato, S. Nakazawa, R. Rahimi, T. Ise, S. Nishida, T. Yoshino, N. Mori, K. Toyota, D. Shiomi, Y. Yakiyama, Y. Morita, M. Kitagawa, K. Nakasuji, M. Nakahara, H. Hara, P. Carl, P. Höfer, T. Takui, *J. Mater. Chem.* **2009**, *19*, 3739–3754; (c) K. Sato, S. Nakazawa, R. Rahimi, S. Nishida, T. Ise, D. Shiomi, K. Toyota, Y. Morita, M. Kitagawa, P. Carl, P. Höfer, T. Takui, Quantum computing using pulse-based electron-nuclear double resonance (ENDOR) molecular spin-qubits, in *Molecular Realizations of Quantum Computing 2007*, M. Nakahara, Y. Ota, R. Rahimi (eds.), World Scientific, Singapore, **2009**, pp. 58–162; (d) Y. Morita, Y. Yakiyama, S. Nakazawa, T. Murata, T. Ise, D. Hashizume, D. Shiomi, K. Sato, M. Kitagawa, K. Nakasuji, T. Takui, *J. Am. Chem. Soc.* **2010**, *132*, 6944–6946.

27. (a) J. F. W. Keana, *Chem. Rev.* **1978**, *78*, 37–64; (b) N. Kocherginsky, H. M. Swartz, *Nitroxide Spin Labels: Reactions in Biology and Chemistry*, CRC Press, Boca Raton, FL, **1995**; (c) M. A. Hemminga, L. J. Berliner (eds.), *ESR Spectroscopy in Membrane Biophysics*, Springer, New York, **2007**; (d) G. I. Likhtenshtein, J. Yamauchi, S. Nakatsuji, A. I. Smirnov, R. Tamura, *Nitroxides: Applications in Chemistry, Biomedicine, and Materials Science*, Wiley-VCH, Weinheim, **2008**.

28. For representative studies on the most recent phenalenyl chemistry, see (a) Y. Morita, T. Ohba, N. Haneda, S. Maki, J. Kawai, K. Hatanaka, K. Sato, D. Shiomi, T. Takui, K. Nakasuji, *J. Am. Chem. Soc.* **2000**, *122*, 4825–4826; (b) J. Inoue, K. Fukui, T. Kubo, S. Nakazawa, K. Sato, D. Shiomi, Y. Morita, K. Yamamoto, T. Takui, K. Nakasuji, *J. Am. Chem. Soc.* **2001**, *123*, 12702–12703; (c) Y. Morita, T. Aoki, K. Fukui, S. Nakazawa, K. Tamaki, S. Suzuki, A. Fuyuhiro, K. Yamamoto, K. Sato, D. Shiomi, A. Naito, T. Takui, K. Nakasuji, *Angew. Chem. Int. Ed.* **2002**, *41*, 1793–1796; (d) M. E. Itkis, X. Chi, A. W. Cordes, R. C. Haddon, *Science* **2002**, *296*, 1443–1445; (e) S. K. Pal, M. E. Itkis, F. S. Tham, R. W. Reed, R. T. Oakley, R. C. Haddon, *Science* **2005**, *309*, 281–284; (f) S. Nishida, Y. Morita, K. Fukui, K. Sato, D. Shiomi, T. Takui, K. Nakasuji, *Angew. Chem. Int. Ed.* **2005**, *44*, 7277–7280; (g) S. K. Mandal, S. Samanta, M. E. Itkis, D. W. Jensen, R. T. Reed, R. T. Oakley, F. S. Tham, B. Donnadieu, R. C. Haddon, *J. Am. Chem. Soc.* **2006**, *128*, 1982–1994; (h) S. Suzuki, Y. Morita, K. Fukui, K. Sato, D. Shiomi, T. Takui, K. Nakasuji, *J. Am. Chem. Soc.* **2006**, *128*, 2530–2531; (i) S. K. Pal, M. E. Itkis, F. S. Tham, R. W. Reed, R. T. Oakley, B. Donnadieu, R. C. Haddon, *J. Am. Chem. Soc.* **2007**, *129*, 7163–7174; (j) Y. Morita, S. Suzuki, K. Fukui, S. Nakazawa, H. Kitagawa, H. Kishida, H. Okamoto, A. Naito, A. Sekine, Y. Ohashi, M. Shiro, K. Sasaki, D. Shiomi, K. Sato, T. Takui, K. Nakasuji, *Nat. Mater.* **2008**, *7*, 48–51; (k) Y. Morita, S. Nishida, J. Kawai, T. Takui, K. Nakasuji, *Pure Appl. Chem.* **2008**, *80*, 507–517; (l) S. K. Pal, M. E. Itkis, F. S. Tham, R. W. Reed, R. T. Oakley, R. C. Haddon, *J. Am. Chem. Soc.* **2008**, *130*, 3942–3951; (m) R. C. Haddon, A. Sarkar, S. K. Pal, X. Chi, M. E. Itkis, F. S. Tham, *J. Am. Chem. Soc.* **2008**, *130*, 13683–13690; (n) P. Bag, M. E. Itkis, S. K. Pal, B. Donnadieu, F. S. Tham, H. Park. J. A. Schlueter, T. Siegrist, R. C. Haddon, *J. Am. Chem. Soc.* **2010**, *132*, 2684–2694.

29. (a) Y. Morita, S. Nishida, Phenalenyls, cyclopentadienyls, and other carbon-centered radicals, in *Stable Radicals: Fundamental and Applied Aspects of Odd-Electron Compounds*, R. Hicks (ed.), Wiley, Chichester, **2010**, pp. 81–145; (b) Y. Morita, S. Suzuki, K. Sato, T. Takui, *Nat. Chem.* **2011**, *3*, 197–204.

30. For reviews on the radical anion system of fullerenes, see (a) S. S. Eaton, G. R. Eaton, *Appl. Magn. Reson.* **1996**, *11*, 155–170; (b) D. V. Konarev, R. N. Lyubovskaya, *Russ. Chem. Rev.* **1999**, *68*, 19–38; (c) C. A. Reed, R. D. Bolskar, *Chem. Rev.* **2000**, *100*, 1075–1120;

(d) D. V. Konarev, R. N. Lyubovskaya, N. V. Drichko, E. I. Yudanova, Y. M. Shul'ga, A. L. Litvinov, V. N. Semkin, B. P. Tarasov, *J. Mater. Chem.* **2000**, *10*, 803–818; (e) T. Sternfeld, M. Rabinovitz, Reduction of carbon-rich compounds, in *Carbon-Rich Compounds*, M. M. Haley, R. R. Tykwinski (eds.), Wiley-VCH, Weinheim, **2006**, pp. 599–611.

31. For review articles of neutral radical adducts of fullerenes, see (a) B. L. Tumanskii, *Russ. Chem. Bull.* **1996**, *45*, 2267–2278; (b) J. R. Morton, F. Negri, K. F. Preston, *Acc. Chem. Res.* **1998**, *31*, 63–69.

32. B. Tumanskii, O. Kalina, *Radical Reactions of Fullerenes and their Derivatives*, Kluwer Academic Publishers, Dordrecht, **2001**.

33. (a) Q. Xie, E. Pérez-Cordero, L. Echegoyen, *J. Am. Chem. Soc.* **1992**, *114*, 3978–3980; (b) Y. Ohsawa, T. Saji, *J. Chem. Soc. Chem. Commun.* **1992**, 781–782.

34. (a) P.-M. Allemand, G. Srdanov, A. Koch, K. Khemani, F. Wudl, Y. Rubin, F. Diederich, M. M. Alvarez, S. J. Anz, R. L. Whetten, *J. Am. Chem. Soc.* **1991**, *113*, 2780–2781; (b) U. Bilow, M. Jansen, *J. Chem. Soc. Chem. Commun.* **1994**, 403–404; (c) M. M. Khaled, R. T. Carlin, P. C. Trulove, G. R. Eaton, S. S. Eaton, *J. Am. Chem. Soc.* **1994**, *116*, 3465–3474; (d) W. C. Wan, X. Liu, G. M. Sweeney, W. E. Broderick, *J. Am. Chem. Soc.* **1995**, *117*, 9580–9581; (e) W. Schütz, J. Gmeiner, A. Schilder, B. Gotschy, V. Enkelmann, *J. Chem. Soc. Chem. Commun.* **1996**, 1571.

35. (a) Q. Zhu, D. E. Cox, J. E. Fischer, *Phys. Rev. B* **1995**, *51*, 3966–3969; (b) M. Kosaka, K. Tanigaki, T. Tanaka, T. Atake, A. Lappas, K. Prassides, *Phys. Rev. B* **1995**, *51*, 12018–12021; (c) G. Oszlányi, G. Bortel, G. Faigel, M. Tegze, L. Gránásy, S. Pekker, P. W. Stephens, G. Bendele, R. Dinnebier, G. Mihály, A. Jánossy, O. Chauvet, L. Forró, *Phys. Rev. B* **1995**, *51*, 12228–12232; (d) G. Oszlányi, G. Bortel, G. Faigel, L. Gránásy, G. M. Bendele, P. W. Stephens, L. Forró, *Phys. Rev. B* **1996**, *54*, 11849–11852.

36. (a) D. V. Konarev, S. S. Khasanov, A. Otsuka, G. Saito, *J. Am. Chem. Soc.* **2002**, *124*, 8520–8521; (b) D. V. Konarev, S. S. Khasanov, G. Saito, A. Otsuka, Y. Yoshida, R. N. Lyubovskaya, *J. Am. Chem. Soc.* **2003**, *125*, 10074–10083.

37. D. V. Konarev, S. S. Khasanov, I. I. Vorontsov, G. Saito, M. Y. Antipin, A. Otsuka, R. N. Lyubovskaya, *Chem. Commun.* **2002**, 2548–2549.

38. (a) P. Bhyrappa, P. Paul, J. Stinchcombe, P. D. W. Boyd, C. A. Reed, *J. Am. Chem. Soc.* **1993**, *115*, 11004–11005; (b) P. Paul, Z. Xie, R. Bau, P. D. W. Boyd, C. A. Reed, *J. Am. Chem. Soc.* **1994**, *116*, 4145–4146; (c) P. D. W. Boyd, P. Bhyrappa, J. Stinchcombe, R. D. Bolskar, Y. Sun, C. A. Reed, *J. Am. Chem. Soc.* **1995**, *117*, 2907–2914; (d) P. C. Trulove, R. T. Carlin, G. R. Eaton, S. S. Eaton, *J. Am. Chem. Soc.* **1995**, *117*, 6265–6272; (e) M. Baumgarten, A. Gügel, L. Gherghel, *Adv. Mater.* **1996**, *5*, 458–461; (f) T. F. Fässler, A. Spiekermann, M. E. Spahr, R. Nesper, *Angew. Chem. Int. Ed. Engl.* **1997**, *36*, 486–488; (g) P. Paul, K.-C. Kim, D. Sun, P. D. W. Boyd, C. A. Reed, *J. Am. Chem. Soc.* **2002**, *124*, 4394–4401; (h) D. V. Konarev, S. S. Khasanov, G. Saito, I. I. Vorontsov, A. Otsuka, R. N. Lyubovskaya, Y. M. Antipin, *Inorg. Chem.* **2003**, *42*, 3706–3708.

39. (a) C. Janiak, S. Mühle, H. Hemling, K. Köhler, *Polyhedron* **1996**, *15*, 1559–1563; (b) M. C. B. L. Shohoji, M. L. T. M. B. Franco, M. C. R. L. R. Lazana, S. Nakazawa, K. Sato, D. Shiomi, T. Takui, *J. Am. Chem. Soc.* **2000**, *122*, 2962–2963; (c) S. C. Drew, J. F. Boas, J. R. Pilbrow, P. D. W. Boyd, P. Paul, C. A. Reed, *J. Phys. Chem. B* **2003**, *107*, 11353–11359; (d) S. Nakazawa, K. Sato, D. Shiomi, M. L. T. M. B. Franco, M. C. R. L. R. Lazana, M. C. B. L. Shohoji, K. Itoh, T. Takui, *Inorg. Chim. Acta.* **2008**, *361*, 4031–4037.

40. (a) S. Liu, Y.-J. Lu, M. M. Kappes, J. A. Ibers, *Science* **1991**, *254*, 408–410; (b) H.-B. Bürgi, E. Blanc, D. Schwarzenbach, S. Liu, Y.-J. Lu, M. M. Kappes, J. A. Ibers, *Angew. Chem. Int. Ed. Engl.* **1992**, *31*, 640–643.

41. For an article on benzyl radical adducts, see P. J. Krusic, E. Wasserman, P. N. Keizer, J. R. Morton, K. F. Preston, *Science* **1991**, *254*, 1183–1185.

42. For literature on alkyl radical adducts, see (a) J. R. Morton, K. F. Preston, P. J. Krusic, S. A. Hill, E. Wasserman, *J. Phys. Chem.* **1992**, *96*, 3576–3578; (b) J. R. Morton, K. F. Preston, P. J. Krusic, S. A. Hill, E. Wasserman, *J. Am. Chem. Soc.* **1992**, *114*, 5454–5455; (c) J. R. Morton, K. F. Preston, P. J. Krusic, E. Wasserman, *J. Chem. Soc. Perkin Trans. 2*, **1992**, 1425–1429; (d) P. J. Krusic, D. C. Roe, E. Johnston, J. R. Morton, K. F. Preston, *J. Phys. Chem.* **1993**, *97*, 1736–1738; (e) P. N. Keizer, J. R. Morton, K. F. Preston, P. J. Krusic, *J. Chem. Soc. Perkin Trans. 2*, **1993**, 1041–1045.

43. For literature on fluoroalkyl radical adducts, see (a) P. J. Fagan, P. J. Krusic, C. N. McEwen, J. Lazar, D. H. Parker, N. Herron, E. Wasserman, *Science* **1993**, *262*, 404–407; (b) J. R. Morton, K. F. Preston, *J. Phys. Chem.* **1994**, *98*, 4993–4997; (c) J. R. Morton, F. Negri, K. F. Preston, G. Ruel, *J. Phys. Chem.* **1995**, *99*, 10114–10117; (d) J. R. Morton, F. Negri, K. F. Preston, G. Ruel, *J. Chem. Soc. Perkin Trans. 2*, **1995**, 2141–2145.

44. (a) H. Iikura, S. Mori, M. Sawamura, E. Nakamura, *J. Org. Chem.* **1997**, *62*, 7912–7913; (b) Y. Matsuo, E. Nakamura, *J. Am. Chem. Soc.* **2005**, *127*, 8457–8466.

45. (a) J. C. Hummelen, B. Knight, J. Pavlovich, R. González, F. Wudl, *Science* **1995**, *269*, 1554–1556; (b) M. Keshavarz-K., R. González, R. G. Hicks, G. Srdanov, V. I. Srdanov, T. G. Collins, J. C. Hummelen, C. Bellavia-Lund, J. Pavlovich, F. Wudl, K. Holczer, *Nature* **1996**, *383*, 147–150; (c) C. M. Brown, L. Cristofolini, K. Kordatos, K. Prassides, C. Bellavia, R. González, M. Keshavarz-K., F. Wudl, A. K. Cheetham, J. P. Zhang, W. Andreoni, A. Curioni, A. N. Fitch, P. Pattison, *Chem. Mater.* **1996**, *8*, 2548–2550; (d) W. Andreoni, A. Curioni, K. Holczer, K. Prassides, M. Keshavarz-K., J.-C. Hummelen, F. Wudl, *J. Am. Chem. Soc.* **1996**, *118*, 11335–11336; (e) C. Bellavia-Lund, R. González, J. C. Hummelen, R. G. Hicks, A. Sastre, F. Wudl, *J. Am. Chem. Soc.* **1997**, *119*, 2946–2947; (f) C. Bellavia-Lund, M. Keshavarz-K., T. Collins, F. Wudl, *J. Am. Chem. Soc.* **1997**, *119*, 8101–8102; (g) A. Gruss, K.-P. Dinse, A. Hirsch, B. Nuber, U. Reuther, *J. Am. Chem. Soc.* **1997**, *119*, 8728–8729; (h) K. Hasharoni, C. Bellavia-Lund, M. Keshavarz-K., G. Srdanov, F. Wudl, *J. Am. Chem. Soc.* **1997**, *119*, 11128–11129; (i) F. Rachdi, L. Hajji, H. Dollt, M. Ribet, T. Yildirim, J. E. Fischer, C. Goze, M. Mehring, A. Hirsch, B. Nuber, *Carbon* **1998**, *36*, 607–611; (j) U. Reuther, A. Hirsch, *Carbon* **2000**, *38*, 1539–1549; (k) O. Vostrowsky, A. Hirsch, *Chem. Rev.* **2006**, *106*, 5191–5207.

46. (a) C. Corvaja, M. Maggini, M. Prato, G. Scorrano, M. Venzin, *J. Am. Chem. Soc.* **1995**, *117*, 8857–8858; (b) F. Arena, F. Bullo, F. Conti, C. Corvaja, M. Maggini, M. Prato, G. Scorrano, *J. Am. Chem. Soc.* **1997**, *119*, 789–795; (c) I. A. Nuretdinov, V. P. Gubskaya, V. V. Yanilkin, V. I. Morozov, V. V. Zverev, A. V. Il'yasov, G. M. Fazleeva, N. V. Nastapova, D. V. Il'matova, *Russ. Chem. Bull.* **2001**, *50*, 607–613.

47. (a) N. Mizuochi, Y. Ohba, S. Yamauchi, *J. Phys. Chem. A* **1999**, *103*, 7749–7752; (b) M. Mazzoni, L. Franco, C. Corvaja, G. Zordan, E. Menna, G. Scorrano, M. Maggini, *ChemPhysChem* **2002**, *6*, 527–531; (c) A. Polimeno, M. Zerbetto, L. Franco, M. Maggini, C. Corvaja, *J. Am. Chem. Soc.* **2006**, *128*, 4734–4741.

48. (a) J. Janata, J. Gendell, C.-Y. Ling, W. Barth, L. Backes, H. B. Mark Jr., R. G. Lawton, *J. Am. Chem. Soc.* **1967**, *89*, 3056–3058; (b) A. Ayalon, M. Rabinovitz, P.-C. Cheng, L. T.

Scott, *Angew. Chem. Int. Ed. Engl.* **1992**, *31*, 1636–1637; (c) P. W. Rabideau, Z. Marcinow, R. Sygula, A. Sygula, *Tetrahedron Lett.* **1993**, *34*, 6351–6354; (d) A. Ayalon, A. Sygula, P.-C. Cheng, M. Rabinovitz, P. W. Rabideau, L. T. Scott, *Science* **1994**, *265*, 1065–1067; (e) M. Baumgarten, L. Gherghel, M. Wagner, A. Weitz, M. Rabinovitz, P.-C. Cheng, L. T. Scott, *J. Am. Chem. Soc.* **1995**, *117*, 6254–6257; (f) G. Zilber, V. Rozenshtein, P.-C. Cheng, L. T. Scott, M. Rabinovitz, H. Levanon, *J. Am. Chem. Soc.* **1995**, *117*, 10720–10725; (g) T. Sato, A. Yamamoto, T. Yamabe, *J. Chem. Phys. A* **2000**, *104*, 130–137; (h) T. Sato, A. Yamamoto, H. Tanaka, *Chem. Phys. Lett.* **2000**, *326*, 573–579; (i) I. Aprahamian, D. Eisenberg, R. E. Hoffman, T. Sternfeld, Y. Matsuo, E. A. Jackson, E. Nakamura, L. T. Scott, T. Sheradsky, M. Rabinovitz, *J. Am. Chem. Soc.* **2005**, *127*, 9581–9587; see also Ref. 67.

49. T. Sternfeld, M. Rabinovitz, Reduction of carbon-rich compounds, in *Carbon-Rich Compounds*, M. M. Haley, R. R. Tykwinski (eds.), Wiley-VCH, Weinheim, **2006**, pp. 586–596.

50. (a) A. Weitz, E. Shabtai, M. Rabinovitz, M. S. Bratcher, C. C. McComas, M. D. Best, L. T. Scott, *Chem. Eur. J.* **1998**, *4*, 234–239; (b) I. Aprahamian, R. E. Hoffman, T. Sheradsky, D. V. Preda, M. Bancu, L. T. Scott, M. Rabinovitz, *Angew. Chem. Int. Ed.* **2002**, *41*, 1712–1715; (c) I. Aprahamian, D. V. Preda, M. Bancu, A. P. Belanger, T. Sheradsky, L. T. Scott, M. Rabinovitz, *J. Org. Chem.* **2006**, *71*, 290–298.

51. (a) Y. Morita, S. Nishida, T. Kobayashi, K. Fukui, K. Sato, D. Shiomi, T. Takui, K. Nakasuji, *Org. Lett.* **2004**, *6*, 1397–1400; (b) S. Nishida, Y. Morita, T. Kobayashi, K. Fukui, A. Ueda, K. Sato, D. Shiomi, T. Takui, K. Nakasuji, *Polyhedron* **2005**, *24*, 2200–2204; (c) Y. Morita, A. Ueda, S. Nishida, K. Fukui, T. Ise, D. Shiomi, K. Sato, T. Takui, K. Nakasuji, *Angew. Chem. Int. Ed.* **2008**, *47*, 2035–2038; (d) A. Ueda, K. Ogasawara, S. Nishida, K. Fukui, K. Sato, T. Takui, K. Nakasuji, Y. Morita, *Aust. J. Chem.* **2010**, *63*, 1627–1633.

52. K. Fukui, Y. Morita, S. Nishida, T. Kobayashi, K. Sato, D. Shiomi, T. Takui, K. Nakasuji, *Polyhedron* **2005**, *24*, 2326–2329.

53. A. Ueda, S. Nishida, K. Fukui, T. Ise, D. Shiomi, K. Sato, T. Takui, K. Nakasuji, Y. Morita, *Angew. Chem. Int. Ed.* **2010**, *49*, 1678–1682.

54. T. J. Seiders, E. L. Elliott, G. H. Grube, J. S. Siegel, *J. Am. Chem. Soc.* **1999**, *121*, 7804–7813.

55. (a) R. C. Haddon, L. T. Scott, *Pure Appl. Chem.* **1986**, *58*, 137–142; (b) R. C. Haddon, *Acc. Chem. Res.* **1988**, *21*, 243–249; (c) R. C. Haddon, *J. Am. Chem. Soc.* **1990**, *112*, 3385–3389.

56. (a) F. van Bolhuis, C. T. Kiers, *Acta Crystallogr. Sect. B* **1978**, *34*, 1015–1016; (b) R. West, J. A. Jorgenson, K. L. Stearley, J. C. Calabrese, *J. Chem. Soc. Chem. Commun.* **1991**, 1234–1235.

57. Aromaticity of fullerene derivatives is extensively discussed by using a chemical term, *spherical aromaticity*; see (a) M. Bühl, A. Hirsch, *Chem. Rev.* **2001**, *101*, 1153–1183; (b) Z. Chen, R. B. King, *Chem. Rev.* **2005**, *105*, 3613–3642.

58. (a) P. von R. Schleyer, C. Maerker, A. Dransfeld, H. Jiao, N. J. R. van E. Hommes, *J. Am. Chem. Soc.* **1996**, *118*, 6317–6318; (b) P. von R. Schleyer, *Chem. Rev.* **2001**, *101*, 1115–1117; (c) Z. Chen, C. S. Wannere, C. Corminboeuf, R. Puchta, P. von R. Schleyer, *Chem. Rev.* **2005**, *105*, 3842–3888.

59. (a) V. Gogonea, P. von R. Schleyer, P. R. Schreiner, *Angew. Chem. Int. Ed.* **1998**, *37*, 1945–1948; (b) Y. Morita, J. Kawai, K. Fukui, S. Nakazawa, K. Sato, D. Shiomi, T. Takui, K. Nakasuji, *Org. Lett.* **2003**, *5*, 3289–3291; see also Ref. 28h.

60. W. T. Borden (ed.), *Diradicals*, Wiley, New York, **1982**.

61. A. Sygula, S. D. Karlen, R. Sygula, P. W. Rabideau, *Org. Lett.* **2002**, *4*, 3135–3137.

62. (a) D. Döhnert, J. Koutecký, *J. Am. Chem. Soc.* **1980**, *102*, 1789–1796; (b) Y. Jung, M. Head-Gordon, *ChemPhysChem* **2003**, *4*, 522–525 and references cited therein.

63. (a) R. Pettit, *Chem. Ind. (Lond.)* **1956**, 1306–1307; (b) D. H. Reid, *Chem. Ind. (Lond.)* **1956**, 1504–1505; (c) D. H. Reid, *Quart. Rev.* **1965**, *19*, 274–302.

64. K. Goto, T. Kubo, K. Yamamoto, K. Nakasuji, K. Sato, D. Shiomi, T. Takui, M. Kubota, T. Kobayashi, K. Yakushi, J. Ouyang, *J. Am. Chem. Soc.* **1999**, *121*, 1619–1620.

65. S. Nishida, Y. Morita, A. Ueda, T. Kobayashi, K. Fukui, K. Ogasawara, K. Sato, T. Takui, K. Nakasuji, *J. Am. Chem. Soc.* **2008**, *130*, 14954–14955.

66. T. J. Seiders, K. K. Baldridge, G. H. Grube, J. S. Siegel, *J. Am. Chem. Soc.* **2001**, *123*, 517–525.

67. Theoretical calculations suggest that the pristine corannulene also shows a decrease in the inversion barrier on increased negative charge; see (a) A. Sygula, P. W. Rabideau, *J. Mol. Struct. (THEOCHEM)* **1995**, *333*, 215–226; (b) C. Bruno, R. Benassi, A. Passalacqua, F. Paolucci, C. Fontanesi, M. Marcaccio, E. A. Jackson, L. T. Scott, *J. Phys. Chem. B* **2009**, *113*, 1954–1962.

68. H. Dodziuk (ed.), *Strained Hydrocarbons*, Wiley-VCH, Weinheim, **2009**.

69. A corannulene-based bowl-shaped *o*-semiquinone radical with an alkali metal cation has more recently been synthesized, and its three-dimensional electronic spin structure and dynamic behavior have been quantitatively evaluated; see A. Ueda, K. Ogasawara, S. Nishida, T. Ise, T. Yoshino, S. Nakazawa, K. Sato, T. Takui, K. Nakasuji, Y. Morita, *Angew. Chem. Int. Ed.* **2010**, *49*, 6333–6337.

70. For reviews of hoop- or belt-shaped aromatic compounds, see (a) A. Schröder, H.-B. Mekelburger, F. Vögtle, *Top. Curr. Chem.* **1994**, *172*, 179–201; (b) U. Girreser, D. Giuffrida, F. H. Kohnke, J. P. Mathias, D. Philp, J. F. Stoddart, *Pure Appl. Chem.* **1993**, *65*, 119–125; (c) L. T. Scott, *Angew. Chem. Int. Ed.* **2003**, *42*, 4133–4135; (d) R. Gleiter, B. Hellbach, S. Gath, R. J. Schaller, *Pure Appl. Chem.* **2006**, *78*, 699–706; (e) K. Tahara, Y. Tobe, *Chem. Rev.* **2006**, *106*, 5274–5290; (f) Y. Matsuo, E. Nakamura, Cyclophenacene cut out of fullerene, in *Functional Organic Materials*, T. J. J. Müller, U. H. F. Bunz (eds.), Wiley-VCH, Weinheim, **2007**, pp. 59–80; (g) T. Yao, H. Yu, R. J. Vermeij, G. J. Bodwell, *Pure Appl. Chem.* **2008**, *80*, 533–546. See also Ref. 11e.

71. For overviews of helicene chemistry, see (a) R. H. Martin, *Angew. Chem. Int. Ed. Engl.* **1974**, *13*, 649–660; (b) H. Hopf, *Classics in Hydrocarbon Chemistry*, Wiley-VCH, Weinheim, **2000**, pp. 321–330; (c) A. Rajca, M. Miyasaka, Synthesis and characterization of novel chiral conjugated materials, in *Functional Organic Materials*, T. J. J. Müller, U. H. F. Bunz (eds.), Wiley-VCH, Weinheim, **2007**, pp. 547–581; (d) Non-Kekulé or open-shell polynuclear aromatic hydrocarbons possessing helical π-conjugated structure are designed, and their band structures are calculated; see K. Fukui, *An Electron-Nuclear Magnetic Resonance Study of Highly Symmetric Organic Open-Shell Systems with Homo- and Hetero-Atomic π-Conjugation and Their High-Spin Assemblages*, PhD thesis, Osaka City Univ. (Japan), **2001**; (e) In 2010, Rajca and coworkers reported the preparation and chiroptical properties of a chiral radical cation of a thiophene-based helicene derivative. Unfortunately, isolation as the solid state is not achieved because of the low stability; see J. K. Zak, M. Miyasaka, S. Rajca, M. Lapkowski, A. Rajca, *J. Am. Chem. Soc.* **2010**, *132*, 3246–3247; (f) As the first chiral neutral radical with highly spin-delocalized nature, a

phenalenyl-fused helicenic neutral radical has been designed, synthesized, and isolated as a stable solid in air (Y. Morita and A. Ueda, unpublished results).

72. (a) D. Ajami, O. Oeckler, A. Simon, R. Herges, *Nature* **2003**, *426*, 819–821; (b) R. Herges, *Chem. Rev.* **2006**, *106*, 4820–4842; (c) R. Herges, *Nature* **2007**, *450*, 36–37; (d) Z. S. Yoon, A. Osuka, D. Kim, *Nat. Chem.* **2009**, *1*, 113–122. (d) In 2010, Osuka and coworkers synthesized and isolated a stable zwitterionic radical species by metalation of a Möbius aromatic hexaphyrin derivative and demonstrated a delocalization nature of electronic spin over a doubly twisted π-conjugation system, see H. Rath, S. Tokuji, N. Aratani, K. Furukawa, J. M. Lim, D. Kim, H. Shinokubo, A. Osuka, *Angew. Chem. Int. Ed.* **2010**, *49*, 1489–1491.

73. (a) M. Sawamura, Y. Kuninobu, M. Toganoh, Y. Matsuo, M. Yamanaka, E. Nakamura, *J. Am. Chem. Soc.* **2002**, *124*, 9354–9355; (b) M. Toganoh, Y. Matsuo, E. Nakamura, *J. Am. Chem. Soc.* **2003**, *125*, 13974–13975; (c) E. Nakamura, *J. Organomet. Chem.* **2004**, *689*, 4630–4635; (d) Y. Matsuo, K. Tahara, E. Nakamura, *J. Am. Chem. Soc.* **2006**, *128*, 7154–7155; (e) Y. Matsuo, K. Tahara, T. Fujita, E. Nakamura, *Angew. Chem. Int. Ed.* **2009**, *48*, 6239–6241; (f) T. Fujita, Y. Matsuo, E. Nakamura, *Chem. Asian. J.* **2010**, *5*, 835–840.

74. The colloquial chemical term "π-bowl complex" was coined by Hirao (see Ref. 18c,f,h).

75. (a) E. M. Pérez, M. Sierra, L. Sánchez, M. R. Torres, R. Viruela, P. M. Viruela, E. Ortí, N. Martín, *Angew. Chem. Int. Ed.* **2007**, *46*, 1847–1851; (b) S. S. Gayathri, M. Wielopolski, E. M. Pérez, G. Fernández, L. Sánchez, R. Viruela, E. Ortí, D. M. Guldi, N. Martín, *Angew. Chem. Int. Ed.* **2009**, *48*, 815–819.

76. K. Inoue, S.-i. Ohkoshi, H. Imai, Chiral molecule-based magnets, in *Magnetism: Molecules to Materials*, J. S. Miller, M. Drillon (eds.), Wiley-VCH, Weinheim, **2005**, Vol. 5, pp. 41–70.

# CHAPTER 5

# EXPERIMENTAL AND CALCULATED PROPERTIES OF FULLERENE AND NANOTUBE FRAGMENTS

DEREK R. JONES, PRAVEEN BACHAWALA, and JAMES MACK

## 5-1. INTRODUCTION

The concept of nanotechnology has been around since the late 1950s;[1] however, scientists were unable to fully utilize this potential until the discovery and development of new instrumentation and new materials. Development of the scanning tunneling microscope in 1982 enabled scientists for the first time to view individual atoms on a metal surface.[2] Soon thereafter, the discovery of fullerenes[3] and subsequently nanotubes[4] provided the organic materials necessary to start making the dream of nanotechnology a reality. Although fullerenes and nanotubes have properties that are well suited for various nanotechnological fields, the ability to tailor these properties for specific tasks have been slow to develop. The chemistry of fullerenes suffers from the formation of various complex mixtures, while the synthesis of consistently reproducible single-walled nanotubes is still a major hurdle. Until these problems are corrected, the implementation of these molecules in nanotechnology will continue to be limited.

The ability to synthesize and modify independent fullerene and nanotube fragments is necessary in order to build targeted structures for nanotechnological purposes. Over the years, reports of the various properties of fullerene fragments have surfaced in the chemical literature. Corannulene (1), a fullerene fragment that spans exactly one-third of fullerene[60] (2), has been shown to possess some of the same properties of fullerenes but on a smaller scale (Figure 5-1). The focus of this chapter is to cover the photophysical and electrochemical properties of corannulene

*Fragments of Fullerenes and Carbon Nanotubes: Designed Synthesis, Unusual Reactions, and Coordination Chemistry*, First Edition. Edited by Marina A. Petrukhina and Lawrence T. Scott. © 2012 John Wiley & Sons, Inc. Published 2012 by John Wiley & Sons, Inc.

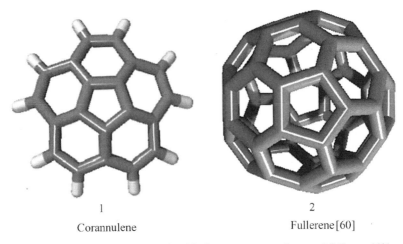

1                        2

Corannulene                Fullerene[60]

**Figure 5-1.** The structural relationship between corannulene and fullerene[60].

and other fullerene fragments and to discuss the range of our ability to predict these properties via computational methods.

## 5-2. COMPUTATIONAL CONSIDERATIONS

Vast improvements have been made in computational chemistry through the development of theoretical models, and of computers that can conduct these calculations at ever faster rates.[5–7] Improvements have led to the ability to calculate properties of potential synthetic targets with great accuracy. In the area of fullerene fragments, many of the computational efforts have focused on optimizing the structures of these unique molecules. More recently, computational efforts that focus on the photophysical and electronic properties of fullerene fragments have appeared in the chemical literature.[8–16] The ability to accurately calculate the molecular orbitals of these molecules is significant in the study of fluorescence, phosphorescence, reductions, oxidations, and complexation. If these properties can be accurately predicted, chemists can focus their synthetic efforts on targets that have the desired properties needed in nanotechnology.

### 5-2-1. Ionization Potential

Electron affinity and ionization potential are important properties for the development of electronically active organic materials. Electron affinity provides information about a molecule's potential as a conducting material, whereas ionization potential provides information about a molecule's potential as an insulating material. Various levels of theory can calculate these properties for fullerene fragments with a high level of accuracy. The electron affinity and ionization

**TABLE 5-1. Comparison of Calculated and Experimental Ionization Potentials (eV) of Corannulene (1) and 1,6-Dimethylcorannulene (3)**

|  | 1 | 3 |
| --- | --- | --- |

| Method | Corannulene | 1,6-Dimethylcorannulene |
| --- | --- | --- |
| Cyclic voltammetry | 8.37 | 8.57 |
| Photoemission spectroscopy | 8.50 | 8.22 |
| MP2/cc-pVDZ | 8.77 | 8.37 |
| Koopman's theorem[a] | 7.99 | 7.66 |

[a]Calculation performed at the MP2/cc-pVDZ level.

potential of corannulene have been calculated and measured in a variety of ways. The ionization potential of corannulene was measured to be 8.37 eV by photoelectron spectroscopy (PES) and 8.50 eV by cyclic voltammetry (CV).[10] 1,6-Dimethylcorannulene (3) gives a measured ionization potential of 8.59 eV by CV and 8.22 eV when measured by PES. Calculations performed at the MP2/cc-pVDZ level give a calculated ionization potential of 8.77 eV for corannulene and 8.38 eV for 1,6-dimethylcorannulene. Using the same level of theory and a Koopman's theorem approach (i.e., estimating the first ionization energy as the negative of the highest occupied molecular orbital energy), the ionization energy for corannulene is calculated to be 7.99 and 7.66 eV for 1,6-dimethylcorannulene. Therefore, in terms of ionization potential, calculations at the MP2/cc-pVDZ level slightly overestimate the values for corannulene and 1,6-dimethylcorannulene compared to experimental values, whereas the ionization potentials calculated by using Koopman's theorem are underestimated (Table 5-1).

## 5-2-2. Electron Affinity

The electron affinity (EA) of corannulene has been measured to be 0.5 ($\pm$.1) eV by collision-induced dissociation (CID) mass spectrometry.[11] Using the same methodology, dibenzocorannulene has a measured EA of 1.00 ($\pm$0.2) eV and diindenochrysene has an EA of 1.16 ($\pm$0.2) eV.[12] Calculating an accurate electron affinity is more challenging than calculating the ionization potential. Because an ionization potential represents the removal of an electron to form a cationic species, empty orbital contributions can be investigated in such a way that Koopman's theorem is applicable. In the case of electron affinity, which represents the addition of an electron, electron correlation must be accounted for. This requires that a self-consistent field method be employed in order to accurately calculate this property. Several trends can be observed from the calculations of electron affinity shown in

**TABLE 5-2. Comparison between Calculated and Experimental Electron Affinities (eV) of Corannulene (1)**

| Method | Electron Affinity |
| --- | --- |
| Collision-induced dissociation | +0.50 |
| Electrochemistry | +0.80 |
| Koopman's theorem [(HF/6-311G(d,p)] | +0.16 |
| UB3LYP/6-311G(d,p) | +0.50 |
| UB3LYP/631G+(d,p) | +0.63 |
| UB3LYP/6-311G(d,p) | +0.83 |
| UB3LYP/6-31G+(d,p) | +0.96 |
| UMP2/6-31G(d) | +0.13 |
| UMP3/6-31G(d) | +0.23 |
| UMP2/6-31+G(d) | −0.26 |
| UMP2/6-31+G(d) | −0.21 |
| CBS-4M | −1.38 |
| CBS-4MB | −1.60 |

Table 5-2.[10,13] Incorporation of diffuse functions provides less agreement with the experimentally measured EA's. It would be expected that inclusion of diffuse functions would lead to a better prediction of the EA of the radical anion, but this does not seem to be the case. The same trend is observed for fullerene[60] as well. At the B3LYP/6-311G(d,p) level an EA of 2.52 eV is calculated. The same calculation at the B3LYP/6-31G$^+$(d,p) level gives a calculated EA of 2.99 eV. Comparison of the calculated values to the experimental value of 2.65 eV[14,15] again shows that the incorporation of diffuse functions does not provide an accurate description of the electron affinity of curved aromatic compounds.

Calculated EAs at the MP2 and MP3 levels are less accurate than at the DFT level. Incorporation of diffuse functions gives the same trend as that observed at the DFT level. Calculations conducted at the CBS-4M level predict corannulene to have an EA of -1.38 eV; an EA of -1.60 eV is predicted when calculations are performed at the CBS-4MB level.[16]

Fullerene[60] has a triply degenerate low-lying LUMO with the ability to accept up to six electrons.[17] Similarly, corannulene has a doubly degenerate low-lying LUMO with the ability to accept up to four electrons.[18] As will be discussed in more detail later in this chapter (Section 5-5), corannulene has been chemically reduced 4 times by alkali metals, but the electrochemical reductions have not been measured past the third reduction state. Reduction potentials of corannulene were calculated at various levels of theory to determine how accurately these values compared to the experimental results and predict the reduction potential necessary to achieve the fourth reduction state. As shown in Table 5-3, as the reduction state is increased, only a few levels of theory are in good agreement with experiment. For some levels of theory, as the reduction state increases, the calculated potential decreases. Calculations performed at the B3PW91/6-31G level give values closest to

**TABLE 5-3. Comparison of Predicted Electron Reduction Potentials (eV) of Corannulene with Experimental Values**

| Method | Solvent | Neutral to Anion | Anion to Dianion | Dianion to Trianion | Trianion to Tetraanion |
|---|---|---|---|---|---|
| MP2/6-31G* | DMF | −3.04 | −3.44 | −4.44 | −4.24 |
| MP3/6-31G* | DMF | −3.14 | −3.84 | −4.64 | −5.04 |
| UMP2/6-31+G* | DMF | −2.74 | −3.01 | | |
| UMP3/6-31+G* | DMF | −3.94 | −1.24 | | |
| HF/Midix | DMF | −3.84 | −4.24 | −5.04 | −5.24 |
| B3PW91/6-31G* | DMF | −2.34 | −3.04 | −3.74 | −4.44 |
| B3PW91/6-31+G* | DMF | −2.04 | −2.44 | −2.24 | −0.74 |
| B3LYP/6-31+G* | DMF | −2.04 | −2.54 | −2.14 | −0.84 |
| B3PW91/Midi | DMF | −2.24 | −3.04 | −3.84 | −4.54 |
| Experimental | TMAB/DMF | −1.88 | −2.42 | −3.01 | — |

experimental results. Again, incorporation of diffuse functions seem to give results that are in poorer agreement with experiment than do calculations performed without diffuse functions.[16]

## 5-2-3. Dipole Moment

Another property that is unique to many fullerene fragments is the ability to complex to a variety of guests.[19–26] Corannulene has a measured dipole moment of 2.07 D.[27] Various levels of theory have been used to calculate the dipole moment of corannulene. AM1 and PM3 methods calculate a dipole moment of 2.80 and 2.43 D,[28] respectively, for corannulene. As expected, *ab initio* calculations are more accurate at predicting the dipole moment of corannulene.[29,30] Table 5-4 shows the various self-consistent field (SCF) and DFT methods that have been used to calculate the dipole moment of corannulene. This shows that using the correct basis set with the correct functional can provide accurate dipole moments of corannulene.

## 5-2-4. Spectroscopy

Computational chemistry has been important in the prediction of spectra of various molecules. Vibrational modes (i.e., the infrared spectrum) of corannulene are very well reproduced at the DFT level of theory.[8,9] Calculations performed at the B3LYP/cc-pVTZ level give vibrational modes that are in good agreement with corannulene's infrared spectrum. A scaling factor of 0.975 gives the best correlation between the calculated and experimental vibrational modes, which is consistent with the scaling factors used for other polyaromatic hydrocarbons, such as naphthalene. Calculated at the same level of theory (B3LYP/cc-pVTZ), the Raman spectral data are consistent with the experimental Raman spectral data. As shown in Figure 5-2, the shifts are in good agreement with the experimental values; however, the intensities are not as accurate.

**TABLE 5-4.  Calculated Dipole Moments of Corannulene at Various Levels of Theory Compared against Measured Dipole Moment of Corannulene**

| Functional | Dipole Moment (D) | Deviation from Experimentally Measured Dipole Moment |
|---|---|---|
| PM3 | 2.43 | +0.36 |
| AM1 | 2.80 | +0.73 |
| HF/3-21G(2d,p) | 2.69 | +0.62 |
| HF/DZV(2d,p) | 2.09 | +0.02 |
| HF/DZV(2df,2p) | 2.07 | 0.00 |
| B3LYP[a] | 1.88 | −0.19 |
| BLYP[a] | 1.76 | −0.31 |
| MP2[a] | 2.45 | +0.38 |
| BP86[a] | 2.00 | −0.07 |
| PBE[a] | 1.97 | −0.10 |
| SVWN[a] | 2.10 | +0.03 |
| PW91[a] | 1.94 | −0.13 |
| X3LYP[a] | 1.89 | −0.18 |
| B97[a] | 2.00 | −0.07 |
| BMK[a] | 2.14 | +0.07 |
| M06-2X[a] | 2.19 | +0.12 |
| Experimental | 2.07 | — |

[a]Calculated using the cc-pVDZ basis set.

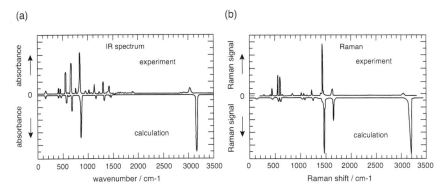

**Figure 5-2.** (a) Infrared spectrum compared with calculated values at B3LYP/cc-pVTZ level of theory; (b) Raman spectrum compared with calculated values at B3LYP/cc-pVTZ level of theory. (Reprinted from Huisken et al., *ChemPhysChem* **2008**, *9*, 2085–2091.)

## 5-3. HYDROGEN COMPLEX OF FULLERENE FRAGMENTS

One of the more intriguing properties of corannulene is its ability to complex hydrogen. Catalysts designed to complex hydrogen have been highly sought after because of their potential use as hydrogen storage materials for the hydrogen economy. Because corannulene has an inherent dipole moment, it has the potential to complex hydrogen through a dipole-induced dipole mechanism. Various calculations have been performed that predict the ability of corannulene to adsorb hydrogen.[31–35] Because weak van der Waals interactions are expected to be responsible for the adsorption of hydrogen onto the surface of corannulene, many of these calculations are conducted at the Moller–Plesset level.

Calculations at the MP2/6-311++G(3df,2p) level predict that hydrogen has an interaction energy of 2.81 kcal/mol [basis set superposition error (BSSE)-uncorrected] with corannulene. By comparison, calculations performed at the MP2/6-311G (d,2p) level predict an interaction energy of 3.05 kcal/mol. Surprisingly, calculations performed at the SVWN5/6-31G level give a calculated interaction energy of 2.9 kcal/mol, in good agreement with the more expensive Moller–Plesset level. The calculated binding energy at the MP2(full)/6-311++G(3df,2p) level was found to be 1.38 kcal/mol, compared to the 1.2 kcal/mol found at SVWN5/6-31G level.[32] For comparison, the encapsulation energy of endohedral fullerene $H_2@C_{60}$ is predicted to be 2.36, 6.94, and 6.07 kcal/mol at the MP2/6-31G$^{**}$, MP2/6-311G$^{**}$, and MP2/6-311(2d,2p) levels of theory, respectively (all values are BSSE-corrected; the uncorrected values are 3.99, 8.63, and 10.62 kcal/mol).[36] Calculations at the SVWN5/6-31G$^*$ level give a binding energy of 5.4 kcal/mol (BSSE-corrected) for endohedral fullerene $H_2@C60$, which is in very good agreement with the values obtained at the MP2/6-31G$^{**}$, MP2/6-311G$^{**}$, and MP2/6-311G(2d,2p) levels.

Various other corannulene derivatives and fullerene fragments have also been studied computationally to determine the tendency of these fragments to adsorb hydrogen molecules. Tetraindenocorannulene (**4**) and pentaindenocorannulene (**5**) have been calculated to have binding energies for hydrogen of 3.0 and 2.9 kcal/mol, respectively. Substituting two nitrogen atoms for carbon in corannulene (**6**) gives a calculated binding energy to hydrogen of 2.6 kcal/mol. If four nitrogen atoms are substituted for four carbon atoms in corannulene (**7**), a calculated binding energy of 2.9 kcal/mol is predicted. Calculations conducted using six nitrogen atoms in pentaindenocorannulene give a binding energy of 3.2 kcal/mol compared to 2.9 kcal/mol for the nonsubstituted pentaindenocorarnnulene (**8**). When a boron atom is substituted for one of the hub carbons on corannulene (**9**), the binding energy greatly decreases, leading to a metastable state (Figure 5-3).[32]

Because a hydrogen storage material cannot consist of only one molecule, a cooperative effect of many molecules must be exhibited. For this reason, molecular dynamics (MD) calculations have been conducted in order to calculate the viability of corannulene to behave as a true hydrogen storage material (Table 5-5). Because of the high computational demand of MD calculations, the MD was performed at the B3LYP/6-311G(d,p) level. MD calculations predict that for crystalline corannulene the highest percent hydrogen that could be adsorbed is 3.03% at a pressure of 198 bar at 173 K. Figure 5-4 shows the linear relationship of temperature and pressure calculated for hydrogen adsorption on the surface of corannulene. At 273 K and pressure of 72 bar, the MD simulations predicts 0.79% hydrogen storage. This value is in very good agreement with the experimental values of 0.80% hydrogen storage at 72 bar pressure and 273 K.[31] It is clear from calculations that a cooperative interaction between two molecules provides greater hydrogen adsorption potential than does the sum of the

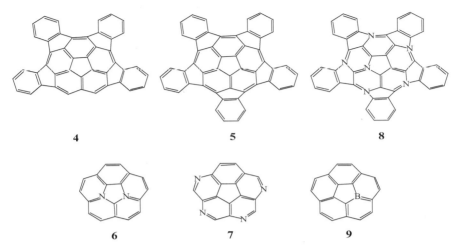

**Figure 5-3.** Structures of tetraindenocorannulene (**4**), pentaindenocorannulene (**5**), diazacorannulene (**6**), tetraazacorannulene (**7**), hexaazapentaindenocorannulene (**8**), and borocorannulene (**9**).

binding energy of each molecule individually. Calculations conducted at the SVWN level predict a binding energy of 1.8 kcal/mol for hydrogen bond to benzene and a binding energy of 2.9 kcal/mol to pentaindenocorannulene. A calculated sandwich complex of hydrogen between benzene and pentaindenocorannulene has a calculated binding energy of greater than 6 kcal/mol. Further, a sandwich of corannulene and pentaindenocorannulene gives a cooperative binding energy of greater than 7 kcal/mol (Figure 5-5).[36] Both of the aforementioned sandwich complexes are predicted to have a larger binding energy for hydrogen than for fullerene[60].

The calculated percent hydrogen adsorption is predicted to be affected significantly by the interlayer distance (ILD) between corannulene units. As the spacing between corannulene units becomes progressively larger, it is calculated that more hydrogen will be adsorbed (Figure 5-6); the optimum intermolecular distance between corannulene units is approximately 8 Å.[28]

Molecular dynamics simulations predict that a 1,5-corannulenylcyclophane (**10**) whose corannulene units are 7.2 Å apart (Figure 5-7) will be able to store 6.5% hydrogen by weight (i.e., the percent currently targeted by the US Department of Energy) at 294 bar of pressure at 273 K. This level of $H_2$ uptake would require a pressure of 361 bar for corannulene placed at 6.0 Å apart.[35]

## 5-4. PHOTOPHYSICAL PROPERTIES OF FULLERENE FRAGMENTS

### 5-4-1. Corannulene and Cyclopentacorannulene

#### *Photophysical Properties of Corannulene and Cyclopentacorannulene.*
The fluorescence spectrum of corannulene has been reported in various solvents.[37] It has been observed that the phosphorescence and fluorescence of corannulene have

**TABLE 5-5. Hydrogen Uptake by Crystalline Corannulene**

| Temperature (K) | Gas-Phase Volume ($\text{Å}^3$) | Number of Sorbents | Number of $H_2$ Unadsorbed | Pressure (bar) | Number of $H_2$ Adsorbed | $H_2$ Uptake (wt%) |
|---|---|---|---|---|---|---|
| 173 | $26.52 \times 23.72 \times 52$ | 32 | 275 | 198 | 125 | 3.03 |
| | $26.52 \times 23.72 \times 72$ | 32 | 283 | 147 | 117 | 2.84 |
| 273 | $26.52 \times 23.72 \times 92$ | 32 | 327 | 210 | 73 | 1.79 |
| | $26.52 \times 23.72 \times 122$ | 32 | 338 | 164 | 62 | 1.53 |
| | $53.08 \times 47.44 \times 57$ | 128 | 187 | 48 | 113 | 0.7 |
| 300 | $26.52 \times 23.74 \times 102$ | 32 | 336 | 214 | 64 | 1.57 |
| | $26.52 \times 23.74 \times 132$ | 32 | 344 | 170 | 56 | 1.38 |
| | $53.08 \times 47.44 \times 67$ | 128 | 207 | 50 | 93 | 0.58 |

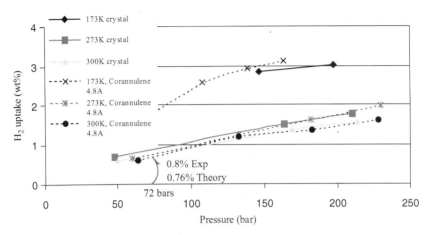

Hydrogen Adsorption - Crystalline Corannulene and Corannulene with 4.8 Å Interlayer Seperation

**Figure 5-4.** Comparison of calculated hydrogen adsorption versus experimental measurements.

lifetimes of 2.6 s and 10.3 ns, respectively.[38] Because corannulene bowl inverts at a rate of approximately 200,000 times per second,[39] the fluorescence and phosphorescence spectra of corannulene are averages of the continuous bowl inversions (Figure 5-8). The absorption and steady-state fluorescence measurements of corannulene and cyclopentacorannlene (**11**) were examined.[40]

Cyclopentacorannulene consists of two additional carbon atoms attached to the outer rim of corannulene forming a cyclopentene ring. The additional carbon atoms extend conjugation and increase rigidity and bowl depth to 1.05 Å (corannulene has a bowl depth of 0.89 Å).[41] Corannulene has an intense absorbance band at 250 nm and a weaker band at 285 nm. By comparison, cyclopentacorannulene has a more intense

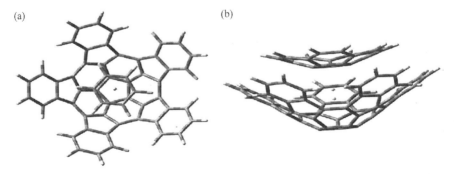

**Figure 5-5.** (a) Top view of benzene-$H_2$-pentaindenocorannulene complex; (b) side view of corannulene-$H_2$-pentaindenocorannulene complex. (Reprinted from Denis et al., *J. Phys. Chem. C* **2008**, *112*, 2791–2796.)

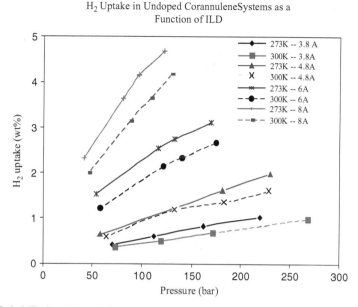

**Figure 5-6.** MD simulation of the precent hydrogen uptake by corannulene as a function of the interlayer distance between the corannulene units.

and broader absorbance band at 285 nm with a shoulder that extends to 360 nm (Figure 5-8). The observed bathochromic shift from corannulene to cyclopentacorannulene can be attributed to the additional π electrons that extend the overall conjugation of the corannulene ring system. The bathochromic shift of the fluorescence spectrum of cyclopentacorannulene compared to corannulene can also be attributed to the additional π electrons. One major drawback to the use of corannulene

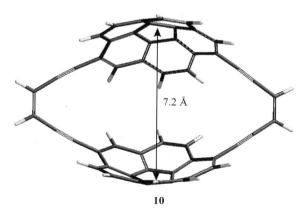

**10**

**Figure 5-7.** 1,5-Corannulene cyclophane.

(a)

(b)

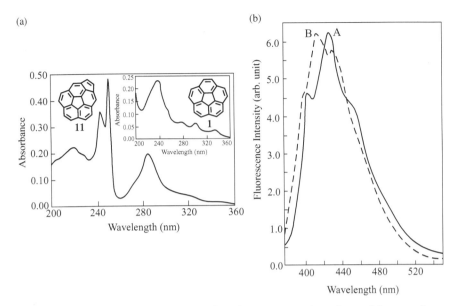

**Figure 5-8.** (a) Absorption spectrum of cyclopentacorannulene in cyclohexane (inset: absorption spectrum of corannulene in cyclohexane); (b) fluorescence spectra of purified samples of (curve A) cyclopentacorannulene and (curve B) corannulene (excitation wavelength = 285 nm). (Reprinted from Rabideau et al., *J. Fluoresc.* **1997**, *7*, 231–236.)

and cyclopentacorannulene as fluorescent materials is their low fluorescence quantum yields of 0.07 and 0.01, respectively.[40]

***Computational Considerations.*** Time-dependent density functional theory (TD-DFT) is routinely used to calculate the electronic excitation of a molecule.[42–46] Corannulene has absorbance bands of 333.75, 284.50 and 249.50 as measured in an argon matrix.[9] Very early intermediate neglect of differential overlap/configuration interaction (INDO/CI) calculations predicted corannulene to have three major transitions at 288, 286, and 252 nm.[47] More recent zero INDO (ZINDO) calculations predict values of allowed transitions of 289, 287, 259, and 251 nm. B3LYP/cc-pVTZ calculations predict transitions of 309, 286, and 236 nm.

## 5-4-2. Monosubstituted Corannulene Derivatives

Although corannulene and cyclopentacorannulene suffer from very low fluorescence quantum yields, the substitution of corannulene (and other fullerene fragments) in a manner that keeps the conjugation of the system intact is an important concept to consider. The monoaddition of substituents onto a corannulene framework does not disrupt the overall conjugation, which allows the absorbance and fluorescence to be shifted to longer wavelengths. The ability of corannulene-based derivatives with extended π conjugation to be redshifted is

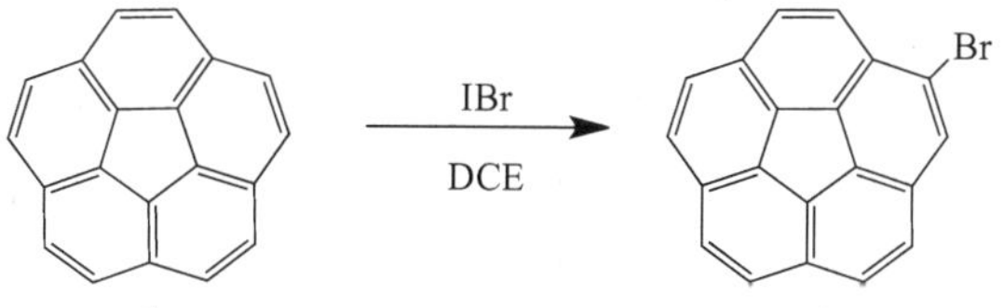

**Scheme 5-1.** Monobromination of corannulene.

advantageous over fullerenes, since linking two fullerenes with an ethynyl group leads to no visible redshift.[48]

**Synthesis of Monosubstituted Corannulene Derivatives.** Corannulene can be monobrominated by using 3 eq of IBr in dry 1,2-dichloroethane to give **12** in 90% yield (Scheme 5-1).[49] This key intermediate can then be transformed into a number of different monosubstituted corannulene derivatives that have increased conjugation, leading to longer absorbance wavelengths, different emission spectra, and increased quantum yields (Scheme 5-2).

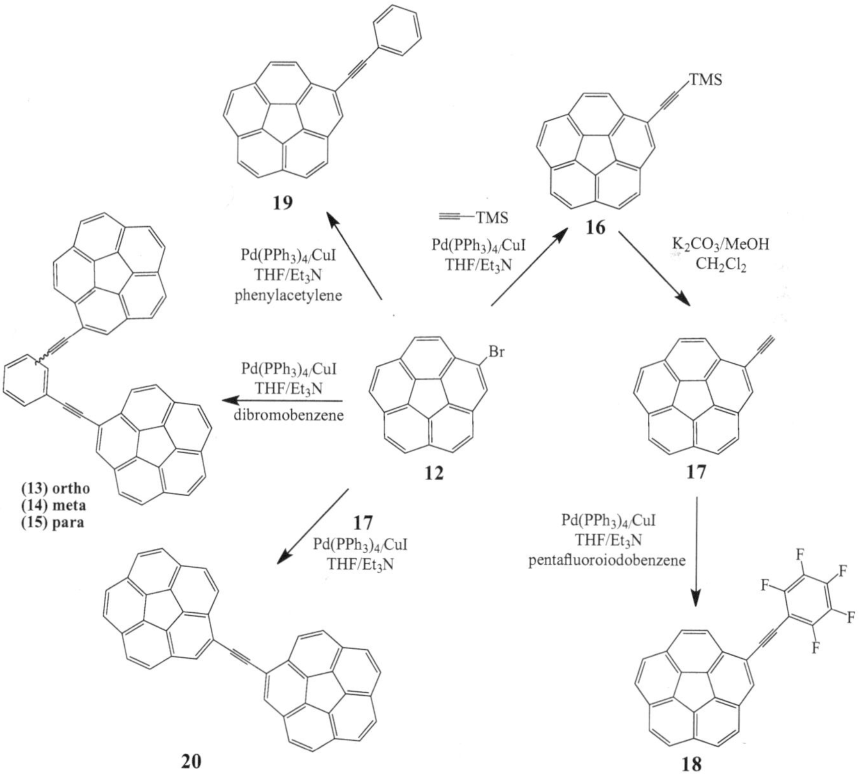

**Scheme 5-2.** Synthesis of monosubsituted corannulene derivatives.

***Computational Considerations.*** Extending to a more conjugated corannulene system with the addition of ethynyl groups, TD-DFT calculations are in better agreement with the observed experimental transitions than ZINDO calculations. Although the calculated excitations do not match the experimental values exactly, the calculations are able to predict the type of effect that a substituent will have on the wavelength of absorbance, as well as predict the effect that the placement of the substituent around the corannulene system will have.[50]

The TD-DFT calculations were shown to be in good agreement with the experimental absorption spectra of the *ortho* (**13**), *meta* (**14**), and *para* (**15**), di(corannulenylethynyl)benzene molecules.[49] Compound **1** calculated at the TD-B3LYP/6-31-G(d) level gives a calculated maximum long-wavelength absorption of 293 nm, compared to the observed wavelength of 299 nm. The *meta* isomer (**14**) has strong observed absorbance bands at 300 and 365 nm, with a broad shoulder that trails past 400 nm. TD-B3LYP/6-31G(d) calculations predict electronic transitions of 384, 365, 306, 305, and 300 nm. The *para* isomer (**15**) gives observed absorbance bands at wavelengths of 250, 302, and 371 and calculated excitations at 244, 303, and 401.

***Photophysical Properties of Monosubstituted Derivatives.*** For each of the molecules in Scheme 5-2, the $\pi-\pi^*$ transition energy is lowered, resulting in a redshift in absorbance when compared to corannulene (Table 5-6). Compounds **16** and **17** are redshifted by 10 nm, which can be attributed to the addition of the ethynyl linkage.[51] With substitution of the hydrogen atom on compound **17** by a perfluorophenyl (**18**), a phenyl (**19**), or a corannulene moiety (**20**), a larger redshift is observed. Compounds **13–15** display two absorption maxima between 250–254 and 299–302 nm, with compound **15** extending a shoulder to 430 nm.[49] This is a significant redshift in absorbance when compared to that of the parent corannulene. Surprisingly, all of these compounds give similar emission spectra. Although these compounds do not give longer-wavelength emissions, they do give fluorescence

**TABLE 5-6. UV–Visible Absorption and Emission Maxima and Fluorescence Quantum Yields**

| Compound | Absorption | Emission | $\Phi$ |
|---|---|---|---|
| 1 | 254, 289 | 417, 432 | 0.07 |
| 13[a] | 254, 299 | 420, 440 | 0.08 |
| 14[a] | 254, 301 | 418, 435 | —[b] |
| 15[a] | 250, 302, 371 | 420, 443 | 0.60 |
| 16[c] | 256, 295 | 407 sh, 425 | 0.14 |
| 17[c] | 261, 297 | 409, 429 | 0.08 |
| 18[c] | 255, 300 | 411 sh, 431 | 0.26 |
| 19[c] | 254, 297 | 409, 429 | 0.31 |
| 20[d] | 249, 302 | 418, 443 | 0.57 |

[a]Measurements recorded in $CH_2Cl_2$, compounds excited at 300 nm.
[b]Quantum yield not published.
[c]Measurements recorded in cyclohexane, compounds excited at 342 nm.
[d]Measurements recorded in cyclohexane, compounds excited at 372 nm.

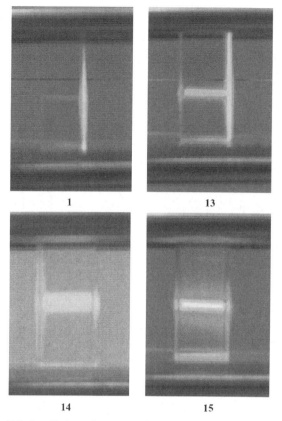

**Figure 5-9.** Irradiation of compounds **1** and **13–15** with 405-nm laser.

quantum yields higher than that of corannulene; compounds **15**, **18**, **19**, **20** give pronounced increases in fluorescence quantum yields compared to the parent corannulene.[49,51] The best results in quantum yields are achieved by compounds **15** and **20**. However, compound **20** is unstable and decomposes within just a few days, even on refrigeration.[51] Compound **15**, on the other hand, is stable to an open atmosphere and has shown no signs of degradation. Furthermore, excitation of compound **20** with a 405-nm laser beam gives bright blue fluorescence, allowing for viability as a blue emitter (Figure 5-9). The increased fluorescence and luminescent properties of these molecules suggest that with the right substituent it should be possible to generate stable intense fluorophores with increased quantum yields.

### 5-4-3. Highly Substituted Corannulene

*Synthesis of Penta- and Decasubstituted Corannulene.* Corannulene can be chlorinated to the penta- and decasubstituted moieties (Scheme 5-3). These

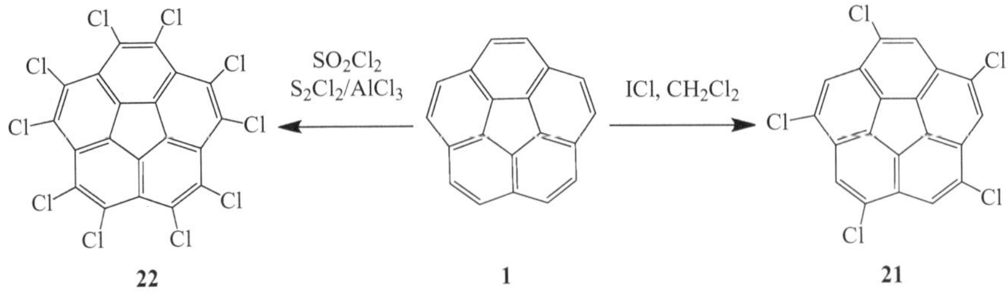

**Scheme 5-3.** Halogenation of corannulene.

highly halogenated species can then be functionalized further, thus providing a starting point for further synthetic transformation and added conjugation (Schemes 5-4 and 5-5).[52–55]

### *Photophysical Properties of Pentasubstituted Corannulene.*

The conversion of **21** into a variety of pentasubstituted corannulene derivatives allows for study of the photophysical properties and uncovers a unique cooperative interaction of functional groups around the rim of the bowl.[54] Transforming 21 into **23** by adding straight-chain hydrocarbons produces a redshift in the $\pi$–$\pi$ absorption band of around 9–10 nm. This is similar to the shift seen when alkyl groups are added to a benzene ring. Since the alkyl groups are weakly electron-donating, this suggests that the overall electron density of the corannulene system is slightly increased. When **21** is converted to **24** (pentaphenyl), the overall redshift is dramatically larger (+ 31 nm) and indicates an increase in conjugation. This effect is lost when *ortho* substituents are added to the phenyl ring, causing them to twist out of the plane and losing this $\pi$-system overlap. This is similar to the effect seen when comparing compounds **13** and **15**. Compound **25**, with phenyl and alkyl thiols, gives a significant redshift compared to the parent corannulene.[54] With conversion of **21** into the polyether-substituted corannulene **26**, a large redshift is observed corresponding to the strong electron-donating property of the alkoxy over alkyl groups.[54] Compound **27** can help determine the effects of arylalkynyl substitution as well as general aryl effects.[52]

### *Photophysical Properties of Decasubstituted Corannulene Derivatives.*

The conversion of **22** into decasubstituted corannulene derivatives can help determine the properties of this new class of corannulene-based molecules. The conversion of **1** to **28** with the addition of 10 alkyl groups leads to a redshift in the absorption of the four transitions of 8, 14, 21, and 18 nm, respectively.[53] This result is similar to those for **23** but are slightly more redshifted, based on the more weakly electron-donating alkyl groups. A similar redshift is observed in the fluorescence spectra of these compounds as well.[53] Decasubstitution reduces the overall bowl depth through steric effects and allows for fine-tuning of the electronic properties of the $\pi$ system according to the nature of the substituents.[56] One example of this is compound **29**, which adds 10 methylsulfide groups around the corannulene rim, which, in turn, allows for an overall

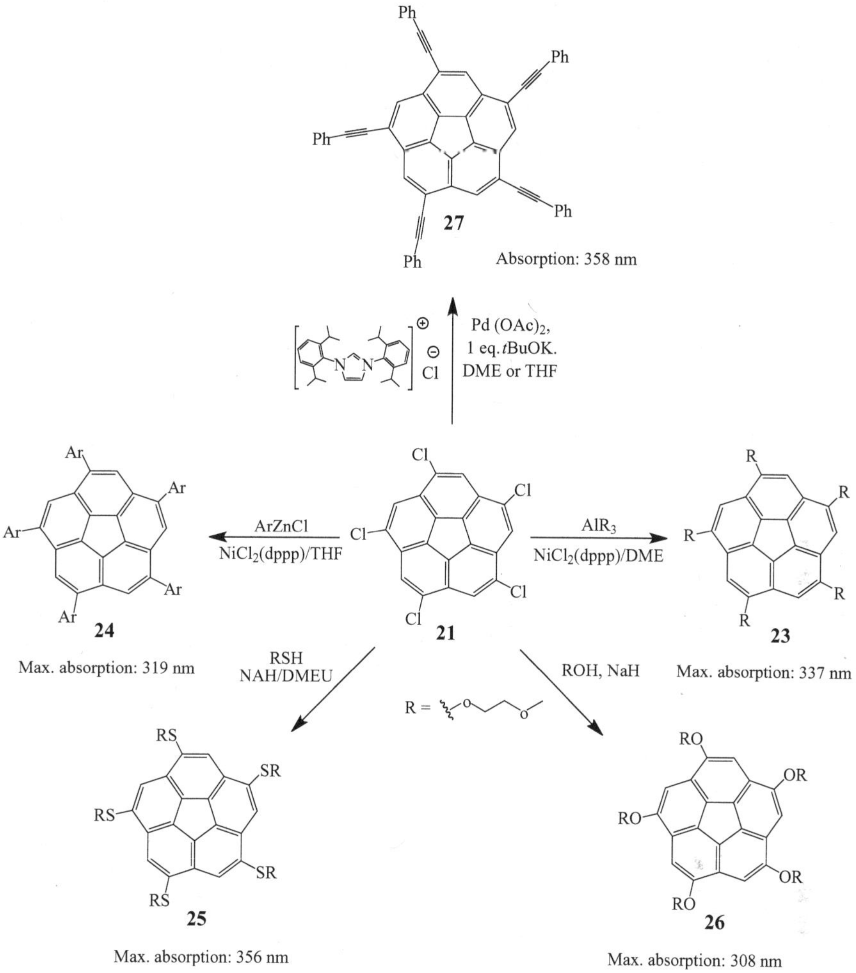

**Scheme 5-4.** Synthesis of pentasubsituted corannulene derivatives.

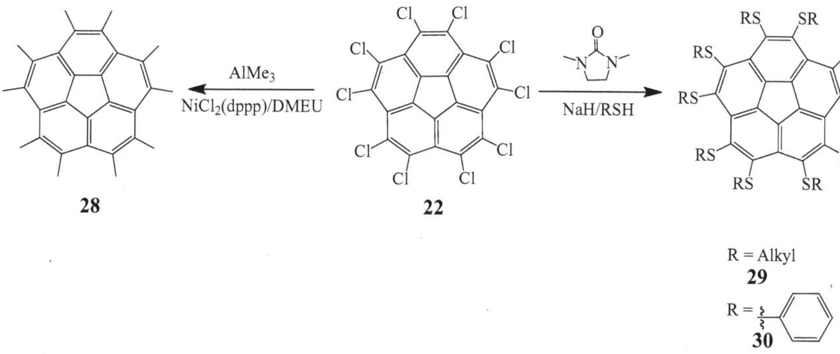

**Scheme 5-5.** Synthesis of decasubsituted corannulene derivatives.

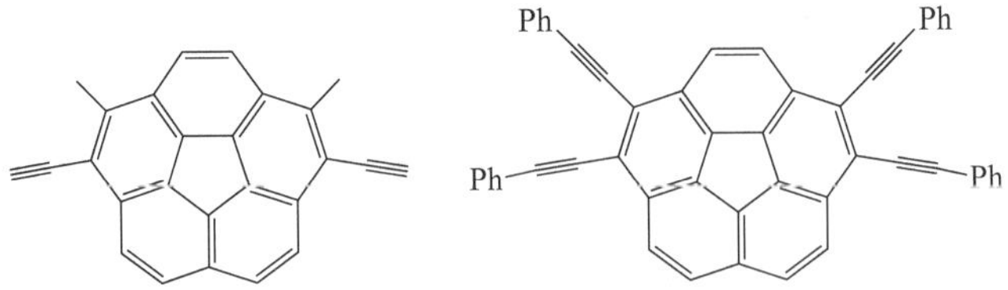

31                                    32

**Figure 5-10.** Other multiethynylcorannulene derivatives.

increase of electron density to the corannulene nucleus. Absorption spectra of **30** give significant redshift compared to **1** (200, 375, and 410 nm), which is comparable to fullerene[60]. This can be accredited to the LUMO extending beyond the core corannulene system to include the adjacent sulfur atoms.

***Photophysical Properties of Other Corannulene Derivatives.*** Compounds **31** and **32** are similar to other ethynyl corannulenes (Figure 5-10), **19**, and **27**. Each molecule gives a redshift, with compound **32** having wavelength absorptions longer than those of **31**. Additionally, the two compounds give exceptional fluorescence quantum yields of 75% and 83%, respectively.[52]

## 5-5. REDUCTIONS AND OXIDATIONS OF FULLERENE FRAGMENTS

### 5-5-1. Electrochemistry of Corannulene and Corannulene Derivatives

Fullerenes and corannulene share many similar properties, one of which is their electrochemistry. Fullerene[60] has a triply degenerate low-lying LUMO and has been electrochemically reduced 6 times, with reduction potentials of −0.98, −1.37, −1.87, −2.35, −2.85, and 3.26.[57–59] Corannulene has a doubly degenerate low-lying LUMO that can take up to four electrons. The electrochemistry of corannulene was performed shortly after its initial synthesis.[60] The first electron reduction potential was observed at −1.88 eV and the second reduction potential, at −2.36 eV. In addition, a color change associated with each reduction was reported. The first reduction occurs with an associated green color, whereas the second reduction has a distinct red color. Each of these changes is reversible and can be observed under chemical and electrochemical conditions. More recently it has been shown that the triply reduced state can be observed by conducting the electrochemistry under extremely dry conditions and using liquid dimethylamine as the solvent. Under these conditions corannulene has reduction potentials of −1.83, −2.28, and −3.12.[16] On attempting the oxidation of corannulene, researchers have observed the formation of a polymer-like substance that completely coats and insulates the electrode.

Both pentamethylcorannulene and decamethylcorannulene have greatly enhanced electron densities compared to corannulene, which affects their oxidation and reduction potentials. Electrochemical reduction potential studies for both molecules indicate differences of 250 and 160 mV, respectively for the radical anion and dianion.

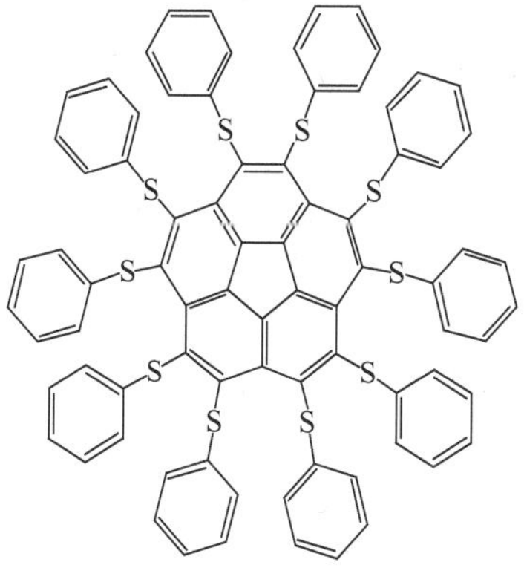

**30**

**Figure 5-11.** Structure of decakis(phenylthio)corannulene.

There is also a decrease in ionization potentials, as expected. Cyclic voltammetry of the decakis(thiophenyl)corannulene (**30**) (Figure 5-11) displays four waves. Values for its first reduction potential are on par with that of fullerene[60]. This outcome could be a combined effect of the electron-withdrawing ability of phenyl groups and the negative-charge stabilizing behavior of the sulfur atoms.

## 5-5-2. Reduction of Corannulenes with Alkali Metals

Although electrochemical conditions have been able to achieve only the triply reduced state, using the correct alkali metal allows for the reduction of corannulene to the tetraanion.[18, 61–63] Lithium wire completely reduces corannulene to the fourth reduction state. Similar to what is observed under electrochemical conditions, each reduction corresponds to a distinct color change. In addition to manifestation of the color changes associated with the first two reductions, the third reduction state gave a purple color and the fourth reduction state gave a brown color. Since the reductions were observed chemically, the fourth electrochemical reduction potential has not been reported and is still absent from the literature. In addition to lithium, other alkali metals such as Na, K, Rb, and Cs are capable of reducing corannulene.[64] Both the dianionic and tetranionic reduction stages have been detected from the alkali metals except from sodium, which is able to reduce corannulene only to the dianion, even after prolonged contact for several days. Substituted corannulene derivatives can also be reduced by alkali metals to the tetraanion. Monosubstituted corannulene derivatives *tert*-butylcorannulene and isopropylcorannulene were examined, and each was shown to be fully reduced to the tetraanion.[65] When two corannulene units are tethered together by an alkyl chain, alkali metals are able to reduce this molecule to the corresponding octaanion (total charge) with Li, K, Rb, and Cs; Na only reduces the molecule to the tetranion stage (total charge).[64]

Higher substituted corannulene molecules have reduction properties similar to those of the parent corannulene. 1,3,5,7,9-Penta-*tert*-butylcorannulene undergoes facile reduction with lithium and resembles corannulene in all its reduction states.[66]

## 5-6. CONCLUSION

Corannulene and other fullerene fragments have considerable potential for uses in a variety of applications. Corannulene has fluorescent and electrochromic properties and shows the ability to store hydrogen gas. These properties can be directly applied to organic light-emitting diode (OLED) technology as well as to the hydrogen economy. Improvements in the syntheses of these molecules can potentially make these structures viable building blocks in the development of nanotechnology. As bigger and more diverse fullerene fragments are synthesized, more properties of these unique molecules will be discovered. The development of corannulene-based materials will potentially be the bridge between current problems and possible solutions.

## REFERENCES

1. R. P. Feynman, There is plenty of room at the bottom http://www.zyvex.com/nanotech/feynman.html (accessed 1/15/10).
2. G. Binnig, H. Rohrer, *Helv. Phys. Acta* **1982**, *55*, 726–735.
3. H. W. Kroto, J. R. Heath, S. C. O'Brien, R. F. Curl, R. E. Smalley, *Nature* **1985**, *318*, 162–163.
4. S. Iijima, *Nature* **1991**, *354*, 56–58.
5. C. J. Cramer, *Essentials of Computational Chemistry: Theories and Models*, Wiley, Hoboken, NJ, **2004**.
6. F. Jensen, *Introduction to Computational Chemistry*, Wiley, Hoboken, NJ, **2007**.
7. K. I. Ramachandran, G. Deepa, K. Namboori, *Computational Chemistry and Molecular Modeling: Principles and Applications*, Springer, London, **2008**.
8. A. S. Tiwary, A. K. Mukherjee, *THEOCHEM* **2008**, *859*, 107–112.
9. G. Rouille, C. Jaeger, M. Steglich, F. Huisken, T. Henning, G. Theumer, I. Bauer, H.-J. Knoelker, *ChemPhysChem* **2008**, *9*, 2085–2091.
10. T. J. Seiders, K. K. Baldridge, J. S. Siegel, R. Gleiter, *Tetrahedron Lett.* **2000**, *41*, 4519–4522.
11. G. Chen, R. G. Cooks, E. Corpuz, L. T. Scott, *J. Am. Soc. Mass Spectrom.* **1996**, *7*, 619–627.
12. U. Chen, S. Ma, R. G. Cooks, H. E. Bronstein, M. D. Best, L. T. Scott, *J. Mass Spectrom.* **1997**, *32*, 1305–1309.
13. L. D. Betowski, M. Enlow, L. Riddick, D. H. Aue, *J. Phys. Chem. A* **2006**, *110*, 12927–12946.
14. L. S. Wang, J. Conceicao, C. Jin, R. E. Smalley, *Chem. Phys. Lett.* **1991**, *182*, 5–11.
15. S. H. Yang, C. L. Pettiette, J. Conceicao, O. Cheshnovsky, R. E. Smalley, *Chem. Phys. Lett.* **1987**, *139*, 233–238.

16. C. Bruno, R. Benassi, A. Passalacqua, F. Paolucci, C. Fontanesi, M. Marcaccio, E. A. Jackson, L. T. Scott, *J. Phys. Chem. B* **2009**, *113*, 1954–1962.

17. Q. Xie, E. Perez-Cordero, L. Echegoyen, *J. Am. Chem. Soc.* **1992**, *114*, 3978–3980.

18. A. Ayalon, M. Rabinovitz, P. C. Cheng, L. T. Scott, *Angew. Chem.* **1992**, *104*, 1691–1692; see also *Angew. Chem. Int. Ed. Engl.* **1992**, *1631* (1612), 1636–1697.

19. A. Y. Rogachev, M. A. Petrukhina, *J. Phys. Chem. A* **2009**, *113*, 5743–5753.

20. H. Choi, C. Kim, K.-M. Park, J. Kim, Y. Kang, J. Ko, *J. Organomet. Chem.* **2009**, *694*, 3529–3532.

21. W. Xiao, D. Passerone, P. Ruffieux, K. Aiet-Mansour, O. Groening, E. Tosatti, J. S. Siegel, R. Fasel, *J. Am. Chem. Soc.* **2008**, *130*, 4767–4771.

22. B. M. Wong, *J. Comput. Chem.* **2008**, *30*, 51–56.

23. M. A. Petrukhina, *Angew. Chem. Int. Ed.* **2008**, *47*, 1550–1552.

24. A. Y. Rogachev, Y. Sevryugina, A. S. Filatov, M. A. Petrukhina, *Dalton Trans.* **2007**, 3871–3873.

25. R. J. Angelici, B. Zhu, S. Fedi, F. Laschi, P. Zanello, *Inorg. Chem.* **2007**, *46*, 10901–10906.

26. J. S. Siegel, K. K. Baldridge, A. Linden, R. Dorta, *J. Am. Chem. Soc.* **2006**, *128*, 10644–10645.

27. F. J. Lovas, R. J. McMahon, J.-U. Grabow, M. Schnell, J. Mack, L. T. Scott, R. L. Kuczkowski, *J. Am. Chem. Soc.* **2005**, *127*, 4345–4349.

28. J. Mack, unpublished results.

29. R. Peverati, K. K. Baldridge, *J. Chem. Theory Comput.* **2008**, *4*, 2030–2048.

30. K. K. Baldridge, J. S. Siegel, *Theor. Chem. Acc.* **1997**, *97*, 67–71.

31. L. G. Scanlon, P. B. Balbuena, Y. Zhang, G. Sandi, C. K. Back, W. A. Feld, J. Mack, M. A. Rottmayer, J. L. Riepenhoff, *J. Phys. Chem. B* **2006**, *110*, 7688–7694.

32. P. A. Denis, *J. Phys. Chem. C* **2008**, *112*, 2791–2796.

33. Y. Zhang, L. G. Scanlon, M. A. Rottmayer, P. B. Balbuena, *J. Phys. Chem. B* **2006**, *110*, 22532–22541.

34. L. G. Scanlon, W. A. Feld, P. B. Balbuena, G. Sandi, X. Duan, K. A. Underwood, N. Hunter, J. Mack, M. A. Rottmayer, M. Tsao, *J. Phys. Chem. B* **2009**, *113*, 4708–4717.

35. Y. Zhang, L. G. Scanlon, P. B. Balbuena, *Theor. Comput. Chem.* **2007**, *18*, 127–166.

36. Z. Slanina, P. Pulay, S. Nagase, *J. Chem. Theory Comput.* **2006**, *2*, 782–785.

37. S. A. Tucker, W. E. Acree, Jr., J. C. Fetzer, R. G. Harvey, M. J. Tanga, P. C. Cheng, L. T. Scott, *Appl. Spectrosc.* **1993**, *47*, 715–722.

38. J. F. Verdieck, W. A. Jankowski, *Mol. Lumin., Int. Conf.* **1969**, 829–836.

39. L. T. Scott, M. M. Hashemi, M. S. Bratcher, *J. Am. Chem. Soc.* **1992**, *114*, 1920–1921.

40. J. Dey, A. Y. Will, R. A. Agbaria, P. W. Rabideau, A. H. Abdourazak, R. Sygula, I. M. Warner, *J. Fluoresc.* **1997**, *7*, 231–236.

41. T. J. Seiders, K. K. Baldridge, G. H. Grube, J. S. Siegel, *J. Am. Chem. Soc.* **2001**, *123*, 517–525.

42. M. E. Casida, *Comput. Meth. Catal. Mater. Sci.* **2009**, 33–59.

43. P. Elliott, F. Furche, K. Burke, *Rev. Comput. Chem.* **2009**, *26*, 91–165.

44. F. Furche, D. Rappoport, *Theor. Comput. Chem.* **2005**, *16*, 93–128.

45. T. A. Niehaus, N. H. March, *Theor. Chem. Acc. 125*, 427–432.

46. M. van Faassen, K. Burke, *Phys. Chem. Chem. Phys.* **2009**, *11*, 4437–4450.

47. J. Feng, J. Li, Z. Wang, M. C. Zerner, *Int. J. Quantum Chem.* **1990**, *37*, 599–607.

48. T. Tanaka, K. Komatsu, *J. Chem. Soc., Perkin Trans. 1* **1999**, 1671–1676.

49. J. Mack, P. Vogel, D. Jones, N. Kaval, A. Sutton, *Org. Biomol. Chem.* **2007**, *5*, 2448–2452.

50. Y.-T. Wu, D. Bandera, R. Maag, A. Linden, K. K. Baldridge, J. S. Siegel, *J. Am. Chem. Soc.* **2008**, *130*, 10729–10739.

51. C. S. Jones, E. Elliott, J. S. Siegel, *Synlett* **2004**, 187–191.

52. Y.-T. Wu, J. S. Siegel, *Chem. Rev.* **2006**, *106*, 4843–4867.

53. T. J. Seiders, E. L. Elliott, G. H. Grube, J. S. Siegel, *J. Am. Chem. Soc.* **1999**, *121*, 7804–7813.

54. G. H. Grube, E. L. Elliott, R. J. Steffens, C. S. Jones, K. K. Baldridge, J. S. Siegel, *Org. Lett.* **2003**, *5*, 713–716.

55. T. J. Seiders, K. K. Baldridge, E. L. Elliott, G. H. Grube, J. S. Siegel, *J. Am. Chem. Soc.* **1999**, *121*, 7439–7440.

56. K. K. Baldridge, K. I. Hardcastle, T. J. Seiders, J. S. Siegel, *Org. Biomol. Chem.* **2010**, *8*, 53–55.

57. Q. Xie, F. Arias, L. Echegoyen, *J. Am. Chem. Soc.* **1993**, *115*, 9818–9819.

58. L. Echegoyen, L. E. Echegoyen, *Acc. Chem. Res.* **1998**, *31*, 593–601.

59. F. Zhou, C. Jehoulet, A. J. Bard, *J. Am. Chem. Soc.* **1992**, *114*, 11004–11006.

60. J. Janata, J. Gendell, C.-Y. Ling, W. E. Barth, L. Backes, H. B. Mark, Jr., R. G. Lawton, *J. Am. Chem. Soc.* **1967**, *89*, 3056–3058.

61. M. Baumgarten, L. Gherghel, M. Wagner, A. Weitz, M. Rabinovitz, P.-C. Cheng, L. T. Scott, *J. Am. Chem. Soc.* **1995**, *117*, 6254–6257.

62. A. Weitz, E. Shabtai, M. Rabinovitz, M. S. Bratcher, C. C. McComas, M. D. Best, L. T. Scott, *Chem. Eur. J.* **1998**, *4*, 234–239.

63. M. Baumgarten, L. Gherghel, M. Wagner, A. Weitz, M. Rabinovitz, P.-C. Cheng, L. T. Scott, *J. Am. Chem. Soc.* **1995**, *117*, 6254–6257.

64. E. Shabtai, R. E. Hoffman, P.-C. Cheng, E. Bayrd, D. V. Preda, L. T. Scott, M. Rabinovitz, *J. Chem. Soc., Perkin Trans. 2* **2000**, 129–133.

65. A. Ayalon, A. Sygula, P.-C. Cheng, M. Rabinovitz, P. W. Rabideau, L. T. Scott, *Science* **1994**, *265*, 1065–1067.

66. A. Weitz, M. Rabinovitz, P.-C. Cheng, L. T. Scott, *Synth. Met.* **1997**, *86*, 2159–2160.

# CHAPTER 6

# COORDINATION PREFERENCES OF BOWL-SHAPED POLYAROMATIC HYDROCARBONS

ALEXANDER S. FILATOV and MARINA A. PETRUKHINA

## 6-1. INTRODUCTION

The bowl-shaped polyaromatic hydrocarbons constitute a new and unique family of polyarenes that is now confirmed to exhibit system-dependent preferences for metal binding. By selecting a proper metal unit and by tuning the geometry and electronics of the bowl, one can modulate the overall binding mode. The careful choice of metal complexes allows the direction of metal binding inside or outside of the bowl or engaging both faces of π bowls in coordination.

## 6-2. MONONUCLEAR METAL COMPLEXES

Transition metal complexes of fullerenes have attracted considerable attention since 1991, when Fagan et al. reported the synthesis and structure of the first complex, $[Pt(PPh_3)_2(\eta^2\text{-}C_{60})]$.[1] A great number of *exo*-bound transition metal complexes of fullerenes, in which a metal coordinates to the junction of two 6-membered rings of the fullerene surface[2–6] or to the five-membered ring of pentaalkyl- or pentaaryl-derivatized fullerenes,[7–13] have been synthesized since then. Compared to the closed-surface buckyballs, open geodesic polyaromatic hydrocarbons that constitute fragments of fullerenes[14–21] (and therefore are referred to as "buckybowls" or "π bowls") represent even more interesting π ligands from a coordination viewpoint.[22] Buckybowls have multisite coordination possibilities, namely, convex and concave interior polyaromatic faces, as well as edge and rim carbon atoms capped by hydrogen atoms. They share with fullerenes the convex three-dimensional unsaturated carbon

*Fragments of Fullerenes and Carbon Nanotubes: Designed Synthesis, Unusual Reactions, and Coordination Chemistry*, First Edition. Edited by Marina A. Petrukhina and Lawrence T. Scott.
© 2012 John Wiley & Sons, Inc. Published 2012 by John Wiley & Sons, Inc.

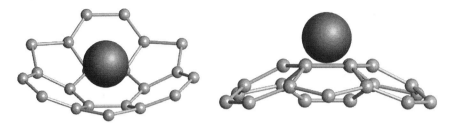

**Figure 6-1.** Schematic representation of *endo/exo*-coordination of a buckybowl.

surface, but, in contrast, have an open concave surface that is readily accessible. The study of relative preferences of convex and concave faces of buckybowls for binding metal centers (Figure 6-1) has been a focus of considerable attention more recently because of its fundamental and practical importance.

The controlled positioning of metal centers inside the bowls is expected to provide a direct route toward the inclusion complexes of fullerenes and nanotubes, which still remain challenging synthetic targets. In addition, the coordination of metal centers to the outside of the bowls should find applications in surface activation and functionalization of fullerenes and nanotubes, thus further stimulating the use of curved and strained polyarenes in materials chemistry. Substantial efforts have been directed toward studying the reactivity and ligating properties of buckybowls using various computational techniques.[23–35] However, only a limited number of metal complexes of buckybowls have been isolated and structurally characterized by single-crystal X-ray diffraction to date. The latter technique is required to reveal which face of a bowl is directly involved in metal binding, as spectroscopic methods cannot clearly address this critical issue. For example, Bohme and coworkers studied the reactivity of several small neutral ligands with the iron–corannulene complex using a selected-ion flowtube (SIFT) apparatus.[36] By comparing that with the bare $Fe^+$ ion reactivity, they found that the presence of corannulene led to enhancements in reactivity of up to five orders of magnitude in room temperature reactions. A few years later, Duncan et al. investigated[37] the gas-phase corannulene complexes of transition (Ti, V, Cr, Fe) and actinide (U) metals, as well as their oxides produced by covaporization of materials by a laser plasma source. The time-of-flight mass spectra revealed that metals yield mono- and diligand complexes in the form of $[M(C_{20}H_{10})_n]^+$ ($n = 1$ or 2), while their oxides efficiently produce monoadducts. Neither study could provide experimental hints for elucidation of binding modes of corannulene or for its *exo/endo* coordination preference.

In general, the smallest $C_{5v}$- and $C_{3v}$-symmetric subunits of the $C_{60}$-fullerene, namely, corannulene and sumanene (Scheme 6-1), have served as the primary models for theoretical and experimental coordination studies.

For corannulene, both $\sigma$- and $\pi$-metal complexes have been reported. It was found that the halogen-substituted corannulene core can undergo standard organometallic reactions on the rim, similar to those with planar aromatic systems. In 2005, Lee and Sharp[38] showed that oxidative addition of bromocorannulene to $Ni(COD)_2/2PEt_3$ and

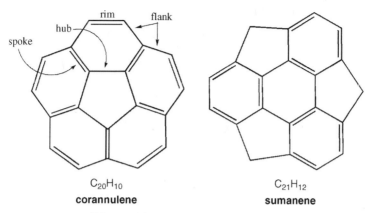

$C_{20}H_{10}$
**corannulene**

$C_{21}H_{12}$
**sumanene**

**Scheme 6-1.** Subunits of $C_{60}$-fullerene.

$Pt(PEt_3)_4$ opens a route to $\sigma$-bonded organometallic corannulene derivatives (see Chapter 8 for details). In 2009, these systems were expanded by Siegel[39] and others[40] to include reactions with multisubstituted halo- and ethynylcorannulenes.

The first $\pi$-metal complex of corannulene, $[(\eta^6\text{-}C_{20}H_{10})Ru(C_5Me_5)]^+$, was isolated and spectroscopically characterized in Siegel's group back in 1997.[23] NMR evidence was also subsequently reported for the formation of $\eta^6$-coordinated complexes of corannulene and tetramethylcorannulene with $[Ir(C_5Me_5)]^{2+}$.[41] Interestingly, this iridium cation showed no reactivity toward the $C_{60}$-fullerene. Generally, many attempts to isolate corannulene complexes with metal units that were reactive toward the surface of $C_{60}$ or planar aromatic hydrocarbons have been unsuccessful,[41,42] clearly illustrating the striking differences in their coordination preferences. Furthermore, the lability of corannulene complexes in solutions may have hindered the first crystallization attempts. Thus, it was not until 2004 that Angelici and coworkers finally succeeded in structural characterization of $\eta^6$-coordinated corannulene complexes for the first time (Figure 6-2).[43,44]

The X-ray study revealed the dramatic impact that transition metals can have on bowl-shaped polyaromatic ligands. The observed differences in bond lengths of the six-membered rings of $C_{20}H_{10}$ involved in metal coordination approach 0.05 Å. Moreover, binding of two ruthenium atoms to corannulene in $[(Cp^*Ru)_2(\mu_2\text{-}\eta^6{:}\eta^6\text{-}C_{20}H_{10})]X_2$ ($X = PF_6^-$ and $SbF_6^-$) caused significant structural changes in the bowl shape and almost completely flattened out its curved polyaromatic surface. This effect is noteworthy, since metal ion binding may likewise change the shape of other nonplanar carbon surfaces, such as the walls or caps of carbon nanotubes. The family of $\eta^6$-corannulene metal complexes was later expanded to include several new members. The more recently prepared complexes of the type $[(\eta^6\text{-arene})M(\eta^6\text{-}C_{20}H_{10})]X_2$ ($M = Ru$, $Os$; $X = BF_4^-$, $PF_6^-$, or $SbF_6^-$; arene = $C_6HMe_5$, $C_6Me_6$, $C_6EtMe_5$, cymene)[31] showed an increased stability compared to analogous $\eta^6$-corannulene complexes of $[Cp^*Ru]^+$, $[Cp^*Ir]^{2+}$, and $[(COE)_2M]^+$ ($M = Rh$ or $Ir$).[30]

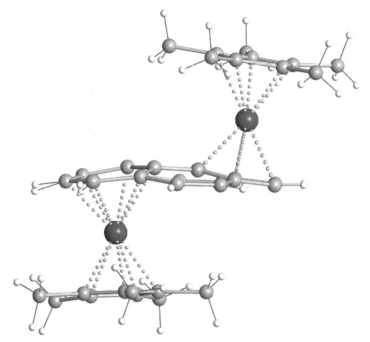

**Figure 6-2.** Molecular structure of $[(Cp^*Ru)_2(\mu_2\text{-}\eta^6{:}\eta^6\text{-}C_{20}H_{10})]^{2+}$.

Interestingly, NMR investigations demonstrated that, in solution, a metal unit is moving around the curved surface of corannulene. The first observations for such migration of a metal complex were reported by Angelici and coworkers in 2003, but the mechanism could not be elucidated at that time.[41] It was found that the reaction of tetramethylcorannulene with $[Cp^*Ir]^{2+}$ initially affords the product where the cation is coordinated to a nonmethylated ring. Within a short period of time, the isomerization occurs and the more stable isomer, in which the metal unit is bound to one of the methylated rings, is detected. In 2006, Siegel and coworkers ruled out the possibility for migration along the corannulene surface via the intermolecular dissociation/association mechanism on the basis of their NMR and calculation data. They favored the inter-ring migration of cationic metal units by walking through the central five-membered ring (hub migration) over the migration from one arene ring to the next along the outer edge of corannulene. Importantly, the single-crystal X-ray diffraction studies revealed that all $\eta^6$-corannulene complexes having a single metal bound to a bowl exhibit exclusive convex metal coordination (Figure 6-3).

A similar trend has been seen for $C_{20}H_{10}$ complexation by $Ag^+$ cations studied in solution.[45] In three silver(I)-based extended networks built on $\eta^2$ and $\eta^1$ binding of $Ag^+$ ions to the rim sites of corannulene, a metal was found at the outside of the bowl (Figure 6-4). These facts illustrated a strong general preference of the convex face of corannulene for metal coordination, and thus for some time thwarted the idea of using buckybowls to access inclusion metal complexes.

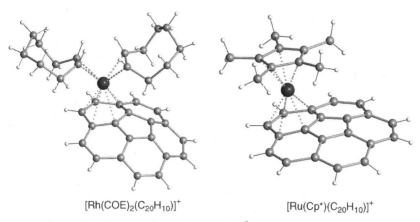

[Rh(COE)$_2$(C$_{20}$H$_{10}$)]$^+$                    [Ru(Cp*)(C$_{20}$H$_{10}$)]$^+$

**Figure 6-3.** *Exo*-bound transition metal η$^6$-complexes of corannulene.

In this regard, the first selective concave coordination reported by Hirao and coworkers in 2007[46] has been a breakthrough. *Endo*-metal coordination has been successfully accomplished for sumanene[18,19] (see Chapter 7 for details), which has a deeper and more rigid bowl than does corannulene (Figure 6-5). Prior to this work, which has been expanded to include the chiral cyclopentadienyl [Fe(Cp-(*S*)-*sec*-Bu]$^+$ complex[47] and the more flexible [RuCp]$^+$ product showing a dynamic inversion behavior,[48] no selective coordination of metal ions to the concave face of a π bowl had been observed experimentally.

This discovery serves as the first step toward the elusive inclusion complexes of buckybowls and has several important implications. It confirms that bowl-shaped polyarenes indeed are excellent multisite models for revealing trends in binding of

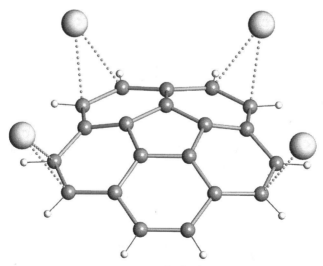

**Figure 6-4.** *Exo*-bound [Ag$_4$(η$^2$:η$^2$:η$^2$:η$^1$-C$_{20}$H$_{10}$)]$^{4+}$ cation.

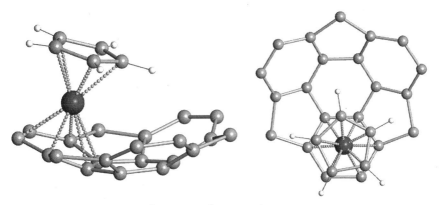

**Figure 6-5.** *Endo*-bound $[(\eta^5\text{-}C_5H_5)Fe(\eta^6\text{-}C_{21}H_{12})]^+$ cation: side (a) and top (b) views.

metal centers to curved and strained π-carbon surfaces. It proves that despite all prior examples of the preferential metal coordination to the convex surfaces of buckybowls, their inside concave carbon face can also be engaged in metal binding. It stimulates the expansion of this research to π bowls with larger and more strained polyaromatic surfaces, which, in contrast to corannulene and sumanene, have not yet been broadly studied. For the latter two small bowls, all known coordination modes identified by the X-ray diffraction studies are summarized in Table 6-1.

One additional important outcome of this work should also be emphasized. In contrast to solution preparation methods used to access the above-mentioned corannulene complexes, Hirao's group took advantage of the solid-state synthesis to successfully place a cyclopentadienyliron unit in the sumanene bowl. Since large polyarenes can be expected to show limited solubility, new methods are clearly needed to engage buckybowls in coordination and organometallic reactions and to

**TABLE 6-1. Coordination Modes of π Bowls in Structurally Characterized Mononuclear Metal Complexes**

| Complex | $C_{20}H_{10}$ | Ref. | Complex | $C_{21}H_{12}$ | Ref. |
|---------|----------------|------|---------|----------------|------|
| $[RuCp^*]^+$ | $\mu_2\text{:}\eta^6\text{:}\eta^6$ | 43,44 | $[FeCp]^+$ | *endo*-$\eta^6$ | 46 |
| | *exo*-$\eta^6$ | 44 | | | |
| $[Rh(COE)_2]^+$ | *exo*-$\eta^6$ | 30 | $[Fe(Cp\text{-}(S)\text{-}sec\text{-}Bu]^+$ | *endo*-$\eta^6$ | 47 |
| $[Rh(COE)_2(P(i\text{-}Pr)_3)]^+$ | *exo*-$\eta^6$ | 30 | $[RuCp]^+$ | *endo*-$\eta^6$ | 48 |
| $[Ir(COE)_2]^+$ | *exo*-$\eta^6$ | 30 | | | |
| $AgBF_4$ | $\mu_4\text{:}\eta^2\text{:}\eta^2\text{:}\eta^2\text{:}\eta^1$ | 45 | | | |
| $AgClO_4$ | $\mu_2\text{:}\eta^2\text{:}\eta^2$ | 45 | | | |
| $Ag(SO_3CF_3)$ | $\mu_2\text{:}\eta^2\text{:}\eta^2$ | 45 | | | |
| $[Ru(arene^a)]^{2+}$ | *exo*-$\eta^6$ | 31 | | | |
| $[Os(cymene)]^{2+}$ | *exo*-$\eta^6$ | 31 | | | |

$^a$Arene $= C_6Me_6$, $C_6HMe_5$, $C_6EtMe_5$.

access their new metalated products. Moreover, the highly strained large π bowls are still available in microscale quantities only, and that additionally limits their solution chemistry. In this regard, our approach based on codeposition of the complementary reactive units in a solvent-free environment offers a useful alternative to solution coordination methods. This technique has proved to be very efficient for the preparation of metal complexes of buckybowls in the single-crystalline form, suitable for direct X-ray diffraction studies.

## 6-3. DINUCLEAR METAL COMPLEXES

For effective utilization of intermolecular metal–π-arene interactions, which in solutions may be thwarted by lability of complexes, solvent and ion-templating effects, or solvent competition for coordination,[45,49,50] we introduced a microscale gas-phase coordination technique that excludes the use of solvents. It is based on codeposition of volatile complementary building units under reduced pressure.[51] Several experimental variables such as reaction temperature, temperature gradient, and the ratio of reactants affect the reaction outcome and allow variations in the product composition. Careful control of these parameters permits the one-step synthesis of products with desired stoichiometries as well as their deposition in the form of single crystals from the gas phase. This technique can utilize a variety of volatile organic donors spanning the range from planar molecules with isolated multiple bonds[52] to conjugated planar polyarenes,[53] to curved unsaturated π bowls, discussed here. As metal-containing units, a number of reactive and volatile complexes can generally be used, but, in order to facilitate the formation of directional metal–π-arene interactions, strong Lewis acidic complexes having one or multiple metal sites open for coordination are preferred. The synthetic availability, strong avidity for axial coordination, and ease of effective modulation of electronic properties of dimetal core complexes[54] make them a primary choice for the solvent-free coordination studies of buckybowls. Specifically, dimetal trifluoroacetates and trifluoromethyl-substituted benzoates have been chosen because of the well-known ability of the trifluoromethyl group to provide substantial Lewis acidity along with increased volatility to the resulting metal complexes (Scheme 6-2). Importantly, an excess of volatile metal units can be readily created in the gas phase, and that allowed us to test the coordination limits of several buckybowls ranging from corannulene to hemifullerene in metal binding reactions.

***Corannulene.*** The first crystalline complexes of corannulene were successfully prepared and structurally characterized by single-crystal X-ray diffraction in our laboratory back in 2003.[55] Synthesis of the complexes was accomplished by sublimation–deposition reactions of $[Rh_2(O_2CCF_3)_4]$ and $C_{20}H_{10}$ at 160°C. Different ratios of starting reagents in the solid state (1 : 1 and 3 : 2) resulted in the formation of two different products. The former exhibits an extended one-dimensional structure having alternating dirhodium units and corannulene molecules and built on directional $\eta^2$-coordination (Figure 6-6). Both faces of $C_{20}H_{10}$ are involved in

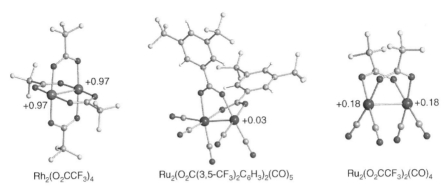

$Rh_2(O_2CCF_3)_4$           $Ru_2(O_2C(3,5-CF_3)_2C_6H_3)_2(CO)_5$           $Ru_2(O_2CCF_3)_2(CO)_4$

**Scheme 6-2.** Selected dimetal complexes with the computed charges on metal atoms.

coordination, where the average Rh−C separation is slightly longer for the convex side (2.580(4) Å for Rh−$C_{concave}$ and 2.643(4) Å for Rh−$C_{convex}$). Only rim carbon atoms of a bowl are engaged in these interactions.

The X-ray diffraction analysis of the second product revealed a two-dimensional network consisting of large rhomboid cells built of six dimetal units and six corannulene molecules [Figure 6-7(a)]. As in the case of 1D polymeric chain, each corannulene bowl uses its rim carbon atoms ($\eta^2$-mode) at both the convex and concave faces for Rh binding. One of the Rh−$C_{convex}$ bond lengths is shorter (2.548(3) Å), while the other one is longer (2.636(3) Å) than the Rh−$C_{concave}$ bond (2.570(3) Å).

The observed $\eta^2$-rim coordination of dirhodium units does not perturb the geometric parameters of the polyarene, as the corannulene core is only slightly flattened in both cases (0.839, 0.861, and 0.875 Å[56,57] in 1D, 2D, and free corannulene, respectively). At the same time, the Rh−Rh bond distance (averaged at 2.43 Å in both complexes) is substantially elongated compared to that of 2.38 Å in the unligated dirhodium tetratrifluoroacetate.[58] The exclusive coordination of Rh atoms at the rim of corannulene shows some similarities with the complexation of planar polycyclic

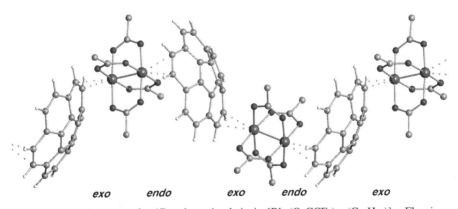

*exo*      *endo*        *exo*        *endo*        *exo*

**Figure 6-6.** A fragment of a 1D polymeric chain in $[Rh_2(O_2CCF_3)_4 \cdot (C_{20}H_{10})]_\infty$. Fluorine atoms are omitted for clarity.

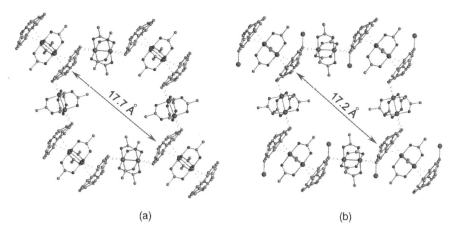

(a)                                            (b)

**Figure 6-7.** Fragments of the 2D layers in $[(Rh_2(O_2CCF_3)_4)_3 \cdot (C_{20}H_{10})_2]_\infty$ (a) and $[(Rh_2(O_2CCF_3)_4)_3 \cdot (C_{20}H_9Br)_2]_\infty$ (b). Fluorine and hydrogen atoms are omitted for clarity.

aromatic hydrocarbons (PAHs), where the $\eta^2$-rim coordination was seen previously.[53] From Hückel calculations, we have noticed that the electrophilic dirhodium units tend to coordinate to CC bonds of PAHs having the highest $\pi$-bond order.[53] We have demonstrated that the direction of this interaction depends on the topology of the frontier molecular orbitals (FMOs) for aromatic ligands and dimetal complexes.[35,52]

Density functional theory calculations (PBE0//LANL2DZ(Rh)/6-311G**) showed only a 2.2 kcal/mol energetic preference of rim over spoke binding of $[Rh_2(O_2CCF_3)_4]$, thus making the spoke complex of corannulene a feasible experimental target. However, in order to access it, one would need to exploit Lewis acids much softer than dirhodium(II,II) tetra(trifluoroacetate). In this regard, a one-end diruthenium(I,I) unit, $[Ru_2(O_2C(3,5-CF_3)_2C_6H_3)_2(CO)_5]$, has been chosen (Scheme 6-2). It has a proven record of reactivity toward planar aromatic systems[59] and is expected to act as a very soft Lewis acid, due to the smallest charge on Ru(I) atom among the selected dimetal units ($+0.03$ vs. $+0.97$ on Rh(II) in $[Rh_2(O_2CCF_3)_4]$). The experimental results on codeposition of corannulene with the above-mentioned diruthenium unit confirmed our expectations, as the first $\eta^1$-hub-coordinated complex of $C_{20}H_{10}$ was isolated along with the cocrystallized $\eta^2$-rim-bound isomer (Figure 6-8).[60]

The compelling experimental evidence for a bonding interaction between the Ru(I) center and the interior site of $C_{20}H_{10}$ stems from the fact that the pyramidalization of the complexed hub atom increases by nearly 30%, while the pyramidalization of the rest of the C atoms remains indistinguishable from that of uncoordinated corannulene. The $\pi$-orbital axis vector (POAV) analysis by Haddon[61] is commonly used as a quantitative evaluation method for the degree of pyramidalization for trigonal C atoms. As reference points, the planar trigonal atom is taken with a POAV angle of $0.0°$, whereas the highly pyramidalized carbon atoms of the $C_{60}$-fullerene have POAV angles of $11.6°$. The averaged pyramidalization angle of the hub C atoms of the corannulene core is $8.3 \pm 0.1°$. On $\eta^1$-binding, the POAV angle of the coordinated

(a)　　　　　　　　　　　　　　　　(b)

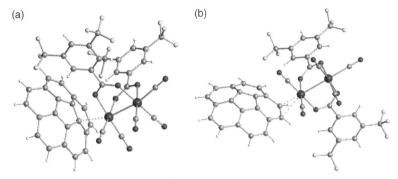

**Figure 6-8.** The $\eta^1$-hub (a) and $\eta^2$-rim-coordinated (b) complexes of $[Ru_2(O_2C(3,5-CF_3)_2C_6H_3)_2(CO)_5]$ with $C_{20}H_{10}$.

hub C atom is increased to 10.6°. Consequently, the bowl depth is also increased from 0.875(2) Å in free corannulene to 0.940(4) Å in the $\eta^1$-hub ruthenium(I) complex. In contrast, the $\eta^2$-rim complexation of the same Ru(I) unit does not significantly perturb the corannulene core geometry. Cocrystallization of the hub and rim isomers of $[Ru_2(O_2C(3,5-CF_3)_2C_6H_3)_2(CO)_5 \cdot (C_{20}H_{10})]$ suggests that they should be close in energy. In fact, the same small energetic preference of ~2.5 kcal/mol [PBE0// LANLDZ(Ru)/6-311G**] for the rim-bound structure over the hub complex was computed, similarly to that calculated for the dirhodium tetra(trifluoroacetate) corannulene system.

All $\eta^2$-complexes discussed above revealed that coordination properties of $C_{20}H_{10}$ are more reminiscent of those for planar polyarenes and contrast the reactivity of $C_{60}$ in its exohedral metal complexes. Thus, the hub corannulene Ru(I) complex, while representing the first example of metal coordination to the bowl interior, has also demonstrated some degree of similarity between the convex carbon surfaces of buckybowls and buckyballs in metal binding reactions. Interestingly, when $[Ru_2(O_2C(3,5-CF_3)_2C_6H_3)_2(CO)_5]$ reacted with $C_{60}$, a complex with an $\eta^2$-bound metal to the (6 : 6) site was obtained.[32] The bonding energy of the $C_{20}H_{10}$ complex was shown to be 5 times smaller than that of the $C_{60}$-fullerene complex, and such a remarkable difference was attributed to noticeably weaker acceptor abilities of corannulene in metal binding reactions compared to that of fullerene.

In addition to the dimetal units described above, we have tested the $[Ru_2(O_2CCF_3)_2(CO)_4]$ complex that has both Ru(I) centers open for coordination and a slightly increased charge on the metal atoms (+0.18, Scheme 6-2) compared to $[Ru_2(O_2C(3,5-CF_3)_2C_6H_3)_2(CO)_5]$. It is less volatile than $[Rh_2(O_2CCF_3)_4]$, and that allows creation of excess organic substrates over diruthenium units in the gas phase. This resulted in the formation of the first biscorannulene transition metal complex, $[Ru_2(O_2CCF_3)_2(CO)_4 \cdot (C_{20}H_{10})_2]$ (Figure 6-9).[62]

The central diruthenium(I,I) core of the bisadduct shown in Figure 6-9 has two terminally $\eta^2$-rim-coordinated $C_{20}H_{10}$ bowls with metals bound to the *exo* faces of

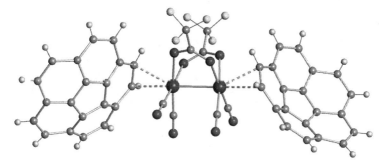

**Figure 6-9.** The bisadduct, $[Ru_2(O_2CCF_3)_2(CO)_4 \cdot (C_{20}H_{10})_2]$.

corannulene. The average Ru−C distance (2.51 Å) is noticeably shorter than those in the aforementioned 1D and 2D dirhodium(II,II) complexes (averaged at 2.60 Å). Similarly, there is a little perturbation of the corannulene core on ruthenium complexation with a small bowl depth flattening effect (0.860(1) vs. 0.875(2) Å in free corannulene) and a noticeable elongation of the Ru−Ru bond distance (from 2.627 to 2.665 Å). DFT calculations (PBE0//LANL2DZ(Ru)/6-311G$^{**}$) show that the $\eta^2$-*exo*-rim corannulene complex is favored over the $\eta^2$-*endo*-rim-bound one by 2 kcal/mol.[62] Although the energetic difference between these two isomers is small, we have not seen experimental evidence of *endo* coordination in this system.

Summing up the coordination preferences of corannulene in reactions with electrophilic dimetal units, several coordination modes have been experimentally identified to date (Figure 6-10). In the structurally characterized complexes, Rh(II) and Ru(I) centers approach the bowls from both the *exo* and *endo* sides, interacting with the rim CC bonds or interior hub C atom of corannulene. The preferential mode of coordination is $\eta^2$-rim for all dimetal units except one case, where the Ru(I) center is bound to the interior bowl part in an $\eta^1$-fashion.

For the next step, we have extended these coordination studies to buckybowls having substituents on the rim or larger surface areas than corannulene (Scheme 6-3).

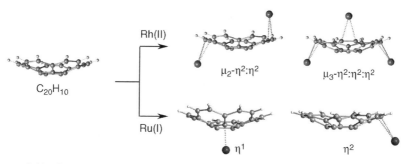

**Figure 6-10.** Coordination of $C_{20}H_{10}$ in the structurally characterized Rh(II) and Ru(I) complexes. Only one metal atom of each dimetal unit is shown.

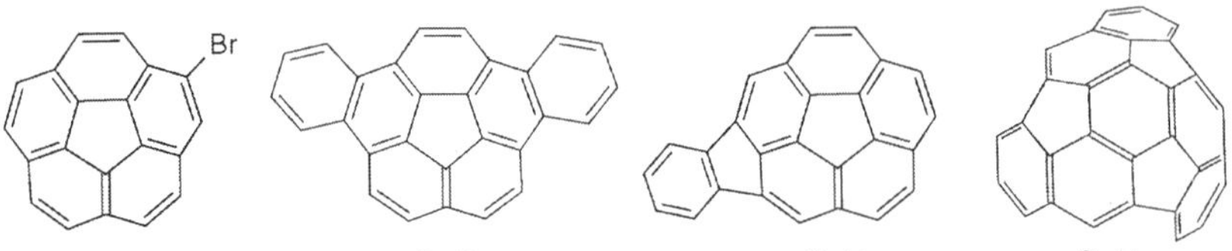

$C_{20}H_9Br$
monobromo-
corannulene

$C_{28}H_{14}$
dibenzo[$a,g$]-
corannulene

$C_{26}H_{12}$
monoindeno-
corannulene

$C_{30}H_{12}$
$C_3$-hemi-
fullerene

**Scheme 6-3.** Examples of surface extension.

***Monobromocorannulene.*** Rim substitution of hydrogen atoms by electron-donating groups is known to enhance π-donation properties of buckybowls.[63] In this regard, Siegel et al.[63] have developed an efficient synthesis of monobromocorannulene, $C_{20}H_9Br$, by reacting corannulene with molecular bromine in the presence of iron trichloride as a catalyst. DFT calculations (PBE0//6-31G$^*$) show that the frontier molecular orbitals of $C_{20}H_{10}$ and $C_{20}H_9Br$ are very similar, although the FMOs of the latter are no longer degenerate.[64] The additional electron density provided by the bromine atom to the corannulene core is distributed predominantly over the rim carbon atoms and does not affect the interior carbon sites. Thus, although bromocorannulene, is a stronger Lewis base, it is a very close electronic analog of corannulene. Consequently, it was not surprising that deposition reactions afforded two products with the $[Rh_2(O_2CCF_3)_4]$-to-$C_{20}H_9Br$ ratio of 1 : 1 (1D polymer) and 3 : 2 (2D layer), similar to those of corannulene.[64] Both 1D coordination polymers with $C_{20}H_{10}$ and $C_{20}H_9Br$ have extended structures built on weak η²-coordination with the identically positioned rim CC bonds of the corannulene core. However, there is some difference in these otherwise very similar 1D structures. In the corannulene-based chain, each dirhodium unit is coordinated to both convex (*exo*) and concave (*endo*) surfaces of the bowl (Figure 6-6), while there is an exclusive coordination of dirhodium units to only convex or concave surfaces of $C_{20}H_9Br$ (Figure 6-11). There is also a noticeably greater flattening of the bowls in the bromocorannulene complex ($\Delta = 0.054$ Å) in comparison with the corannulene analog ($\Delta = 0.036$ Å). The 2D networks with $C_{20}H_{10}$ and $C_{20}H_9Br$ also exhibit very similar solid-state structures

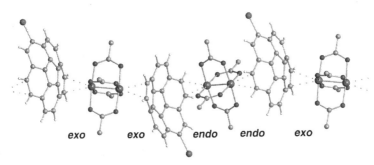

*exo*   *exo*   *endo*   *endo*   *exo*

**Figure 6-11.** A fragment of a 1D infinite chain in $[Rh_2(O_2CCF_3)_4 \cdot (C_{20}H_9Br)]_\infty$. Fluorine atoms are omitted.

(Figure 6-7). The estimated diagonal pore size in the structure with bromocorannulene is slightly decreased compared with the corannulene case (17.2 vs. 17.7 Å), reflecting its enhanced π-donating abilities.

***Dibenzo[a,g]corannulene.*** To access buckybowls with larger surfaces, the benzoannulation synthetic procedures have been developed by Scott[65,66] and Mehta.[67] Dibenzoannulation at *a* and *g* positions[65] was shown to flatten the bowl depth of the corannulene core by ~0.4 Å.[68] The gas-phase reaction of dibenzo[*a,g*]corannulene (Scheme 6-3) with [$Rh_2(O_2CCF_3)_4$] affords the product with the composition of [$Rh_2$] : ($C_{28}H_{14}$) = 3 : 2.[68] Although the same composition was also observed in the 2D corannulene-based network, the dibenzocorannulene complex assembles the reacting moieties in a complex 1D infinite polymer that features the unprecedented complexation of two transition metals to the concave surface of a geodesic polyarene (Figure 6-12).

Again, coordination is seen exclusively at the rim carbon atoms of a bowl, as in all dirhodium complexes described above. While the average Rh–C distances are similar to those in other Rh(II) complexes, the Rh–C contact to the peripheral benzene ring of $C_{28}H_{14}$ is slightly longer than those to the central corannulene core. Binding of three dirhodium units to one bowl results in a decrease of the bowl depth from 0.830 to 0.770 Å, where the overall effect (Δ = 0.060 Å) is greater than that observed in the corannulene complexes.

***Monoindenocorannulene.*** Indenoannulation, which was developed by Scott and de Meijere[69–72] and by Siegel,[73] increases the curvature of the corannulene core, due to an additional five-membered ring in the polyarene framework.[74] Unlike corannulene and dibenzocorannulene, where the bowl-to-bowl interconversion is fast at room temperature,[75–77] monoindenocorannulene, $C_{26}H_{12}$ (Scheme 6-3),[78] exists as a static bowl[73] with both planar (indeno-site) and nonplanar (corannulene core) parts. Its first transition metal complex, [($Rh_2(O_2CCF_3)_4$)$_2$ · ($C_{26}H_{12}$)], was

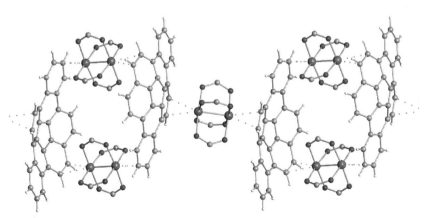

**Figure 6-12.** A fragment of a 1D infinite chain in [($Rh_2(O_2CCF_3)_4$)$_3$ · ($C_{28}H_{14}$)$_2$]$_\infty$. The $CF_3$ groups are omitted.

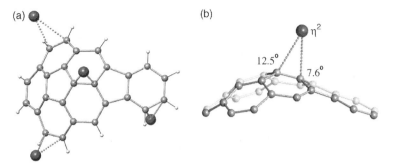

**Figure 6-13.** (a) A tetradentate $\mu_4$-$\eta^2$:$\eta^2$:$\eta^2$:$\eta^2$ coordination of $C_{26}H_{12}$ in [(Rh$_2$(O$_2$CCF$_3$)$_4$)$_2$ · (C$_{26}$H$_{12}$)]$_\infty$, depicting one metal atom of each dimetal unit; (b) side view of indenocorannulene showing the spoke site coordination along with pyramidalization angles.

recently synthesized by gas-phase deposition.[79] The X-ray diffraction study revealed the formation of a 2D organometallic network based on a rare tetra-bridged coordination of a $\pi$ bowl [Figure 6-13(a)]. The 2 : 1 stoichiometry of [Rh$_2$] to the polyarene is unique among structurally characterized complexes of buckybowls, including corannulene, monobromocorannulene, and dibenzocorannulene, where products with [Rh$_2$] : L = 1 : 1 and/or 3 : 2 were obtained. Importantly, in addition to $\eta^2$-rim binding, one Rh(II) center interacts exclusively with interior carbon atoms on the convex bowl surface, exhibiting an $\eta^2$-coordination type to the spoke bond previously observed only in closed all-carbon buckyballs. The latter unique coordination of Rh(II) accentuates the pyramidalization of the C atoms of monoindenocorannulene [Figure 6-13(b)]. Pyramidalization of the complexed hub C atom is increased to 12.5°, surpassing the curvature of C$_{60}$ and reaching that of the highly curved pentaindenocorannulene (the average for all hub C atoms is 12.6°)[74] having four more five-membered rings than C$_{26}$H$_{12}$. As a consequence, the bowl depth of C$_{26}$H$_{12}$ is significantly increased in comparison with its free form ($\Delta = 0.08$ Å), setting the record for how metal coordination may increase the curvature of a buckybowl.

DFT calculations (PBE0//def2-TZVPP(Rh)/TZVP) show[75] that indenoannulation at the corannulene core results in a significant perturbation of the electronic structure of C$_{26}$H$_{12}$ in comparison with that of C$_{20}$H$_{10}$. The $\Delta E_{HOMO-LUMO}$ gap of C$_{26}$H$_{12}$ is significantly reduced compared to C$_{20}$H$_{10}$ (3.86 vs. 4.73 eV), rendering it to be a softer ligand. The bond order calculations reveal that the preferred coordination sites of C$_{26}$H$_{12}$ are the rim of the corannulene core (nonplanar part), followed by the interior spoke, and then the rim of the indeno-site (planar part). This calculated trend is nicely followed by the average experimental Rh–C bond distances in the solid state structure of [(Rh$_2$(O$_2$CCF$_3$)$_4$)$_2$ · (C$_{26}$H$_{12}$)]: 2.567 (rim) < 2.687 (spoke) < 2.715 Å (indeno-site). A similar trend was observed for the rim and spoke dirhodium adducts with corannulene.[35] DFT calculations explicitly support the conclusion that the $\eta^2$-coordination to the interior part of the bowl should have the greatest effect on the bowl depth.[79]

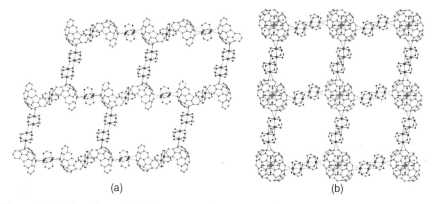

(a)                                                    (b)

**Figure 6-14.** Two views of the 3D structure in $[(Rh_2(O_2CCF_3)_4)_3 \cdot (C_{30}H_{12})]_\infty$, perpendicular (a) and along (b) the direction of 1D polymer based on alternating $[Rh_2]$ and $C_{30}H_{12}$ units. "Dimers of dimers" extend the structure into the other two dimensions. The $CF_3$ groups and H atoms are omitted.

***Hemifullerene.*** The largest and deepest buckybowl used in metal binding studies so far is the $C_3$-symmetric hemifullerene, $C_{30}H_{12}$ (Scheme 6-3).[80,81] First attempts to prepare the hemifullerene transition metal complex by treating it with $[Pt(C_2H_4)(PPh_3)]$ resulted in the insertion of platinum into the peripheral CC bond of one of the five-membered rings.[82] The feasibility of such an unusual aryl-aryl bond breaking has been attributed to the relief of strain at the edge of this highly curved aromatic hydrocarbon. The first coordination complex of $C_{30}H_{12}$ was obtained in our laboratory in the single crystalline form by co-deposition.[83] Under gas-phase conditions, dirhodium tetra(trifluoroacetate) binds to hemifullerene providing a complex with the $[Rh_2]:(C_{30}H_{12}) = 3:1$ composition. Its X-ray diffraction analysis revealed the formation of a complex three-dimensional network in the solid state (Figure 6-14). Similar to the complex with indenocorannulene, four metal centers are bound to each $C_{30}H_{12}$-bowl resulting in its tetrabridged coordination. While three rhodium atoms approach the hemifullerene ligand from the convex side ($\eta^2$-mode), one is bound to its concave face in an $\eta^1$ fashion. Despite the large available surface area, only rim carbon atoms of $C_{30}H_{12}$ are involved in metal coordination. In this complex, only one of the three dirhodium units has the $C_{30}H_{12}$ molecules coordinated to both of its open axial positions. The other two are involved in Rh$\cdots$O intermolecular interactions, thus forming the "dimer of dimers" core structures. The formation of such dimers is not surprising, due to a very limited volatility of $C_{30}H_{12}$ and thus the prevailing concentrations of dirhodium units in the gas phase. The "dimer of dimers" type of structure has been seen in deposition reactions when an excess of dimetal units over ligands was used.[84]

Several structurally characterized complexes with $[Rh_2(O_2CCF_3)_4]$ provide a good set for evaluation of metal binding effects on $\pi$ bowls having different curvature and surface areas. A comparison of the bowl depths in the free and coordinated forms shows that the effect varies depending on the coordination site of a buckybowl (Table 6-2). In general, the $\eta^2$-rim coordination flattens a bowl. In contrast, the interior $\eta^2$-spoke

**TABLE 6-2. Coordination Modes of π Bowls in Structurally Characterized [Rh₂(O₂CCF₃)₄] Complexes**

| Formula | Side View,[a] Bowl Depth in Free Form (Å) | Coordination Mode,[b] Bowl Depth in Complexed Form (Å) |
|---|---|---|
| $C_{20}H_{10}$ | 0.875(2) | $\mu_2$-$\eta^2$:$\eta^2$ 0.839(3) (Ref. 55) $\mu_3$-$\eta^2$:$\eta^2$:$\eta^2$ 0.861(3) (Ref. 55) |
| $C_{20}H_9Br$ | 0.862 (calc) | $\mu_2$-$\eta^2$:$\eta^2$ 0.808(2) (Ref. 64) $\mu_3$-$\eta^2$:$\eta^2$:$\eta^2$ – [c] (Ref. 64) |
| $C_{28}H_{14}$ | 0.830(3) | $\mu_3$-$\eta^2$:$\eta^2$:$\eta^2$ 0.770(3) (Ref. 68) |
| $C_{26}H_{12}$ | 1.065(4) | $\mu_4$-$\eta^2$:$\eta^2$:$\eta^2$:$\eta^2$ 1.148(6) (Ref. 79) |

$C_{30}H_{12}$

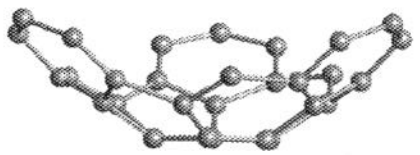

2.41(4)

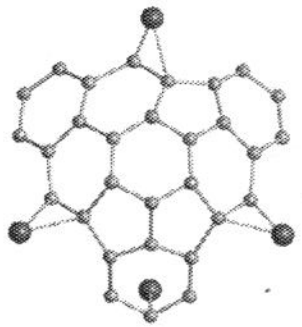

$\mu_4$-$\eta^2$:$\eta^2$:$\eta^2$:$\eta^1$ — [c] (Ref. 83)

$C_{40}H_{50}$

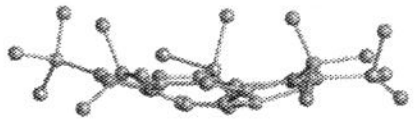

0.72(5)

No binding (Ref. 85)

—

[a] H atoms are removed.

[b] Only one metal atom of each dimetal unit is shown.

[c] Severe disorder of $C_{20}H_9Br$ and $C_{30}H_{12}$ precludes obtaining accurate bowl depth values in the X-ray structures of their complexes.

coordination leads to a noticeable increase of the bowl depth, with an effect similar to that provided by an annulation of one additional pentagon to a bowl.[79] Overall, the solid-state effects of $\eta^2$ coordination are not as pronounced as in some $\eta^6$-complexes, where almost complete flattening of the corannulene molecule has been observed.[31] Rather than affecting the bowl as a whole, $\eta^2$-complexation is more delicate and site-specific, causing the subtle modification of a particular surface area.

It is noteworthy that multiple metal binding to a bowl is readily achieved under gas-phase reaction conditions (Table 6-2). For example, two and three metal centers are bound to corannulene and bromocorannulene in the isolated Rh(II) complexes to cause the bowl to function in the $\mu_2$-$\eta^2$:$\eta^2$- and $\mu_3$-$\eta^2$:$\eta^2$:$\eta^2$-bridging modes, respectively. The latter tridentate mode is also realized in the dibenzo[$a,g$]corannulene complex, but the large concave surface area of $C_{28}H_{14}$ allows coordination of two dirhodium units at the *endo* face of the bowl. Four metal centers coordinate to monoindenocorannulene and hemifullerene ligands in their Rh(II) complexes, resulting in rare tetradentate $\mu_4$-$\eta^2$:$\eta^2$:$\eta^2$:$\eta^{2(1)}$-bridging coordination of the $\pi$ bowls. It is noteworthy that when the rim sites of the corannulene core are blocked by five bulky *tert*-butyl groups in $C_{40}H_{50}$, no metal complexation is observed on codeposition with the same dirhodium complex.[85]

## 6-4. TRINUCLEAR METAL COMPLEXES: CURVATURE TRADEOFFS

Coordination of mono- and dinuclear metal complexes to bowl-shaped polyarenes allows identification of their preferential binding sites. The use of trinuclear metal complexes permits analysis of the solid-state interactions between planar and nonplanar molecular surfaces. As a planar trimetal complex, the highly Lewis acidic perfluoro-*ortho*-phenylenemercury, [Hg$_3$],[86] has been chosen (Figure 6-15).

We opted for this unit for its known ability to form stable complexes with a great number of planar single-ring and polycyclic aromatic hydrocarbons.[87–100] DFT calculations (B3LYP//Huzinaga–Dunning basis sets for all atoms) show a positively charged electrostatic potential surface at the center of the [Hg$_3$] unit,[101] thus making it an excellent electrophilic probe for solid-state interactions with the negatively charged surfaces of buckybowls. The complementarity of electronic properties and incompatibility of molecular geometries of these reacting partners allow evaluation of the mutual structural influences of bowl-shaped polyarenes and a planar polynuclear metal unit on their attractive interactions. Curving of the trimercury unit to match the convex surface of a $\pi$ bowl is required for the formation of a stable metal–organic complex. In addition, flattening of the bowl-shaped polyarene is also anticipated, and both effects should change the strain energy of interacting partners leading to "mutual curvature adaptations" at the molecular level. Several bowl-shaped polyarenes having different surface size, rim functionalization, and curvature have been tested in this work.

Reaction of corannulene and [Hg$_3$] in solution affords the [Hg$_3$ · $C_{20}H_{10}$] complex.[102] In the solid state, the extended binary stacks, in which the [Hg$_3$] units alternate with the corannulene bowls, are formed (Figure 6-16). The trimercury unit is placed over the five-membered ring of the convex surface of corannulene providing the first example of multiple metal binding to the interior of a buckybowl surface.

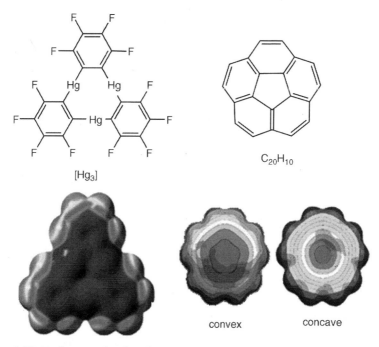

**Figure 6-15.** Perfluoro-*ortho*-phenylenemercury and corannulene (top) along with their complementary electrostatic potential surfaces (bottom).

The sandwiched corannulene molecule is engaged in a number of interactions using its five-membered ring of the convex (hub C atoms, 3.14–3.51 Å) and concave (flank C atoms, 3.17–3.40 Å) faces. These contacts are among the shortest for the previously reported [Hg$_3$] complexes with planar aromatic hydrocarbons.[87–100] The observed strong interactions in the solid-state structure of [Hg$_3 \cdot$ C$_{20}$H$_{10}$] also have a pronounced flattening effect on bowl depth (from 0.875 to 0.754 Å). On complex formation, the perfluoro-*ortho*-phenylenemercury unit, which is planar in the solid state,[88] embraces the curved surface of corannulene with a bending angle $\gamma$ equal to 11.2° (Figure 6-16).

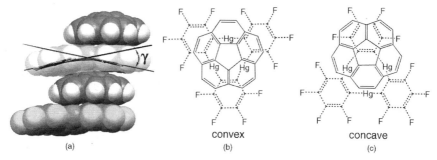

**Figure 6-16.** Space-filling view of the [Hg$_3 \cdot$ C$_{20}$H$_{10}$] stack showing the bending angle ($\gamma$) (a) and schematic representations of the top views to the convex (b) and concave (c) faces of corannulene.

Such a strong deformation allows for better arene–perfluoroarene interactions between the concave surface of $C_{20}H_{10}$ and the perfluorobenzene ring of [$Hg_3$]. The favorable arene–fluoroarene interactions and steric hindrance that blocks access to the five-membered ring of $C_{20}H_{10}$ from the inside cause the sliding of [$Hg_3$] along the concave face of corannulene [Figure 6-16 (b)].

Since packing forces along the 1D stacks in [$Hg_3 \cdot C_{20}H_{10}$] may have additional influence on the solid-state interactions between π bowls and planar trimercury units, an attempt to break these columns into discrete subunits was undertaken. Addition of alkyl substituents to the corannulene core is expected to enhance its donating ability and thus to provide stronger interactions with metal units. At the same time, methyl groups at the rim of corannulene add some steric hindrance and may block the coordination of [$Hg_3$] to the concave face of the bowl. Pursuing this line of thought, we set out to test 1,2,5,6-tetramethylcorannulene[63] and isolated its crystalline product with [$Hg_3$] from solution.[103] The X-ray diffraction study revealed that [$Hg_3 \cdot C_{24}H_{18}$] crystallizes in a unit cell that is only slightly different from that of the corannulene analog. The cell volume is increased by approximately 150 $Å^3$, due to the additional space occupied by four methyl groups. In contrast to our expectations, the methylated corannulene also forms 1D extended binary stacks, as was found in the corannulene case. The sandwiched $C_{24}H_{18}$ bowls are engaged in similar types of interaction with [$Hg_3$], although, as expected, the [$Hg_3 \cdot C_{24}H_{18}$] adduct displays shorter Hg$\cdots$C distances to the convex and longer ones to the concave bowl surfaces. The reduced curvature of $C_{24}H_{18}$ in comparison with $C_{20}H_{10}$ and the weakening of the intermolecular arene–fluoroarene interactions lead to a noticeably smaller deformation of the planar [$Hg_3$] moiety ($\gamma = 6.6°$). Similar to that of corannulene, the bowl depth of tetramethylcorannulene is decreased from 0.850 to 0.721 Å.

Analysis of the systems described above shows the importance of $\pi–\pi_F$ interactions in bending the planar trimercury unit over the nonplanar polyarenes. Increasing the surface area of the latter should strengthen these intermolecular interactions by creating additional sites for $\pi–\pi_F$ contacts. For this purpose, monoindenocorannulene (Scheme 6-3) has been selected. Indenocorannulene has a deeper molecular bowl than do corannulenes, and thus is expected to impose greater curvature to the planar trimercury unit. The complexation reaction between the equimolar quantities of [$Hg_3$] and indenocorannulene in dichloromethane provided [$Hg_3 \cdot C_{26}H_{12}$] in quantitative yield. Single-crystal X-ray diffraction analysis revealed the existence of two crystallographically independent extended stacks **A** and **B** (Figure 6-17).[102]

In its free form, the monoindenocorannulene bowls are packed into one-dimensional extended stacks, and thus the overall strength of interactions between the [$Hg_3$] and $C_{26}H_{12}$ units can be nicely illustrated by a noticeable shortening of the distance between the bowls in the solid-state structure of [$Hg_3 \cdot C_{26}H_{12}$]. This difference in stacking distance between **A** and **B** correlates well with the variation in bowl depth of the corresponding $C_{26}H_{12}$ molecules (Figure 6-17). Coordination flattens the indenocorannulene surface compared to its free form (1.056 Å); the bowl is less curved in **A** (1.008 Å) than in **B** (1.047 Å). The substantially smaller bowl deformations than in the complexes with shallower π bowls are attributed to the shift

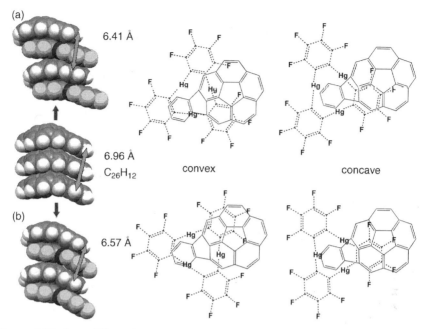

**Figure 6-17.** Space-filling views of two independent [Hg$_3$ · C$_{26}$H$_{12}$] stacks along with free C$_{26}$H$_{12}$ (left), and top views to the convex (center) and concave (right) bowl surfaces: (a) **A** and (b) **B**.

of primary Hg· · ·C interactions away from the corannulene core (nonplanar part) to the peripheral indeno group (planar part). In stacks **A**, there is a strong bonding interaction between the Hg atom and the spoke bond connecting two 5-membered rings on the convex face. The Hg· · ·C$_{spoke}$ distance (3.052(2) Å) is the shortest in the series of [Hg$_3$] complexes with all other planar[87–100] or nonplanar arenes. This strong and directional interaction leads to the difference in curvature of the C$_{26}$H$_{12}$ bowls in both stacks.

In **A** and **B**, both convex and concave faces of C$_{26}$H$_{12}$ are engaged in arene–fluoroarene interactions, in contrast to corannulene and tetramethylcorannulene cases. This fact can be attributed to the availability of larger surface area attained by the indenoannulation of the corannulene core in C$_{26}$H$_{12}$. The shortest distances between the centroids are 3.84 and 3.33 Å with the convex and 3.88 and 3.97 Å with the concave surfaces of C$_{26}$H$_{12}$ in **A** and **B**, respectively. The best surface match of interacting subunits and the strongest $\pi$–$\pi_F$ interactions with the convex face of C$_{26}$H$_{12}$ in **B** lead to a large distortion of the [Hg$_3$] unit ($\gamma = 21.8°$, Figure 6-18), which is the greatest bending found among all trimercury complexes with buckybowls. For comparison, slightly weaker $\pi$–$\pi_F$ interactions in **A** provide a smaller bending angle (15.0°), which is still greater than those observed in the corannulene and tetra-methylcorannulene complexes (10.3° and 6.6°, Table 6-3). It is worth mentioning that in **A**, where the directional Hg· · ·C$_{spoke}$ interaction takes place, the corannulene core

$\gamma = 21.8°$

**Figure 6-18.** Bending of [Hg$_3$] over the surface of C$_{26}$H$_{12}$ in stack **B**.

**TABLE 6-3. Bowl Depth and Bending Angles in [Hg$_3$] Complexes of Buckybowls**

| Buckybowl | Bowl Depth, (Å) | | | Bending Angle of [Hg$_3$] Unit, $\gamma$ ° |
|---|---|---|---|---|
| | Free | Complex | $\Delta$ | |
| Corannulene | 0.870 | 0.754 | 0.116 | 11.2 |
| Tetramethylcorannulene | 0.850 | 0.721 | 0.129 | 6.6 |
| Indenocorannulene | 1.056 | (**A**) 1.008 | 0.048 | 15.0 |
| | | (**B**) 1.047 | 0.009 | 21.8 |

of C$_{26}$H$_{12}$ is affected noticeably more strongly, whereas the [Hg$_3$] unit is less deformed than in **B**, where the above-mentioned Hg$\cdots$C interactions could not be clearly identified. This fact corroborates the importance of $\pi-\pi_F$ interactions for matching the surfaces of the planar trimercury unit and those of $\pi$ bowls.

## 6-5. POLYNUCLEAR METAL COMPLEXES

In 2007 we used electrophilic copper(I) carboxylate complexes as probes for intermolecular copper(I)−arene substrate interactions. In the course of these studies, the first cyclic hexanuclear copper(I) carboxylate, [Cu$_6$(O$_2$C(3,5-F)$_2$C$_6$H$_3$)$_6$], and its adduct with planar coronene, [Cu$_6$(O$_2$C(3,5-F)$_2$C$_6$H$_3$)$_6$ · C$_{24}$H$_{12}$], have been synthesized.[104] DFT calculations (PBE0//LANL2DZ(Cu)/3-21G$^*$) show the latter to be a cocrystallization product held by strong electrostatic ion−dipole interactions without covalent contributions to bonding. The use of bulkier substituents on the benzene rings of the carboxylate groups resulted in the isolation of a new complex, [Cu(O$_2$C (3,5-CF$_3$)$_2$C$_6$H$_3$)], having a remarkable infinite double-helix structure in the solid state.[105] The latter forms reactive copper(I) fragments with the nuclearities of two, four, or six in the gas phase on sublimation with various planar polyaromatic hydrocarbons.[105] Its reaction with corannulene affords the first copper buckybowl adduct, [Cu$_6$(O$_2$C(3,5-CF$_3$)$_2$C$_6$H$_3$)$_6$ · (C$_{20}$H$_{10}$)$_2$] (Figure 6-19).[106]

In the solid state, each Cu$_6$ cluster is bound to two bowls from the *exo* (3.1−3.2 Å) and to the other two from the *endo* (2.8−3.3 Å) sides. The Cu$\cdots$C interactions are

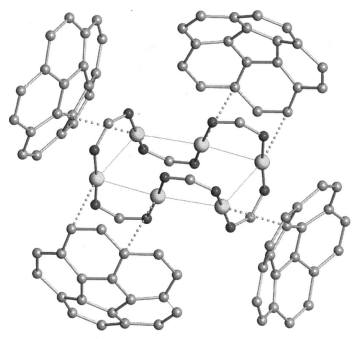

**Figure 6-19.** Fragment of a 2D layer in $[Cu_6(O_2C(3,5-CF_3)_2C_6H_3)_6 \cdot (C_{20}H_{10})_2]$ showing a hexanuclear Cu(I) core enclosed by four corannulene molecules. The benzoate groups and H atoms are omitted for clarity.

most probably purely electrostatic, similar to those found in the complex with coronene.[104] This is consistent with the facts that on coordination, the bowl depth of $C_{20}H_{10}$ is only slightly reduced to 0.824(2) Å from 0.875(2) Å in free corannulene, while the pyramidalization angles of the C atoms remained essentially the same.

We have also prepared and structurally characterized the first $\pi$ complex of corannulene with palladium.[107] As mentioned above, previous attempts to synthesize metal complexes of buckybowls using $[Pt(C_2H_4)(PPh_3)]$ led to the $\sigma$-insertion product.[82] Reaction of bis(benzonitrile)palladium(II) dichloride with corannulene in benzene, on the other hand, provides a $[Pd_6Cl_{12} \cdot (C_{20}H_{10})_2 \cdot (C_6H_6)_3]$ complex (Figure 6-20).

In the solid state, the $Pd_6$ cluster interacts with two benzene molecules and the convex faces of two corannulene bowls, providing additional experimental evidence for the *exo*-coordination preference of $C_{20}H_{10}$ on metal binding. It is noteworthy that the $Pd_6Cl_{12}$ cluster binds to $C_6H_6$ in an $\eta^6$-fashion, while exhibiting an $\eta^1$-mode with respect to $C_{20}H_{10}$, again illustrating the different coordination properties of planar and nonplanar aromatic systems. One of the Pd atoms approaches the interior hub C atom of $C_{20}H_{10}$ with the $Pd\cdots C_{hub}$ bond length of 3.085(3) Å. While this $Pd\cdots C_{hub}$ separation is well beyond the conventional bonding range found in typical palladium olefin complexes (2.1–2.3 Å),[108,109] it is shorter than those in the previously reported Pd complexes with planar PAHs and fullerenes (3.2–3.4 Å).[110,111] Complexation does not affect the corannulene core, with the pyramidalization angles of the hub

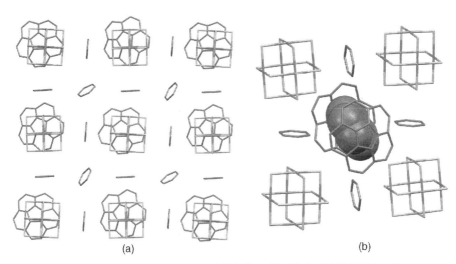

**Figure 6-20.** Fragment of a 3D structure of $[Pd_6Cl_{12} \cdot (C_{20}H_{10})_2 \cdot (C_6H_6)_3]$ (a) and an encapsulation of one benzene molecule, rendered as a space-filling model (b).

C atoms ($8.3° \pm 0.1°$) and the bowl depth [$0.885(5)$ Å] being essentially unchanged. These results contrast to the $\eta^1$-hub complex of Ru(I) discussed above, where a noticeable increase in the degree of pyramidalization of the hub C atom was observed.[60] This fact indicates that the interaction between the $Pd_6$ cluster and corannulene lacks a covalent contribution and is mostly electrostatic. Overall, the complex exhibits a 3D structure formed by extended 2D sheets composed of binary stacks of alternating $Pd_6Cl_{12}$ clusters and benzene molecules and expanded to the third dimension through Pd$\cdots$C interactions with the corannulene bowls. Interestingly, the concave faces of $C_{20}H_{10}$ that are free from metal interaction are utilized for the encapsulation of a benzene molecule in the square cage formed by $Pd_6Cl_{12}$ and benzene [Figure 6-20(b)].

## 6-6. CONCLUDING REMARKS AND FUTURE PROSPECTS

System-dependent preference for metal binding of bowl-shaped polyaromatic hydrocarbons has been confirmed. The formation of Pd(II) and Ru(I)[60,62] complexes, along with $\eta^6$-bound Rh,[30] Ir,[30] Ru,[31] and $\eta^2$-coordinated Ag[45] products, experimentally reveals the general preference of the convex face of $C_{20}H_{10}$ for metal coordination. However, the new *endo*-bound sumanene complexes[46–48] show that the concave face of a $\pi$ bowl can also be engaged in "dishing up" a metal. While studies of mono- and dinuclear metal complexes revealed reactive coordination sites of curved polyarenes,[22] the utilization of electrophilic trimercury units provided a unique opportunity to probe an interface between the geometrically mismatched molecular surfaces.[102–103] In the solid state, significant mutual geometry adaptations of both partners, referred to as "curvature tradeoffs," have been revealed as a result of reinforced intermolecular interactions. The use of molecular aromatic bowls as templates for bending planar molecules should be further utilized in storage and release of system strain energy by manipulating the geometries of reacting units.

Despite all the more recent developments, the control of reactivity of nonplanar polyarenes in metal binding reactions still presents a challenge, especially considering that highly strained and thus the most interesting large $\pi$ bowls are still not readily available. This severely hinders further progress in coordination and reactivity investigations of carbon-rich and curved aromatic systems. For larger bowls, novel preparative methods should be sought and developed to achieve controlled metal binding to a specific site on nonplanar polyaromatic surfaces. As a result, novel aspects of their reactivity will be unraveled and novel metallated buckybowl products will be synthesized, and that should greatly stimulate the use and applications of carbon-rich molecules in materials chemistry.

## ACKNOWLEDGMENTS

We thank the National Science Foundation for financial support of this research and for funds to purchase single crystal and powder X-ray diffractometers. The contribution to this work by all students, postdoctoral associates, and collaborators listed in the references cited is gratefully acknowledged.

## REFERENCES

1. P. J. Fagan, J. C. Calabrese, B. Malone, *Science* **1991**, *252*, 1160–1161.

2. P. J. Fagan, J. C. Calabrese, B. Malone, *Acc. Chem. Res.* **1992**, *25*, 134–142.

3. A. L. Balch, M. M. Olmstead, *Chem. Rev.* **1998**, *98*, 2123–2165.

4. K. Lee, H. Song, J. T. Park, *Acc. Chem. Res.* **2003**, *36*, 78–86.

5. L.-C. Song, G.-A. Yu, F.-H. Su, Q.-M. Hu, *Organometallics* **2004**, *23*, 4192–4198.

6. B. K. Park, G. Lee, K. H. Kim, H. Kang, C. Y. Lee, M. A. Miah, J. Jung, Y.-K. Han, J. T. Park, *J. Am. Chem. Soc.* **2006**, *128*, 11160–11172.

7. M. Sawamura, H. Iikura, E. Nakamura, *J. Am. Chem. Soc.* **1996**, *118*, 12850–12851.

8. M. Sawamura, Y. Kuninobu, E. Nakamura, *J. Am. Chem. Soc.* **2000**, *122*, 12407–12408.

9. E. Nakamura, M. Sawamura, *Pure Appl. Chem.* **2001**, *73*, 355–359.

10. Y. Matsuo, A. Iwashita, E. Nakamura, *Organometallics* **2008**, *27*, 4611–4617.

11. Y. Matsuo, Y. Kuninobu, A. Muramatsu, M. Sawamura, E. Nakamura, *Organometallics* **2008**, *27*, 3403–3409.

12. M. W. Bouwkamp, A. Meetsma, *Inorg. Chem.* **2009**, *48*, 8–9.

13. Y. Matsuo, K. Tahara, T. Fujita, E. Nakamura, *Angew. Chem. Int. Ed.* **2009**, *48*, 6239–6241.

14. L. T. Scott, *Pure Appl. Chem.* **1996**, *68*, 291–300.

15. P. W. Rabideau, A. Sygula, *Acc. Chem. Res.* **1996**, *29*, 235–242.

16. L. T. Scott, H. E. Bronstein, D. V. Preda, R. B. M. Ansems, M. S. Bratcher, S. Hagen, *Pure Appl. Chem.* **1999**, *71*, 209–219.

17. L. T. Scott, *Angew. Chem. Int. Ed.* **2004**, *43*, 4994–5007.

18. H. Sakurai, T. Daiko, T. Hirao, *Science* **2003**, *301*, 1878–1882.

19. H. Sakurai, T. Daiko, H. Sakane, T. Amaya, T. Hirao, *J. Am. Chem. Soc.* **2005**, *127*, 11580–11581.

20. V. M. Tsefrikas, L. T. Scott, *Chem. Rev.* **2006**, *106*, 4868–4884.

21. Y.-T. Wu, J. S. Siegel, *Chem. Rev.* **2006**, *106*, 4843–4867.

22. A. S. Filatov, M. A. Petrukhina, *Coord. Chem. Rev.* **2010**, *254*, 2234–2246.

23. T. J. Seiders, K. K. Baldridge, J. M. O'Connor, J. S. Siegel, *J. Am. Chem. Soc.* **1997**, *119*, 4781–4782.

24. A. L. Chistyakov, I. V. Stankevich, *J. Organomet. Chem.* **2000**, *599*, 18–27.

25. M. V. Frash, A. C. Hopkinson, D. K. Bohme, *J. Am. Chem. Soc.* **2001**, *123*, 6687–6695.

26. R. C. Dunbar, *J. Phys. Chem. A* **2002**, *106*, 9809–9819.

27. U. D. Priyakumar, G. N. Sastry, *Tetrahedron Lett.* **2003**, *44*, 6043–6046.

28. Y. Kameno, A. Ikeda, Y. Nakao, H. Sato, S. Sakaki, *J. Phys. Chem. A* **2005**, *109*, 8055–8063.

29. A. K. Kandalam, B. K. Rao, P. Jena, *J. Phys. Chem. A* **2005**, *109*, 9220–9225.

30. J. S. Siegel, K. K. Baldridge, A. Linden, R. Dorta, *J. Am. Chem. Soc.* **2006**, *128*, 10644–10645.

31. B. Zhu, A. Ellern, A. Sygula, R. Sygula, R. J. Angelici, *Organometallics* **2007**, *26*, 1721–1728.

32. A. Y. Rogachev, Y. Sevryugina, A. S. Filatov, M. A. Petrukhina, *Dalton Trans.* **2007**, 3871–3873.

33. R. Peverati, K. K. Baldridge, *J. Chem. Theory Comput.* **2008**, *4*, 2030–2048.

34. A. Sygula, S. Saebo, *Int. J. Quantum Chem.* **2009**, *109*, 65–72.

35. A. Y. Rogachev, M. A. Petrukhina, *J. Phys. Chem. A* **2009**, *113*, 5743–5753.

36. D. Caraiman, G. K. Koyanagi, L. T. Scott, D. V. Preda, D. K. Bohme, *J. Am. Chem. Soc.* **2001**, *123*, 8573–8582.

37. T. M. Ayers, B. C. Westlake, D. V. Preda, L. T. Scott, M. A. Duncan, *Organometallics* **2005**, *24*, 4573–4578.

38. H. B. Lee, P. R. Sharp, *Organometallics* **2005**, *24*, 4875–4877.

39. R. Maag, B. H. Northrop, A. Butterfield, A. Linden, O. Zerbe, Y. M. Lee, K.-W. Chi, P. J. Stang, J. S. Siegel, *Org. Biomol. Chem.* **2009**, *7*, 4881–4885.

40. H. Choi, C. Kim, K.-M. Park, J. Kim, Y. Kang, J. Ko, *J. Organomet. Chem.* **2009**, *694*, 3529–3532.

41. C. M. Alvarez, R. J. Angelici, A. Sygula, R. Sygula, P. W. Rabideau, *Organometallics* **2003**, *22*, 624–626.

42. M. W. Stoddart, J. H. Brownie, M. C. Baird, H. L. Schmider, *J. Organomet. Chem.* **2005**, *690*, 3440–3450.

43. P. A. Vecchi, C. M. Alvarez, A. Ellern, R. J. Angelici, A. Sygula, R. Sygula, P. W. Rabideau, *Angew. Chem., Int. Ed.* **2004**, *43*, 4497–4500.

44. P. A. Vecchi, C. M. Alvarez, A. Ellern, R. J. Angelici, A. Sygula, R. Sygula, P. W. Rabideau, *Organometallics* **2005**, *24*, 4543–4552.

45. E. L. Elliott, G. A. Hernandez, A. Linden, J. S. Siegel, *Org. Biomol. Chem.* **2005**, *3*, 407–413.

46. T. Amaya, H. Sakane, T. Hirao, *Angew. Chem. Int. Ed.* **2007**, *46*, 8376–8379.

47. H. Sakane, T. Amaya, T. Moriuchi, T. Hirao, *Angew. Chem. Int. Ed.* **2009**, *48*, 1640–1643.

48. T. Amaya, W.-Z. Wang, H. Sakane, T. Moriuchi, T. Hirao, *Angew. Chem. Int. Ed.* **2010**, *49*, 403–406.

49. L. Pirondini, A. G. Stendardo, S. Geremia, M. Campagnolo, P. Samori, J. P. Rabe, R. Fokkens, E. Dalcanale, *Angew. Chem. Int. Ed.* **2003**, *42*, 1384–1387.

50. J. R. Nitschke, D. Schultz, G. Bernardinelli, D. Gerard, *J. Am. Chem. Soc.* **2004**, *126*, 16538–16543.

51. M. A. Petrukhina, *Coord. Chem. Rev.* **2007**, *251*, 1690–1698.

52. A. S. Filatov, A. Y. Rogachev, M. A. Petrukhina, *Cryst. Growth Des.* **2006**, *6*, 1479–1484.

53. F. A. Cotton, E. V. Dikarev, M. A. Petrukhina, *J. Am. Chem. Soc.* **2001**, *123*, 11655–11663.

54. F. A. Cotton, C. A. Murillo, R. A. Walton, *Multiple Bonds between Metal Atoms*, 3rd ed., Springer Science and Business Media, New York, **2005**, p. 818.

55. M. A. Petrukhina, K. W. Andreini, J. Mack, L. T. Scott, *Angew. Chem. Int. Ed.* **2003**, *42*, 3375–3379.

56. J. C. Hanson, C. E. Nordman, *Acta Crystallogr.* **1976**, *B32*, 1147–1153.

57. M. A. Petrukhina, K. W. Andreini, J. Mack, L. T. Scott, *J. Org. Chem.* **2005**, *70*, 5713–5716.

58. F. A. Cotton, E. V. Dikarev, X. Feng, *Inorg. Chim. Acta* **1995**, *237*, 19–26.

59. Y. Sevryugina, A. V. Olenev, M. A. Petrukhina, *J. Cluster Sci.* **2005**, *16*, 217–229.

60. M. A. Petrukhina, Y. Sevryugina, A. Y. Rogachev, E. A. Jackson, L. T. Scott, *Angew. Chem. Int. Ed.* **2006**, *45*, 7208–7210.

61. R. C. Haddon, *J. Am. Chem. Soc.* **1990**, *112*, 3385–3389.

62. M. A. Petrukhina, Y. Sevryugina, A. Y. Rogachev, E. A. Jackson, L. T. Scott, *Organometallics* **2006**, *25*, 5492–5495.

63. T. J. Seiders, E. L. Elliott, G. H. Grube, J. S. Siegel, *J. Am. Chem. Soc.* **1999**, *121*, 7804–7813.

64. A. S. Filatov, M. A. Petrukhina, *J. Organomet. Chem.* **2008**, *693*, 1590–1596.

65. H. A. Reisch, M. S. Bratcher, L. T. Scott, *Org. Lett.* **2000**, *2*, 1427–1430.

66. V. M. Tsefrikas, S. Arns, P. M. Merner, C. C. Warford, B. L. Merner, L. T. Scott, G. J. Bodwell, *Org. Lett.* **2006**, *8*, 5195–5198.

67. G. Mehta, P. V. V. S. Sarma, *Chem. Commun.* **2000**, 19–20.

68. M. A. Petrukhina, K. W. Andreini, V. M. Tsefrikas, L. T. Scott, *Organometallics* **2005**, *24*, 1394–1397.

69. H. A. Wegner, L. T. Scott, A. de Meijere, *J. Org. Chem.* **2003**, *68*, 883–887.

70. H. A. Wegner, H. Reisch, K. Rauch, A. Demeter, K. A. Zachariasse, A. de Meijere, L. T. Scott, *J. Org. Chem.* **2006**, *71*, 9080–9087.

71. A. de Meijere, B. Stulgies, K. Albrecht, K. Rauch, H. A. Wegner, H. Hopf, L. T. Scott, L. Eshdat, I. Aprahamian, M. Rabinovitz, *Pure Appl. Chem.* **2006**, *78*, 813–830.

72. J. M. Quimby, L. T. Scott, *Adv. Synth. Catal.* **2009**, *351*, 1009–1013.

73. Y.-T. Wu, T. Hayama, K. K. Baldridge, A. Linden, J. S. Siegel, *J. Am. Chem. Soc.* **2006**, *128*, 6870–6884.

74. E. A. Jackson, B. D. Steinberg, M. Bancu, A. Wakamiya, L. T. Scott, *J. Am. Chem. Soc.* **2007**, *129*, 484–485.

75. A. Borchardt, A. Fuchicello, K. V. Kilway, K. K. Baldridge, J. S. Siegel, *J. Am. Chem. Soc.* **1992**, *114*, 1921–1923.

76. L. T. Scott, M. M. Hashemi, M. S. Bratcher, *J. Am. Chem. Soc.* **1992**, *114*, 1920–1921.

77. T. J. Seiders, K. K. Baldridge, G. H. Grube, J. S. Siegel, *J. Am. Chem. Soc.* **2001**, *123*, 517–525.

78. B. D. Steinberg, E. A. Jackson, A. S. Filatov, A. Wakamiya, M. A. Petrukhina, L. T. Scott, *J. Am. Chem. Soc.* **2009**, *131*, 10537–10545.

79. A. S. Filatov, A. Y. Rogachev, E. A. Jackson, L. T. Scott, M. A. Petrukhina, *Organometallics* **2010**, *29*, 1231–1237.

80. A. H. Abdourazak, Z. Marcinow, A. Sygula, R. Sygula, P. W. Rabideau, *J. Am. Chem. Soc.* **1995**, *117*, 6410–6411.

81. S. Hagen, M. S. Bratcher, M. S. Erickson, G. Zimmermann, L. T. Scott, *Angew. Chem. Int. Ed.* **1997**, *36*, 406–408.

82. R. M. Shaltout, R. Sygula, A. Sygula, F. R. Fronczek, G. G. Stanley, P. W. Rabideau, *J. Am. Chem. Soc.* **1998**, *120*, 835–836.

83. M. A. Petrukhina, K. W. Andreini, L. Peng, L. T. Scott, *Angew. Chem. Int. Ed.* **2004**, *43*, 5477–5481.

84. E. V. Dikarev, K. W. Andreini, M. A. Petrukhina, *Inorg. Chem.* **2004**, *43*, 3219–3224.

85. Y. Sevryugina, A. Y. Rogachev, E. A. Jackson, L. T. Scott, M. A. Petrukhina, *J. Org. Chem.* **2006**, *71*, 6615–6618.

86. P. Sartori, A. Golloch, *Chem. Ber.* **1968**, *101*, 2004–2009.

87. M. Tsunoda, F. P. Gabbai, *J. Am. Chem. Soc.* **2000**, *122*, 8335–8336.

88. M. R. Haneline, M. Tsunoda, F. P. Gabbai, *J. Am. Chem. Soc.* **2002**, *124*, 3737–3742.

89. J. B. King, M. R. Haneline, M. Tsunoda, F. P. Gabbai, *J. Am. Chem. Soc.* **2002**, *124*, 9350–9351.

90. M. R. Haneline, J. B. King, F. P. Gabbai, *Dalton Trans.* **2003**, 2686–2690.

91. M. A. Omary, R. M. Kassab, M. R. Haneline, O. Elbjeirami, F. P. Gabbai, *Inorg. Chem.* **2003**, *42*, 2176–2178.

92. M. R. Haneline, F. P. Gabbai, *Angew. Chem. Int. Ed.* **2004**, *43*, 5471–5474.

93. C. Burress, O. Elbjeirami, M. A. Omary, F. P. Gabbai, *J. Am. Chem. Soc.* **2005**, *127*, 12166–12167.

94. M. Melaimi, F. P. Gabbai, *Adv. Organomet. Chem.* **2005**, *53*, 61–99.

95. I. A. Tikhonova, K. I. Tugashov, F. M. Dolgushin, A. A. Yakovenko, B. N. Strunin, P. V. Petrovskii, G. G. Furin, V. B. Shur, *Inorg. Chim. Acta* **2006**, *359*, 2728–2735.

96. T. J. Taylor, C. N. Burress, L. Pandey, F. P. Gabbai, *Dalton Trans.* **2006**, 4654–4656.

97. T. J. Taylor, C. N. Burress, F. Gabbai, *Organometallics* **2007**, *26*, 5252–5263.

98. O. Elbjeirami, C. N. Burress, F. P. Gabbai, M. A. Omary, *J. Phys. Chem. C* **2007**, *111*, 9522–9529.

99. C. N. Burress, F. P. Gabbai, *Heteroatom. Chem.* **2007**, *18*, 195–201.

100. C. N. Burress, M. I. Bodine, O. Elbjeirami, J. H. Reibenspies, M. A. Omary, F. P. Gabbai, *Inorg. Chem.* **2007**, *46*, 1388–1395.

101. A. Burini, J. P. Fackler, Jr., R. Galassi, T. A. Grant, M. A. Omary, M. A. Rawashdeh-Omary, B. R. Pietroni, R. J. Staples, *J. Am. Chem. Soc.* **2000**, *122*, 11264–11265.

102. A. S. Filatov, E. A. Jackson, L. T. Scott, M. A. Petrukhina, *Angew. Chem. Int. Ed.* **2009**, *48*, 8473–8476.

103. A. S. Filatov, A. K. Greene, E. A. Jackson, L. T. Scott, M. A. Petrukhina, *J. Organomet. Chem.* **2011**, *696*, in press. doi:10.1016/j.jorganchem.2011.01.038

104. Y. Sevryugina, A. Y. Rogachev, M. A. Petrukhina, *Inorg. Chem.* **2007**, *46*, 7870–7879.

105. Y. Sevryugina, M. A. Petrukhina, *Eur. J. Inorg. Chem.* **2008**, *2*, 219–229.

106. Y. Sevryugina, E. A. Jackson, L. T. Scott, M. A. Petrukhina, *Inorg. Chim. Acta* **2008**, *361*, 3103–3108.

107. S. N. Spisak, A. S. Filatov, M. A. Petrukhina, *J. Organomet. Chem.* **2011**, *696*, 1228–1231.

108. J. H. Nelson, H. B. Jonassen, *Coord. Chem. Rev.* **1971**, *6*, 26–73.

109. H. Kurosawa, I. Ikeda, *J. Organomet. Chem.* **1992**, *428*, 289–301.

110. M. M. Olmstead, A. S. Ginwalla, B. C. Noll, D. S. Tinti, A. L. Balch, *J. Am. Chem. Soc.* **1996**, *118*, 7737–7745.

111. M. M. Olmstead, P.-P. Wei, A. S. Ginwalla, A. L. Balch, *Inorg. Chem.* **2000**, *39*, 4555–4559.

# CHAPTER 7

# SUMANENES: SYNTHESIS AND COMPLEXATION

TOSHIKAZU HIRAO and TORU AMAYA

## 7-1. INTRODUCTION

Nonplanar polyaromatic carbon molecules including fullerenes and carbon nanotubes have been attracting significant interest for their potential as materials and catalysts.[1] In this context, bowl-shaped polyaromatic hydrocarbons (here, we use the term "π bowls") are now considered as another group of key materials in the science of nonplanar π-conjugated carbon systems.[2] As a polar cap fragment of fullerene $C_{60}$, there are two representative key subunits for π bowls. (1) is a $C_{5v}$ symmetric "corannulene (**2**, $C_{20}H_{10}$)" and (2) a $C_{3v}$ symmetric "sumanene (**1**, $C_{21}H_{12}$)" (Figure 7-1). Most of the attention in π-bowl chemistry has focused on **2** since the first synthesis in 1966.[3] Subsequent development of a practical synthetic method for accessing **2** in 1991 resulted in a breakthrough in this area.[4] On the other hand, most of the research based on sumanene (**1**) began with its first synthesis in 2003,[5] with the exception of the earlier theoretical studies.[6] The characteristic structural feature of **1** depends on three $sp^3$-hybridized carbon atoms at the benzylic positions, which is in sharp contrast to the structure of **2**, the rim of which is covered with five aromatic rings. These benzylic positions of **1** undergo further functionalization via the corresponding radicals, cations, anions, carbenes, and other species. In such context, sumanene is a more attractive key molecule for creating new bowl-shaped compounds. This chapter focuses on the research of sumanene, including its synthesis, structural characterization, derivatization, complexation, and electrical materials application to date. As an aside, *sumanene* was named by Mehta et al. after "suman" which means flower in Sanskrit because the shape of **1**, with the ring edges on the rim resembling petals, is reminiscent of a flower.[7]

*Fragments of Fullerenes and Carbon Nanotubes: Designed Synthesis, Unusual Reactions, and Coordination Chemistry*, First Edition. Edited by Marina A. Petrukhina and Lawrence T. Scott.
© 2012 John Wiley & Sons, Inc. Published 2012 by John Wiley & Sons, Inc.

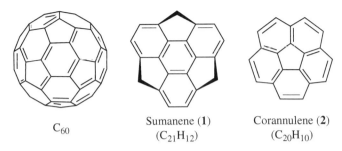

**Figure 7-1.** $C_{60}$, sumanene (**1**, $C_{21}H_{12}$), and corannulene (**2**, $C_{20}H_{10}$).

## 7-2. SYNTHESIS OF SUMANENE

Early attempts for the synthesis of **1** are based on approaches from planar aromatic compounds. For example, tribromomethyltriphenylene **3** undergoes flash vacuum pyrolysis (FVP) to give monocyclized and dicyclized products **4** and **5**, respectively, without formation of desired sumanene [Scheme 7-1(a)].[7] McGlinchey et al. have proposed another approach using trithia compound **6** as a precursor for **1** in a few steps

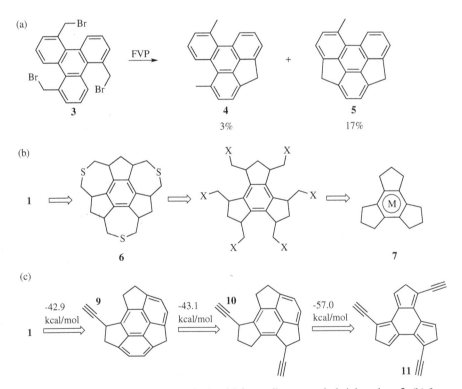

**Scheme 7-1.** (a) Attempt at the synthesis of **1** from tribromomethyltriphenylene **3**; (b) from $\eta^6$-trindane metal complex **7**; (c) by using theoretical calculation.

from $\eta^6$-trindane metal complex **7** as a starting material [Scheme 7-1(b)].[8] A theoretical calculation-based strategy is also proposed as shown in Scheme 7-1 (c).[9] Release of the strained energy is designed in all steps for cyclization, but there is no report proving this approach experimentally.

These results suggest that a route from planar compounds is unlikely, due to the strain energy. Our strategy lies in constructing a three-dimensional framework based mostly on tetrahedral $sp^3$ carbons, which leads to the required $\pi$-conjugated structure by oxidative aromatization.[5] The related strategy to corannulene has been reported by Barth and Lawton in 1966.[3] In the synthetic route shown in Scheme 7-2, the key intermediate is hexahydrosumanene **10**, which is prepared from the isomer **9** by transferring the alkene bridges. This isomeric transformation from *syn*-**9** to **10** is estimated to be 51.4 kcal/mol exothermic by density functional theory (DFT) calculation [B3LYP/6-31G(d)*], while that from *anti* isomer of **9** is calculated to be 37.4 kcal/mol endothermic.

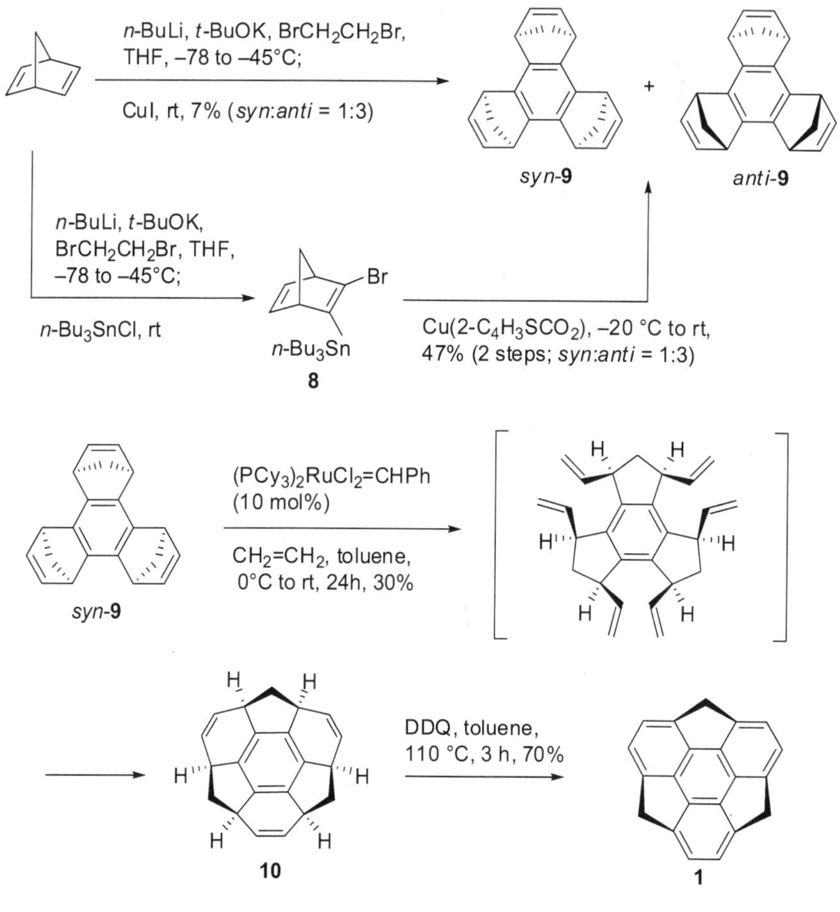

**Scheme 7-2.**    Synthetic scheme of sumanene (**1**).

Trimerization of norbornadiene by a modified procedure based on the organo-copper-mediated cyclization yields *syn*- and *anti*-benzotris(norbornadiene)s **9**.[10] Alternatively, stepwise transmetallation via an organotin compound increases the yield.[11] Alkene bridge exchange of *syn*-**9** gives **10** in 30% yield via the Ru-catalyzed tandem ring-opening metathesis (ROM) and ring-closing metathesis (RCM) reaction under an atmospheric pressure of ethylene. In contrast, the tandem ROM-RCM reaction of *anti*-**9** does not give **10**. Finally, DDQ oxidation of **10** affords sumanene (**1**) in 70% yield (Scheme 7-2).[5]

## 7-3. STRUCTURE OF SUMANENE

The molecular structure of **1** by X-ray crystal structure analysis is illustrated in the ORTEP diagrams in Figure 7-2.[12] Bond alternation is observed in the hub six-membered ring to give the bond lengths of r1a and r1b as 1.381 and 1.431 Å, respectively. The magnitude of this bond alternation is slightly smaller than that in triphenylene (1.41 and 1.47 Å),[13] or that in $C_{60}$ (1.398 and 1.455 Å).[14] In contrast, the bond lengths in the flank six-membered ring (r1a, r2, and r3a) are almost identical (1.38 to 1.40 Å) except for a significantly elongated rim bond (r4: 1.43 Å). As shown in the side view [Figure 7-2(b)], sumanene possesses a bowl-shaped structure. Haddon's π-orbital axis vector (POAV) analysis[15] reveals that the six hub carbons are pyramidalized to an extent of 8.8°. By comparison, POAV angles of all the carbons of $C_{60}$ and the five carbons at the hub of corannulene **2** are 11.6° and 8.2°, respectively.[16] On the other hand, the bowl depth, defined as the distance between a plane of a hub benzene ring and an aromatic carbon rim, is observed as 1.11 Å, which is deeper than that of corannulene (0.87 Å).[17] Bowl-shaped π aromatics sometimes favor a stacking structure in a concave–convex fashion.[2a,18] Such an arrangement is also observed in the intermolecular packing structure of **1** [Figure 7-2(d)]. The stacking distance is approximately 3.86 Å, which is relatively larger than that of ordinary π stacking. Columnar stacking of **1** occurs along a crystallographic threefold axis [Figure 7-2(c)]. Such a stacking arrangement is also observed with hemifullerene ($C_{30}H_{12}$)[18d] and circumtrindene ($C_{36}H_{12}$),[18b] in sharp contrast with corannulenes.[17] Each layer of the column is stapled in a staggered fashion, which might be due to a repulsive effect of the three methylene units and/or maximization of the HOMO-LUMO overlap. Every column is oriented in the same direction,

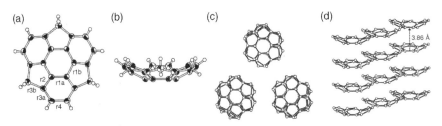

**Figure 7-2.**    ORTEP drawings and packing structure of **1**: (a,c) top views and (b, d) side views.

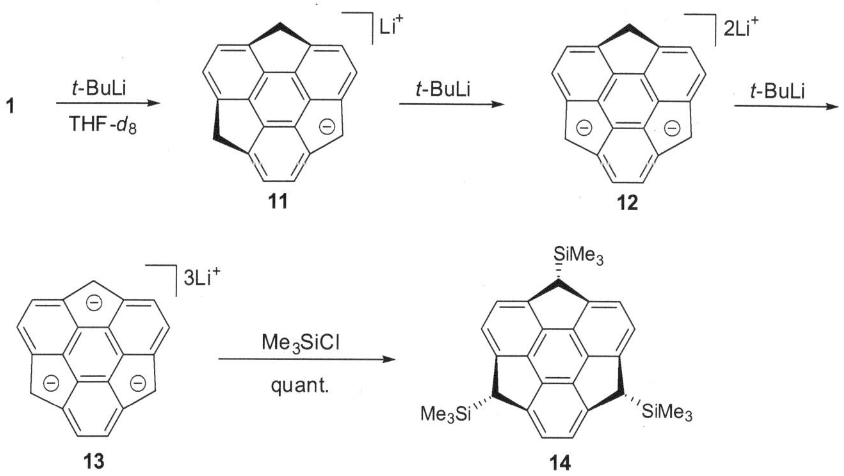

**Scheme 7-3.** Generation of mono-, di-, and tri-anions **11–13** and synthesis of tris(trimethylsilyl)sumanene **14**.

giving polar crystals [Figure 7-2(d)]. Such bowl-in-bowl stacks with polarity are also found in crystals of circumtrindene[18b] and hemifullerene.[18d]

## 7-4. STEPWISE ANION GENERATION AT BENZYLIC POSITIONS OF SUMANENE

As described in Section 7-1, sumanene is characterized by its benzylic positions. In this regard, the generation of the anionic species can be monitored by an NMR tube reaction in THF-$d_8$ (Scheme 7-3).[12] Benzylic protons are kinetically deprotonated with the treatment of $t$-BuLi so that mono-, di-, and tri-anions are selectively generated by careful control of the amount of $t$-BuLi. Difference in chemical shifts between the *exo/endo* benzylic protons increases in the order of the number of anions (dianion **12** > monoanion **11** > **1**). The reaction of trianion **13** with an electrophile is demonstrated by trapping with Me₃SiCl, in which *exo*-introduced tris(trimethylsilyl) derivative **14** is obtained with perfect selectivity, probably due to the steric demand.

## 7-5. BOWL-TO-BOWL INVERSION OF SUMANENE

Bowl-to-bowl inversion is one of the characteristic behaviors for some flexible π bowls. This intriguing dynamics has been exclusively studied with corannulene and its derivatives. Scott and coworkers demonstrated for the first time that the bowl-to-bowl inversion of the corannulene derivative **15** occurs rapidly with the activation barrier of 10.2 kcal/mol [Figure 7-3(a)].[19] Further elaboration of the π-bowl structure

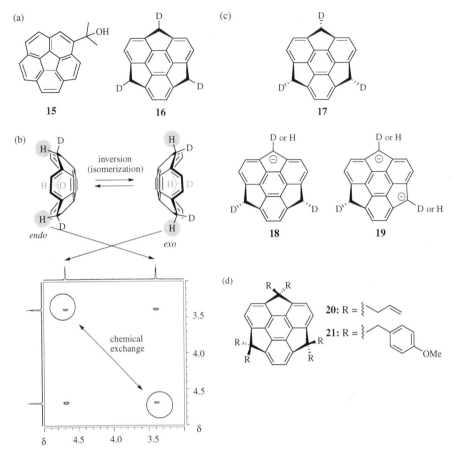

**Figure 7-3.** (a) Corannulene derivative **15** and trideuteriosumanene **16**; (b) schematic representation of the bowl-to-bowl inversion and a selected region of the EXSY spectrum of **16** in CDCl$_3$; (c) *exo*-trideuteriosumanene **17**, monoanion **18**, and dianion **19**; (d) hexaallylated and hexa-*p*-methoxybenzylated sumanene **20** and **21**.

gives more detailed insight into the inversion dynamics; for example, Sygula and Rabideau found that the introduction of the rigid five-membered ring to the rim of corannulene decelerated the inversion.[20] Siegel and coworkers investigated the relationship between the inversion energy barrier and the bowl depth.[21]

On the other hand, a variable-temperature $^1$H NMR experiment implies that **1** inverts much more slowly than does corannulene.[5] Theoretical study using the DFT method suggests that the inversion barrier of **1** is 16.9 kcal/mol (B3LYP/cc-pVTZ// B3LYP/cc-pVDZ, Gaussian 94),[6b] and more recent calculation shows 19.0 kcal/mol [B3LYP/6-311+G(2d,p)//B3LYP/6-31G(d,p), Gaussian 03].[22] Control of the dynamics is envisioned not only to contribute to the enantioselective synthesis of sumanene-based curved molecules but also to afford a new building block for molecular machines, molecular devices, and molecular switches.

Trideuteriosumanene **16** can be used to measure bowl-to-bowl inversion [Figure 7-3(a)]. In this molecule, the bowl-to-bowl inversion is equivalent to the isomerization between the diastereomers [Figure 7-3(b)]. Accordingly, estimation of the isomerization rate using 2D EXSY NMR experimentation gives the inversion rate. The activation energy ($\Delta G^{\ddagger}$, kcal/mol) is 20.4 kcal/mol in CDCl$_3$,[23] which is approximately twice as high as that of the corannulene derivative **15**. Use of CD$_2$Cl$_2$ and THF-$d_8$ shows similar values. On the other hand, the use of methylated benzene-based solvents (toluene-$d_8$, $p$-xylene-$d_{10}$, and mesitylene-$d_{12}$) inparts a tendency to decrease the energy barrier (19.7–19.9 kcal/mol). The difference between the highest and the lowest ones reaches 0.7 kcal/mol, and the inversion of **16** can be accelerated up to 3 times by simply exchanging the solvent (0.066 s$^{-1}$ for CDCl$_3$ and 0.21 s$^{-1}$ for $p$-xylene-$d_{10}$ at 318K).

Because the inversions for monoanion **18** and dianion **19** [Figure 7-3(c)] are too slow for detection of the crosspeaks in a 2D EXSY experiment, the inversion barriers for them are obtained by following the equilibration of the inversion in the corresponding anion species of *exo*-trideuteriosumanene **17**. The barriers for **18** and **19** are 21.8 and 21.5 kcal/mol in THF-$d_8$, which correspond to 1.5 and 1.2 kcal/mol higher than **16**, respectively.[23]

The inversion barriers of hexaallylated and hexa-$p$-methoxybenzylated sumanene **20** and **21** in CDCl$_3$ are 19.2 and 18.2 kcal/mol, which are 1.2 and 2.2 kcal/mol lower than that of **16**, respectively [Figure 7-3(d)]. In terms of the inversion rate, these values are approximately 7 and 35 times as fast as that of **16** at 318K.[23]

## 7-6. EXTENDED π CONJUGATION OF SUMANENE

Condensation of the benzylic anions of **1** with various aldehydes affords the corresponding π-expanded derivatives when aqueous 30% NaOH solution is used as a base in the presence of tetra-$n$-butylammonium bromide and a minimum amount of THF. Eight derivatives **22a–h** with extended π conjugation have been successfully synthesized in 57% to a quantitative yield (Scheme 7-4).[24] These derivatives are mixtures of $C_3$-symmetric and unsymmetric diastereomers. Among them, the terthiophene derivative **22h** shows a remarkable redshifted absorption with a small bandgap, which is rationalized by DFT calculation.

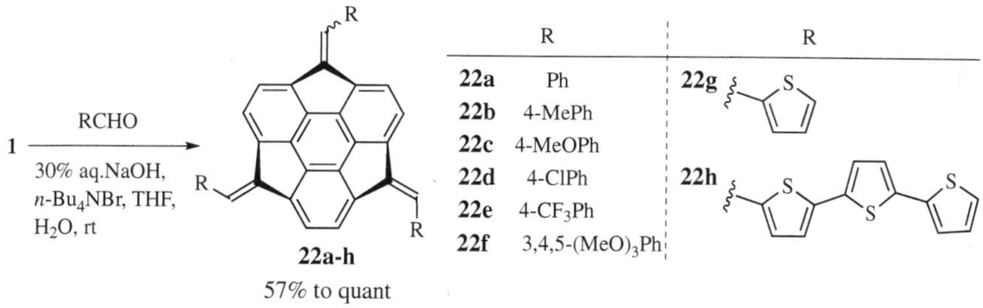

|  | R |  | R |
|---|---|---|---|
| **22a** | Ph | **22g** | (thienyl) |
| **22b** | 4-MePh | | |
| **22c** | 4-MeOPh | **22h** | (terthienyl) |
| **22d** | 4-ClPh | | |
| **22e** | 4-CF$_3$Ph | | |
| **22f** | 3,4,5-(MeO)$_3$Ph | | |

**Scheme 7-4.** Synthesis of π-expanded derivatives **22a–h**.

## 7-7. SYNTHESIS OF NAPHTHOSUMANENES

To synthesize deeper $\pi$ bowls, FVP has proved to be an effective tool for forming curved bonds.[2b] In fact, highly strained $\pi$ bowls, such as hemifullerene $(C_{30}H_{12})$,[25] circumtrindene $(C_{36}H_{12})$,[26] and even $C_{60}$,[27] are synthesized using this methodology. On the other hand, nonpyrolytic synthetic strategy has been required for the elaborate synthesis of this class of molecules because reaction conditions of FVP are severe. Furthermore, a number of employable functional groups is limited. More recently, Scott and coworkers reported the synthesis of the highly strained bowl $C_{50}H_{20}$ from corannulene using intramolecular Mizoroki–Heck reactions as a key step.[28]

A nonpyrolytic synthetic strategy for naphthosumanenes **23**, **24**, and **25** is outlined retrosynthetically in Scheme 7-5, where sumanene is used as a starting material. In this strategy, the key reaction is intramolecular dehydrative benzannulation of the aldehyde via 1,2-addition of the benzylic anions. Such aldehyde arises from the corresponding bromide **26** through the Suzuki–Miyaura coupling reaction with 2-formylphenylboronic acid. Bromide **26** is prepared from **1** under aromatic bromination conditions. According to this strategy, **23**, **24**, and **25** are successfully synthesized in good yields, except that the selectivity for di- and tribromination is very low.[29] Benzylic dideuteration of **23** makes it possible to measure the bowl-to-bowl inversion barrier as described above. Such monitoring experimentation by $^1$H NMR

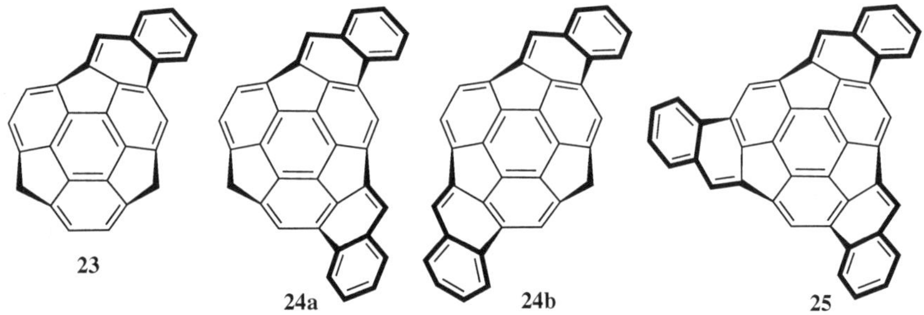

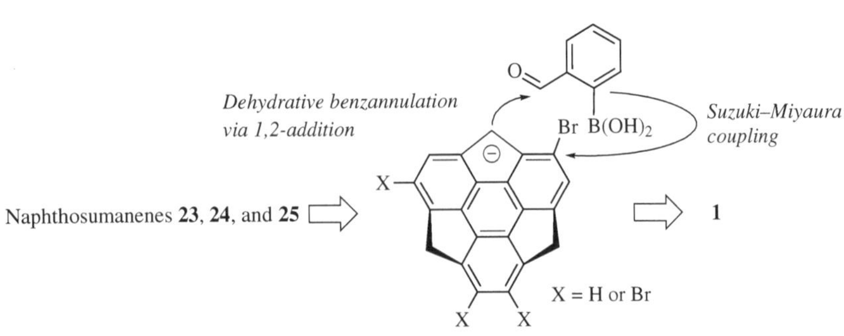

**Scheme 7-5.** Strategy for naphthosumanenes **23**–**25**.

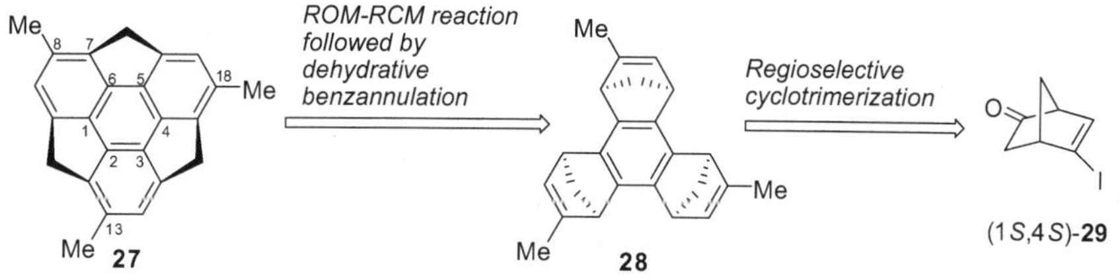

**Scheme 7-6.** Strategy for chiral $C_3$ symmetric **27**.

provides the barrier 32.2 kcal/mol,[29] which is almost 10 kcal/mol higher than that of **1**. This is approximately reproduced by DFT calculation [31.4 kcal/mol, based on B3LYP/6-311+G(2d,p)//B3LYP/6-31G(d,p)].[22,29]

## 7-8. SYNTHESIS OF CHIRAL SUMANENE

Many $\pi$ bowls possess "bowl chirality" derived from their three-dimensional geometry, similar to chiral fullerenes and carbon nanotubes.[30,31] Methods for controlling the bowl chirality can potentially be applied to the related chiral fullerenes and carbon nanotubes as well. In addition, chiral aromatic compounds are expected to contribute to a variety of applications such as asymmetric molecular recognition, homochiral crystal organic materials, and chiral ligands for organometallic catalysis.

A $C_3$-symmetric $(C)$-$(M)$-8,13,18-trimethylsumanene (**27**) is selected as a target compound.[32] The strategy related to the synthesis of sumanene relies on conversion from an $sp^3$ chirality to the bowl chirality (Scheme 7-6). The *syn*-benzonorbornene derivative **28** is a key intermediate, which can be prepared in a few steps by regioselective cyclotrimerization[33] of a iodonorbornene derivative [(1S,4S)-**29**] with $sp^3$ chirality. According to this strategy, **27** is successfully synthesized from **28** via ROM-RCM reactions followed by dehydrogenative aromatization.[32] The enantiomeric excess of **27** is determined to be 90% enantiomeric excess (ee), based on the diastereomeric analysis by *exo*-selective introduction of (S)-Ph(CF$_3$)(OMe)CCO groups at three benzylic positions before racemization, because this benzylic substitution can prevent bowl-to-bowl inversion.

## 7-9. SYNTHESIS OF HETERASUMANENES

### 7-9-1. Trithiasumanene

Trithiasumanene **31** is synthesized from tris(chlorovinyl)trithiophene **30** by FVP (Scheme 7-7).[18c] It is the first synthesis of a $\pi$ bowl with heteroatoms. X-ray crystallographic analysis of **31** shows a shallow $\pi$ bowl-shaped structure. The molecules in a crystal are stacked in a concave–convex fashion with intermolecular S–S contacts.

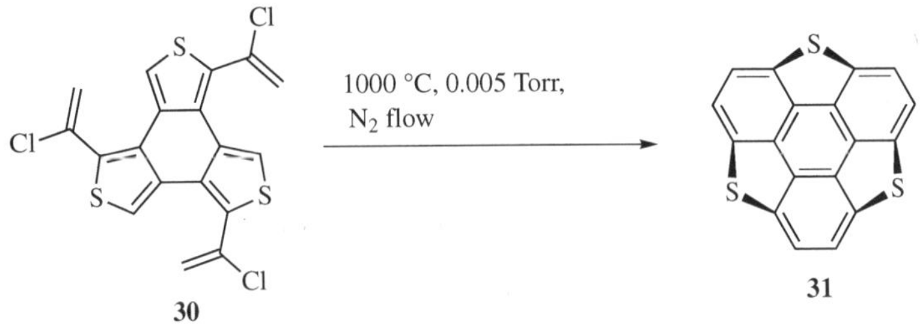

**Scheme 7-7.** Synthesis of trithiasumanene **31**.

## 7-9-2. Trisilasumanene

Trisilasumanene **35** is synthesized by applying the sequential intramolecular sila-Friedel–Crafts reaction as a key step (Scheme 7-8).[34] Tribromide **32** is dilithiated and trapped by $Ph_2SiCl_2$, followed by reduction with $LiAlH_4$ to give the hydrosilane. Intramolecular sila-Friedel–Crafts reaction is performed in the action of $Ph_3CB$ $(C_6F_5)_4$ as a hydride abstraction agent. The thus-obtained **33** is again lithiated and trapped with $Ph_2SiH_2$ to afford the hydrosilane **34**. The further intramolecular sila-Friedel–Crafts reaction yields trisilasumanene **35**. X-ray structural analysis of **35** indicates that the main framework is almost planar. Trisilasumanene **35** shows a blue fluorescence in solution and solid states.

## 7-9-3. Heterasumanenes Having Different Heteroatom Functionalities

Trimethylsilyl groups are first introduced at the two $\alpha$ carbons of the dibenzothio-phene moiety in triphenylenothiophene **36**. Regioselective lithiation of **37** and trapping with $Me_2SiCl_2$ afford silolotriphenylenothiophene **38**. Again, regioselective

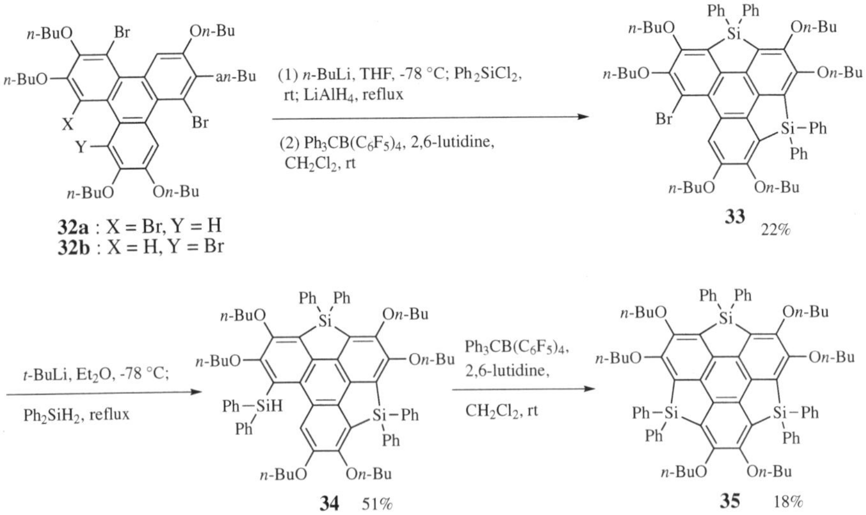

**Scheme 7-8.** Synthesis of trisilasumanene **35**.

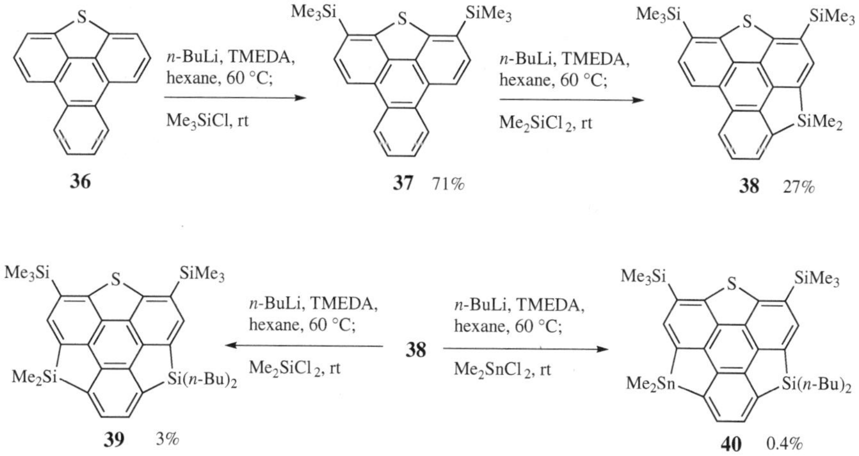

**Scheme 7-9.**  Synthesis of heterasumanenes **39** and **40**.

lithiation of **38** followed by trapping with $Me_2SiCl_2$ affords heterasumanene **39** with a sulfur and two silicones in the sumanene skeleton. Using $Me_2SnCl_2$ instead of $Me_2SiCl_2$ gives heterasumanene **40** with a sulfur, a silicon, and a stannane in the sumanene skeleton, which is the first example of a heterasumanene having three different heteroatom functionalities (Scheme 7-9).[35] X-ray structural analysis of **39** shows a shallow bowl structure, where the bowl depth is 0.23 Å.

## 7-10. COORDINATION STUDY OF SUMANENE

### 7-10-1. Concave-Selective Complexation

"How metals bind to curved carbon π-surfaces" has attracted continuous interest since the discovery of fullerenes and carbon nanotubes. In the coordination chemistry of π bowls, there is an intriguing issue as to the preference for metal binding to the concave surface versus the convex one,[36] which was first addressed by the computational study of hemifullerene ($C_{30}H_{12}$) in 1993.[37] Controlled positioning of metal centers inside the bowls is expected to provide a direct route to the inclusion complexes of fullerenes and nanotubes. On the other hand, the coordination of metal centers to the outside of the bowls should permit applications in surface activation and functionalization of fullerenes and nanotubes. To date, some coordination complexes of π bowls (mainly corannulene) have been prepared and characterized.[18d,38] A convex binding is demonstrated by the X-ray crystal structures in some monometallated corannulene complexes, including $\eta^1$-, $\eta^2$-, and $\eta^6$-coordination modes.[36,38j,l−o] More recent theoretical studies on the complexes of corannulene or its derivatives using DFT have also been carried out to indicate the preferential convex binding to transition metals.[39] Both convex and concave bindings of Ru(II) to corannulene or tetramethylcorannulene with $\eta^6$ coordinations have been reported.[38f,j] Furthermore, tri- and tetra-metallation of corannulenes and hemifullerene ($C_{30}H_{12}$) have been achieved under

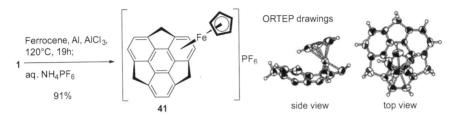

**Scheme 7-10.**    Synthesis of [CpFe(sumanene)]PF$_6$ (**41**) and its ORTEP drawings.

gas-phase deposition reactions, where metal centers are bound in both convex and concave faces with $\eta^2$ binding.[18d,38e,i,p] On the other hand, coordination chemistry of sumanene with transition metals was limited to the computational study[40] prior to the complexation with [cyclopentadienyl (Cp) iron]$^+$ as described below.

The metallation of **1** is performed by ligand exchange of a Cp group of ferrocene in the presence of aluminum powder and aluminum chloride under solvent-free conditions.[41] The counterion is then replaced by hexafluorophosphate to yield [CpFe (sumanene)]PF$_6$ (**41**) in 91% yield as a sole isomer (Scheme 7-10). It is fully characterized by fast-atom bombardment (FAB) mass spectrometry, $^1$H and $^{13}$C NMR spectroscopy, and X-ray crystallography. NMR study shows a coordination at a flank benzene ring. Furthermore, the concave binding of the CpFe$^+$ in solution is revealed by 1D NOE experiment, where bowl-to-bowl inversion is not observed. X-ray crystallographic analysis unambiguously confirms the $\eta^6$ binding of the CpFe$^+$ to a flank benzene ring of the concave face of sumanene in the solid state (Scheme 7-10). Shallowing is slightly induced at the coordinated benzene ring. This selective concave coordination is the first example of a $\pi$-bowl ligand. The redox properties of **41** have been elucidated through an electrochemical study. Complex **41** exhibits Fe(II)/Fe(I) reduction, which displays features of partial chemical reversibility, coupled to decomposition of the corresponding Fe(I) species to the fragments including ferrocene.[42]

### 7-10-2. Chiral Complex

Chiral $\pi$-bowl complexes are desirable as model structures of chiral endohedral metallofullerenes, asymmetric organometallic catalysts, or molecular recognition units. Synthesis of an Fe(II) complex of **1** having the Cp ligand with a chiral *s*-butyl group is performed in a manner similar to that used for the preparation of **41**. Complex **42** also exhibits concave-face selective coordination at a flank benzene ring in solution (Figure 7-4), where rotation of the Cp ring is restricted. Magnetic and optical desymmetrization in the sumanene ligand are shown with complex **42**.[43] This is the first optically active complex with a $\pi$-bowl ligand.

### 7-10-3. Bowl-to-Bowl Inversion of CpRu Sumanene

The selectivity for *endo*-complexation might reflect kinetic and/or thermodynamic control in the complexation, thermodynamic control in "bowl-to-bowl inversion," or both. Corannulene and sumanene ligands themselves are known to exhibit

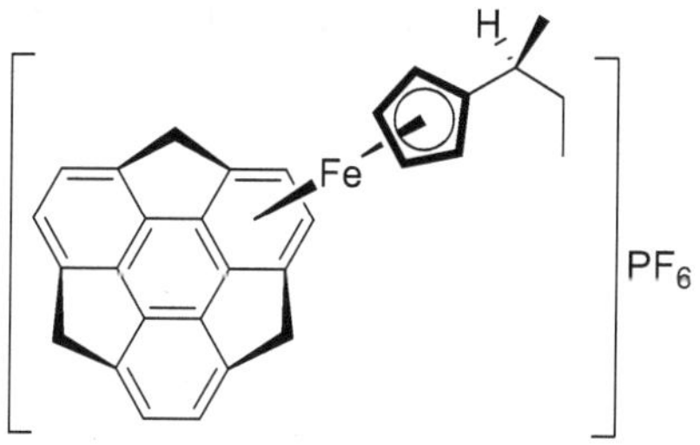

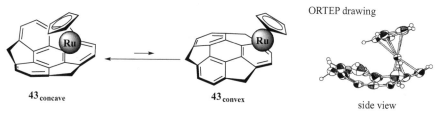

**42**

**Figure 7-4.** Chiral π-bowl complex **42**.

bowl-to-bowl inversion.[19,23] The ring-to-ring migration of $\eta^6$-binding $Cp^*Ir^{2+}$, $Cp^*Ru^+$, $(COE)_2Ir^+$, and $(COE)_2Rh^+$ ($Cp^* = \eta^5\text{-}C_5Me_5$, COE = cyclooctene) is found on the corannulene ligands.[38d,n] However, bowl-to-bowl inversion has not been elucidated with π-bowl transition metal complexes.[44]

The $CpRu^+$ sumanene complex **43** is synthesized similarly.[45] Unlike the scenarios for **41** and **42**, a pair of the two species is observed in the $^1$H NMR spectra of [CpRu (sumanene)]$PF_6$ (**43**). The major and minor species are assigned to the concave-bound complex **43**$_{concave}$ and the convex-bound isomer **43**$_{convex}$, respectively. The ratio of the concave to convex complex depends on solvents to some degree. Furthermore, 2D EXSY experiments reveal that they isomerize through the bowl-to-bowl inversion (Scheme 7-11), where the inversion barrier from **43**$_{convex}$ to **43**$_{concave}$ is 16.7 kcal/mol in $CD_2Cl_2$. Linear van't Hoff plots give a positive $\Delta H$ and a negative $\Delta S$. From a mixture of the diastereomers, the single crystal of **43**$_{concave}$, which is the major species, is obtained. In the X-ray crystallographic analysis, ORTEP diagrams clearly show the concave coordination (Scheme 7-11). Once the crystal of **43**$_{concave}$ is dissolved in solvents, both isomers of **43** appear again. These results indicate that the preference of the concave or convex isomer depends on the solvent and temperature under thermodynamic control.

## 7-11. MATERIALS APPLICATION

Polyaromatic carbon molecules have attracted great interest as electrical materials.[46] In this context, π bowls are also of potential utility as electrically active materials.

ORTEP drawing

**43**$_{concave}$        **43**$_{convex}$      side view

**Scheme 7-11.** Bowl-to-bowl inversion of [CpRu(sumanene)]$PF_6$ (**43**) and ORTEP drawing of concave complex **43**$_{concave}$.

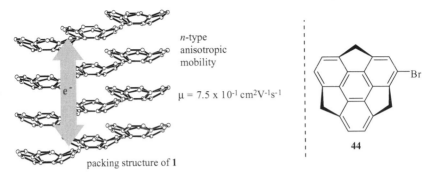

**Figure 7-5.**   Schematic illustration of electron transporting in the crystal of **1** Br-sumanene **44**.

As described above, X-ray crystallographic analysis of **1** shows the 1D columnar π stacking in a concave–convex fashion (Figure 7-2). Each layer is stapled in a staggered fashion probably due to repulsive effect of the three methylene units and/or maximization of the HOMO-LUMO overlap. The charge carrier mobility of a needle-like crystal of **1** is investigated using the time-resolved microwave conductivity (TRMC) method, which can minimize frequently problematic factors in time-of-flight (TOF) and field effect transistor (FET) methods, such as impurity, chemical and physical defects, and organic–electrode interfaces because the nanometer-scale mobility of charge carriers generated by laser pulse irradiation is quantified under oscillating microwave without electrodes.[47] Cocrystal with a small amount (∼2%) of Br-sumanene **44** (Figure 7-5) is employed to determine the charge carrier species because the high electron affinity of the bromo group is known to give effective negative-charge carrier traps. A significant drop in conductivity is found in the cocrystal, supporting the theory that the major charge carrier is electrons in the π-stacking structure. The minimum estimate of the intra-columnar electron mobility of **1** along the stacking axis is calculated to be $7.5 \times 10^{-1} \, \text{cm}^2 \, \text{V}^{-1} \, \text{s}^{-1}$ (Figure 7-5).[48] This value is comparable to that of a single crystal of $C_{60}$ ($0.5 \pm 0.2 \, \text{cm}^2 \, \text{V}^{-1} \, \text{s}^{-1}$ measured by TOF),[49] one of the representative compounds showing *n*-type organic field effect transistor (OFET) properties. Furthermore, the anisotropic difference reaches 9.2 times in the needle-like crystal.

## 7-12. CONCLUSIONS AND OUTLOOK

Synthesis of sumanene (**1**) has opened up an unexplored research area in π-bowl chemistry. Characteristic benzylic positions are able to undergo facile functionalization via the corresponding anions to give the stereoselective substituted compounds, the π-extended derivatives, and the larger π bowls. The dynamically flexible aspect based on bowl-to-bowl inversion is considered to afford a molecular switch or machine. The crystal with columnar bowl-in-bowl stacking exhibits high electron transport ability with anisotropy. The next stage may include the application of the polarity of such crystals. Complexation with $CpFe^+$ has demonstrated selective formation of the first concave-bound complex, which is expected to lead to the inclusion complexes of π bowls. Thus, some characteristic features of sumanenes themselves are becoming apparent. In future, π bowls such as sumanene are expected to provide novel electrical materials, organometallic catalysts, and so on. Research based on sumanene is believed to contribute to the progress in this area.

## REFERENCES

1. (a) L. Dai, A. W. H. Mau, *Adv. Mater.* **2001**, *13*, 899–913; (b) B. C. Thompson, J. M. J. Fréchet, *Angew. Chem.* **2008**, *120*, 62–82; *Angew. Chem. Int. Ed.* **2008**, *47*, 58–77; (c) Q. Cao, J. A. Rogers, *Adv. Mater.* **2009**, *21*, 29–53; (d) K. Woan, G. Pyrgiotakis, W. Sigmund, *Adv. Mater.* **2009**, *21*, 2233–2239.

2. (a) Y.-T. Wu, J. S. Siegel, *Chem. Rev.* **2006**, *106*, 4843–4867; (b)V. M. Tsefrikas, L. T. Scott, *Chem. Rev.* **2006**, *106*, 4868–4884.

3. W. E. Barth, R. G. Lawton, *J. Am. Chem. Soc.* **1966**, *88*, 380–381.

4. L. T. Scott, M. M. Hashemi, D. T. Meyer, H. B. Warren, *J. Am. Chem. Soc.* **1991**, *113*, 7082–7084.

5. H. Sakurai, T. Daiko, T. Hirao, *Science* **2003** *301*, 1878.

6. (a) G. N. Sastry, E. D. Jemmis, G. Mehta, S. R. Shah, *J. Chem. Soc., Perkin Trans. 2* **1993**, 1867–1871; (b) U. D. Priyakumar, G. N. Sastry, *J. Phys. Chem. A* **2001**, *105*, 4488–4494.

7. G. Mehta, S. R. Shah, K. Ravikumar, *J. Chem. Soc., Chem. Commun.* **1993**, 1006–1008.

8. H. K. Gupta, P. E. Lock, M. J. McGlinchey, *Organometallics* **1997**, *16*, 3628–3634.

9. U. D. Priyakumar, G. N. Sastry, *Tetrahedron Lett.* **2001**, *42*, 1379–1381.

10. R. Durr, O. De Lucchi, S. Cossu, V. Lucchini, *Chem. Commun.* **1996**, 2447–2448.

11. G. Borsato, O. De Lucchi, F. Fabris, L. Groppo, V. Lucchini, A. Zambon, *J. Org. Chem.* **2002**, *67*, 7894–7897.

12. H. Sakurai, T. Daiko, H. Sakane, T. Amaya, T. Hirao, *J. Am. Chem. Soc.* **2005**, *127*, 11580–11581.

13. K. K. Baldridge, J. S. Siegel, *J. Am. Chem. Soc.* **1992**, *114*, 9583–9587.

14. K. Hedberg, L. Hedberg, D. S. Bethune, C. A. Brown, H. C. Dorn, R. D. Johnson, M. De Vries, *Science* **1991**, *254*, 410–412.

15. R. C. Haddon, *J. Am. Chem. Soc.* **1987**, *109*, 1676–1685.

16. R. C. Haddon, *J. Am. Chem. Soc.* **1990**, *112*, 3385–3389.

17. J. C. Hanson, C. E. Nordman, *Acta Cryst.* **1976**, *B32*, 1147–1153.

18. (a) A. Sygula, H. E. Folsom, R. Sygula, A. H. Abdourazak, Z. Marcinow, F. R. Fronczek, P. W. Rabideau, *J. Chem. Soc., Chem. Commun.* **1994**, 2571–2572; (b) D. M. Forkey, S. Attar, B. C. Noll, R. Koerner, M. M. Olmstead, A. L. Balch, *J. Am. Chem. Soc.* **1997**, *119*, 5766–5767; (c) K. Imamura, K. Takimiya, Y. Aso, T. Otsubo, *Chem. Commun.* **1999**, 1859–1860; (d) M. A. Petrukhina, K. W. Andreini, L. Peng, L. T. Scott, *Angew. Chem.* **2004**, *116*, 5593–5597; *Angew. Chem. Int. Ed.* **2004**, *43*, 5477–5481.

19. L. T. Scott, M. M. Hashemi, M. S. Bratcher, *J. Am. Chem. Soc.* **1992**, *114*, 1920–1921.

20. (a) A. Sygula, A. H. Abdourazak, P. W. Rabideau, *J. Am. Chem. Soc.* **1996**, *118*, 339–343; (b) Z. Marcinow, A. Sygula, A. Ellern, P. W. Rabideau, *Org. Lett.* **2001**, *3*, 3527–3529.

21. T. J. Seiders, K. K. Baldridge, G. H. Grube, J. S. Siegel, *J. Am. Chem. Soc.* **2001**, *123*, 517–525.

22. T. Amaya, H. Sakane, T. Nakata, T. Hirao, *Pure Appl. Chem.* **2010**, *82*, 969–978.

23. T. Amaya, H. Sakane, T. Muneishi, T. Hirao, *Chem. Commun.* **2008**, 765 767.

24. T. Amaya, K. Mori, H.-L. Wu, S. Ishida, J. Nakamura, K. Murata, T. Hirao, *Chem. Commun.* **2007**, 1902–1904.

25. (a) A. H. Abdourazak, Z. Marcinow, A. Sygula, R. Sygula, P. W. Rabideau, *J. Am. Chem. Soc.* **1995**, *117*, 6410–6411; (b) S. Hagen, M. S. Bratcher, M. S. Erickson, G. Zimmermann, L. T. Scott, *Angew. Chem.* **1997**, *109*, 407–409; *Angew. Chem. Int. Ed.* **1997**, *36*, 406–408; (c) G. Mehta, G. Panda, P. V. V. S. Sarma, *Tetrahedron Lett.* **1998**, *39*, 5835–5836.

26. (a) L. T. Scott, M. S. Bratcher, S. Hagen, *J. Am. Chem. Soc.* **1996**, *118*, 8743–8744; (b) R. B. M. Ansems, L. T. Scott, *J. Am. Chem. Soc.* **2000**, *122*, 2719–2724.

27. L. T. Scott, M. M. Boorum, B. J. McMahon, S. Hagen, J. Mack, J. Blank, H. Wegner, A. de Meijere, *Science* **2002**, *295*, 1500–1503.

28. E. A. Jackson, B. D. Steinberg, M. Bancu, A. Wakamiya, L. T. Scott, *J. Am. Chem. Soc.* **2007**, *129*, 484–485.

29. T. Amaya, T. Nakata, T. Hirao, *J. Am. Chem. Soc.* **2009**, *131*, 10810–10811.

30. C. Thilgen, F. Diederich, *Chem. Rev.* **2006**, *106*, 5049–5135.

31. X. Peng, N. Komatsu, S. Bhattacharya, T. Shimawaki, S. Aonuma, T. Kimura, A. Osuka, *Nat. Nanotechnol.* **2007**, *2*, 361–365 and references cited therein.

32. S. Higashibayashi, H. Sakurai, *J. Am. Chem. Soc.* **2008**, *130*, 8592–8593.

33. S. Higashibayashi, H. Sakurai, *Chem. Lett.* **2007**, *36*, 18–19.

34. S. Furukawa, J. Kobayashi, T. Kawashima, *J. Am. Chem. Soc.* **2009**, *131*, 14192–14193.

35. M. Saito, T. Tanikawa, T. Tajima, J. D. Guo, S. Nagase, *Tetrahedron Lett.* **2010**, *51*, 672–675.

36. M. A. Petrukhina, *Angew. Chem.* **2008**, *120*, 1572–1574; *Angew. Chem. Int. Ed.* **2008**, *47*, 1550–1552.

37. R. Faust, K. P. C. Vollhardt, *J. Chem. Soc., Chem. Commun.* **1993**, 1471–1473.

38. (a) T. J. Seiders, K. K. Baldridge, J. M. O'Connor, J. S. Siegel, *J. Am. Chem. Soc.* **1997**, *119*, 4781–4782; (b) R. M. Shaltout, R. Sygula, A. Sygula, F. R. Fronczek, G. G. Stanley, P. W. Rabideau, *J. Am. Chem. Soc.* **1998**, *120*, 835–836; (c) D. Caraiman, G. K. Koyanagi, L. T. Scott, D. V. Preda, D. K. Bohme, *J. Am. Chem. Soc.* **2001**, *123*, 8573–8582; (d) C. M. Alvarez, R. J. Angelici, A. Sygula, R. Sygula, P. W. Rabideau, *Organometallics* **2003**, *22*, 624–626; (e) M. A. Petrukhina, K. W. Andreini, J. Mack, L. T. Scott, *Angew. Chem.* **2003**, *115*, 3497–3501; *Angew. Chem. Int. Ed.* **2003**, *42*, 3375–3379; (f) P. A. Vecchi, C. M. Alvarez, A. Ellern, R. J. Angelici, A. Sygula, R. Sygula, P. W. Rabideau, *Angew. Chem.* **2004**, *116*, 4597–4600; *Angew. Chem. Int. Ed.* **2004**, *43*, 4497–4500; (g) E. L. Elliott, G. A. Hernández, A. Linden, J. S. Siegel, *Org. Biomol. Chem.* **2005**, *3*, 407–413; (h) M. A. Petrukhina, K. W. Andreini, V. M. Tsefrikas, L. T. Scott, *Organometallics* **2005**, *24*, 1394–1397; (i) for a perspective account, see M. A. Petrukhina, L. T. Scott, *Dalton. Trans.* **2005**, 2969–2975; (j) P. A. Vecchi, C. M. Alvarez, A. Ellern, R. J. Angelici, A. Sygula, R. Sygula, P. W. Rabideau, *Organometallics* **2005**, *24*, 4543–4552; (k) T. M. Ayers, B. C. Westlake, D. V. Preda, L. T. Scott, M. A. Duncan, *Organometallics* **2005**, *24*, 4573–4578; (l) M. A. Petrukhina, Y. Sevryugina, A. Y. Rogachev, E. A. Jackson, L. T. Scott, *Angew. Chem.* **2006**, *118*,

7366–7368; *Angew. Chem. Int. Ed.* **2006**, *45*, 7208–7210; (m) M. A. Petrukhina, Y. Sevryugina, A. Y. Rogachev, E. A. Jackson, L. T. Scott, *Organometallics* **2006**, *25*, 5492–5495; (n) J. S. Siegel, K. K. Baldridge, A. Linden, R. Dorta, *J. Am. Chem. Soc.* **2006**, *128*, 10644–10645; (o) B. Zhu, A. Ellern, A. Sygula, R. Sygula, R. J. Angelici, *Organometallics* **2007**, *26*, 1721–1728; (p) M. A. Petrukhina, *Coord. Chem. Rev.* **2007**, *251*, 1690–1698.

39. (a) R. C. Dunbar, *J. Phys. Chem. A* **2002**, *106*, 9809–9819; (b) T. J. Seiders, K. K. Baldridge, J. M. O'Connor, J. S. Siegel, *Chem. Commun.* **2004**, 950–951; (c) A. K. Kandalam, B. K. Rao, P. Jena, *J. Phys. Chem. A* **2005**, *109*, 9220–9225.

40. Y. Kameno, A. Ikeda, Y. Nakao, H. Sato, S. Sakaki, *J. Phys. Chem. A* **2005**, *109*, 8055–8063.

41. T. Amaya, H. Sakane, T. Hirao, *Angew. Chem.* **2007**, *119*, 8528–8531; *Angew. Chem. Int. Ed.* **2007**, *46*, 8376–8379.

42. P. Zanello, S. Fedi, F. F. de Biani, G. Giorgi, T. Amaya, H. Sakane, T. Hirao, *Dalton Trans.* **2009**, 9192–9197.

43. H. Sakane, T. Amaya, T. Moriuchi, T. Hirao, *Angew. Chem.* **2009**, *121*, 1668–1671; *Angew. Chem. Int. Ed.* **2009**, *48*, 1640–1643.

44. The possibility of a very fast bowl-to-bowl inversion of $[(Cp^*Ru)_2(\mu_2\text{-}(\eta^6{:}\eta^6\text{-}C_{20}H_6)]$ $[PF_6]_2$ and $[(Cp^*Ru)_2(\mu_2\text{-}(\eta^6{:}\eta^6\text{-}C_{20}H_6Me_4)][PF_6]_2$ has been suggested; see Ref. 38j.

45. T. Amaya, W.-Z. Wang, H. Sakane, T. Moriuchi, T. Hirao, *Angew. Chem.* **2010**, *122*, 413–416; *Angew. Chem. Int. Ed.* **2010**, *49*, 403–406.

46. J. E. Anthony, *Angew. Chem.* **2008**, *120*, 460–492; *Angew. Chem. Int. Ed.* **2008**, *47*, 452–483.

47. (a) S. Seki, Y. Yoshida, S. Tagawa, K. Asai, K. Ishigure, K. Furukawa, M. Fujiki, N. Matsumoto, *Philos. Mag. B* **1999**, *79*, 1631–1645; (b) F. C. Grozema, L. D. A. Siebbeles, J. M. Warman, S. Seki, S. Tagawa, U. Scherf, *Adv. Mater.* **2002**, *14*, 228–231; (c) A. Saeki, S. Seki, T. Sunagawa, K. Ushida, S. Tagawa, *Philos. Mag.* **2006**, *86*, 1261–1276; (d) Y. Yamamoto, T. Fukushima, W. Jin, A. Kosaka, T. Hara, T. Nakamura, A. Saeki, S. Seki, S. Tagawa, T. Aida, *Adv. Mater.* **2006**, *18*, 1297–1300; (e) A. Saeki, S. Seki, T. Takenobu, Y. Iwasa, S. Tagawa, *Adv. Mater.* **2008**, *20*, 920–923.

48. T. Amaya, S. Seki, T. Moriuchi, K. Nakamoto, T. Nakata, H. Sakane, A. Saeki, S. Tagawa, T. Hirao, *J. Am. Chem. Soc.* **2009**, *131*, 408–409.

49. E. Frankevich, Y. Maruyama, H. Ogata, *Chem. Phys. Lett.* **1993**, *214*, 39–44.

# CHAPTER 8

# σ-BONDED TRANSITION METAL COMPLEXES OF POLYCYCLIC AROMATIC CARBON COMPOUNDS

PAUL R. SHARP

## 8-1. INTRODUCTION

In the following sections, the broad view is taken that a fullerene or carbon nanotube fragment is any polycyclic aromatic carbon (PAC) compound whose skeleton can be derived from a fullerene or carbon nanotube (Figure 8-1). Benzene does not qualify, but naphthalene, anthracene, phenanthrene, pyrene, perylene, and larger six-membered ring systems do. The nonfused polycyclics biphenyl and pentafulvalene are also included when they are part of a metallacycle complex. Five-membered rings are of special interest as these are critical in the curvature of fullerenes, but only systems that contain all $sp^2$-hybridized carbon atoms and that are neutral will be considered. Thus, indene does not qualify, while acenaphthylene does. It is the σ-bonded transition metal organometallic chemistry of these fragments that holds our interest.

One reason for studying these complexes is the vast differences in the chemistry of the transition metal–carbon bond depending on the type of carbon group bonded to the metal center. Even for relatively simple σ-bonded $sp^3$-hybridized carbon groups there can be large differences. Methyl, ethyl, $t$-butyl, and neopentyl complexes can show differing chemistry ranging from α-, β-, or γ-hydride elimination to reductive coupling. Steric, electronic, and structural features of the alkyl group determine the chemistry on a particular metal center. The discovery, investigation, and understanding of this chemistry has led to amazing advances in areas ranging from catalysis to electronics.[1–3]

*Fragments of Fullerenes and Carbon Nanotubes: Designed Synthesis, Unusual Reactions, and Coordination Chemistry*, First Edition. Edited by Marina A. Petrukhina and Lawrence T. Scott.
© 2012 John Wiley & Sons, Inc. Published 2012 by John Wiley & Sons, Inc.

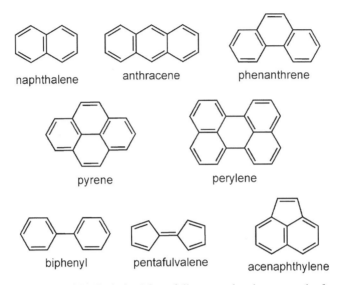

**Figure 8-1.** Selected PACs derived from fullerene and carbon nanotube fragments.

In a similar way, a phenyl group and its extended analogs—naphthyl, anthracenyl, and other PACs σ-bonded to metal centers—should display different chemical reactivity depending on the steric, electronic, and structural features of the group. However, while there have been extensive studies of σ-bonded phenyl complexes, there have been only sporadic (but growing) reports of analogous PAC complexes and little investigation of their properties. This is remarkable considering the importance of PACs.

Relatively recent additions to the PAC family are graphene[4] and fullerenes, including carbon nanotubes.[5-10] Metals are intimately involved in their chemistry and potential technology. Metal atoms or particles catalyze nanotube growth, and the metal usually remains bonded to the edge of the tube probably through σ- and π-M−C bonds.[11-13] The metal is usually dissolved off the end of the tube, but with a good understanding of metal-PAC chemistry it should be possible to utilize the reactivity of the metal−PAC bond in a more productive fashion. Electronic device fabrication requires good electrical connections between a metal contact and a nanotube or a graphene sheet.[14] *Good electrical contact* very likely means strong bonding interactions between the metal and the nanotube or graphene sheet.

Metal−PAC σ-bond chemistry is a rich area as shown by reported reactions distinctive to PAC σ-bonded complexes, including reversible C−C bond formation,[15] metal migrations along the edge of the PAC,[16,17] PAC dimerizations,[18-20] alkyne cycloadditions,[16,18-20] colloidal metal-catalyzed organometallic reactions,[19,20] a chiral PAC synthesis,[21] photochemical halide eliminations,[22] metal-directed PAC brominations,[23] and metal-stabilized PAC radicals.[21] Research in this area, from our laboratories and others, is described in the following sections. An earlier review of our work in this area has appeared previously.[24]

## 8-2. METAL–PAC COMPLEX SYNTHESIS

We begin by examining the concern of synthetic chemists: synthesis. For this purpose, metal σ-bonded metal–PAC complexes are divided into two categories: $\eta^1$ complexes (PAC-yl complexes) and $\eta^2$ complexes (PAC metallacycles). The $\eta^1$ complexes contain a single M–C σ bond, while $\eta^2$ complexes contain two M–C σ bonds to a single PAC. Cluster complexes containing three or more metal atoms bonded to each other and to the PAC are considered separately.

### 8-2-1. $\eta^1$ Complexes

*$\eta^1$ Complexes by Metallation (C–H Activation).* The preparation of compounds by transition metal C–H activation of hydrocarbons has been, and continues to be, an active area of research in both organic and organometallic chemistry.[25–41] Aromatic hydrocarbons are often encountered in C–H activation chemistry, and a number of PAC-yl complexes have been obtained from simple polycyclic aromatic hydrocarbons (PAHs). The common use of PAH radical anions as strong soluble reducing agents resulted in the early appearance of PAH C–H activation in C–H activation research. Metal complex reduction with the radical anions produced highly reactive unsaturated species that inserted into the C–H bond of naphthalene, anthracene, and phenanthrene.[42–45]

An important issue for PAH C–H activation chemistry is regioselectivity. Naphthalene has been the most widely studied, and the metallation site appears to depend on whether C–H activation is under kinetic or thermodynamic control (Figure 8-2). In cases where hydride **2** is in equilibrium with π complex **1** (reversible C–H activation), the 2 position is metallated, suggesting a thermodynamic preference for position 2.[42–44,46–49] In contrast, if there is no equilibrium with the π complex, activation at both positions 1 and 2 is observed (**2** and **3**), suggesting little kinetic differentiation.[44,50] Similar conclusions have been reached in cyclometallation reactions of heteroatom-functionalized naphthalenes and related PACs, although there seems to be a kinetic preference for the 1 position in these reactions (see below).

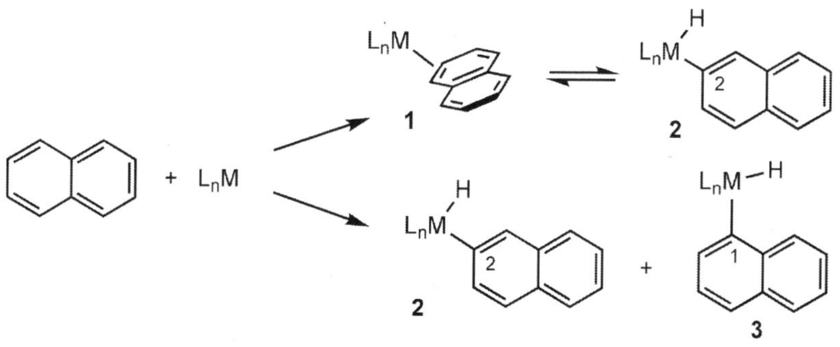

**Figure 8-2.** Naphthalene C–H bond activation scheme.

Equilibria between π complexes and PAC-yl hydrides were investigated for a number of PACs with the Cp*Rh(PMe₃) fragment.[49] Only π complexes are observed for anthracene, phenanthrene, pyrene, and fluoranthene, while equilibrium mixtures are found for naphthalene, perylene, and triphenylene. The different behavior is attributed to changes in the stability of the π complex relative to the hydride complex where bond localization for π-complex formation causes a smaller resonance energy loss for some PACs than for others. The smaller loss of resonance energy gives more stable π complexes and a lower tendency to undergo metal insertion into a C—H bond. These results suggest that low-valence metal insertions into C—H bonds could be problematic for large PAHs where π-complex formation would likely not significantly disrupt the resonance energy.

Other examples of naphthalene C—H activation do not give hydride complexes but rather lose the hydrogen atom as a proton. [Pt(NH₃)(2-naphthyl)Cl₄]²⁻ was obtained along with HCl in the reaction of PtCl₄(NH₃)₂ with naphthalene.[51] Whether this is a thermodynamic product or kinetic product is unknown. Manganese metallation of naphthalene and methoxy naphthalenes by deprotonation with [(tmeda)Na(tmp)(CH₂SiMe₃)Mn(tmp)] (tmp = 2,2,6,6-tetramethylpiperidide) occurs selectively at position 2.[52] This manganese reagent and related reagents[53–55] have shown unusual selectivity in arene metallations. Studies with PAHs other than naphthalene may reveal unusual selectivity for PAHs as well.

### η¹ *Complexes by Cyclometallation.*

If a hydrocarbon is functionalized with a metal-bonding heteroatom group, C—H activation becomes more facile and in a process referred to as *cyclometallation* the heteroatom group coordinates to the metal center and a nearby carbon-bonded hydrogen atom is displaced as a proton by the metal center. A large number of PAC-yl complexes have been obtained in this way, mostly with Pd(II), Pt(II), and Ir(III). Here, only selected examples relevant to PAC chemistry will be discussed. For information on cyclometallation in general the reader is referred to reviews.[25,29,56]

The favorable photochemical properties of Ir(III) and Pt(II) cyclometallated complexes for applications such as OLEDs and other photonic devices have been a more recent stimulus in this area.[57–61] Complexes incorporating various PACs attached to heteroatom groups have been reported. In cases where regioselectivity of the cyclometallation is an issue, factors such as the identity of the metal, the steric and electronic properties of the metal center ligands, and the length of the linker group, as well as the kinetic and thermodynamic reactivity of the PAH itself, might be expected to be involved. In addition, ring formation in these reactions will make reverse reactions less favorable, and equilibria, as observed for PAH's above, will be less likely. Kinetic products should therefore be more readily obtained.

Multiple metallation sites are possible with 1-naphthylene, 9-phenanthrenylene, and 1-pyrene derivatives (Figure 8-3). In these systems the heteroatom group is attached at a *peri*-position and there are two possible metallation sites. One site is the adjacent *peri*-position, and the other is the *vicinal*-position. The cyclometallation ring size differs by one atom, with *peri*-metallation giving the larger ring size. Metallations at both positions have been reported, often with one or the other

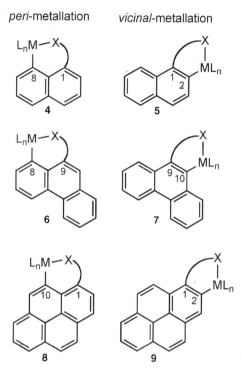

peri-metallation        vicinal-metallation

**Figure 8-3.** Cyclometallation products for 1-functionalized naphthalenes and related PAHs.

product predominating.[62–72] An exception is when the heteroatom is directly bonded to the ring and metallation at the *peri*-position to give a five-membered ring is exclusively observed.[73–75] This is readily rationalized by ring size; metallation at the vicinal position would give a strained four-membered ring.

For one of the naphthalene systems the ratio of the two products was shown to be strongly dependent on reaction conditions.[63] It was concluded that, at least under some conditions, the *peri*-metallation product **4** is kinetically favored, while the alternate product (**5**) is thermodynamically most stable. A related naphthalene system gives only product **5**,[65] whereas the analogous phenanthrene system gives a mixture of the two possible metallation products **6** and **7**.[68] In this case, the difference was attributed to steric destabilization of **7** by the additional benzo ring of the phenanthrene system.

When the heteroatom is linked to naphthalene at the 2 position, the same-size ring is formed by metallation at either position 1 or 3 (Figure 8-4, **10** and **11**). In at least one case, a kinetic preference for 1-metallation (**10**) and a thermodynamic preference for 3-metallation (**11**) is observed. Control of experimental conditions allows selective isolation of either product.[76] Similar control with multiply functionalized naphthalenes selectively gives multiple palladation products.[77] In other cases, only 1-metallation[78] or 3-metallation products are reported, but changes in the reaction

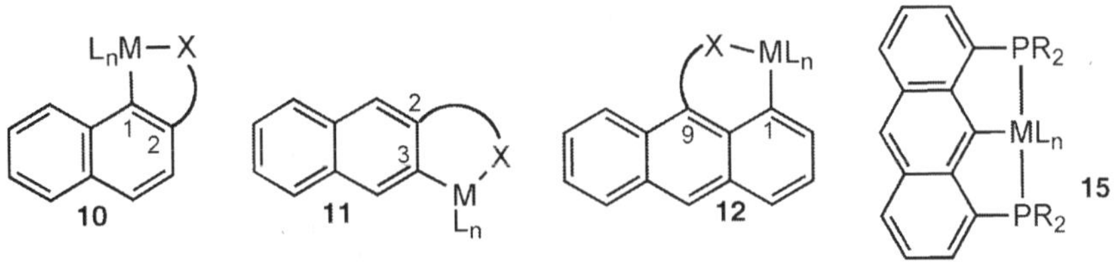

**Figure 8-4.** Metallation products.

conditions could result in the alternative metallation product.[65,79] Blocking the 3 position with a methyl group results in exclusive metallation at the 1 position.[80]

Anthracene with a heteroatom linked to the central 9 position can give only one metallation product (Figure 8-4, **12**). Two metallation sites are possible, but with a symmetric anthracene the resulting products are identical.[68,81] With heteroatoms linked to both the 9 and 10 positions, double metallation can give two isomeric products. This is found for 9,10-bis(diphenylphosphine)anthracene (Figure 8-5), where **13** is the kinetic product, which can be thermally converted to **14**.[82] Double metallation should also be possible with anthracenes substituted at both the 1 and 8 positions. However, what is observed experimentally is single metallation at the central 9 position and formation of "pincer" complexes **15** (Figure 8-4) for Ir(III), Ni (II), Pd(II), and Pt(II).[83–86]

$\eta^1$ **Complexes by Oxidative Addition.** Oxidative addition of carbon–halogen bonds to low-valence metal centers has been effectively applied to the synthesis of PAC-yl complexes. The majority of these complexes are from group 10, where reactive M(0) fragments are readily accessible. Single-metal complexes formed by oxidative addition are known for 1-naphthyl,[87–89] 2-naphthyl,[87] 9-anthracenyl,[23] 9-phenanthryl,[89] 1-acenaphthylenyl,[90] 2-pyrenyl,[91] 9-dibenzanthracene,[17] and 1-corannulenyl (**16**, Figure 8-6).[92] With PACs having more than one carbon–halogen bond, multiple oxidative additions are possible and multimetal complexes have been

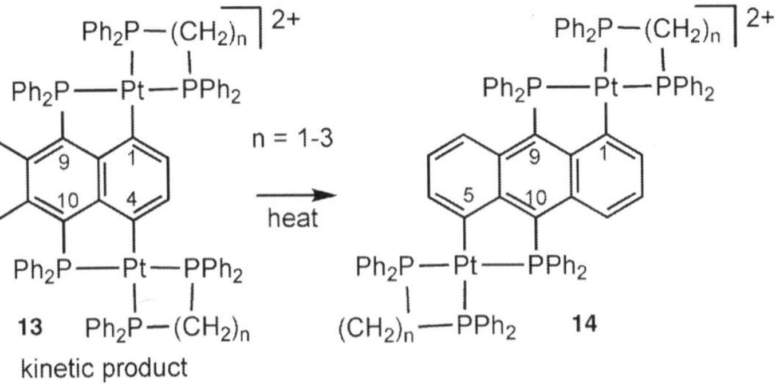

**Figure 8-5.** Isomerization of kinetic product **13**.

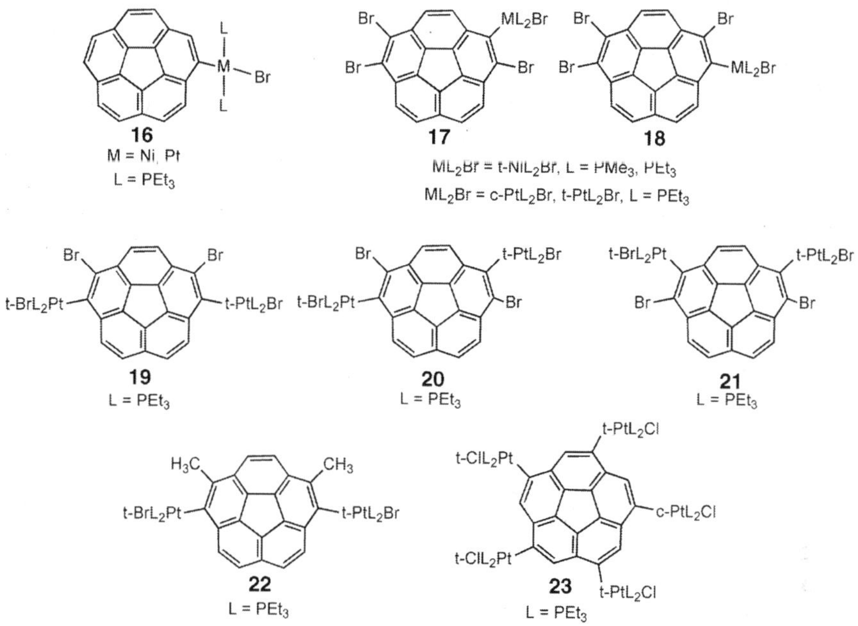

**Figure 8-6.** Corannulene complexes formed by C–X oxidative addition.

reported for the larger systems of anthracene,[93] phenanthrene,[94] pyrene,[91] perylene,[95] hexabenzocoronene,[96] and corannulene (**17–23**, Figure 8-6).[92,97,98]

The curvature of corannulene sets this PAC apart, and some unusual features are observed for its σ-bonded metal complexes. One of these is the inequivalence of the phosphine ligands of edge(rim)-bonded *trans*-M(PEt₃)₂X units. In related planar systems the *trans*-phosphine ligands lie above and below the PAC plane and are equivalent. However, the bowl shape of corannulene results in one phosphine ligand being inside the bowl while the other is outside the bowl (Figure 8-7).[92] Variable-temperature (VT) NMR spectroscopy indicates exchange of the inequivalent phosphine ligands, most likely brought about by bowl inversion, but also possible is rotation about the Pt–C bond, although this appears sterically unlikely. Exchange by phosphine dissociation is not indicated, at least for Pt, as $^{31}P$–$^{195}Pt$ coupling is retained throughout the process.

In **23**, a *cis*-phosphine geometry for one of the five Pt centers persists, even under conditions that would normally give a *trans*-phosphine geometry at all Pt centers.[97,98] This appears to result from a steric limit for space inside the bowl. In the one *cis*-Pt (PEt₃)₂Cl unit, the phosphine ligand *cis* to the Pt-bonded corannulene carbon atom is outside the bowl and the second phosphine ligand is *trans* to the carbon atom and points away from the bowl. The smaller Cl atom points inside the bowl sharing the space with the inward-pointing phosphine ligands of the four *trans*-Pt(PEt₃)₂Cl units. Interestingly, substitution of the chloride ligands in **23** with *p*-nitrophenylacetelyde ligands causes all five Pt centers to go *trans*. To accommodate the additional

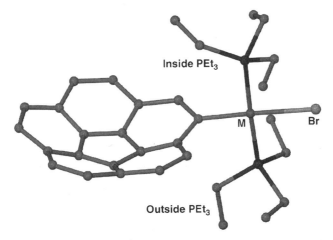

**Figure 8-7.** Structure of *trans*-M(corannulenyl)(PEt₃)₂Br (**16**) illustrating the inside and outside phosphine ligand environments.

phosphine ligand inside the bowl, the corannulene curvature appears to flatten from that in **23**.[97]

Another feature of the corannulene system is observed in the oxidative addition reactions of 1,2,7,8-tetrabromocorannulene.[92] Selective oxidative addition of one C–Br bond, giving **17** and **18**, or two, giving **19**, **20**, and **21**, can be achieved (Figure 8-6). For this to be possible, the second addition must be slower than the first, implying electronic communication across the corannulene ring system. Although it is doubtful that such selectivity is restricted to the corannulene system, it has not been reported for other multiply halogenated PACs. The remaining C–Br bonds in **19–21** are unreactive, probably as a result of steric hindrance from the adjacent Pt centers.

In contrast to pentachlorocorannulene, oxidative addition reactions of 1,3,6-tribromopyrene with three or more Pt(PEt₃)₄ gives only double addition as a mixture of the 1,3 and the 1,6 products.[91] The reluctance of the third C–Br bond to oxidatively add was attributed to sterics. However, the C–Br bond is not adjacent to the Pt center as in the **19–20** reaction, and conditions were milder than those used for penta-chlorocorannulene [room temperature (rt) vs. toluene reflux]. With more forcing conditions it may be possible to prepare the 1,3,6-triplatinated pyrene complex.

**η¹ Complexes by Metathesis.** The reaction of a main group PAC reagent with a transition metal halide complex is probably the most common method for the synthesis of PAC-yl complexes and how the earliest examples of PAC-yl complexes were prepared.[99,100] Li, Mg, Sn, B, and Hg reagents have all been used, Li most frequently. Single metal complexes of 1-naphthyl,[80,99–118] 2-naphthyl,[119–121] 9-anthryl,[99,100,110,112,118,122,123] 9-phenanthryl,[99,100,122] 1-pyrenyl,[91,112,122] and 2-pyrenyl[119,121] have been reported. Multi-Au(I) complexes, of interest for their photophysical properties, have also been reported for 2,6- and 2,7-naphthdiyl,[119,121] 1,8-naphthdiyl,[124] 7,2-pyrendiyl,[119,121] and 1,6-and 1,8- pyrendiyl.[91] Unfortunately,

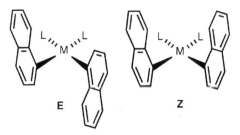

**Figure 8-8.** Atropisomers of a square planar bis(1-naphthyl) complex.

one of the more interesting PACs, corannulene, is reluctant to form a Li or Mg reagent. Attempts to prepare lithiocorannulenyl or the Grignard from bromocorannulene yielded deeply colored solutions (suggesting reduction) from which only corannulene could be isolated.[22] Corannulene deprotonation with BuLi also failed, giving instead BuLi addition across a rim double bond.[125]

An advantage of the metathesis method over others (see text above) is that a variety of metal halide complexes in both high and low oxidation states can be used. For example, Au(I) PAC-yl complexes, some of them analogous to Pt(II) complexes prepared by oxidative addition, have been prepared from Li,[91,112,120,123] Sn,[124] or B[119,121] reagents. Another advantage of the metathesis method is that multiple reactions with a single-metal center are possible giving di-yl complexes[101,105,111,114,122] and even a tetra-yl complex, $[Fe(1\text{-naphthyl})_4]^{-2}$.[103] Very early in the development of this chemistry it was realized that PAC group steric features are important in the stability of the complexes. Five-coordinate $RhL_2(Br)$ (1-naphthyl)$_2$ (L = $PPr^n_3$, $PEt_2Ph$) shows unusually high stability attributed to steric shielding of the metal center by the "*ortho*"-benzo group of the naphthyl ligands.[101]

The steric footprint of PAC-yl ligands can also result in hindered rotation about the M−C bond. In complexes with two -yl ligands that are not symmetric (e.g., 1-naphthyl, 9-phenanthryl, 2-pyrenyl, 5-acenaphthenyl), stable rotational isomers or atropisomers are possible, as illustrated for a square planar 1-naphthyl complex in Figure 8-8.[101,105,111,122,126] Spectroscopic data generally indicate the formation of a single isomer, and crystallographic data show that, in the case of Pt(NBD) (2-ethoxynaphth-1-yl)$_2$, it is the E atropisomer.[126,127] Other examples of steric effects on PAC complexes are discussed in Section 8-3.

## 8-2-2. Other Syntheses

Several other synthetic approaches to PAC-yl complexes have been reported. Decarbonylation of naphthoyl complexes, obtained from naphthoyl chlorides and $NaMn(CO)_5$ or $Na[CpFe(CO)_2]$, yields the corresponding naphthyl complexes.[128,129] Aldehyde H-atom abstraction from, and decarbonylation of, 1-naphthaldehyde is observed in its reaction with $Co(CH_3)(PMe_3)_4$, yielding $Co(1\text{-naphthyl})(CO)(PMe_3)_3$ (Figure 8-9).[130] This complex is also obtained from 1-naphthalenemethanol. Oxidative addition of 1-naphthalenesulfonyl chloride to Vaska's complex, *trans*-IrCl(CO)

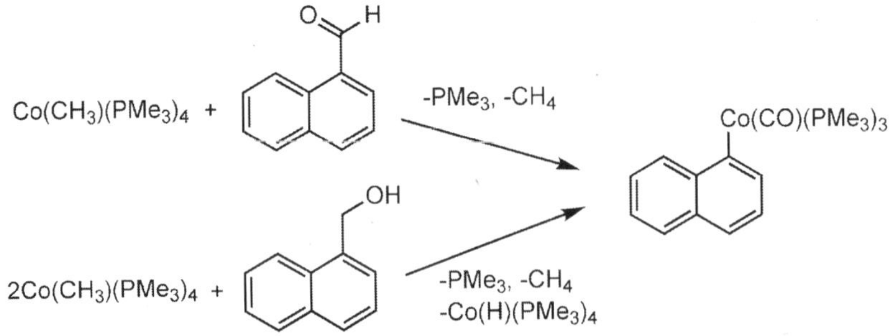

**Figure 8-9.** Formation of a Co 1-naphthyl complex.

(PPh$_3$)$_2$, is reported to give the 1-naphthalenesulfinate complex Ir(1-naphthyl-SO$_2$) Cl$_2$(CO)(PPh$_3$)$_2$, which expels SO$_2$ to give 1-naphthyl complexes.[131] Metal insertions into P−C bonds have also yielded 1-naphthyl complexes.[132,133]

## 8-2-3. Metallacycle Complexes

*Metallacycle complex* here refers to a complex with the metal center bonded to a PAC through two M−C σ bonds. The smallest is a three-membered ring, which in PAC systems is a resonance form of an-yne complex. This is illustrated for 2,3-naphthyne complexes **24** in Figure 8-10. Analogous benzyne complexes are well known,[134] and some examples of PAC-yne complexes have also been reported.[135−137] Larger ring sizes of 4, 5, and 6 are also possible and known. A major challenge in the synthesis of these complexes is the availability of a suitable PAC derivative for reaction with metal complexes. Ideally, two appropriate carbon atoms must have reactive substituents (e.g., halides) that can be used to form the M−C bond by reactions similar to those

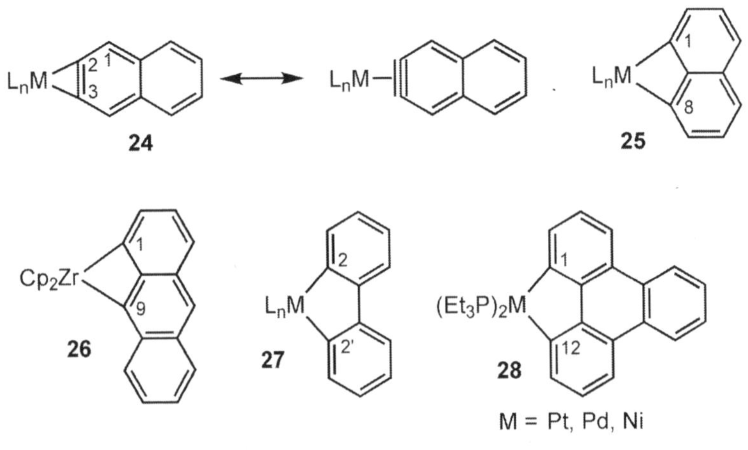

**Figure 8-10.** Selected three-, four-, and five-membered PAC metallacycle complexes.

discussed above for $\eta^1$ complexes. If only one substituted site is available on the PAC, then some type of C−H activation must be employed to form the second M−C bond. Examples of both categories are found below.

***Three-Membered Metallacycles.*** Only Zr and Ni three-membered PAC metallacycles are known (see below), and only the Ni complexes have been isolated and fully characterized. Many more transition metals are known to form benzyne complexes,[134] and the dearth of PAC analogs is most likely due to a lack of attempts to prepare and isolate them.

The three-membered Zr PAC metallacycle is $Cp_2Zr$(1,2-naphthyne) and adducts. The generation of $Cp_2Zr$(1,2-naphthyne) follows a method developed for analogous benzyne complexes.[138,139] Thermolysis of $Cp_2Zr$(Me)(1-naphthyl), prepared from 1-lithionaphthyl and $Cp_2ZrCl$(Me), results in hydrogen abstraction from the naphthyl 2 position and methane elimination. Resulting $Cp_2Zr$(1,2-naphthyne) is unstable under the reaction conditions but can be trapped by $PMe_3$ to give an adduct,[140] by nitriles[135] to give cycloaddition products, or by $B(OEt)_3$ through B−O bond addition across a Zr−C bond.[141]

Nickel 2,3-naphthyne complexes $NiL_2$(2,3-naphthyne) (**24**, $L_2 = 2PEt_3$, dcpe) are prepared by Na/Hg reduction of $NiBrL_2$(3-bromonaphth-2-yl).[136] Again, this is a method used to prepare analogous benzyne complexes.[142,143] Attempts to extend this method to Ni and Pt acenaphthylene complexes have been only partially successful. Treatment of $NiBr(PEt_3)_2$(1-bromoacenaphthylen-2-yl) with Na/Hg gives a product that appears to be $NiBr(PEt_3)_2$(1,2-acenaphthylyne), but workup in the presence of the amalgam gives the Hg complexes **29** (M = Ni, Figure 8-11). The analogous Pt complex **29** (M = Pt) is prepared similarly.[90]

Coordinative trapping of free-ynes is another method for the synthesis of yne complexes. Ni(dcpe)(9,10-phenanthryne) [dcpe = bis(dicyclohexylphosphino)ethane] was prepared by ethylene displacement from Ni(dcpe)($C_2H_4$) with phenanthryne generated *in situ* by dehydrohalogenation of 9-bromophenanthrene.[137] However, this complex could not be isolated in pure form. Similar trapping of free -ynes is probably involved in Pd-catalyzed cyclotrimerization of -yne precursors and alkynes.[144]

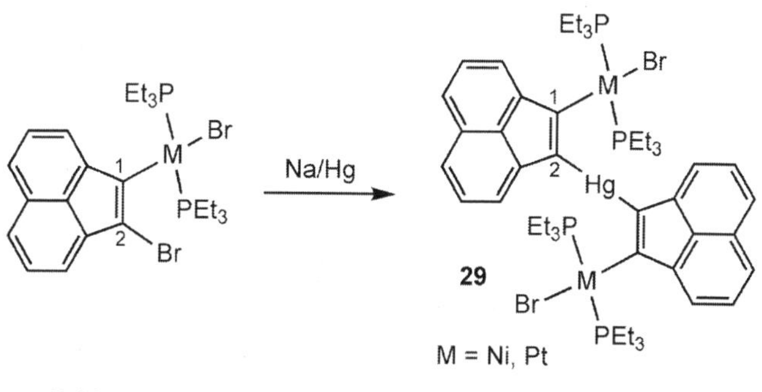

**Figure 8-11.** Attempted preparation of $NiBr(PEt_3)_2$(1,2-acenaphthylyne).

***Four-Membered Metallacycles.*** Owing to the ready availability of the 1,8-naphthylendiyl dianion in the form of Li[145] and Mg[146] reagents, a fair number of four-membered metallacycles based on naphthalene have been reported. Simple treatment of metal dihalide complexes with the dianion gives metallacycles **25** [Figure 8-10, $ML_n = TiCp_2$, $ZrCp^*_2$, $RhCp^*(PPh_3)$, $IrCp^*(PPh_3)$, $Ni(PEt_3)_2$, Pd $(PEt_3)_2$, $Pt(PEt_3)_2$].[19,20,147] For some metal dihalide complexes this reaction fails and the naphthalene dimer, perylene, is observed.[148] Thermolysis of a few of the isolated metallacycles also yields perylene (see text below), suggesting that metallacycles may form but decompose readily by coupling of the naphthalene fragments.

An alternative preparation of platinacycle **25** [$ML_n = Pt(PEt_3)_2$] is the reduction of $Pt(PEt_3)_2(Br)$(8-bromonaphth-1-yl).[19] Attempts to prepare an analogous Ni complex by this pathway yielded perylene, but trapping experiments with alkynes suggest the transient existence of nickelacycle **25**.[18] Decomposition of a digold 1,8-naphthylendiyl is proposed to give **25** [$ML_n = Au(dppe)^+$].[124] Finally, decomposition of $Cp_2Zr(Me)$(9-anthryl) at 25°C results in methane elimination and formation of four-membered zirconacycle **26** (Figure 8-10). This contrasts with the 1-naphthyl analog, $Cp_2Zr(Me)$(1-naphthyl), where hydrogen abstraction occurred at a 2 position yielding a three-membered zirconacycle (see discussion above).

***Five- and Six-Membered Metallacycles.*** Deprotonation of biphenyls[149] and triphenylene[150] with *n*-BuLi occurs selectively in the bay region, yielding dilithio reagents that are ideal for the preparation of five-membered metallacycle complexes. Treatment of metal dihalide complexes with 2,2'-dilithiobiphenyldiyl and 1,12-dilithiotriphenylendiyl yields metallacycle complexes **27**[151–158] and **28**[15,19,20] (Figure 8-10). Metallacycles (**27**) have also been prepared from Sn reagents[159–161] by decarbonylation of fluorenone,[162] dechalcogenation of dibenzochalcophenes,[163–165] double C−H activation of biphenyl,[166] benzyne coupling,[137,167] and insertion of low-valence metal fragments into a strained C−C bond of biphenylene.[168–174] Consecutive insertions into two biphenylene molecules by the $Pt(PEt_3)_2$ fragment gives spirocyclic $Pt(PEt_3)_2$(diphenyldiyl)$_2$.[175] More facile metal center insertions into biphenylene occur on coordination of $Mn(CO)_3^+$ to one of the six-membered rings.[176] Insertion chemistry has been successfully utilized in catalytic conversions of biphenylene.[170,174,175,177–181]

Finally, in the five-membered metallacycles a series of unusual pentafulvalene-based cobaltacycles have been prepared from **30** (Figure 8-12).[182] The pentafulvalendiyl complexes **34** (L = py, Me⁻) are stable only at subambient temperatures. These fullerene fragment complexes are structural isomers of 1,8-naphthalendiyl complexes (**25**).

There are only two reported six-membered PAC metallacycle complexes (Figure 8-13), and both are prepared by low-valence metal fragment insertion into a C−C bond of a five-membered PAC ring. Cobalt complex **35** is obtained from the reaction of a $C_{60}$ derivative with $CpCo(CO)_2$.[183] Similarly, **36** is isolated from the reaction of $Pt(PPh_3)_4$ with the corresponding five-membered ring compound.[184]

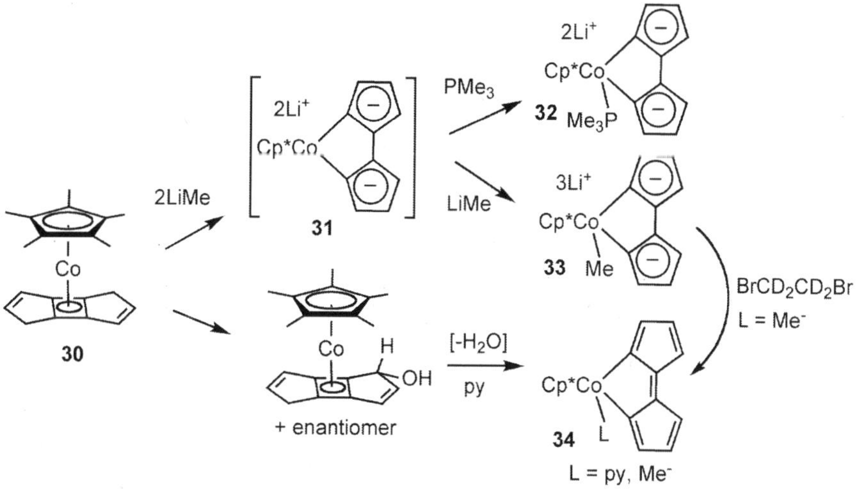

Figure 8-12. Synthesis of pentafulvalendiyl cobaltacycles **34**.

## 8-2-4. Cluster Complexes

Isolated Os and Ru cluster complexes containing σ-bonded PAC fragments give additional insight into the types of interactions that may exist when PACs interact with metal surfaces. The majority of the complexes are formed from heteroatom substituted PACs where the heteroatom presumably initially coordinates to the cluster complex [$Ru_3(CO)_{12}$, $Os_3(CO)_{12}$, or $Os_3(MeCN)_2(CO)_{10}$], and this is followed by C–H activation chemistry in a cyclometallation type reaction.[185–191] In a few instances the carbon–heteroatom bond is also cleaved, giving cluster complexes with heteroatom-free PAC fragments. The loss and addition of metal atoms can also be observed. Only in the case of acenaphthylene have complexes been prepared from the reaction of a heteroatom-free PAH (Figure 8-14).[192] Isolated complexes

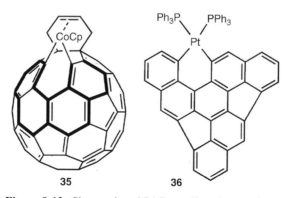

Figure 8-13. Six-membered PAC metallacycle complexes.

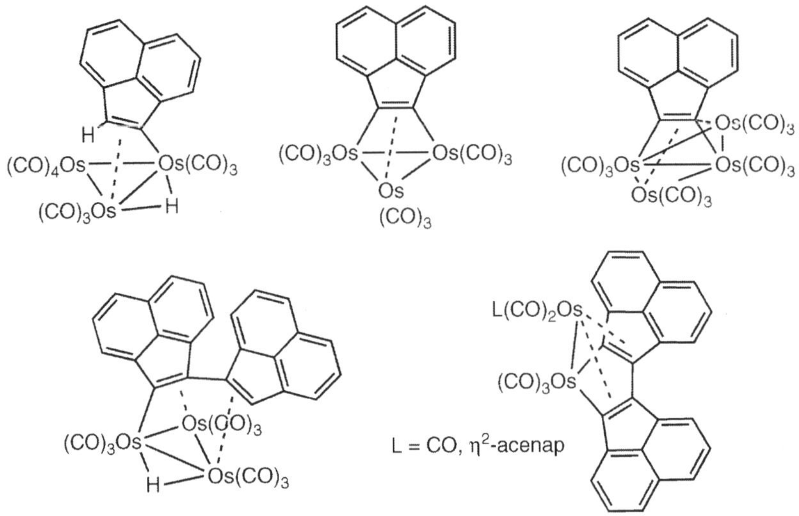

**Figure 8-14.** Cluster products from the reaction of acenaphthylene with [$Os_3(CO)_{10}(CH_3CN)_2$].

include -yl, -yne, and metallacycle PAC cluster complexes usually with additional PAC π interactions with the cluster metal atoms.

## 8-3. METAL–PAC REACTIONS

### 8-3-1. Metal Migrations

In reactions that can be considered analogous to metal center migrations along alkyl chains (Figure 8-15), metal centers in σ-bonded PAC metal complexes have been observed to migrate along the edge of the PAC.

Migrations have been implicated in metal-catalyzed organic syntheses with PACs but the mechanistic pathways can be controversial and may lack direct evidence.[193–198] We prepared and studied a system where a 1,4 migration or 1,4 shift[39] of the metal center along the edge of the PAC is directly observable [Figure 8-16(a)].[17,199]

The migration occurs on thermolysis of **37** to **38** (L = PEt$_3$; M = Pt, X = Cl, Br, I, OTf; M = Pd, X = Br). Solvent studies and D$_2$O addition indicate that the migration rate increases in more polar solvents and that acidic species are not involved (no

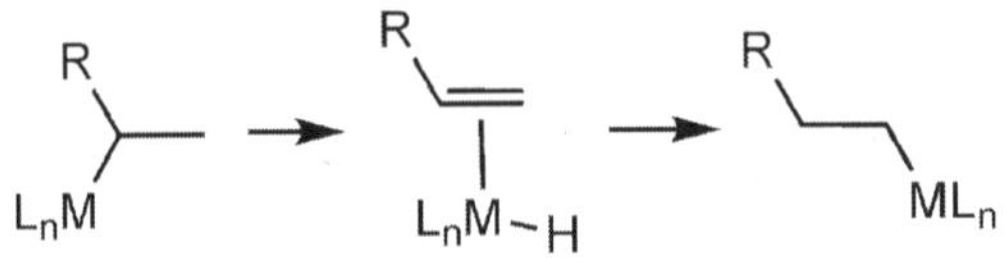

**Figure 8-15.** Metal migration along a C–C chain.

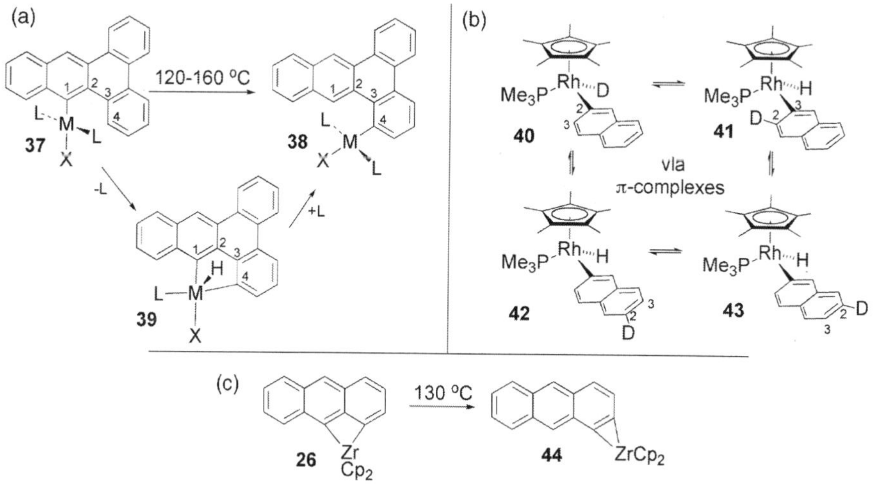

**Figure 8-16.** Metal migrations along the edge of PACs.

deuterium exchange with $D_2O$). The effect of the anionic group X on the migration process for M = Pt shows an increasing rate for the first-order reaction in the order Cl < Br < I < OTf. DFT calculations[140] on Pt model complexes support the intermediacy of the hydride **39**, while previous calculations on Pd model complexes suggest a transition state with a similar structure.[197] An analog of hydride complex **39** is known.[175] DFT studies suggest that similar metal migrations can also occur for other ring configurations, if the hydrogen atom at the destination position is accessible to the metal center.[197]

Another type of migration involves π complexes of naphthalene [Figure 8-16(b)].[48] Reversible insertion of Rh into the C–H bonds from π complexes (Figure 8-2) with migration of the Rh center across the face of the π-bonded naphthalene equilibrates complexes **40–43**. The naphthalene remains bonded to the Rh center throughout the migration process.

Metal migration can also occur in metallacycle complexes [Figure 8-16(c)].[16] Thermolysis of four-membered metallacycle **26** yields-yne complex **44**, which is trapped by $PMe_3$ or PhCCPh. The mechanistic details of this migration are unknown, but it is unimolecular and does not incorporate deuterium from the $C_6D_6$ solvent.

## 8-3-2. Metallacycle Reductive Elimination and Cycloadditions

Unimolecular reductive elimination reactions are not likely for four- and five-membered PAC metallacycles, as formation of strained three- or four-membered rings is unfavorable. In fact, the reverse (C–C oxidative addition) is observed in the synthesis of five-membered metallacycles by metal fragment insertion into a C–C bond of biphenylene (see discussion above). Even for six-membered metallacycles, the two known examples were prepared by metal fragment insertion into five-membered PAC rings.[183,184] However, the Pd-catalyzed synthesis of five-membered

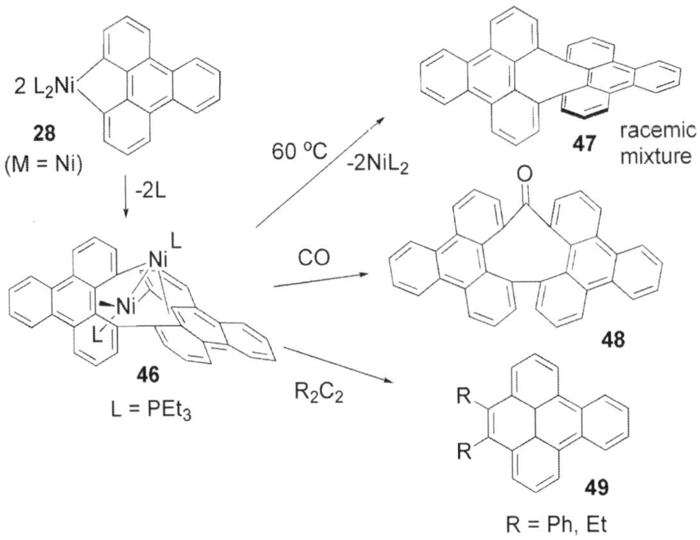

**Figure 8-17.** Reactions of nickelacycles **28** and **46**.

ring PAHs probably via six-membered PAC palladacycles suggests that reductive elimination from six-membered PAC metallacycles can be favorable.[200,201]

Reductive elimination from three- or four-membered metallacycles can occur by coupling of the PAC fragment through bimolecular or more complex pathways. Bimolecular elimination is observed for five-membered nickelacycles. This was first established for 2,2′-biphenydiyl complex **27** [Figure 8-10, $ML_n$ = Ni(PEt$_3$)$_2$], which decomposes to a chiral dimer that gives tetraphenylene on heating.[168] 1,12-Triphenylendiyl complex **28** (M = Ni) shows similar reactivity, giving dimer **46** and then tetraphenylene **47** (Figure 8-17).[15] Tetraphenylene (**47**) is chiral because of a helical twist resulting from the rigid nature of the triphenylene unit. Interestingly, CO cycloaddition gives carbonyl (**48**), which retains the C—C bond of dimer **46**, while alkyne cycloaddition gives **49** from C—C bond rupture in **46**.

The combination of 2,2′-biphenylendiyl reductive coupling with the ability of Ni (0) fragments to insert into biphenylene allows catalytic conversion of biphenylenes into tetraphenylenes via nickelacycles.[168,177] Other metal catalysts are known,[174] and for Pt(PEt$_3$)$_4$, the mechanism does not involve a dimer but rather insertion of the Pt center into two biphenylenes to give Pt(PEt$_3$)$_2$(2,2′-biphenyldiyl)$_2$, which then undergoes a stepwise reductive elimination of tetraphenylene.[173,175]

Careful consideration of a likely mechanism for Ni dimer formation from five-membered nickelacycles **27** and **28** led us to a chiral synthesis of tetraphenylenes. We realized that it should be possible to control the reaction to obtain enantiomerically pure tetraphenylenes both for fused systems such as **28** and for appropriately substituted 2,2′-biphenyldiyl nickelacycles (**50**, Figure 8-18). The chirality-forming step is the first C—C bond coupling that gives chiral dimer **54/55**. Control of the "twistin" or "twistout" by a chiral phosphine (L) selects the enantiomer (**54** or **55**) that

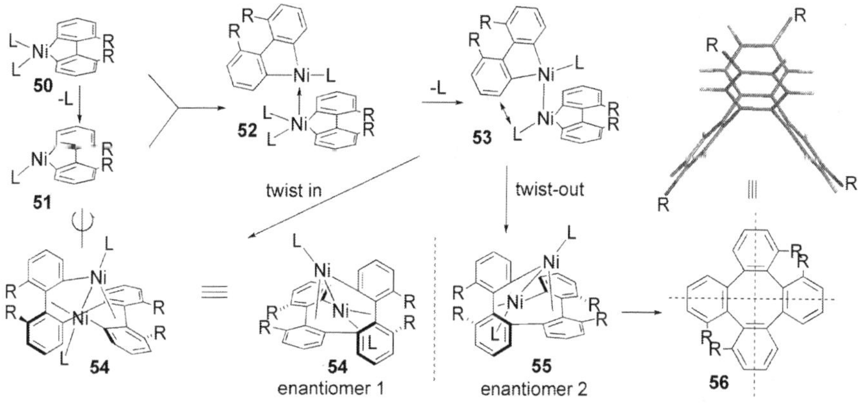

**Figure 8-18.** Chiral tetraphenylene formation from five-membered nickelacycles.

reductively eliminates each enantiomer of tetraphenylene (**56**). The selectivity does not depend on the R group but only on the interaction of the chiral phosphine with the edge of the biphenyl unit in **53**. One substitution pattern on **56** is shown, but any substitution that destroys the mirror symmetry of the tetraphenylene or that gives a fused tetraphenylene such as **47** gives chiral tetraphenylenes.

Biphenylene (**57**) (R = OMe)[202] was combined with chiral phosphine **58**[203] and Ni (COD)$_2$ (Figure 8-19). (We thank Strem Chemical for a gift of chiral phosphine **58**.) Complete conversion of **57** (R = OMe) to a 70% yield of chiral **56** (R = OMe) was achieved.[21] Chiral high-performance liquid chromatography (HPLC) analysis of **56** indicated an enantiomeric excess of 85%. This synthesis compares favorably with the racemic synthesis of **56** (R = OMe) using classic stoichiometric Cu(II) coupling chemistry, where the racemic mixture was laboriously resolved into the enantiomers.[204]

Pd and Pt metallacycles are more stable than their Ni analogs but also decompose to give coupled products (Figure 8-20). However, the reactions follow pathways very different from those of Ni. Autocatalytic kinetics are observed in the decomposition reactions of Pt and Pd metallacycles **25** and **28** (M = Pd).[19,20] The kinetic behavior is caused by colloidal metal catalysis, where increasing amounts of colloid catalyst are formed by decomposition of the ML$_2$ fragment.

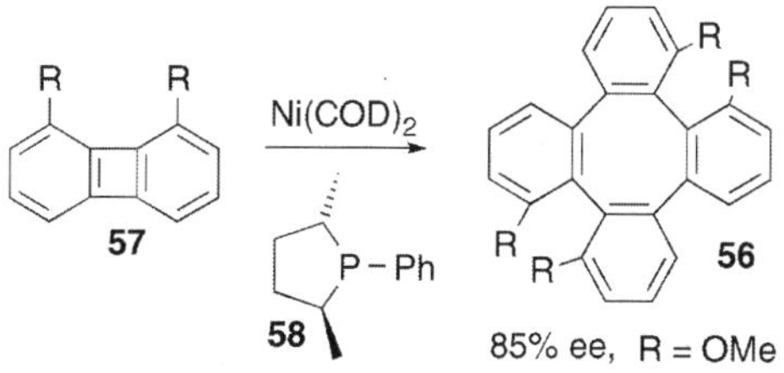

**Figure 8-19.** A chiral synthesis of tetraphenylene (**56**).

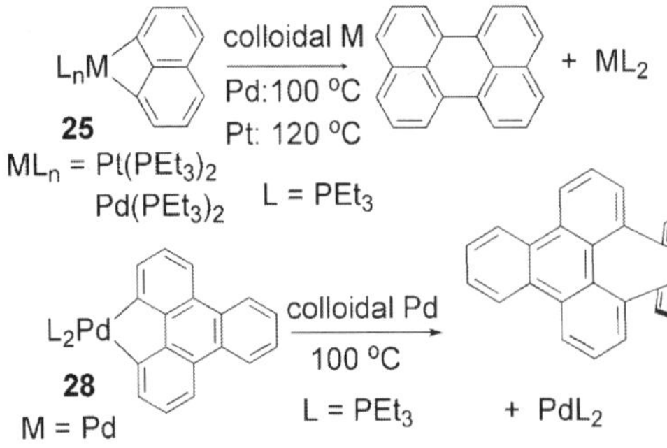

**Figure 8-20.** Colloidal metal-catalyzed decomposition of Pd and Pt metallacycles.

Alkyne addition diverts decomposition to cycloaddition products **59** and **60** (Figure 8-21). For Pt, this is very efficient, and only small amounts of decomposition coupling products are formed. For Pd, it is less efficient, but increasing the alkyne concentration increases the yield of cycloaddition product, indicating competitive processes.

Another difference between the Pt and Pd reactions is that in the presence of alkyne the Pt(PEt$_3$)$_2$ fragment is trapped as the alkyne complex Pt(PEt$_3$)$_2$($\eta^2$-PhCCPh) (**61**). This shuts down the colloidal Pt growth but in the presence of traces of O$_2$ "oxidative deligation" recovers colloidal Pt growth (see "colloid formation step" in Figure 8-21). The involvement of colloidal platinum in the cycloaddition is visually undetectable, and the amount of Pt in the colloid is very small (<μM). More recent work shows similar behavior for the decomposition and alkyne cycloaddition reactions of Pt (PEt$_3$)$_2$(2,2'-biphenyldiyl) (**27**).[205]

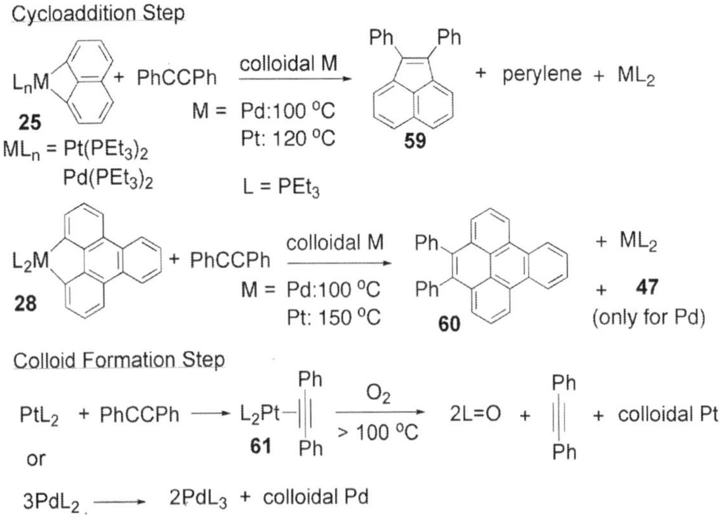

**Figure 8-21.** Alkyne cycloaddition and colloid formation steps.

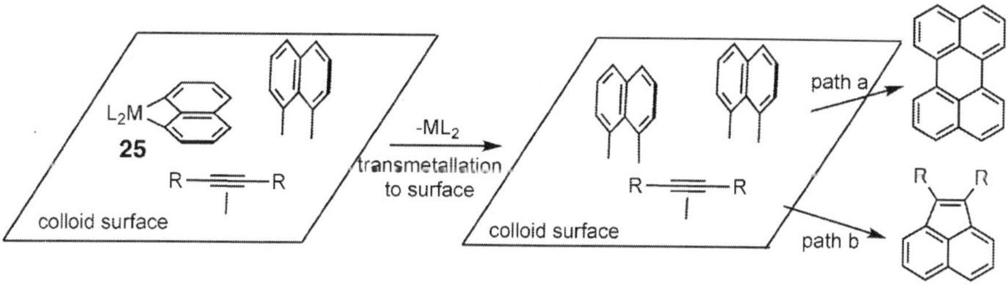

**Figure 8-22.** Model for colloid-catalyzed reactions.

Insight into the reaction mechanism was gained by examining the ability of colloidal Pt to catalyze the reaction of the Pd complexes and vice versa.[20] Colloidal Pt catalyzes the decomposition and alkyne cycloaddition of palladacycle **25** (M = Pd), but colloidal Pd does not catalyze the decomposition or alkyne coupling of platina-cycle **25** (M = Pt). This result, and known surface chemistry of these metals, allowed us to develop a model for the catalytic process illustrated in Figure 8-22 for **25**. A key step is metallacycle transfer to the colloid surface. This transmetallation reaction is feasible only if equal or stronger metal–carbon bonds are formed on the surface. This is expected to be true for a transfer from the palladacycle to a Pt surface (increasing bond enthalpies down a group) but not from the platinacycle to a Pd surface.

A prediction of this model is that once the metallacycle has transferred to the surface, subsequent reactions are that of the catalyst and not the metallacycle complex. Thus, the reaction product distribution of palladacycle **25** should shift toward the distribution for platinacycle **25** (mostly path b) when only colloidal Pt catalyzes the reaction. This is indeed observed. Another conclusion that can be drawn from these observations is that the transmetallation must be rate-limiting, at least in the colloidal Pt catalyzed reactions of platinacycle **25**. This follows from the fact that colloidal Pt catalyzes the palladacycle reactions at a lower temperature (100°C) than do the platinacycle reactions (120°C), indicating that once the metallacycle has transferred to the Pt surface, subsequent reactions are fast, even at the lower temperature.

Cycloaddition reactions of 2,2'-biphenyldiyl metallacycles with CO, isocyanides, and alkynes have been reported to give fluorene, iminofluorenes, and phenan-threnes.[152,168,172] Again, catalytic cycloaddition processes have been developed using biphenylene.[174,179,180,206,207]

## 8-3-3. Reactions at the PAC

In the reaction chemistry described above the metal–C bond is directly involved in the chemistry. More recently, examples have been found where reaction occurs at the PAC and the M–C bond remains intact. Figure 8-23 shows the remarkable bromination chemistry of 9-anthracenyl complex **62**.[23] $Br_2$ addition does not give a Pt(IV) complex or cleavage of the Pt–C bond as is usually observed for Pt(II) organometallic complexes. Instead, ring bromination product **63** is obtained. Even more remarkable

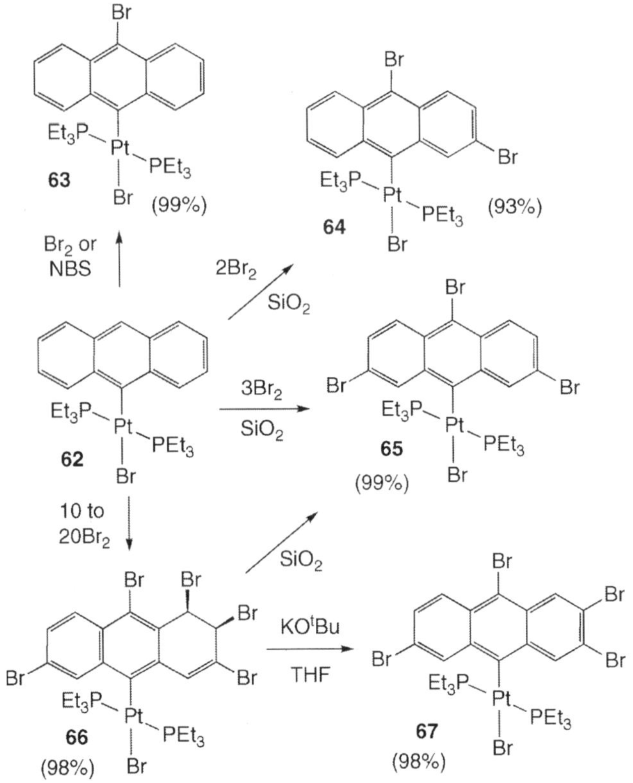

**Figure 8-23.** Bromination chemistry of 9-anthracenyl complex **62**.

is that continued $Br_2$ addition results in further selective ring bromination (**64** and **65**), finally giving highly ring-halogenated **66**. Further $Br_2$ addition has no effect. Complex **66** is readily rearomatized with silica gel or $KOBu^t$ to give **65** or **67**, respectively. (Polybromination of anthracene gives mixtures of products.[208]) All of these complexes are strongly luminescent. Pt(II) center bromination resistance is not limited to the 9-anthryl complexes and has also been observed in **17–21** (Figure 8-6). The resistance of these complexes to reaction at the metal center is attributed to "axial shielding,"[209] where the *peri*-hydrogen atoms and/or a *vicinal*-Br atom on each side of the metal center block access and inhibit the formation of higher coordination numbers. Axial shielding has also been observed in Pt(II) mesityl and *ortho*-halogenated phenyl complexes.[209–211]

Reactivity at the anthryl unit is also observed in pincer complexes **15** (Figure 8-24). Photolysis of **15** (M = Pt) in the presence of $O_2$ yields oxidation product **69** most likely through $O_2$ addition across the central ring in undetected intermediate **68**.[44] Similarly, facile alkyne addition occurs with **15** for all three metals Ni, Pd, and Pt to give **70**.[86] And finally, we have observed H/D exchange of complex **62** with $D_2O$ presumably through protonation (Figure 8-25).[199] Attempts to isolate the presumed protonation product **71** or analogs have not been successful.

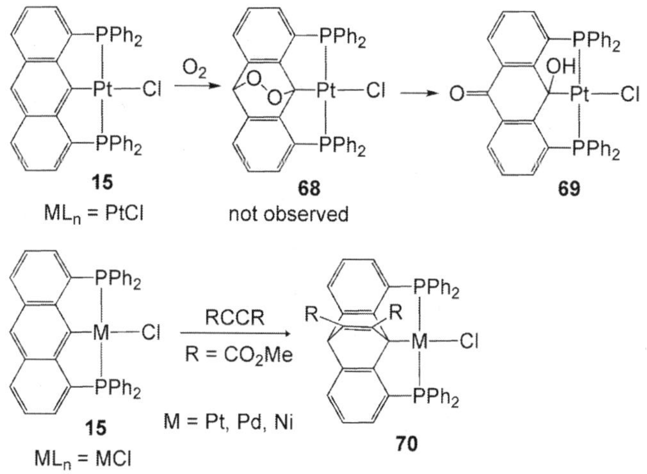

**Figure 8-24.** Anthracene-centered reactions of pincer complexes **15**.

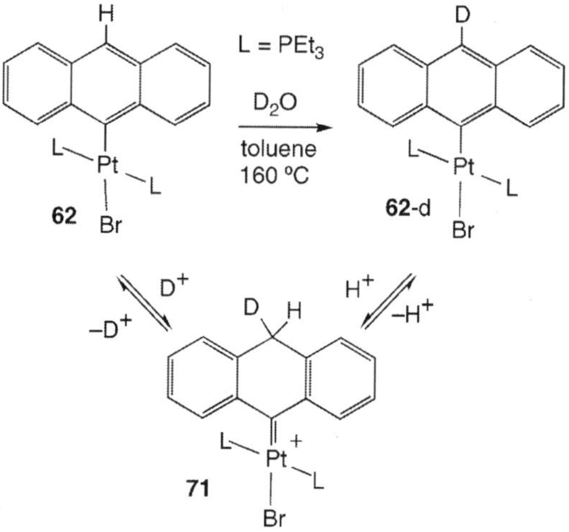

**Figure 8-25.** Deuterium exchange of the ring proton of **62**.

## 8-4. SUMMARY AND CONCLUSIONS

The chemistry of transition metal complexes with the metal $\sigma$-bonded to a polycyclic aromatic carbon compound (PAC) has been reviewed and discussed. Synthetic procedures (C–H activation, oxidative addition, metathesis, etc.) first for $\eta^1$ complexes and then for $\eta^2$-complexes (metallacycle complexes) have been presented, followed by an examination of the reaction chemistry involving the M–C bond (coupling and cycloaddition reactions) and reaction chemistry at the PAC where the M–C bond remains intact.

Transition metal σ-bonded PAC complexes can be prepared by a number of routes, with each showing limitations and advantages. For $\eta^1$ complexes C$-$H activation of often readily available PAHs is attractive but appears limited for low-valence metal insertion reactions by the stability of π complexes. C$-$H activation of PAHs with electrophilic metal centers would not suffer from this problem but has been scarcely investigated. Cyclometallation reactions of heteroatom functionalized PAHs have been extensively utilized with indications of some regiochemical control of differing kinetic and thermodynamic products, but the presence of the heteroatom substituent may be undesirable. Oxidative addition and metathesis reactions provide a wide range of useful metal complexes but are limited by the availability of suitable PAC reagents. This is particularly true for metallacycle complexes where two suitably positioned functionalized carbon atoms must be present in the PAC reagent. Selective lithiation of a few PAHs has provided useful reagents, and new metallation reactions may expand the PAH range and regiochemistry.

Even simple $\eta^1$-PAC complexes display novel and interesting chemistry ranging from metal migrations along the edge of PAHs to enhanced and selective reaction chemistry at the bonded PAC. In some cases the PAC provides steric shielding of the metal center, forcing reactivity onto the PAC, but even in cases without metal center shielding PAC-centered reactivity can be observed. The PAC complexes examined thus far are limited, and other interesting reaction chemistry might be expected. The discovery of colloidal metal catalysis of PAC metallacycle fragment coupling and cycloadditions opens a new area for investigation. As metals are intimately involved in the chemistry and technology of fullerenes and graphene (growth, metal contacts, etc.) and metal$-$carbon bonding is most likely involved, metal$-$PAC chemistry has a direct relationship to potential technological development in this area. Thus, metal$-$PAC chemistry not only is academically interesting but also has potential in the manipulation of fullerenes, nanotubes, and their fragments.

## REFERENCES

1. R. H. Crabtree, *The Organometallic Chemistry of the Transition Metals*, 4th ed., Wiley, New York, **2005**.

2. G. O. Spessard, *Organometallic Chemistry*, Prentice-Hall, Upper Saddle River, NJ, **1997**.

3. C. Elschenbroich, *Organometallics: A Concise Introduction*, 2nd rev. ed., VCH, New York, **1992**.

4. K. S. Novoselov, D. Jiang, F. Schedin, T. J. Booth, V. V. Khotkevich, S. V. Morozov, A. K. Geim, *Proc. Natl. Acad. Sci. USA* **2005**, *102*, 10451.

5. D. Koruga, S. R. Hameroff, J. Withers, R. Loutfy, M. Sundareshan, *Fullerene C60: History, Physics, Nanobiology, Nanotechnology*, North-Holland, New York, **1993**.

6. J. E. Baggott, *Perfect Symmetry: The Accidental Discovery of Buckminsterfullerene*, Oxford Univ. Press, New York, **1994**.

7. H. Aldersey-Williams, *The Most Beautiful Molecule: An Adventure in Chemistry*, Aurum Press, London, **1995**.

8. M. S. Dresselhaus, G. Dresselhaus, P. C. Eklund, *Science of Fullerenes and Carbon Nanotubes*, Academic Press, Boston, **1996**.

9. *Carbon Nanotubes: Preparation and Properties*, CRC Press, New York, **1997**.

10. Y. Saito, in *Carbon Nanotubes: Preparation and Properties*, T. W. Ebbessen (ed.), CRC Press, New York, **1997**, p 249.

11. R. T. K. Baker, M. A. Barber, P. S. Harris, F. S. Feates, R. J. Waite, *J. Catal.* **1972**, *26*, 51.

12. J. Gavillet, A. Loiseau, C. Journet, F. Willaime, F. Ducastelle, J.-C. Charlier, *Phys. Rev. Lett.* **2001**, *87*, 275504/1.

13. J.-Y. Raty, F. Gygi, G. Galli, *Phys. Rev. Lett.* **2005**, *95*, 096103.

14. A. A. Kane, T. Sheps, E. T. Branigan, V. A. Apkarian, M. H. Cheng, J. C. Hemminger, S. R. Hunt, P. G. Collins, *Nano Lett.* **2009**, *9*, 3586.

15. T. V. V. Ramakrishna, P. R. Sharp, *Organometallics* **2004**, *23*, 3079.

16. P. R. Sharp, *J. Am. Chem. Soc.* **2000**, *122*, 9880.

17. A. Singh, P. R. Sharp, *J. Am. Chem. Soc.* **2006**, *128*, 5998.

18. J. S. Brown, P. R. Sharp, *Organometallics* **2003**, *22*, 3604.

19. R. A. Begum, N. Chanda, T. V. V. Ramakrishna, P. R. Sharp, *J. Am. Chem. Soc.* **2005**, *127*, 13494.

20. N. Chanda, P. R. Sharp, *Organometallics* **2007**, *26*, 1635.

21. B.-Y. Wang, P. R. Sharp (unpublished results).

22. H. B. Lee, P. R. Sharp (unpublished results).

23. B.-Y. Wang, A. R. Karikachery, Y. Li, A. Singh, H. B. Lee, W. Sun, P. R. Sharp, *J. Am. Chem. Soc.* **2009**, *131*, 3150.

24. P. R. Sharp, *J. Organomet. Chem.* **2003**, *683*, 288.

25. A. D. Ryabov, *Chem. Rev.* **1990**, *90*, 403.

26. B. A. Arndtsen, R. G. Bergman, T. A. Mobley, T. H. Peterson, *Acc. Chem. Res.* **1995**, *28*, 154.

27. R. H. Crabtree, *Chem. Rev.* **1995**, *95*, 987.

28. A. E. Shilov, G. B. Shul'pin, *Chem. Rev.* **1997**, *97*, 2879.

29. P. Steenwinkel, R. A. Gossage, G. van Koten, *Chem. Eur. J.* **1998**, *4*, 759.

30. Y. Guari, S. Sabo-Etienne, B. Chaudret, *Eur. J. Inorg. Chem.* **1999**, *1999*, 1047.

31. W. D. Jones, *Top. Organomet. Chem.* **1999**, *3*, 9.

32. C. Slugovc, I. Padilla-Martinez, S. Sirol, E. Carmona, *Coord. Chem. Rev.* **2001**, *213*, 129.

33. F. Kakiuchi, S. Murai, *Acc. Chem. Res.* **2002**, *35*, 826.

34. J. A. Labinger, J. E. Bercaw, *Nature* **2002**, *417*, 507.

35. V. Ritleng, C. Sirlin, M. Pfeffer, *Chem. Rev.* **2002**, *102*, 1731.

36. W. D. Jones, *Acc. Chem. Res.* **2003**, *36*, 140.

37. M. E. van der Boom, D. Milstein, *Chem. Rev.* **2003**, *103*, 1759.

38. M. Lersch, M. Tilset, *Chem. Rev.* **2005**, *105*, 2471.

39. S. Ma, Z. Gu, *Angew. Chem. Int. Ed.* **2005**, *44*, 7512.

40. K. Godula, D. Sames, *Science* **2006**, *312*, 67.

41. I. V. Seregin, V. Gevorgyan, *Chem. Soc. Rev.* **2007**, *36*, 1173.

42. J. Chatt, J. M. Davidson, *J. Chem. Soc. A* **1965**, 843.

43. U. A. Gregory, S. D. Ibekwe, B. T. Kilbourn, D. R. Russell, *J. Chem. Soc. A* **1971**, 1118.

44. J. Hu, H. Xu, M.-H. Nguyen, J. H. K. Yip, *Inorg. Chem.* **2009**, *48*, 9684.

45. G. P. Pez, C. F. Putnik, S. L. Suib, G. D. Stucky, *J. Am. Chem. Soc.* **1979**, *101*, 6933.

46. W. D. Jones, L. Dong, *J. Am. Chem. Soc.* **1989**, *111*, 8722.

47. S. T. Belt, L. Dong, S. B. Duckett, W. D. Jones, M. G. Partridge, R. N. Perutz, *J. Chem. Soc., Chem. Commun.* **1991**, 266.

48. R. M. Chin, L. Dong, S. B. Duckett, W. D. Jones, *Organometallics* **1992**, *11*, 871.

49. R. M. Chin, L. Dong, S. B. Duckett, M. G. Partridge, W. D. Jones, R. N. Perutz, *J. Am. Chem. Soc.* **1993**, *115*, 7685.

50. P. Diversi, A. Ferrarini, G. Ingrosso, A. Lucherini, G. Uccello-Barretta, C. Pinzino, F. De Biani Fabrizi, F. Laschi, P. Zanello, *Gazz. Chim. Ital.* **1996**, *126*, 391.

51. R. P. Shibaeva, L. P. Rozenberg, R. M. Lobkovskaya, A. E. Shilov, G. B. Shul'pin, *J. Organomet. Chem.* **1981**, *220*, 271.

52. V. L. Blair, W. Clegg, R. E. Mulvey, L. Russo, *Inorg. Chem.* **2009**, *48*, 8863.

53. P. C. Andrikopoulos, D. R. Armstrong, D. V. Graham, E. Hevia, A. R. Kennedy, R. E. Mulvey, C. T. O'Hara, C. Talmard, *Angew. Chem. Int. Ed.* **2005**, *44*, 3459.

54. W. Clegg, S. H. Dale, E. Hevia, L. M. Hogg, G. W. Honeyman, R. E. Mulvey, C. T. O'Hara, *Angew. Chem. Int. Ed.* **2006**, *45*, 6548.

55. V. L. Blair, L. M. Carrella, W. Clegg, B. Conway, R. W. Harrington, L. M. Hogg, J. Klett, R. E. Mulvey, E. Rentschler, L. Russo, *Angew. Chem. Int. Ed.* **2008**, *47*, 6208.

56. G. R. Newkome, W. E. Puckett, V. K. Gupta, G. E. Kiefer, *Chem. Rev.* **1986**, *86*, 451.

57. M. S. Lowry, S. Bernhard, *Chem. Eur. J.* **2006**, *12*, 7970.

58. J. A. G. Williams, in *Photochemistry and Photophysics of Coordination Compounds*, N. Balzani, S. Campagna (eds.), Springer, Berlin-Heidelberg, **2007**, Vol. II, p. 205.

59. J. A. G. Williams, S. Develay, D. L. Rochester, L. Murphy, *Coord. Chem. Rev.* **2008**, *252*, 2596.

60. J. A. G. Williams, A. J. Wilkinson, V. L. Whittle, *J. Chem. Soc., Dalton Trans.* **2008**, 2081.

61. J. A. G. Williams, *Chem. Soc. Rev.* **2009**, *38*, 1783.

62. K. Gehrig, M. Hugentobler, A. J. Klaus, P. Rys, *Inorg. Chem.* **1982**, *21*, 2493.

63. L. Kind, A. J. Klaus, P. Rys, V. Gramlich, *Helv. Chim. Acta* **1998**, *81*, 307.

64. J. Albert, J. M. Cadena, J. R. Granell, X. Solans, M. Font-Bardia, *Tetrahedron: Asym.* **2000**, *11*, 1943.

65. M. Crespo, M. Font-Bardìa, S. Pérez, X. Solans, *J. Organomet. Chem.* **2002**, *642*, 171.

66. H.-F. Klein, R. Beck, U. Flörke, H.-J. Haupt, *Eur. J. Inorg. Chem.* **2003**, 1380.

67. Y. Li, K.-H. Ng, S. Selvaratnam, G.-K. Tan, J. J. Vittal, P.-H. Leung, *Organometallics* **2003**, *22*, 834.

68. M. Crespo, E. Evangelio, *J. Organomet. Chem.* **2004**, *689*, 1956.

69. K. R. J. Thomas, M. Velusamy, J. T. Lin, C.-H. Chien, Y.-T. Tao, Y. S. Wen, Y.-H. Hu, P.-T. Chou, *Inorg. Chem.* **2005**, *44*, 5677.

70. Z. Chen, H. W. Schmalle, T. Fox, H. Berke, *J. Chem. Soc., Dalton Trans.* **2005**, 580.

71. H.-F. Klein, S. Camadanli, R. Beck, U. Florke, *Chem. Commun.* **2005**, 381.

72. K.-H. Tam, M. C. W. Chan, H. Kaneyoshi, H. Makio, N. Zhu, *Organometallics* **2009**, *28*, 5877.

73. I. G. Phillips, P. J. Steel, *J. Organomet. Chem.* **1991**, *410*, 247.

74. M. Pfeffer, N. Sutter-Beydoun, A. De Cian, J. Fischer, *J. Organomet. Chem.* **1993**, *453*, 139.

75. J. Hu, J. H. K. Yip, D.-L. Ma, K.-Y. Wong, W.-H. Chung, *Organometallics* **2008**, *28*, 51.

76. B. J. O'Keefe, P. J. Steel, *Inorg. Chem. Commun.* **1999**, *2*, 10.

77. B. J. O'Keefe, P. J. Steel, *Organometallics* **2003**, *22*, 1281.

78. C. Bonnefous, A. Chouai, R. P. Thummel, *Inorg. Chem.* **2001**, *40*, 5851.

79. S. C. F. Kui, I. H. T. Sham, C. C. C. Cheung, C.-W. Ma, B. Yan, N. Zhu, C.-M. Che, W.-F. Fu, *Chem. Eur. J.* **2007**, *13*, 417.

80. J.-M. Valk, R. van Belzen, J. Boersma, A. L. Spek, G. van Koten, *J. Chem. Soc., Dalton Trans.* **1994**, 2293.

81. J. Albert, R. Bosque, J. Granell, R. Tavera, *J. Organomet. Chem.* **2000**, *595*, 54.

82. J. Hu, J. H. K. Yip, *Organometallics* **2009**, *28*, 1093.

83. M. W. Haenel, D. Jakubik, C. Krüger, P. Betz, *Chem. Ber.* **1991**, *124*, 333.

84. M. W. Haenel, S. Oevers, K. Angermund, W. C. Kaska, H.-J. Fan, M. B. Hall, *Angew. Chem. Int. Ed.* **2001**, *40*, 3596.

85. P. E. Romero, M. T. Whited, R. H. Grubbs, *Organometallics* **2008**, *27*, 3422.

86. C. Azerraf, A. Shpruhman, D. Gelman, *Chem. Commun.* **2009**, 466.

87. R. Tomat, S. Zecchin, G. Schiavon, G. Zotti, *J. Electroanal. Chem.* **1988**, *252*, 215.

88. D. Gelman, S. Dechert, H. Schumann, J. Blum, *Inorg. Chim. Acta* **2002**, *334*, 149.

89. H. B. Lee, PhD thesis, Univ. Missouri, Columbia, **2007**.

90. R. Begum, A. J. James, P. R. Sharp, *Organometallics* **2005**, *24*, 2670.

91. W. Y. Heng, J. Hu, J. H. K. Yip, *Organometallics* **2007**, *26*, 6760.

92. H. B. Lee, P. R. Sharp, *Organometallics* **2005**, *24*, 4875.

93. C. J. Kuehl, S. D. Huang, P. J. Stang, *J. Am. Chem. Soc.* **2001**, *123*, 9634.

94. Y. K. Kryschenko, S. R. Seidel, A. M. Arif, P. J. Stang, *J. Am. Chem. Soc.* **2003**, *125*, 5193.

95. H. Weissman, E. Shirman, T. Ben-Moshe, R. Cohen, G. Leitus, L. J. W. Shimon, B. Rybtchinski, *Inorg. Chem.* **2007**, *46*, 4790.

96. B. El Hamaoui, F. Laquai, S. Baluschev, J. Wu, K. Müllen, *Synth. Met.* **2006**, *156*, 1182.

97. H. Choi, C. Kim, K.-M. Park, J. Kim, Y. Kang, J. Ko, *J. Organomet. Chem.* **2009**, *694*, 3529.

98. R. Maag, B. H. Northrop, A. Butterfield, A. Linden, O. Zerbe, Y. M. Lee, K.-W. Chi, P. J. Stang, J. S. Siegel, *Org. Biomol. Chem.* **2009**, *7*, 4881.

99. J. Chatt, B. L. Shaw, *Chem. Ind. (Lon.)* **1959**, 675.

100. J. Chatt, B. L. Shaw, *J. Chem. Soc.* **1960**, 1718.

101. J. Chatt, A. E. Underhill, *J. Chem. Soc. A* **1963**, 2088.

102. J. Dehand, A. Mauro, H. Ossor, M. Pfeffer, R. H. de A. Santos, J. R. Lechat, *J. Organomet. Chem.* **1983**, *250*, 537.

103. T. A. Bazhenova, R. M. Lobkovskaya, R. P. Shibaeva, A. K. Shilova, M. Gruselle, G. Leny, E. Deschamps, *J. Organomet. Chem.* **1983**, *244*, 375.

104. C. Arlen, F. Maassarani, M. Pfeffer, J. Fischer, *Nouv. J. Chim.* **1985**, 9.

105. F. S. M. Hassan, D. M. McEwan, P. G. Pringle, B. L. Shaw, *J. Chem. Soc., Dalton Trans.* **1985**, 1501.

106. E. Wehman, G. van Koten, M. Knotter, H. Spelten, D. Heijdenrijk, A. N. S. Mak, C. H. Stam, *J. Organomet. Chem.* **1987**, *325*, 293.

107. E. Wehman, G. Van Koten, C. T. Knaap, H. Ossor, M. Pfeffer, A. L. Spek, *Inorg. Chem.* **1988**, *27*, 4409.

108. E. Wehman, G. van Koten, J. T. B. H. Jastrzebski, H. Ossor, M. Pfeffer, *J. Chem. Soc., Dalton Trans.* **1988**, 2975.

109. T. Debaerdemaeker, C. Weisemann, H.-A. Brune, *J. Organomet. Chem.* **1988**, *350*, 91.

110. H. A. Brune, G. Schmidtberg, C. Weisemann, *J. Organomet. Chem.* **1989**, *371*, 121.

111. C. Weisemann, G. Schmidtberg, H.-A. Brune, *J. Organomet. Chem.* **1989**, *362*, 63.

112. V. W. W. Yam, L. P. Chan, T. F. Lai, *J. Chem. Soc., Dalton Trans.* **1993**, 2075.

113. M. D. Janssen, M. A. Corsten, A. L. Spek, D. M. Grove, G. van Koten, *Organometallics* **1996**, *15*, 2810.

114. M. Bouwkamp, D. van Leusen, A. Meetsma, B. Hessen, *Organometallics* **1998**, *17*, 3645.

115. C. M. P. Kronenburg, C. H. M. Amijs, J. T. B. H. Jastrzebski, M. Lutz, A. L. Spek, G. van Koten, *Organometallics* **2002**, *21*, 4662.

116. C. H. M. Amijs, G. P. M. van Klink, M. Lutz, A. L. Spek, G. van Koten, *Organometallics* **2005**, *24*, 2944.

117. C. H. M. Amijs, A. W. Kleij, G. P. M. van Klink, A. L. Spek, G. van Koten, *Organometallics* **2005**, *24*, 2773.

118. K. Onitsuka, M. Yamamoto, T. Mori, F. Takei, S. Takahashi, *Organometallics* **2006**, *25*, 1270.

119. D. V. Partyka, M. Zeller, A. D. Hunter, T. G. Gray, *Angew. Chem. Int. Ed.* **2006**, *45*, 8188.

120. M. Osawa, M. Hoshino, D. Hashizume, *J. Chem. Soc., Dalton Trans.* **2008**, 2248.

121. L. Gao, M. A. Peay, D. V. Partyka, J. B. Updegraff, T. S. Teets, A. J. Esswein, M. Zeller, A. D. Hunter, T. G. Gray, *Organometallics* **2009**, *28*, 5669.

122. C. Weisemann, G. Schmidtberg, H. A. Brune, *J. Organomet. Chem.* **1989**, *365*, 403.

123. V. W.-W. Yam, K.-L. Cheung, S.-K. Yip, N. Zhu, *Photochem. Photobiol. Sci.* **2005**, *4*, 149.

124. N. Meyer, C. W. Lehmann, T. K. M. Lee, J. Rust, V. W. W. Yam, F. Mohr, *Organometallics* **2009**, *28*, 2931.

125. A. Sygula, R. Sygula, F. R. Fronczek, P. W. Rabideau, *J. Org. Chem.* **2002**, *67*, 6487.

126. C. Weisemann, H. A. Brune, *J. Organomet. Chem.* **1986**, *312*, 133.

127. T. Debaerdemaeker, C. Weisemann, H. A. Brune, *Acta Crystallogr., Sect. C: Cryst. Struct. Commun.* **1987**, *43*, 432.

128. G. D. Vaughn, K. A. Krein, J. A. Gladysz, *Organometallics* **2002**, *5*, 936.

129. A. D. Hunter, D. Ristic-Petrovic, J. L. McLernon, *Organometallics* **2002**, *11*, 864.

130. R. Beck, U. Flörke, H.-F. Klein, *Inorg. Chem.* **2009**, *48*, 1416.

131. J. Blum, G. Scharf, *J. Org. Chem.* **1970**, *35*, 1895.

132. K. Kajiyama, A. Nakamoto, S. Miyazawa, T. K. Miyamoto, *Chem. Lett.* **2003**, *32*, 332.

133. T. Mizuta, Y. Iwakuni, T. Nakazono, K. Kubo, K. Miyoshi, *J. Organomet. Chem.* **2007**, *692*, 184.

134. W. M. Jones, J. Klosin, *Adv. Organomet. Chem.* **1998**, *42*, 147.

135. S. L. Buchwald, S. M. King, *J. Am. Chem. Soc.* **1991**, *113*, 258.

136. M. A. Bennett, D. C. R. Hockless, E. Wenger, *Organometallics* **1995**, *14*, 2091.

137. M. A. Bennett, M. R. Kopp, E. Wenger, A. C. Willis, *J. Organomet. Chem.* **2003**, *667*, 8.

138. G. Erker, *J. Organomet. Chem.* **1977**, *134*, 189.

139. S. L. Buchwald, B. T. Watson, J. C. Huffman, *J. Am. Chem. Soc.* **1986**, *108*, 7411.

140. P. R. Sharp (unpublished results).

141. F. M. G. de Rege, W. M. Davis, S. L. Buchwald, *Organometallics* **1995**, *14*, 4799.

142. M. A. Bennett, *Pure Appl. Chem.* **1989**, *61*, 1695.

143. M. A. Bennett, H. P. Schwemlein, *Angew. Chem. Int. Ed. Engl.* **1989**, *28*, 1296.

144. D. Pena, D. Perez, E. Guitian, L. Castedo, *J. Org. Chem.* **2000**, *65*, 6944.

145. W. Neugebauer, T. Clark, P. von Ragué Schleyer, *Chem. Ber.* **1983**, *116*, 3283.

146. M. A. G. M. Tinga, G. Schat, O. S. Akkerman, F. Bickelhaupt, E. Horn, W. J. J. Kooijiman, W. J. J. Smeets, A. L. Spek, *J. Am. Chem. Soc.* **1993**, *115*, 2808.

147. M. A. G. M. Tinga, G. Schat, O. S. Akkerman, F. Bickelhaupt, W. J. J. Smeets, A. L. Spek, *Chem. Ber.* **1994**, *127*, 1851.

148. M. S. Goedheijt, O. S. Akkerman, F. Bickelhaupt, P. W. N. M. van Leeuwen, N. Veldman, A. L. Spek, *Organometallics* **1994**, *13*, 2931.

149. W. Neugebauer, A. J. Kos, P. von Ragué Schleyer, *J. Organomet. Chem.* **1982**, *228*, 107.

150. A. J. Ashe, III, J. W. Kampf, P. M. Savla, *J. Org. Chem.* **1990**, *55*, 5558.

151. S. A. Gardner, H. B. Gordon, M. D. Rausch, *J. Organomet. Chem.* **1973**, *60*, 179.

152. Y. Wakatsuki, O. Nomura, H. Ton, H. Yamazaki, *J. Chem. Soc., Perkin Trans. 2* **1980**, 1344.

153. M. D. Rausch, L. P. Klemann, W. H. Boon, *Synth. React. Inorg. Met.-Org. Chem.* **1985**, *15*, 923.

154. C. Cornioley-Deuschel, A. von Zelewsky, *Inorg. Chem.* **1987**, *26*, 3354.

155. C. B. Blanton, Z. Murtaza, R. J. Shaver, D. P. Rillema, *Inorg. Chem.* **1992**, *31*, 3230.

156. Y.-h. Chen, J. W. Merkert, Z. Murtaza, C. Woods, D. P. Rillema, *Inorg. Chim. Acta* **1995**, *240*, 41.

157. M. R. Plutino, L. M. Scolaro, A. Albinati, R. Romeo, *J. Am. Chem. Soc.* **2004**, *126*, 6470.

158. C. L. Hilton, B. T. King, *Organometallics* **2006**, *25*, 4058.

159. R. Usón, J. Vicente, J. A. Cirac, M. T. Chicote, *J. Organomet. Chem.* **1980**, *198*, 105.

160. R. Hohenadel, H. A. Brune, *J. Organomet. Chem.* **1988**, *350*, 101.

161. H. A. Brune, H. Roth, T. Debaerdemaeker, H. M. Schiebel, *J. Organomet. Chem.* **1991**, *402*, 435.

162. Z. Hou, Y. Wakatsuki, A. Fujita, H. Yamazaki, *Chem. Commun.* **1998**, 669.

163. K. Singh, W. R. McWhinnie, H. L. Chen, M. Sun, T. A. Hamor, *J. Chem. Soc., Dalton Trans.* **1996**, 1545.

164. D. A. Vicic, W. D. Jones, *Organometallics* **1998**, *17*, 3411.

165. D. A. Vicic, W. D. Jones, *J. Am. Chem. Soc.* **1999**, *121*, 7606.

166. C. N. Iverson, R. J. Lachicotte, C. Mueller, W. D. Jones, *Organometallics* **2002**, *21*, 5320.

167. A. L. Keen, M. Doster, S. A. Johnson, *J. Am. Chem. Soc.* **2007**, *129*, 810.

168. J. J. Eisch, A. M. Piotrowski, K. I. Han, C. Krüger, Y. H. Tsay, *Organometallics* **1985**, *4*, 224.

169. Z. Lu, C.-H. Jun, S. R. de Gala, M. Sigalas, O. Eisenstein, R. H. Crabtree, *J. Chem. Soc., Chem. Commun.* **1993**, 1877.

170. C. Perthuisot, W. D. Jones, *J. Am. Chem. Soc.* **1994**, *116*, 3647.

171. Z. Lu, C.-H. J., S. R. de Gala, M. P. Sigalas, O. Eisenstein, R. H. Crabtree, *Organometallics* **1995**, *14*, 1168.

172. C. Perthuisot, B. L. Edelbach, D. L. Zubris, W. D. Jones, *Organometallics* **1997**, *16*, 2016.

173. N. Simhai, C. N. Iverson, B. L. Edelbach, W. D. Jones, *Organometallics* **2001** *20*, 2759.

174. C. Perthuisot, B. L. Edelbach, D. L. Zubris, N. Simhai, C. N. Iverson, C. Muller, T. Satoh, W. D. Jones, *J. Mol. Catal. A: Chem.* **2002**, *189*, 157.

175. B. L. Edelbach, R. J. Lachicotte, W. D. Jones, *J. Am. Chem. Soc.* **1998**, *120*, 2843.

176. X. Zhang, G. B. Carpenter, D. A. Sweigart, *Organometallics* **1999**, *18*, 4887.

177. H. Schwager, S. Spyroudis, K. P. C. Vollhardt, *J. Organomet. Chem.* **1990**, *382*, 191.

178. B. L. Edelbach, D. A. Vicic, R. J. Lachicotte, W. D. Jones, *Organometallics* **1998**, *17*, 4784.

179. C. N. Iverson, W. D. Jones, *Organometallics* **2001**, *20*, 5745.

180. C. Müller, R. J. Lachicotte, W. D. Jones, *Organometallics* **2002**, *21*, 1975.

181. T. Shibata, G. Nishizawa, K. Endo, *Synlett* **2008**, 765.

182. A. G. Myers, M. Sogi, M. A. Lewis, S. P. Arvedson, *J. Org. Chem.* **2004**, *69*, 2516.

183. M. J. Arce, A. L. Viado, Y. Z. An, S. I. Khan, Y. Rubin, *J. Am. Chem. Soc.* **1996**, *118*, 3775.

184. R. M. Shaltout, R. Sygula, A. Sygula, F. R. Fronczek, G. G. Stanley, P. W. Rabideau, *J. Am. Chem. Soc.* **1998**, *120*, 835.

185. W. R. Cullen, S. J. Rettig, T. C. Zheng, *Organometallics* **1995**, *14*, 1466.

186. A. J. Deeming, C. M. Martin, *Chem. Commun.* **1996**, 53.

187. A. J. Deeming, C. M. Martin, *Angew. Chem. Int. Ed.* **1998**, *37*, 1691.

188. M. I. Bruce, P. A. Humphrey, R. Schmutzler, B. W. Skelton, A. H. White, *J. Organomet. Chem.* **2005**, *690*, 784.

189. W.-Y. Wong, F.-L. Ting, Y. Guo, Z. Lin, *J. Cluster Sci.* **2005**, *16*, 185.

190. A. Sharmin, A. Minazzo, L. Salassa, E. Rosenberg, J. B. A. Ross, S. E. Kabir, K. I. Hardcastle, *Inorg. Chim. Acta* **2008**, *361*, 1624.

191. A. J. Arce, F. Cañavera, Y. De Sanctis, J. Ascanio, R. Machado, T. González, *J. Organomet. Chem.* **2009**, *694*, 1834.

192. R. D. Adams, B. Captain, J. L. Smith, *J. Organomet. Chem.* **2003**, *683*, 421.

193. G. Karig, M.-T. Moon, N. Thasana, T. Gallagher, *Org. Lett.* **2002**, *4*, 3115.

194. D. Masselot, J. P. H. Charmant, T. Gallagher, *J. Am. Chem. Soc.* **2006**, *128*, 694.

195. D. L. Davies, S. M. A. Donald, S. A. Macgregor, *J. Am. Chem. Soc.* **2005**, *127*, 13754.

196. D. Garcia-Cuadrado, A. A. C. Braga, F. Maseras, A. M. Echavarren, *J. Am. Chem. Soc.* **2006**, *128*, 1066.

197. A. J. Mota, A. Dedieu, *Organometallics* **2006**, *25*, 3130.

198. M. A. Campo, H. Zhang, T. Yao, A. Ibdah, R. D. McCulla, Q. Huang, J. Zhao, W. S. Jenks, R. C. Larock, *J. Am. Chem. Soc.* **2007**, *129*, 6298.

199. A. Singh, P. R. Sharp (unpublished results).

200. D. Kim, J. L. Petersen, K. K. Wang, *Org. Lett.* **2006**, *8*, 2313.

201. E. A. Jackson, B. D. Steinberg, M. Bancu, A. Wakamiya, L. T. Scott, *J. Am. Chem. Soc.* **2007**, *129*, 484.

202. T. Ooi, M. Takahashi, M. Yamada, E. Tayama, K. Omoto, K. Maruoka, *J. Am. Chem. Soc.* **2004**, *126*, 1150.

203. M. J. Burk, J. E. Feaster, R. L. Harlow, *Organometallics* **1990**, *9*, 2653.

204. H. Y. Peng, C. K. Lam, T. C. W. Mak, Z. Cai, W. T. Ma, Y. X. Li, H. N. C. Wong, *J. Am. Chem. Soc.* **2005**, *127*, 9603.

205. R. Robertson, Jr., P. R. Sharp (unpublished results).

206. B. L. Edelbach, R. J. Lachicotte, W. D. Jones, *Organometallics* **1999**, *18*, 4040.

207. B. L. Edelbach, R. J. Lachicotte, W. D. Jones, *Organometallics* **1999**, *18*, 4660.

208. O. Cakmak, R. Erenler, A. Tutar, N. Celik, *J. Org. Chem.* **2006**, *71*, 1795.

209. A. Klein, H.-D. Hausen, W. Kaim, *J. Organomet. Chem.* **1992**, *440*, 207.

210. R. Uson, J. Fornies, M. Tomas, B. Menjon, R. Bau, K. Suenkel, E. Kuwabara, *Organometallics* **1986**, *5*, 1576.

211. A. Klein, W. Kaim, *Organometallics* **1995**, *14*, 1176.

# CHAPTER 9

# HEMISPHERICAL GEODESIC POLYARENES: ATTRACTIVE TEMPLATES FOR THE CHEMICAL SYNTHESIS OF UNIFORM-DIAMETER ARMCHAIR NANOTUBES

ANTHONY P. BELANGER, KATHARINE A. MIRICA,
JAMES MACK, and LAWRENCE T. SCOTT

## 9-1. INTRODUCTION

Carbon nanotubes show great promise for technological applications in a host of fields, ranging from computer science to medicine.[1] Alongside diamond, graphite, and fullerenes, carbon nanotubes constitute a unique family of carbon allotropes.[2] Conceptually, they can be viewed as rectangular sections of graphene sheets that have been rolled up into cylinders with the overlapping carbons merged. Of course, there are many different ways in which a rectangle could be cut initially from a graphene sheet, each resulting in a different carbon nanotube, and that makes for a great diversity of structures.

All carbon nanotubes can be categorized[2] as either chiral or achiral, depending on the orientation of the six-membered rings along the shaft. The achiral nanotubes have been further classified as either zigzag or armchair, depending on the appearance of the rim, which also depends on the orientation of the six-membered rings along the shaft. Figure 9-1 shows examples from each class. A good sense of the chirality can be obtained by looking down the axis of the tube. If multiple concentric nanotubes share a common axis, the collection is classified as a *multiwalled nanotube* (Figure 9-1).

Carbon nanotubes of all structural classes share certain properties in common. For example, they are all astonishingly strong. Figure 9-2 compares carbon nanotubes to several other familiar materials with respect to their tensile strength, which is

*Fragments of Fullerenes and Carbon Nanotubes: Designed Synthesis, Unusual Reactions,*
*and Coordination Chemistry*, First Edition. Edited by Marina A. Petrukhina and Lawrence T. Scott.
© 2012 John Wiley & Sons, Inc. Published 2012 by John Wiley & Sons, Inc.

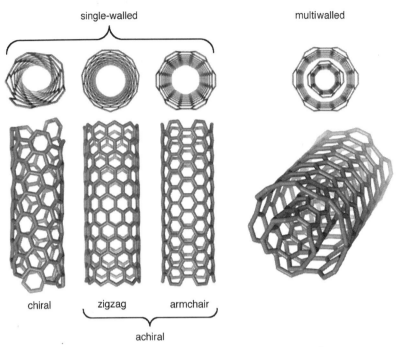

**Figure 9-1.** Different classes of carbon nanotubes.[2]

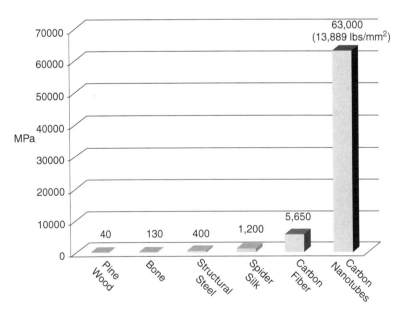

**Figure 9-2.** Tensile strengths of various materials compared to carbon nanotubes.

a measure of the force required to pull the material apart. Structural steel has a tensile strength about 3 times that of bone, a particularly strong biological material, and 10 times that of pine wood. Spider silk, one of Nature's strongest biological materials, exhibits a tensile strength 3 times that of structural steel.[3] None of these materials, however, approaches the incredible strength of carbon nanotubes. Tensile strengths as high as 63,000 MPa have been recorded for carbon nanotubes.[4] To put that in perspective, a carbon nanotube cable with a cross-sectional area of 1 square millimeter ($1 mm^2$) would be capable of supporting an astounding 6300 kg!

In addition to great strength, carbon nanotubes also display considerable flexibility; they are not brittle. Contrasting with these universal mechanical properties, on the other hand, the electronic properties of nanotubes depend critically on subtle variations in structure of the sort pictured in Figure 9-1.[2] Armchair nanotubes, for example, will conduct electricity, whereas most zigzag and chiral nanotubes are semiconductors. This metallic behavior, combined with their small diameters, has made armchair nanotubes particularly attractive candidates to serve as nanowires in molecule-scale electronic devices.[5] The diameter of the thinnest carbon nanotubes ($\sim$1 nm) is approximately one tenth of the width of the narrowest channel that can be carved by state-of-the-art silicon lithography techniques. If carbon nanotubes were used instead of lithography, two-dimensional chips could theoretically occupy just one-hundredth of the area that they currently fill. Research groups worldwide have already begun exploring uses of carbon nanotube as nanowires in electronics of the future.[6]

Unfortunately, current carbon nanotube production methods yield mixtures of all the tube varieties depicted in Figure 9-1. Countless worker-years have been spent on efforts to separate and purify single classes of carbon nanotubes, but only marginal success has been achieved.[7] One particularly fascinating method utilizes DNA, which can wrap around a carbon nanotube and render the tube soluble.[8] Because there are minor variations in the DNA−nanotube interactions, depending on the particular structure of the tube, enriched samples of specific nanotubes can be obtained by passing the DNA−nanotube mixtures through an anion exchange column.

Although interesting, this method and all the other fractionation and purification strategies that have been explored for more than a decade still fail to provide access to pure samples of any one particular type of carbon nanotube. Our approach to solving this problem circumvents the issues associated with separation by relying on rational chemical synthesis. Instead of making a mixture of all types of nanotubes that must be separated, we propose building an endcap template from which one specific variety of nanotube may be grown (Figure 9-3). The template itself will be a highly curved polycyclic aromatic hydrocarbon (e.g., **1**). We envision that once the hemispherical endcap has been synthesized, it will be possible to use an "acetylene-like" feedstock to extend it in an iterative fashion to produce the corresponding single-walled nanotube (SWNT).[9] All tubes produced by this method should be identical in diameter and chirality. Their lengths should be controllable by adjusting the time allowed for the growth process.

We have focused our initial attention on nanotube endcaps that, when extended, will become conductive armchair nanotubes, suitable for use as ultrathin nanowires.

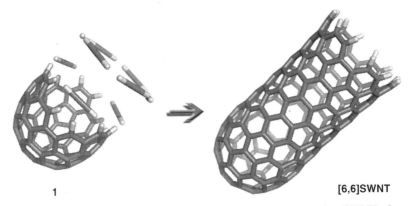

**1**                                                                 **[6,6]SWNT**

**Figure 9-3.** Conceptual process for growing a single-walled nanotube (SWNT) from an endcap template (**1**).

Preliminary accounts of our approaches to the synthesis of various endcaps for [5,5] and [6,6]SWNTs have already appeared.[10] It should be noted that the larger diameter nanotubes can have more than one possible arrangement of five- and six-membered rings in their endcaps, even when consideration is given only to isomers that obey the isolated pentagon rule (IPR).[11] In this chapter, as a case study, we describe our efforts to synthesize the $C_{3v}$ hemispherical geodesic polyarene **1** that constitutes an endcap for the [6,6]SWNT shown in Figure 9-3.

## 9-2. PROGRESS TOWARD THE SYNTHESIS OF A $C_{3v}$ [6,6]SWNT ENDCAP

### 9-2-1. Retrosynthetic Analysis

A threefold rotational symmetry axis runs down the center of the endcap that we have chosen as our target (**1**). Retrosynthetic disconnections in a symmetric manner down the three clefts of this $C_{66}H_{12}$ polyarene lead back to an appealing endcap precursor (**2**). This $C_{66}H_{24}$ hydrocarbon can be viewed as a central benzene ring with three identical corannulene-containing arms radiating from alternate sides. Having curved subunits built into the endcap precursor should make the final step of stitching up the endcap less demanding, because much of the strain will already have been built into the molecule (Figure 9-4).

Our laboratory has considerable experience in synthesizing triply fused benzene rings of this sort by the acid-catalyzed aldol trimerization of cyclic aromatic ketones.[12] For the preparation of hydrocarbon **2** by this strategy, we would need access to the previously unknown acecorannulenone **3** (Figure 9-5). The initial challenge, then, reduces to synthesizing ketone **3**, and we anticipated that **3** should be available from the known alkene, acecorannulylene (**4**).[13]

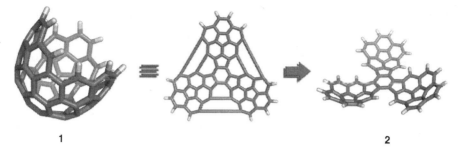

**Figure 9-4.** Retrosynthetic analysis of the $C_{3v}$ [6,6]SWNT endcap (**1**).

## 9-2-2. A Practical Synthesis of Acecorannulylene

Acecorannulylene (**4**) was first reported in 1993 by Rabideau et al.[13] Their synthesis (Scheme 9-1) utilized flash vacuum pyrolysis (FVP) in the final step and was modeled closely after the three-step synthesis of corannulene developed a year earlier by Scott et al.[14] Unfortunately, the FVP precursor (**5**) requires many steps to prepare, and the yield of only 10–15% in the final step made this synthesis unappealing as a starting point for our ambitious venture.

At about the same time, we were investigating an alternative synthesis of acecorannulylene (**4**) that started from corannulene (Scheme 9-2). Friedel–Crafts acylation of corannulene (**6**) with acetyl chloride gives acetylcorannulene (**7**) in good yield. The two-carbon sidechain could then be converted to an α-chlorovinyl group, and subsequent FVP produced acecorannulylene (**4**), presumably by a carbene cyclization analogous to that outlined in Scheme 9-1.[15] Unfortunately, the yield was again low, and this route was abandoned.

In 1999, two new syntheses of acecorannulylene (**4**) were reported in back-to-back publications by Sygula et al.[16] and Siegel et al.[17] Both groups used hexakis(dibromomethyl)-fluoranthene **10** as their final synthetic intermediate and closed the last three rings by reductive cyclizations (Scheme 9-3). The two groups reported complementary routes to **10** and employed different reducing agents in the last step; Sygula and Rabideau used low-valence vanadium, whereas Siegel et al. used low-valence titanium. The most appealing aspect of these syntheses is their avoidance of

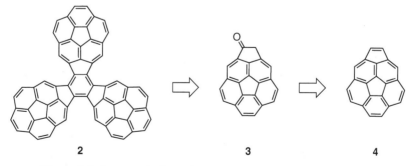

**Figure 9-5.** Retrosynthetic analysis of the $C_{66}H_{24}$ endcap precursor (**2**).

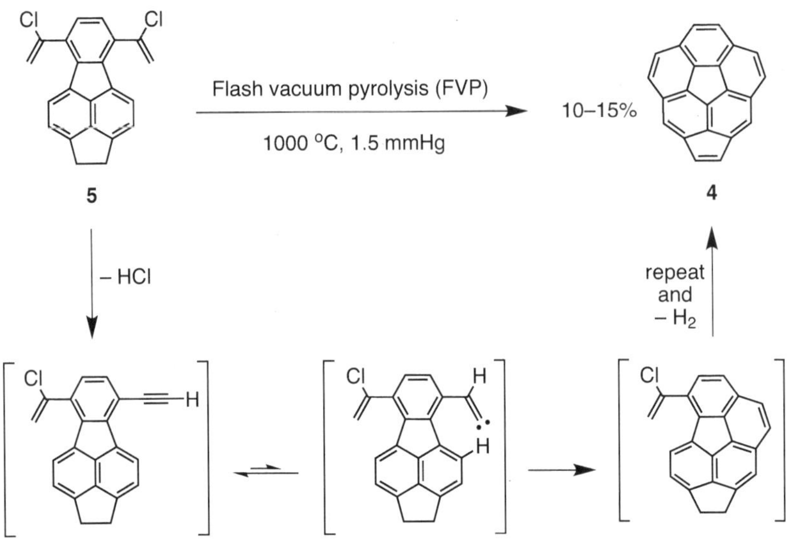

**Scheme 9-1.** Rabideau's pyrolytic synthesis of acecorannulylene (**4**).[13]

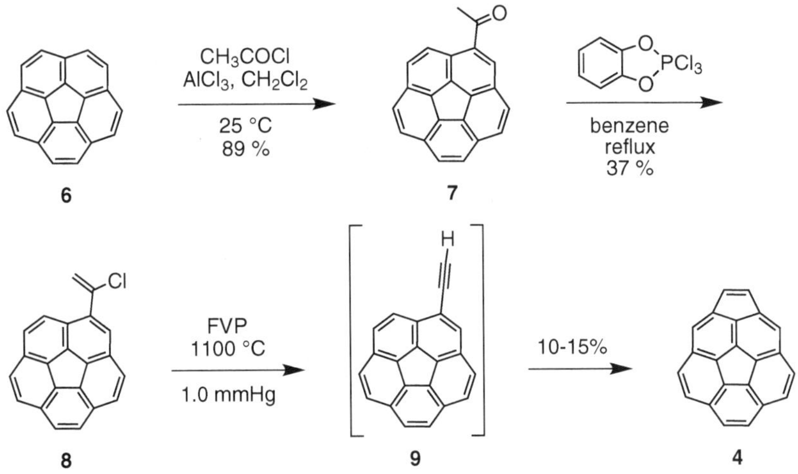

**Scheme 9-2.** Cheng and Scott's pyrolytic synthesis of acecorannulylene (**4**).[15]

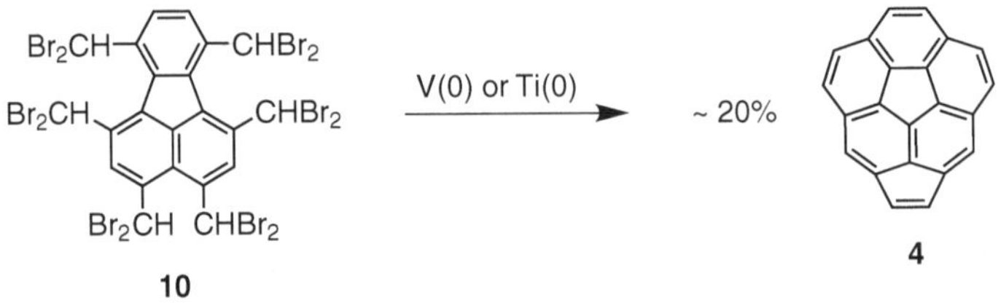

**Scheme 9-3.** Rabideau's and Siegel's nonpyrolytic syntheses of acecorannulylene (**4**). [16,17]

FVP; all of the steps involve ordinary solution-phase chemistry. For our purposes, however, neither synthesis looked practical, as reported. The yield in the final step was only ~20%, regardless of which reducing agent was used, and the quantities produced (60 mg maximum) were far less than what we needed for use as a starting material. Worst of all, we were unable to obtain even a 20% yield of acecorannulylene (**4**) consistently in our preliminary attempts to repeat this tricky reductive cyclization.

No other syntheses of acecorannulylene (**4**) had been reported when we started working on this project. Consequently, we began by evaluating alternative routes. Scheme 9-4 shows two more FVP precursors (**11** and **12**) that are both accessible in high yield from corannulene in two steps; however, neither gave **4** in good yield after isolation and purification.[18,19]

We also prepared and isolated ethynylcorannulene (**9**) and attempted to cyclize it to **4** in solution by a ruthenium-catalyzed isomerization[20] (Scheme 9-5), but we could never get this reaction to work.[19]

Reluctantly, we gave up searching for new routes to **4** and, with some trepidation, turned our attention back to the Rabideau–Siegel solution-phase synthesis (Scheme 9-3). To use this route, we recognized that it would have to be scaled up 10-fold or more and that reproducible conditions for the final reductive cyclization step would have to be worked out. It was our hope, of course, that the yield in the final step might also be improved.

Preparing hexakis(dibromomethyl)fluoranthene **10** is relatively straightforward. The route that we use (Scheme 9-6) is a hybrid of the Rabideau synthesis[16] and the Siegel synthesis.[17] It begins with the benzylic chlorination of mesitylene, a reaction first reported by H. C. Brown in 1939.[21] Using sulfuryl chloride as the limiting reagent, so as not to overchlorinate, we were easily able to access chloromesitylene

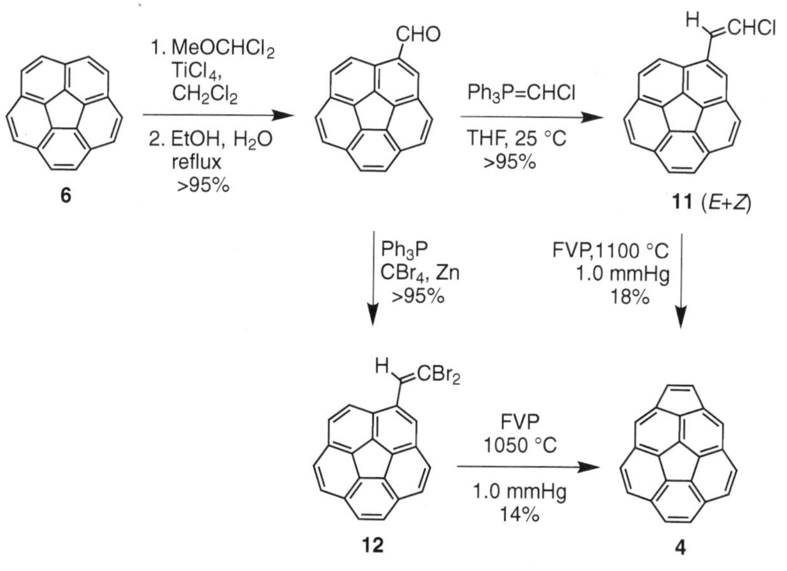

**Scheme 9-4.** More FVP routes to acecorannulylene (**4**).[18,19]

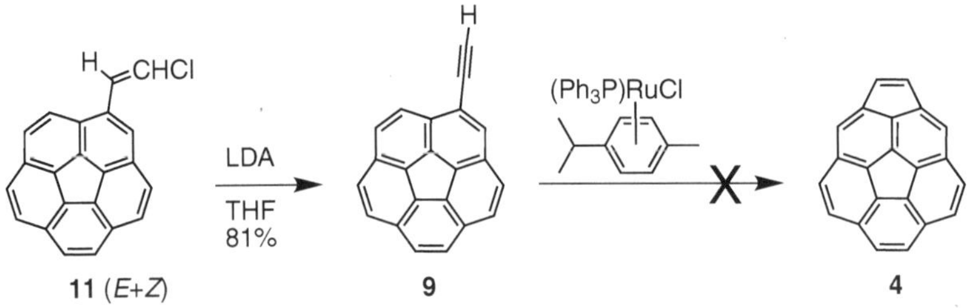

**Scheme 9-5.** An unsuccessful cyclization route to acecorannulylene (**4**).[19]

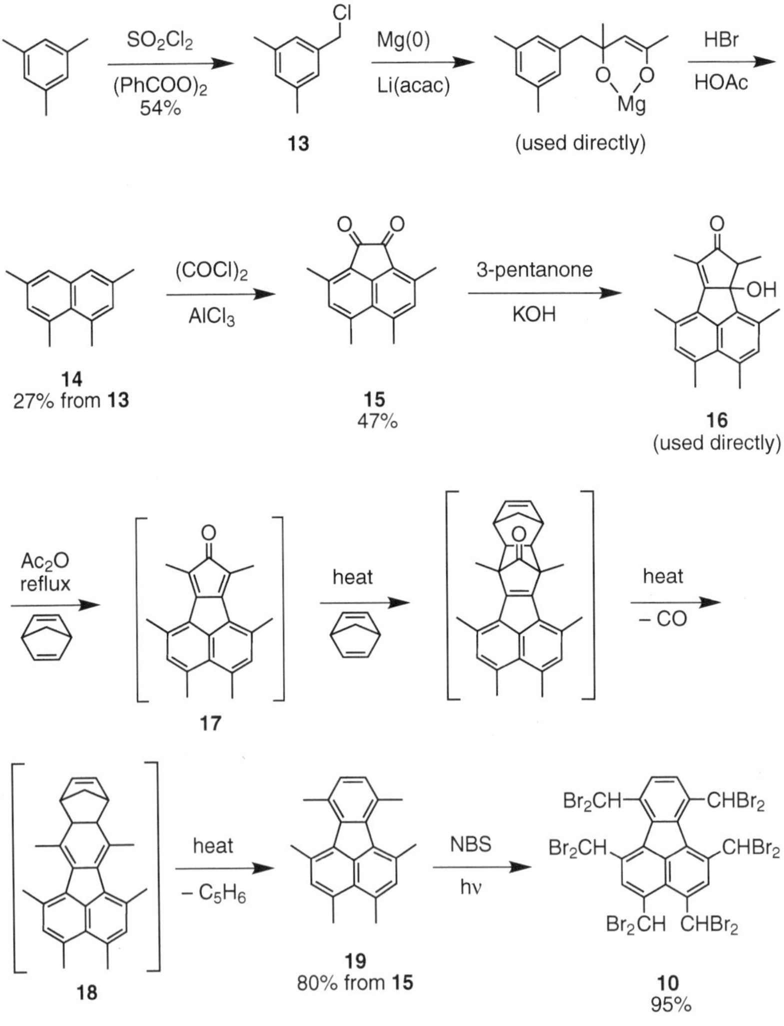

**Scheme 9-6.** Large-scale synthesis of *hexakis*(dibromomethyl)fluoranthene (**10**).[16,17]

(**13**) in large quantities (~80 g per reaction). This benzylic chloride was then converted to the corresponding Grignard reagent in the presence of lithium acetylacetonate, to which it adds *in situ*. Subjecting the crude product of this reaction to HBr in AcOH at reflux produces 1,3,6,8-tetramethylnaphthalene (**14**) in 27% overall yield from **13**.[22] This hydrocarbon is then converted to quinone **15** in 47% yield by a double Friedel–Crafts acylation with oxalyl chloride and aluminum chloride.[16] A double-aldol condensation of **15** with 3-pentanone leads to hydroxy ketone **16**, which gives 1,3,4,6,7,10-hexamethylfluoranthene (**19**) in 80% yield when it is heated at reflux in a mixture of norbornadiene and acetic anhydride. In this remarkable reaction, the acetic anhydride promotes dehydration of **16** to unveil the reactive cyclopentadienone **17**, which is immediately trapped in an inverse electron demand Diels–Alder reaction. The heptacyclic intermediate thus formed loses CO spontaneously,[23] thereby setting the stage for a retro-Diels–Alder reaction (loss of cyclopentadiene) to deliver **19**.[16] Hydrocarbon **19** was then exhaustively benzylically brominated to give **10** in an impressive 95% yield. The reaction stops cleanly at this stage because the three pairs of face-to-face substituents can presumably accommodate bromine atoms only at the positions pointing away from the through-space neighboring substituent, and those become saturated with bromine atoms when the number reaches 12.

In our hands, the procedures of Rabideau and Siegel for the triple reductive cyclization of **10** to acecorannulylene (**4**) do actually work, but we could never improve the yield above 20%, and the heterogeneous reducing agents were difficult to prepare reproducibly. Neither Rabideau nor Siegel had any reason to optimize the yield for this transformation, but we were strongly motivated to do so. Accordingly, we embarked on a serious crusade to make acecorannulylene (**4**) available in gram quantities. After extensive, frustrating experimentation with other reducing agents and conditions, we eventually worked out a procedure that provides acecorannulylene (**4**) in batches of 1.0 g or more as a bright orange solid in 66% isolated yield from **10**, and this opened the gate to let us move forward.

Our success in developing a practical route to acecorannulylene (**4**) depended on several key factors (other than persistence) and is worth detailing:

1. We found that a highly active form of Ti(0) can be conveniently and reproducibly prepared by adding TiCl$_4$ dropwise to a homogeneous THF solution of sodium bis(2-methoxyethoxy)aluminum hydride (commercially available as RedAl™).[19]

2. We found that the yield of intramolecular reductive couplings could be significantly increased by adding the starting material (**10**) slowly to an excess of the Ti(0) under high dilution conditions, thereby diminishing the intervention of undesirable intermolecular reductive couplings. For this purpose, we used the high-dilution apparatus pictured in Figure 9-6.[24]

3. We found that the yield could be further improved by shortening the reaction time from overnight to one hour. Finally, we found that acecorannulylene (**4**) requires some special handling to minimize polymerization and decomposition during work-up, isolation, and storage. This was the first hurdle we had to overcome.[25]

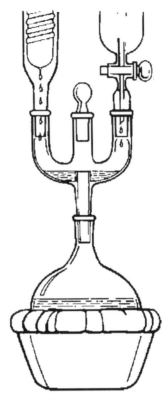

**Figure 9-6.** High-dilution apparatus[24] used for the triple reductive cyclization of **10** to acecorannulylene (**4**) (see Scheme 9-3).

### 9-2-3. Conversion of Acecorannulylene to Acecorannulenone

We expected this step to be easy. In earlier studies on related compounds, we discovered that the epoxide of acenaphthylene (**20a**) isomerizes cleanly and essentially quantitatively to the corresponding ketone (**21a**) simply on contact with silica gel at room temperature (Scheme 9-7).[19] We have subsequently used this chemistry to synthesize substituted acenaphthenones (e.g., **21b**).[12b] DFT calculations (B3LYP/6-31G$^*$) reveal that this isomerization is exothermic by more than 30 kcal/mol, and it looked like an easy way to prepare acecorannulenone (**3**).

The epoxidation of acecorannulylene (**4**) with *meta*-chloroperoxybenzoic acid (MCPBA) is not clean,[25] but the reaction with dimethyldioxirane[26] affords the epoxide of acecorannulylene (**22**) in nearly 100 % yield (Scheme 9-8).[19] From the NMR spectrum of **22**, it is clear that a single diastereomer is formed; we presume that it is the *exo*-epoxide, but we have no direct proof of that stereochemical assignment.

To our surprise, epoxide **22** survives chromatography on silica gel completely unchanged. We initially interpreted this unexpectedly large difference in reactivity

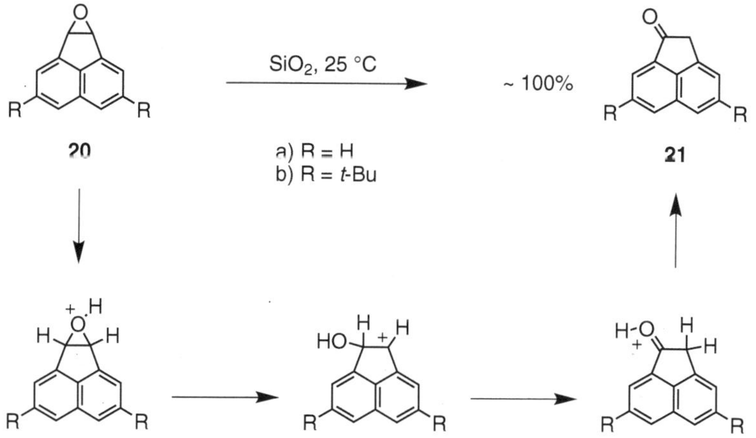

**Scheme 9-7.** A mild epoxide to ketone isomerization.[12b,19]

between **20** and **22** as an indication that the benzylic cation formed by opening epoxide **22** must be significantly less stable than the corresponding benzylic cation in the acenaphthyl system (see the mechanism proposed in Scheme 9-7). Contrary to this interpretation, however, DFT calculations (homodesmic B3LYP/6-31G*) indicate that generation of the benzylic cation from the epoxide should be no more difficult in the acecorannulyl system than in the acenaphthyl system. Further computational probing of the mechanism then revealed that the activation energy for the 1,2 shift of hydrogen (see the mechanism proposed in Scheme 9-7) is nearly 5 kcal/mol higher in the acecorannulyl system than in the acenaphthyl system. We do not fully understand the reason for this large difference in reactivity, and we certainly did not foresee it.

Unwilling to admit defeat, we attempted to effect the isomerization of epoxide **22** to acecorannulenone (**3**) by the action of numerous other acid catalysts. Acetic acid gave no reaction. A two-phase mixture of HCl/HOAc (1:2) and $CH_2Cl_2$ produced small amounts of **3**, but stronger acids and harsher conditions invariably gave multiple products. Reluctantly, we decided to investigate a more classical two-step route involving hydration of the double bond and oxidation of the resulting alcohol (Scheme 9-9).

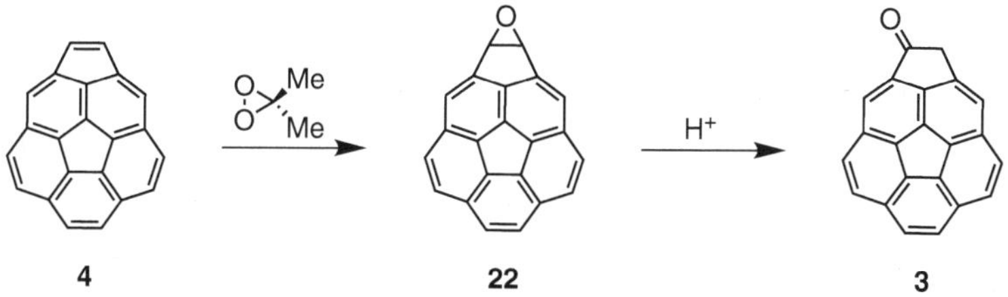

**Scheme 9-8.** The epoxide isomerization route to acecorannulenone (**3**).[19,25]

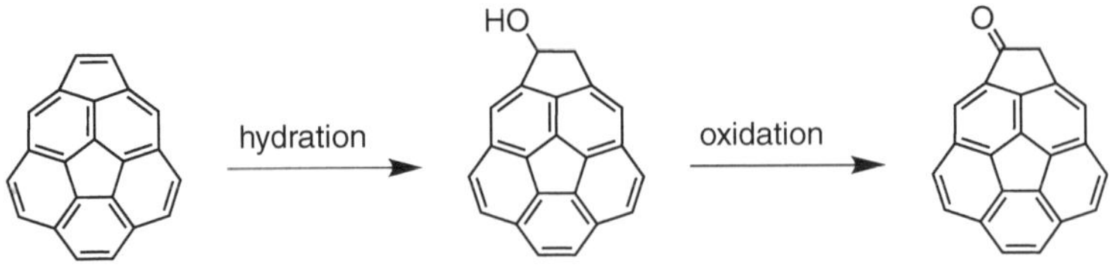

**Scheme 9-9.** Two-step route to acecorannulenone (**3**).[25]

Oxymercuration of **4** with Hg(OAc)$_2$ in aqueous THF followed by demercuration with NaBH$_4$ gives the expected acecorannulenol (**23**) as a mixture of two stereo-isomers, *endo* and *exo* (Scheme 9-10). Because the bowl-to-bowl inversion barrier of the acecorannulene ring system is slow at room temperature,[13b] the two diastereomers could be separated by chromatography and individually analyzed spectroscopically.[25] For synthetic purposes, however, the unseparated mixture of alcohols could simply be oxidized directly with pyridinium chlorochromate (PCC) to give the desired ketone, acecorannulenone (**3**). Unfortunately, neither the oxymercuration–demercuration nor the PCC oxidation worked in high yield in our preliminary investigation of this chemistry, and before we attempted to optimize either step, the prospect of a simpler, alternative one-step route diverted our attention.

It occurred to us that the Wacker oxidation[27] of acecorannulylene (**4**) ought to produce acecorannulenone (**3**) directly in a single step. Our initial attempts to effect this transformation, however, were thwarted by the poor solubility of acecorannu-lylene (**4**) in wet acetonitrile, the standard solvent for such oxidations. Fortunately, we found that by adding THF as a cosolvent, we could obtain acecorannulenone (**3**) directly from acecorannulylene (**4**) in yields as high as 58% (Scheme 9-11).[25] We were never able to improve the yield further, however, even on small-scale reactions, and preparative-scale runs generally gave **3** in 30–35% yield. Nevertheless, consid-ering the directness of this transformation and the fact that hundreds of milligrams of **3** could be obtained this way in a single reaction, we considered it adequate for our purposes and moved on to the aldol cyclotrimerization. We figured that we could always resume optimization experiments later, if this step were ever to become a bottleneck.

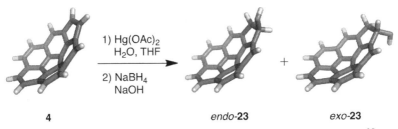

**Scheme 9-10.** Oxymercuration–demercuration of acecorannulylene (**4**).[25]

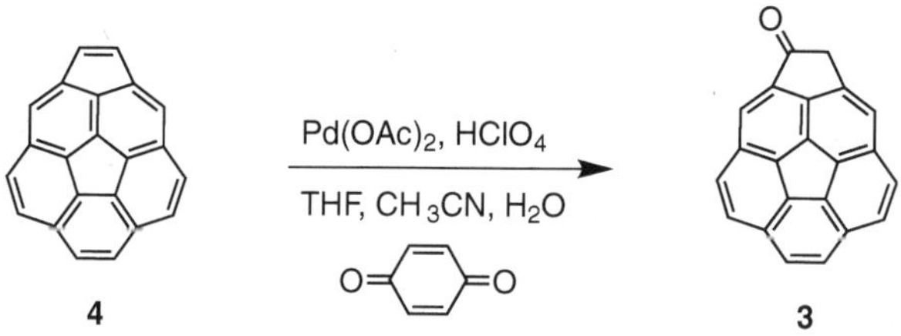

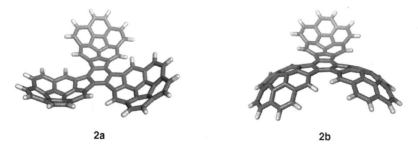

**Scheme 9-11.** The one-step Wacker oxidation route to acecorannulenone (**3**).[25]

## 9-2-4. Acid-Catalyzed Aldol Trimerization of Acecorannulenone

Before discussing the cyclotrimerization of acecorannulenone (**3**), it is worth looking ahead at the desired product (**2**) and noticing that it can exist in two different conformations (Figure 9-7). In conformation **2a**, all three corannulene arms curve in the same direction, whereas in conformation **2b**, one arm curves away from the other two. The former conformer is poised for us to stitch the arms together, as outlined in Figure 9-4, with the curvature already started all in the same direction, but the other is not.

A single bowl-to-bowl inversion of the third arm in **2b** would convert that conformer to **2a**. The energy barrier for such inversions in simple corannulenes is low ($\sim$11 kcal/mol),[28] but bridges between *peri*-positions on corannulene raise the barrier significantly.[13b,29] In the absence of experimental data on inversion barriers for corannulenes that are *peri*-bridged by *ortho*-phenylene units, as in **2**, we take the value measured for acecorannulene (**24**) as a rough estimate for the barrier in **2** (27.7 kcal/mol, Figure 9-8).[13b] A barrier of this height correlates to a bowl-to-bowl inversion half-life of 140 s at 125° C, which is slow on the NMR timescale but still fast enough for the two conformers to interconvert during the course of reactions designed to stitch the arms together at elevated temperatures. DFT calculations (B3LYP/6-31G*) indicate that conformer **2b** should be lower in energy by 5.5 kcal/mol than conformer **2a**.

Initial attempts to catalyze the aldol cyclotrimerization of **3** using a combination of *p*-TsOH·H$_2$O and benzoic acid in hot *ortho*-dichlorobenzene gave no traces of the desired product (**2**). We were surprised by this failure, because very similar conditions were known to promote the aldol cyclotrimerizations of the structurally related acenaphthenone **21a** in good yield (66%).[12c] Other acid catalysts, solvents, temperatures, concentrations, and conditions were tested, but nothing seemed to work.

**2a**                    **2b**

**Figure 9-7.** Conformational isomers of trimer **2** (optimized geometries).

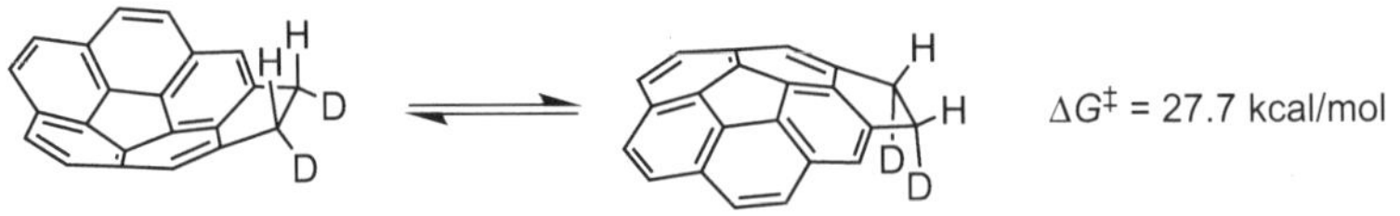

**Figure 9-8.** Bowl-to-bowl inversion of acecorannulene **24**.[13b]

Finally, BBr$_3$ in refluxing 1,1,2,2-tetrachloroethane (147° C)[12c] was found to give clean product formation, albeit only in low yield. The major product, by far, was always the aldol dimer (**25**, Scheme 9-12). Even after extensive optimization, the best we could ever achieve was 6% isolated yield of the desired trimer (**2**).[25] The use of BBr$_3$ as an acid catalyst at higher temperatures in refluxing *ortho*-dichlorobenzene (180°C) also gave trimer **2**, but brominated trimer began to appear as an inseparable contaminant. At room temperature in carbon disulfide, dimer **25** is formed in 96% isolated yield, with no trace of trimer **2**.

For an aldol cyclotrimerization of this sort to work efficiently, as it does for indanone,[12] the crossed aldol condensation of monomer + dimer (e.g., **3** + **25**) to give the acyclic trimer (not shown) must successfully compete with aldol self-condensation of the monomer (e.g., **3** + **3** → **25**). Factors that are known to work against this requirement include insolubility of the initial dimer and preferential dehydration of the initial aldol addition product to a β,γ-unsaturated dimer ketone.[12] Neither complication was anticipated in the present case, however, and neither seems to be the source of the problem encountered. Some other unidentified factor apparently favors aldol self-condensation of the monomer in this case (**3** + **3** → **25**) over the desired crossed aldol of monomer + dimer (**3** + **25**) leading ultimately to **2**. Once the monomer has all been consumed, the product composition is more or less fixed.

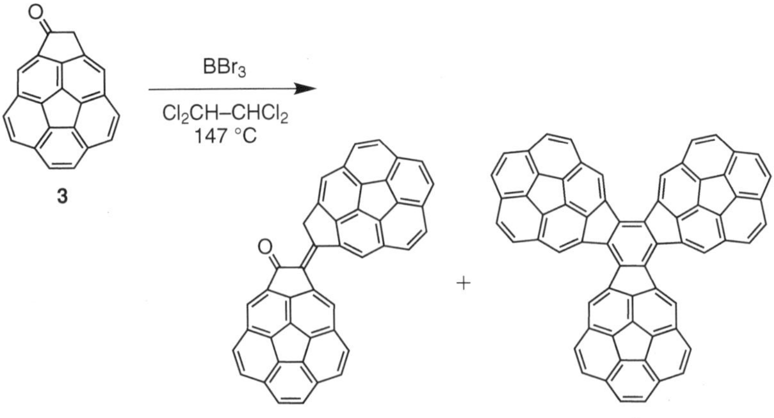

**Scheme 9-12.** Aldol cyclotrimerization of acecorannulenone (**3**).

With access to both the monomer (**3**) and the dimer (**25**), we attempted to effect the crossed aldol by adding **3** slowly to a hot solution of **25** and the acid catalyst, but, not surprisingly, the aldol self-condensation of **3** still dominates the chemistry, and very little trimer (**2**) is formed.

High-resolution mass spectrometry confirms the $C_{66}H_{24}$ formula for trimer **2**, and NMR spectroscopy ($^1$H and $^{13}$C) confirms that the compound adopts the lower-energy conformer **2b**, as predicted. The higher-energy conformer **2a** is calculated to have a propeller-like $C_3$ geometry; however, rapid interconversion of the left- and right-handed propellers by hydrogen atoms passing each other in the fjord regions would be expected to impart time-averaged $C_{3v}$ symmetry. The $^1$H NMR spectrum of conformer **2a** would therefore simplify to just two singlets and two doublets. What we see instead (Figure 9-9) is three low-field singlets (2:2:2 for the three symmetry-related pairs of fjord-region hydrogens: $H_a$, $H_h$, and $H_i$), one high-field singlet for the hydrogens on the ring that is bisected by the pseudosymmetry plane ($H_l$), and eight doublets for the remaining hydrogens (one pair actually collapsed to another singlet, probably $H_d$ and $H_e$), exactly as expected for conformer **2b** with time-averaged $C_s$ symmetry.

As further confirmation of the structural assignment, this $^1$H NMR spectrum obtained experimentally was found to correlate beautifully with the NMR spectrum calculated for **2b** at the B3LYP/6-31G$^*$ level of theory (Table 9.1, referenced to the

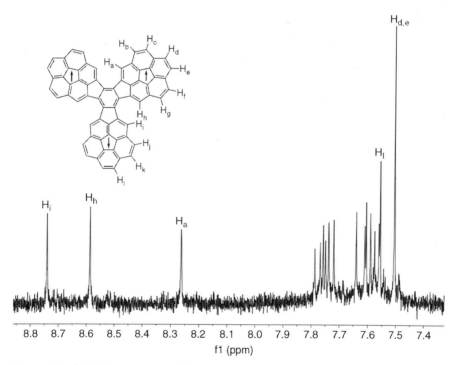

**Figure 9-9.** $^1$H NMR spectrum (400 MHz, 1 : 3 CDCl$_3$ : CS$_2$) of trimer **2b** (arrows in the corannulene arms signify relative bowl curvature).[25]

**TABLE 9-1. Calculated $^1$H NMR Shifts (B3LYP/6-31G$^*$) Compared to Experimental $^1$H NMR Shifts for Trimer 2b**

| B3LYP/6-31G$^*$ (ppm) | Experimental (ppm) | Deviation (ppm)[a] |
|---|---|---|
| 7.54 | 7.51 | 0.03 |
| 7.57 | 7.56 | 0.01 |
| 7.60 | 7.58 | 0.02 |
| 7.64 | 7.59 | 0.05 |
| 7.65 | 7.62 | 0.03 |
| 7.72 | 7.73 | 0.01 |
| 7.75 | 7.75 | 0.00 |
| 7.77 | 7.77 | 0.00 |
| 8.36 | 8.26 | 0.10 |
| 8.60 | 8.59 | 0.01 |
| 8.76 | 8.74 | 0.02 |

*Source*: Ref. 25.
[a]Average deviation $= 0.03$.

calculated chemical shift of benzene). The average deviation of the calculated shifts from those obtained experimentally was only 0.03 ppm (the largest individual variation was 0.10 ppm).

To search for evidence that the bowls in **2b** can invert, even if only slowly, we recorded the $^1$H NMR spectrum of the trimer in $C_2D_2Cl_4$ at several temperatures over the range from room temperature to 125°C (Figure 9-10). As hoped, we see definite

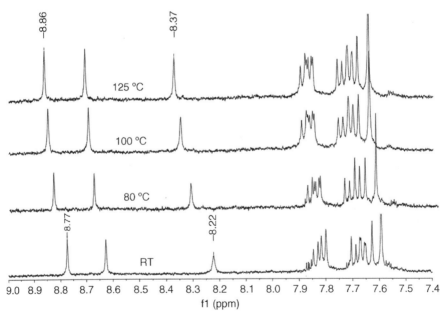

**Figure 9-10.** Variable-temperature $^1$H NMR (500 MHz, $C_2D_2Cl_4$) of trimer **2b**.[25]

signs that the three singlets begin to coalesce; the span over which they appear decreases from 0.55 ppm to 0.49 ppm as the temperature is raised to 125°C. Full coalescence would require a temperature higher than we could access with our NMR spectrometer.

With trimer in hand, we were ready to attempt the crucial sixfold cyclodehydrogenation (Figure 9-4).

## 9-2-5. Attempts to "Stitch up" the Acecorannulenone Trimer (2)

Flash vacuum pyrolysis (FVP) has been successfully employed for the syntheses of more than two dozen geodesic polyarenes,[30] and that might seem like an obvious method to try in this case. The yields in such reactions are always best, however, when radical precursors (usually halogen atoms) are incorporated at strategic positions in the FVP substrate, as the synthesis of circumtrindene (**26**) in Scheme 9-13 illustrates. Direct FVP of unfunctionalized decacyclene (**27a**) affords circumtrindene in only 0.5% yield, and temperatures in excess of 1200°C are required.[31] By contrast, 1,7,13-trichlorodecacyclene (**27b**) is converted to circumtrindene in yields as high as 40% at less severe temperatures.[31b,32]

Faced with the challenge of forming six new bonds intramolecularly, starting from an unfunctionalized hydrocarbon (**2**), we held little hope for FVP in the present case and concentrated instead on solution-phase methods. In particular, we were attracted to the oxidative cyclodehydrogenation chemistry pioneered by Müllen et al., which has been used to convert hexaphenylbenzene to hexabenzocoronene in essentially quantitative yield (Scheme 9-14).[33] It did not escape our notice that six new bonds are formed intramolecularly in this example, starting from an unfunctionalized hydrocarbon, and that all the new rings are six-membered, just as in our case. Müllen and his coworkers have pushed this chemistry to even more breathtaking syntheses of very large PAHs, forming as many as 54 new aryl–aryl bonds in one reaction in an astonishing 62% yield.[34] King and his coworkers have studied the scope, limitations, and mechanism of this spectacular reaction and have introduced some alternative Lewis acidic oxidizing agents and conditions.[35]

We tried everything to cyclize **2** but never saw any evidence for even a trace of the desired $C_{66}H_{12}$ endcap (**1**).[25] Under the normal Müllen conditions at room temperature, we generally recovered only unchanged starting material. Even heating

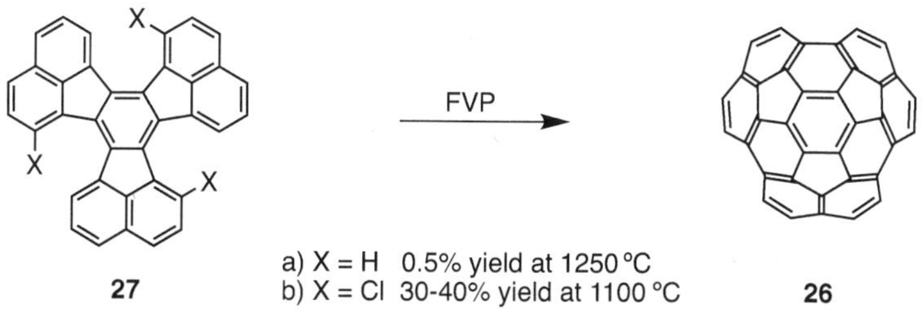

a) X = H    0.5% yield at 1250 °C
b) X = Cl   30-40% yield at 1100 °C

**27**                                                                    **26**

**Scheme 9-13.** FVP syntheses of circumtrindene.[31,32]

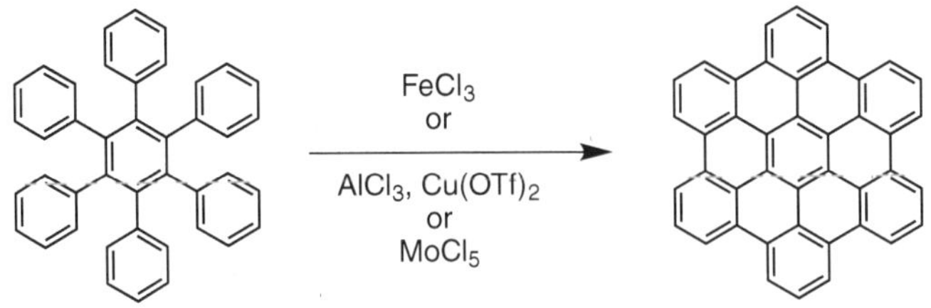

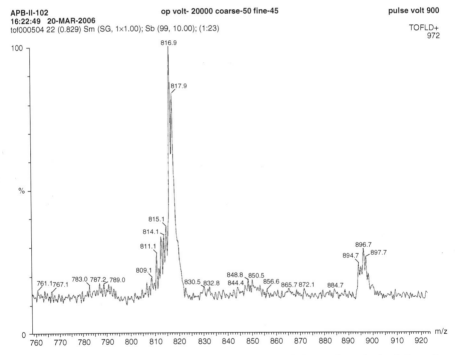

**Scheme 9-14.** Oxidative cyclodehydrogenation of hexaphenylbenzene to hexabenzo-coronene.[33,35]

hydrocarbon **2** at 60°C overnight in a pressure vessel with 18 eq of FeCl$_3$ in CS$_2$/CH$_2$Cl$_2$ (1 : 1) gave back starting material. MoCl$_5$, on the other hand, caused polychlorination of **2**, even at room temperature. Our most promising results were obtained using AlCl$_3$/Cu(OTf)$_2$ (6 eq of each) in CS$_2$ at 60°C for 30 min in a microwave reactor. These conditions gave back mostly unchlorinated **2**, very cleanly, but the mass spectrum of the recovered material showed the first hints of partial cyclodehydrogenation (Figure 9-11). The large peak at *m/z* 816 corresponds to starting material **2**, and the progression of small peaks of slightly lower mass presumably corresponds to minor amounts of partially cyclized material. Unfortunately, none of our attempts to improve

**Figure 9-11.** Mass spectrum (LDI, positive-ion mode) of the product obtained from the reaction of trimer **2b** (*m/z* 816) with AlCl$_3$/Cu(OTf)$_2$ (6 eq of each) in CS$_2$ at 60°C for 30 min in a microwave reactor.[25]

on this result ever gave significantly more cyclization products. The cluster of peaks at $m/z$ 894/896 corresponds to small amounts of brominated **2**, carried over as impurities from the $BBr_3$-catalyzed aldol trimerization.

The resistance of hydrocarbon **2** to oxidative cyclization reagents, even at 60°C, presumably reflects the high ionization potential of corannulene and of curved polyarenes in general, relative to planar polycyclic aromatic hydrocarbons.[36] Whereas corannulenes easily take on one, two, three, and even four electrons to produce anionic states in a reversible manner,[37] the radical cation of corannulene is difficult to form and cannot be generated reversibly even with rapid-scan cyclic voltammetry.[36a,37f]

In light of this intrinsic reactivity of the corannulene nucleus, we decided to turn from cationic/oxidative cyclodehydrogenations to anionic/reductive cyclodehydrogenations. Such reactions are less widely known than their cationic variants, but good precedents can be found.[38] The reduction of [5]helicene with potassium metal in THF-$d_8$, for example, provides important mechanistic insights (Scheme 9-15).[38d] In this experiment, Rabinovitz and coworkers followed the reduction by NMR spectroscopy and were able to record the spectrum of the initial cyclization product, a dianion with two hydrogens attached to $sp^3$ carbon atoms. No signals could be seen for the dianion of [5]helicene, which means that either (1) the cyclization has already occurred at the stage of the monoanion radical or (2) the dianion of [5]helicene cyclizes rapidly as soon as it is formed. Over a period of several days at $-30°C$, or sooner at room temperature, the initial cyclization product loses the original fjord-region hydrogens, and the NMR spectrum for the dianion of benzo[$ghi$]perylene grows in. Bubbling oxygen through the reaction mixture at that stage yields the neutral product of reductive cyclodehydrogenation, benzo[$ghi$]perylene. Whether the hydrogens are lost as $H_2$ or as 2 mol of KH is not known, but they are definitely lost before oxygen is added. The same sequence of spectra is seen when sodium or lithium metal is used in place of potassium.

We were further enticed by the behavior of indenocorannulene (**28**) toward potassium metal.[39] Two monoanion radicals of indenocorannulene come together

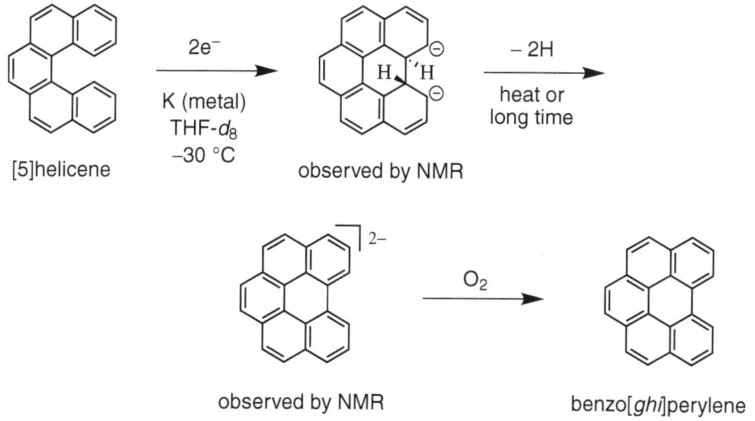

**Scheme 9-15.** Reductive cyclodehydrogenation of [5]helicene with potassium metal.[38d]

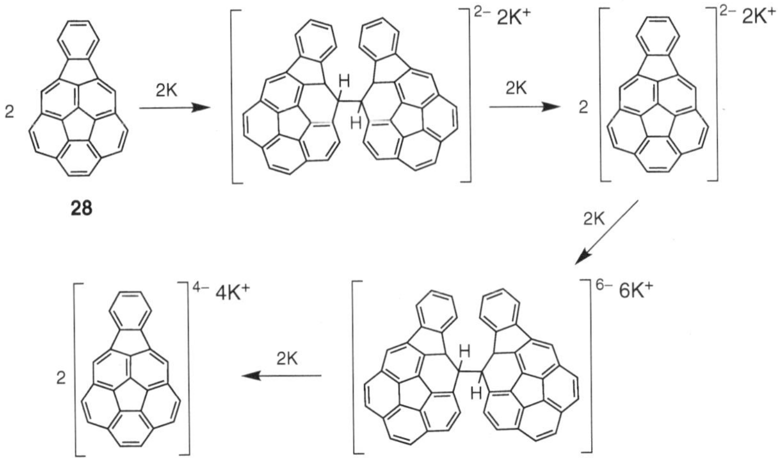

**Scheme 9-16.** Reductive coupling of indenocorannulene with potassium metal.[39]

to form a dimer dianion (Scheme 9-16). On further reduction, the dimer breaks apart into two monomer dianions. A third round of reduction gives a dimer hexaanion, and a fourth reduction gives monomeric tetraanions.

These experiments were all performed at low temperatures, so no hydrogens were lost, but we were excited to note that the reactive position on the indenocorannulene (**28**) corresponds exactly to the positions on the indenocorannulene moieties in our trimer (**2**) where we are seeking to effect C−C bond formation (Scheme 9-17). By raising the temperature, we would expect to see rearomatization by hydrogen loss, as Rabinovitz sees with [5]helicene,[38d] and others have seen in different systems.[38]

We were further encouraged to note that the final three ring closures of **2** all correspond to [5]helicene cyclizations (Scheme 9-18).

Finally, the poor solubility of most highly condensed polyarenes causes serious problems with their isolation, purification, and characterization, as Müllen and others have noted many times.[40] The dianions and more highly reduced derivatives of

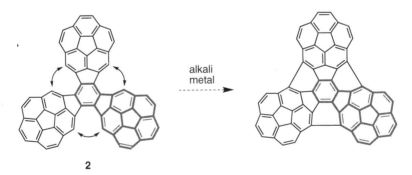

**Scheme 9-17.** Proposed reductive couplings of the indenocorannulene moieties in **2**.

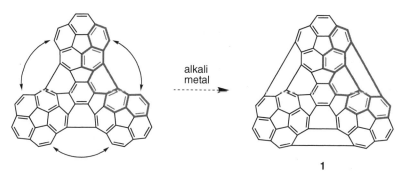

**Scheme 9-18.** Proposed reductive couplings of the [5]helicene moieties in the partially stitched-up **2**.

large polyarenes, on the other hand, generally exhibit good solubility and can be conveniently characterized by NMR.[37] We saw this as another attraction of the reductive cyclodehydrogenation strategy.

Despite all these reasons for optimism, unfortunately, we have (so far) not succeeded in obtaining evidence that **2** can be cyclized to the nanotube endcap **1** with alkali metal reducing agents. Lithium, sodium, and potassium were all tried, with large excesses of the metals being used in refluxing THF for periods of 2–16 h, but mass spectrometric analysis of the crude product mixtures (after oxygen quenches) showed only the gradual disappearance of starting material.

To better control the amount and strength of the reducing agent, we abandoned the use of raw metals and tried using measured amounts of titrated, homogeneous solutions of potassium naphthalenide in THF.[41] Experiments with model systems demonstrated that this protocol works well, especially with microwave heating.[25] Unfortunately, reduction of **2** with ~25 eq of potassium naphthalenide in THF at 90°C in a microwave reactor, followed by an oxygen quench, gave only starting material (*m/z* 816) and what appears to be small amounts of Birch reduction products (*m/z* 818–824).

One possible explanation for the recovery of starting material from these reductions (the optimistic explanation) is that hydrocarbon **2** is reduced too far under the conditions used and goes to a nonreactive polyanionic state that simply sits idle until the workup. No experimental details are available concerning the stage at which anionic cyclodehydrogenations of this sort occur (monoanion radical vs. dianion in simple systems). The behavior of indenocorannulene (**28**, Scheme 9-16) suggests that reduction of the corannulene moieties in **2** all the way to their tetraanions would make a stable species that should resist C—C bond formation. Consequently, we decided to change to another electron source that is not as powerful as potassium naphthalenide. For this purpose, we chose tetrapotassium corannulenide (Cor$^{4-}$/4K$^+$), which can also be prepared as a homogeneous solution[37a–c] and used in controllable stoichiometry. Limited experiments with tetrapotassium corannulenide used in large excess (90°C/microwave, 20 min, followed by oxygen quench) gave back starting material (**2**),[25] but the full range of stoichiometries has yet to be explored.

## 9-3. FUTURE PROSPECTS

For future attempts at anionic cyclodehydrogenation of trimer **2**, it might be profitable to employ an even milder reducing agent than tetrapotassium corannulenide $(Cor^{4-}/4K^{+})$, such as the green monoanion radical of corannulene (monopotassium corannulenide, $Cor^{-}/K^{+}$).[37a–c] This electron source should not overreduce **2**, even if used in excess. Of course, electrochemical reduction would be the most precise way to tune the reduction state of **2**, but these experiments have not been initiated. Stitching up trimer **2** on the surface of a metal catalyst is another alternative, and precedents from the research of Echavarren[42] and Nuckolls[43] provide encouragement that this strategy might yield the target nanotube endcap **1**, already attached to a metal surface and ready for catalytic growth to a [6,6]SWNT.

The rational chemical synthesis of single-walled carbon nanotubes that have predefined diameter and chirality remains a goal that has yet to be attained,[44] but we are confident that success lies within reach.

## ACKNOWLEDGMENTS

We thank the National Science Foundation and the Department of Energy for financial support of this research and for funds to purchase mass spectrometers.

## REFERENCES AND NOTES

1. (a) I. Singh, A. K. Rehni, P. Kumar, M. Kumar, H. Y. Aboul-Enein, *Fullerenes, Nanotubes, Carbon Nanostruct.* **2009**, *17*, 361; (b) B. Perez, A. Merkoci, *Chem. Carbon Nanotubes* **2008**, *2*, 337; (c) N. I. Kovtyukhova, *Chem. Carbon Nanotubes* **2008**, *2*, 299; (d) M. J. O'Connell, *Carbon Nanotubes: Properties and Applications*, CRC Press, Boca Raton, FL, **2006**; (e) M. Meyyappan (ed.), *Carbon Nanotubes: Science and Applications*, CRC Press, Boca Raton, FL, **2005**.

2. M. S. Dresselhaus, G. Dresselhaus, P. Avouris, *Topics in Current Physics*, Vol. 80, Springer, Berlin/Heidelberg, **2001**.

3. F. Vollrath, W. J. Fairbrother, R. J. P. Williams, E. K. Tillinghast, D. T. Bernstein, K. S. Gallagher, M. A. Townley, *Nature* **1990**, *345*, 526.

4. (a) M.-F. Yu, O. Lourie, M. J. Dyer, K. Moloni, T. F. Kelly, R. S. Ruoff, *Science* **2000**, *287*, 637; (b) W. Ding, L. Calabri, K. M. Kohlhaas, X. Chen, D. A. Dikin, R. S. Ruoff, *Exp. Mech.* **2007**, *47*, 25.

5. P. G. Collins, P. Avouris, *Sci. Amer.* **2000**, *283*, 62.

6. Y.-C. Tseng, P. Xuan, A. Javey, R. Malloy, Q. Wang, J. Bokor, H. Dai, *Nano Lett.* **2004**, *4*, 123.

7. (a) R. Martel, *ACS Nano* **2008**, *2*, 2195; (b) R. C. Haddon, J. Sippel, A. G. Rinzler, F. Papadimitrakopoulos, *MRS Bull.* **2004**, *29*, 252.

8. (a) M. Zheng, A. Jagota, E. D. Semke, B. A. Diner, R. S. McLean, S. R. Lustig, R. E. Richardson, N. G. Tassi, *Nat. Mater.* **2003**, *2*, 338; (b) M. Zheng, A. Jagota, M. S. Strano, A. P. Santos, P. Barone, S. G. Chou, B. A. Diner, M. S. Dresselhaus, R. S. McLean, G. B. Onoa, G. G. Samsonidze, E. D. Semke, M. Usrey, D. J. Walls, *Science* **2003**, *302*, 1545.

9. E. H. Fort, P. M. Donovan, L. T. Scott, *J. Am. Chem. Soc.* **2009**, *131*, 16006.

10. (a) B. D. Steinberg, E. A. Jackson, A. S. Filatov, A. Wakamiya, M. A. Petrukhina, L. T. Scott, *J. Am. Chem. Soc.* **2009**, *131*, 10537; (b) T. J. Hill, R. K. Hughes, L. T. Scott, *Tetrahedron* **2008**, *64*, 11360; (c) E. A. Jackson, B. D. Steinberg, M. Bancu, A. Wakamiya, L. T. Scott, *J. Am. Chem. Soc.* **2007**, *129*, 484.

11. (a) T. G. Schmalz, W. A. Seitz, D. J. Klein, G. E. Hite, *Chem. Phys. Lett.* **1986**, *130*, 203; (b) H. W. Kroto, *Nature* **1987**, *329*, 529.

12. (a) M. M. Boorum, L. T. Scott, in *Modern Arene Chemistry*, D. Astruc (ed.), Wiley-VCH, Weinheim, **2002**, p. 20; (b) A. W. Amick, K. S. Griswold, L. T. Scott, *Can. J. Chem.* **2006**, *84*, 1268; (c) A. W. Amick, L. T. Scott, *J. Org. Chem.* **2007**, *72*, 3412.

13. (a) A. H. Abdourazak, A. Sygula, P. W. Rabideau, *J. Am. Chem. Soc.* **1993**, *115*, 3010; (b) A. Sygula, A. H. Abdourazak, P. W. Rabideau, *J. Am. Chem. Soc.* **1996**, *118*, 339. The term *acecorannulylene* for hydrocarbon **4** is used analogously to the naming of acenaphthylene. A more proper name would be cyclopenta[*bc*]corannulene. By analogy to the naming of acenaphthene, the corresponding hydrocarbon with a saturated bridge would be termed *acecorannulene*.

14. L. T. Scott, P.-C. Cheng, M. S. Bratcher, *Proc. 7th Int. Symp. Novel Aromatic Compounds*, Victoria, British Columbia, Canada, July **1992**. Abstract 64. Experimental details for this work were sent to Prof. Rabideau in 1992 as a personal communication and were later published as a full paper: L. T. Scott, P.-C. Cheng, M. M. Hashemi, M. S. Bratcher, D. T. Meyer, H. B. Warren, *J. Am. Chem. Soc.* **1997**, *119*, 10963.

15. P.-C. Cheng, PhD dissertation, Boston College, Chestnut Hill, MA, **1996**.

16. A. Sygula, P. W. Rabideau, *J. Am. Chem. Soc.* **1999**, *121*, 7800.

17. T. J. Seiders, E. L. Elliott, G. H. Grube, J. S. Siegel, *J. Am. Chem. Soc.* **1999**, *121*, 7804.

18. The high-yield formylation of corannulene has been described in D. V. Preda, PhD dissertation, Boston College, Chestnut Hill, MA, **2001**. The other chemistry in Scheme 9-4 has not been published.[19]

19. J. Mack, K. A. Mirica (neé Zharkova), and L. T. Scott (unpublished results).

20. P. M. Donovan, L. T. Scott, *J. Am. Chem. Soc.* **2004**, *126*, 3108 and references cited therein.

21. M. S. Kharasch, H. C. Brown, *J. Am. Chem. Soc.* **1939**, *61*, 2142.

22. P. Boudjouk, W. H. Ohrbom, J. B. Woel, *Synth. Commun.* **1986**, *16*, 401–410.

23. As evidence that the loss of CO precedes the retro-Diels–Alder step in cascade reactions of this type, a byproduct was isolated in a related example, and we determined by X-ray crystallography that it was a Diels–Alder adduct of cyclopentadiene (acting as the dienophile) with the diene corresponding to intermediate **18**.

24. L. T. Scott, C. A. Sumpter, *Org. Synth.* **1990**, *69*, 180.

25. A. P. Belanger, PhD dissertation, Boston College, Chestnut Hill, MA, **2008**.

26. J. K. Crandall, R. Curci, L. D'Accolti, C. Fusco, Dimethyldioxirane, in *Encyclopedia of Reagents for Organic Synthesis*, Wiley, Chichester, **2005**.

27. J. A. Keith, P. M. Henry, *Angew. Chem. Int. Ed.* **2009**, *48*, 9038.

28. L. T. Scott, M. M. Hashemi, M. S. Bratcher, *J. Am. Chem. Soc.* **1992**, *114*, 1920.

29. T. J. Seiders, K. K. Baldridge, G. H. Grube, J. S. Siegel, *J. Am. Chem. Soc.* **2001**, *123*, 517.

30. V. M. Tsefrikas, L. T. Scott, *Chem. Rev.* **2006**, *106*, 4868.

31. (a) L. T. Scott, M. S. Bratcher, S. Hagen, *J. Am. Chem. Soc.* **1996**, *118*, 8743; (b) R. B. M. Ansems, PhD dissertation, Boston College, Chestnut Hill, MA, **2004**.

32. R. B. M. Ansems, L. T. Scott, *J. Am. Chem. Soc.* **2000**, *122*, 2719.

33. X. Feng, J. Wu, V. Enkelmann, K. Müllen, *Org. Lett.* **2006**, *8*, 1145 and references cited therein.

34. C. D. Simpson, J. D. Brand, A. J. Berresheim, L. Przybilla, H. J. Rader, K. Müllen, *Chem.– Eur. J.* **2002**, *8*, 1424.

35. (a) B. T. King, J. Kroulik, C. R. Robertson, P. Rempala, C. L. Hilton, J. D. Korinek, L. M. Gortari, *J. Org. Chem.* **2007**, *72*, 2279; (b) P. Rempala, J. Kroulik, B. T. King, *J. Org. Chem.* **2006**, *71*, 5067; (c) P. Rempala, J. Kroulik, B. T. King, *J. Am. Chem. Soc.* **2004**, *126*, 15002.

36. (a) T. J. Seiders, K. K. Baldridge, J. S. Siegel, R. Gleiter, *Tetrahedron Lett.* **2000**, *41*, 4519; (b) D. Schroder, J. Loos, H. Schwarz, R. Thissen, D. V. Preda, L. T. Scott, D. Caraiman, M. V. Frach, D. K. Böhme, *Helv. Chim. Acta* **2001**, *84*, 1625; (c) H. Becker, G. Javahery, S. Petrie, P. C. Cheng, H. Schwarz, L. T. Scott, D. K. Böhme, *J. Am. Chem. Soc.* **1993**, *115*, 11636.

37. (a) A. Ayalon, M. Rabinovitz, P. C. Cheng, L. T. Scott, *Angew. Chem.* **1992**, *104*, 1691; (b) A. Ayalon, A. Sygula, P.-C. Cheng, M. Rabinovitz, P. W. Rabideau, L. T. Scott, *Science* **1994** 265 (5175), **1065**; (c) M. Baumgarten, L. Gherghel, M. Wagner, A. Weitz, M. Rabinovitz, P.-C. Cheng, L. T. Scott, *J. Am. Chem. Soc.* **1995**, *117*, 6254; (d) A. Weitz, E. Shabtai, M. Rabinovitz, M. S. Bratcher, C. C. McComas, M. D. Best, L. T. Scott, *Chem.– Eur. J.* **1998**, *4*, 234; (e) I. Aprahamian, D. V. Preda, M. Bancu, A. P. Belanger, T. Sheradsky, L. T. Scott, M. Rabinovitz, *J. Org. Chem.* **2006**, *71*, 290; (f) C. Bruno, R. Benassi, A. Passalacqua, F. Paolucci, C. Fontanesi, M. Marcaccio, E. A. Jackson, L. T. Scott, *J. Phys. Chem. B* **2009**, *113*, 1954.

38. (a) D. Tamarkin, D. Benny, M. Rabinovitz, *Angew. Chem.* **1984**, *96*, 594; (b) D. Tamarkin, Y. Cohen, M. Rabinovitz, *Synthesis* **1987**, 196; (c) K. H. Koch, K. Müllen, *Chem. Ber.* **1991**, *124*, 2091; (d) A. Ayalon, M. Rabinovitz, *Tetrahedron Lett.* **1992**, *33*, 2395; (e) R. Benshafrut, R. E. Hoffman, M. Rabinovitz, K. Müllen, *J. Org. Chem.* **1999**, *64*, 644; (f) L. Eshdat, A. Ayalon, R. Beust, R. Shenhar, M. Rabinovitz, *J. Am. Chem. Soc.* **2000**, *122*, 12637; (g) K. Briner, B. M. Mathes, Lithium, *Encyclopedia of Reagents for Organic Synthesis*, Wiley, *Chichester*, **2001**.

39. I. Aprahamian, R. E. Hoffman, T. Sheradsky, D. V. Preda, M. Bancu, L. T. Scott, M. Rabinovitz, *Angew. Chem. Int. Ed.* **2002**, *41*, 1712.

40. See, for example (a) V. S. Iyer, K. Yoshimura, V. Enkelmann, R. Epsch, J. P. Rabe, K. Müllen, *Angew. Chem. Int. Ed.* **1998**, *37*, 2696; (b) O. De Frutos, B. Gomez-Lor, T. Granier, M. A. Monge, E. Gutierrez-Puebla, A. M. Echavarren, *Angew. Chem. Int. Ed.* **1999**, *38*, 204; (c) O. De Frutos, T. Granier, B. Gomez-Lor, J. Jimenez-Barbero, A. Monge, E. Gutierrez-Puebla, A. M. Echavarren, *Chem.–Eur. J.* **2002**, *8*, 2879; (d) D. Wasserfallen, M. Kastler, W. Pisula, W. A. Hofer, Y. Fogel, Z. Wang, K. Müllen, *J. Am. Chem. Soc.* **2006**, *128*, 1334.

41. B. A. Merrill, Potassium naphthalenide, in *Encyclopedia of Reagents for Organic Synthesis*, Wiley, Chichester, **2001**.

42. G. Otero, G. Biddau, C. Sanchez-Sanchez, R. Caillard, M. F. Lopez, C. Rogero, F. J. Palomares, N. Cabello, M. A. Basanta, J. Ortega, J. Mendez, A. M. Echavarren, R. Perez, B. Gomez-Lor, J. A. Martin-Gago, *Nature* **2008**, *454*, 865.

43. K. T. Rim, M. Siaj, S. Xiao, M. Myers, V. D. Carpentier, L. Liu, C. Su, M. L. Steigerwald, M. S. Hybertsen, P. H. McBreen, G. W. Flynn, C. Nuckolls, *Angew. Chem. Int. Ed.* **2007**, *46*, 7891.

44. B. D. Steinberg, L. T. Scott, *Angew. Chem. Int. Ed.* **2009**, *48*, 5400.

# CHAPTER 10

# AROMATIC BELTS AS SECTIONS OF NANOTUBES

GASTON R. SCHALLER and RAINER HERGES

## 10-1. INTRODUCTION

The structure of the carbon nanotube is somewhat similar to a "warp and weft" or woven fabric (see Figure 10-1).

The three conventional methods for producing carbon nanotubes (CNT's) are

1. Vaporization of graphite in the presence of transition metals in an electric arc or by irradiation with a laser
2. Pyrolysis of organic or metal organic precursors
3. Chemical vapor deposition (CVD)

These methods require drastic conditions and provide mixtures of CNTs with various helicities (zigzag, armchair, and chiral), diameters, and lengths. Moreover, the structures of the individual nanotubes are not uniform but usually include a multitude of defects. A number of approaches have been developed to separate CNTs and to isolate monodisperse samples. The application of density gradients obviously is a powerful method.[1] However, only small amounts of specific tubes can be isolated.[2,3] In particular for applications in molecular electronics, nanotubes with homogeneous geometries and thus well-defined properties are needed. Rational chemical synthesis could solve this problem. Considering the fact that the rational synthesis of $C_{60}$ has been accomplished,[4,5] wet chemical synthesis of CNTs seems to be a worthwhile but probably an even more ambitious endeavor. From a strategic perspective, a ring, belt, or hemisphere template has to be synthesized first, which can then be extended to a long tube in a second step.

Consequently, the two major stages of a rational synthesis of nanotubes[6] are (1) synthesis of a template and (2) growing the template toward-CNTs.

*Fragments of Fullerenes and Carbon Nanotubes: Designed Synthesis, Unusual Reactions, and Coordination Chemistry*, First Edition. Edited by Marina A. Petrukhina and Lawrence T. Scott.
© 2012 John Wiley & Sons, Inc. Published 2012 by John Wiley & Sons, Inc.

**Figure 10-1.** Knitting a CNT. (Photomontage by R. Herges and T. Winkler based on the painting "La Tricoteuse" by W. A. Bouguereau (1825–1905). Copyright R. Herges and T. Winkler.)

## 10-2. CNT TEMPLATE SYNTHESIS FOR NANOTUBE GROWTH

Three major approaches for template synthesis have been developed. Figure 10-2 depicts the general strategies for the synthesis of suitable CNT templates:

1. Nanotube endcaps (fullerene hemispheres)
2. Nanotube belts by assembly of (a) concave precursors, or from (b) prefabricated rings by ring enlargement metathesis (REM).

### 10-2-1. Synthesis of Nanotube Endcaps

Nanotubes produced by high-temperature methods usually are capped by fullerene hemispheres. Hence, fullerene hemispheres should be suitable templates for nanotube synthesis, because they define diameter and helicity of the growing nanotube, such as in a CVD process. Meanwhile, a number of bowl-shaped molecules (buckybowls)

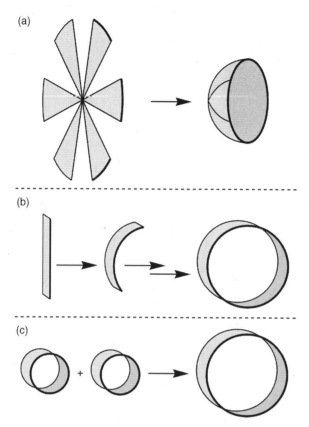

**Figure 10-2.** Strategies for the synthesis of CNT templates: (a) generation of CNT endcaps (fullerene hemispheres); (b) nanotube belts from concave precursors (Section 10.2.1) or (c) from smaller belts by REM (Section 10.2.2).

have been prepared[7] from planar polyaromatic hydrocarbons (PAHs) by flash vacuum pyrolysis (FVP). This high-temperature cyclodehydrogenation reaction usually is a highly endothermic process because alongside with the curvature, a considerable amount of strain is generated.

In 2007, Nuckolls et al. presented a method furnishing fullerene hemispheres from planar PAHs on a ruthenium surface.[8] The planar PAH precursor hexabenzocoronene (**1**) is absorbed on the surface and converted to a $C_{84}$-fullerene hemisphere **2** by intramolecular dehydrogenation, on heating to 600°C (Scheme 10-1).

Whether a subsequent growing process on the ruthenium surface can be accomplished to furnish CNTs remains to be seen.

## 10-2-2. Synthesis of Nanotube Belts from Concave Precursors

To build a beltlike PAH structure, curvature has to be introduced into the usually planar PAHs. Depending on the diameter of the belt, a considerable amount of strain is

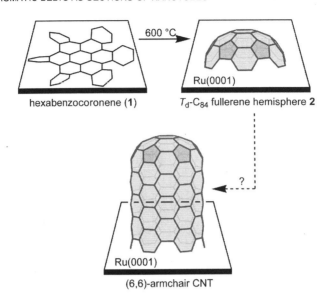

**Scheme 10-1.** Formation of a $C_{84}$ fullerene hemisphere **2** by cyclodehydrogenation of hexabenzocoronene (**1**) on a Ru(0001) surface and the anticipated subsequent growth of an (6,6)-armchair nanotube. (Reproduced from Nuckolls et al., *Angew. Chem.* **2007**, *119*, 8037–8041.)

induced. Any cyclization reaction of a fully benzenoid conjugated linear precursor therefore is prone to failure. The approaches used so far to circumvent this problem are based on the general strategy for building a less strained precursor belt (including $sp^3$ carbons) that is converted into a fully conjugated aromatic belt in a final (usually dehydrogenation) step.

In 1987, Stoddart et al. used concave prefabricated building blocks that were connected by Diels–Alder reactions. Because of the *endo* selectivity of the Diels–Alder reaction, a beltlike nonaromatic structure was formed. However, the full aromatization intended to yield a zigzag nanotube failed.[9,10]

More recently, several other groups have developed similar strategies using either initial Diels–Alder[11,12] or Wittig reactions[13] to assemble the concave precursors.

In 2008, Bertozzi et al. reported the successful synthesis of [9]-, [12]-, and [18]-cycloparaphenylene (Scheme 10-2).[14] The curvature in the precursors is accomplished by using a 3,6-*syn*-dimethoxycyclohexa-1,4-diene moiety (**3**) as a masked aromatic ring. After macrocyclization the marginally strained intermediate **4** is aromatized using lithium naphthalenide as reducing agent, to obtain the desired [*n*]cycloparaphenylenes **5**, **6**, and **7**.

Following this approach, but using cyclohexyl and PtCl$_2$(cod) respectively, to introduce the curvature in the precursor molecule, followed by a cyclization reaction, lead also to various [*n*]cycloparaphenylenes.[15,16]

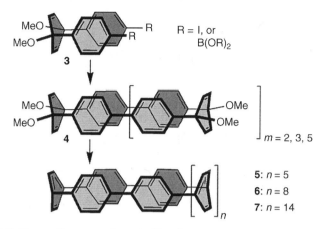

**Scheme 10-2.** Bertozzi's. approach: assembly of concave precursors via Suzuki coupling.

A completely different strategy is under development by Bodwell et al. (Scheme 10-3).[17] They use cyclophanes as the concave precursors. So far, one-half of an aromatic belt (**8**) has already been synthesized that could be viewed as section of an (8,8)-armchair nanotube. The fusion of the two half-pipes **8** would then furnish the fully conjugated nanobelt **9**.

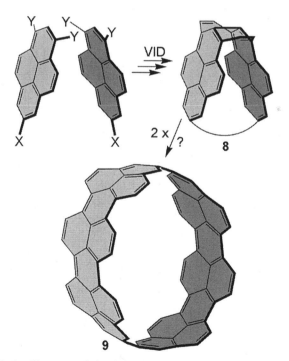

**Scheme 10-3.** Bodwell's approach for the synthesis of aromatic belts using the cyclophane strategy.

### 10-2-3. Synthesis of Nanotube Belts by REM

Carbon nanotubes with small diameters are highly strained because the $sp^2$-carbon atoms are pyramidalized and distorted from their preferred trigonal planar configuration. The most frequently observed nanotubes produced by the high-temperature methods exhibit a diameter of at least 14 Å. The periphery of an (8,8)-armchair nanotube (diameter 13.6 Å) is an [32]annulene. A corresponding beltlike template for nanotube formation or growth, therefore, should be at least that size to avoid excessive strain. Large rings and particularly large annulenes can be prepared by ring enlargement metathesis (REM) of smaller rings. The synthesis of [16]annulene has been achieved by metathesis of [8]annulene.[18,19] However, to build up belts and short tubes, a slightly modified procedure has to be applied. The starting materials must contain at least two flat aromatic plates [e.g., anthracene (**10**) or bisanthene (**11**)], which are connected by at least two quinoid double bonds. REM would then form larger cyclic oligomers of appropriate diameter. In a second independent stage the tube walls have to be closed by cyclodehydrogenation (Scheme 10-4).

The smallest conceivable nanotube is an (2,2)-armchair nanotube **12**, which is extremely strained and therefore has never been observed among the products of high-temperature nanotube production. However, a small section of this nanotube the tetradehydrodianthracene (TDDA) (**13**) is known (Figure 10-3). Because of its high strain it should be a suitable starting compound for preparation of larger rings by ring enlargement metathesis (REM). The driving force for the ring enlargement is the release of strain.

***Synthesis of TDDA: A Promising Building Block.*** Tetradehydrodianthracene (TDDA) (**13**) is accessible in gram quantities using an optimized procedure (Scheme 10-5) developed by Greene et al.[20]

Starting from 9-bromoanthracene (**14**), a photochemically induced [4+4] cycloaddition furnishes 9,10′-dibromodianthracene (**15**). After a twofold base-mediated elimination of hydrobromic acid, TDDA (**13**) is formed as intermediate, that needs to be trapped *in situ* by a 1,3-dipolar cycloaddition using sodium azide as reagent. Otherwise, the base needed for the elimination reacts with TDDA (**13**) by forming nucleophilic addition products such as **16**. Bistriazolinedianthracene **17** reacts with Carpino's reagent to produce *N*-aminobistriazolinedianthracene (**18**), which is subjected to an oxidation with lead tetraacetate to furnish TDDA (**13**) with 42% yield in the final step. The overall yield of this synthesis is 10%.

***Properties, Reactivity, and General Reactions of TDDA.*** TDDA (**13**) belongs to the class of [0_n]paracyclophanes or [*n*]cycloparaphenylenes, which can be written with two mesomeric structures: the quinoid (**13a**) and the benzenoid (**13b**) structure (Figure 10-4).[21] The larger members of this class tend to twist the aromatic subunits out of conjugation, forming a benzenoid structure. In contrast, the smaller members of this class, including TDDA (**13**), adopt a quinoid structure.

The distance between the opposite quinoid double bonds of TDDA (**13**) is 240 pm and hence, significantly below the sum of the van der Waals radii of the $sp^2$-carbon

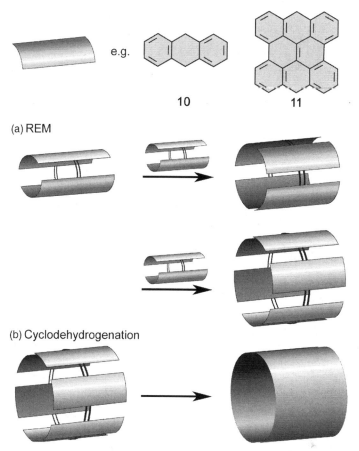

**Scheme 10-4.** Herges' approach for the synthesis of aromatic belts using the REM strategy: (a) REM; (b) cyclodehydrogenation.

atoms (340 pm)[22] (Figure 10-5), resulting in an extremely large *syn*-pyramidalization[23,24] (pyramidalization angle 35°) of the carbon atoms of the quinoid double bonds. The pyramidalization, as well as through-bond and through-space interactions are responsible for its enhanced reactivity. Therefore, TDDA (**13**) reacts with electrophiles,[25] nucleophiles,[26] and dienes (Diels–Alder reaction).[27,28]

**REM of TDDA.** Because of the severe steric shielding of the tetraphenyl-substituted quinoid double bonds, Grubbs and Schrock catalysts failed to induce REM. However, TDDA (**13**) exhibits a remarkable reactivity in photochemically induced cycloadditions. Subsequent thermochemical cycloreversion, driven by release of ring strain, gives the formal metathesis products.

*Molecular Construction Scheme.* TDDA (**13**), in combination with the REM reaction scheme, forms the basic component of a modular construction system (Scheme 10-6):

(a)

(b)

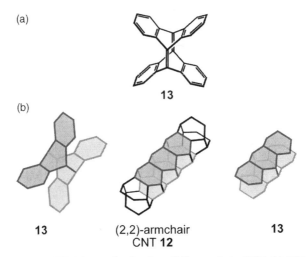

**13**

**13**          (2,2)-armchair          **13**
                 CNT **12**

**Figure 10-3.** (a) TDDA as subunit of an (2,2)-armchair CNT; (b) TDDA (**13**).

1. TDDA reacts with acyclic alkenes to give bianthraquinodimethanes.[29]
2. With cyclic alkenes cyclophane-like bridged bianthraquinodimethanes are formed.[29]
3. REM of TDDA with [*n*]annulenes gives bianthraquinodimethanes, bridged by conjugated chains. Subsequent addition of TDDA to one of the double bonds in the bridging chain gives fully conjugated molecular belts in which the anthracene units are bridged by double bonds and diene units in alternating sequence.[29]
4. Dimerization or cyclic oligomerization of TDDA gives beltlike aromatics in which the anthracene units are connected by double bonds.[30]

*Acyclic and Cyclic Alkenes.* TDDA (**13**) exhibits an unusually high reactivity in photochemically induced [2+2]cycloadditions. The highly strained cyclobutane products in most cases undergo spontaneous cycloreversion, forming the formal methathesis products.

On irradiation at room temperature, ethene readily reacts with TDDA (**13**), forming an isolable cyclobutane derivative **19**, which, on heating, gives bianthraquinodimethane **20** (Scheme 10-7). The steric repulsion of the *peri*-hydrogen atoms prevents a planar conformation of **20**. According to calculations $C_{2h}$ *anti*-isomer **20b** is favored by $-9.3$ kcal/mol relative to the *syn*-isomer **20a** with $C_{2v}$ symmetry.[31]

In the corresponding reaction with propene, the cyclobutane could not be isolated but underwent spontaneous cycloreversion.[32]

Cyclic alkenes react with TDDA (**13**), forming cyclophane-like bridged bianthraquinodimethanes (Scheme 10-8).[33] Because of the cyclophane-like bridging, compounds **21–27** are fixed in the *syn*-pyramidalized conformation. Surprisingly, a reaction with cyclohexene was not observed.

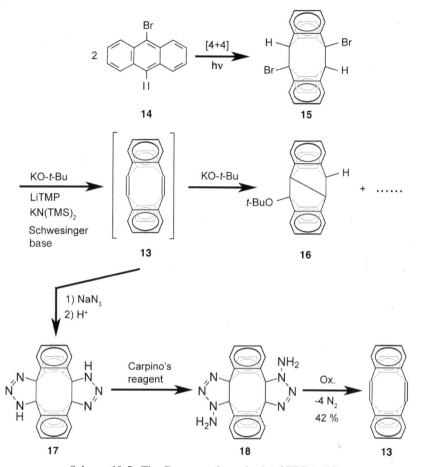

**Scheme 10-5.** The Greene et al. synthesis of TDDA (**13**).

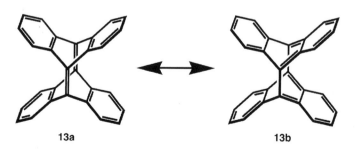

**Figure 10-4.** Quinoid (**13a**) and benzenoid (**13b**) structures of TDDA (**13**).

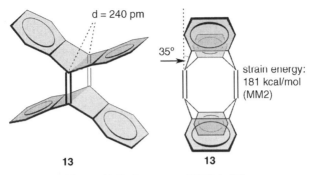

**Figure 10-5.** Structure of TDDA (**13**).

*[n]Annulenes.* On reaction of TDDA (**13**) with [*n*]annulenes (conjugated systems), fully conjugated beltlike bianthraquinodimethane systems are formed.[29,34]

The smallest neutral [*n*]annulene is cyclobutadiene. It can be generated *in situ*[35,36] and subjected to a REM[37,38] with TDDA (**13**) to furnish **28** [Scheme 10-9(a)]. A more

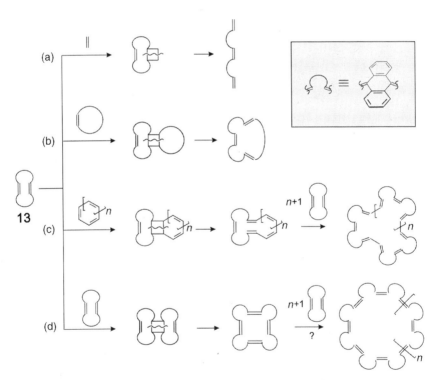

**Scheme 10-6.** REM construction system for furnishing molecular rings, belts, and tubes, depending on the applied reaction partner for TDDA (**13**): (a) acyclic alkene; (b) cyclic alkene; (c) [*n*]annulene; (d) TDDA (**13**).

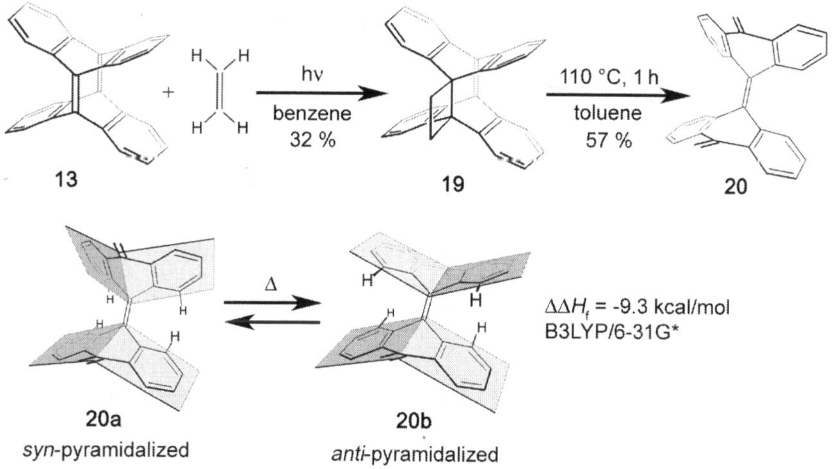

**Scheme 10-7.** REM of TDDA (**13**) with ethene. [From R. Herges and H. Hopf (eds.), *Modern Cyclophane Chemistry*, 2004. Copyright Wiley-VCH Verlag GmbH & Co. KGaA. Reproduced with permission.]

convenient detour to the same target **28** is the use of the easily accessible synthetic equivalents of cyclobutadiene: α-pyrone (**29**) or 1,2-diazine (**30**). In both cases, the dienes react with TDDA (**13**) in a thermochemical [4+2]cycloaddition. The Diels–Alder products **31** and **32** thus obtained eliminate $CO_2$ and $N_2$, respectively, to give **33**. A subsequent electrocyclic ring opening of **33** yields the beltlike structure **28**, with $C_s$ symmetry [Scheme 10-9(b)]. Compound **28** contains 12 π electrons in its annulene periphery. It represents a subunit of an (3,3)-armchair CNT.

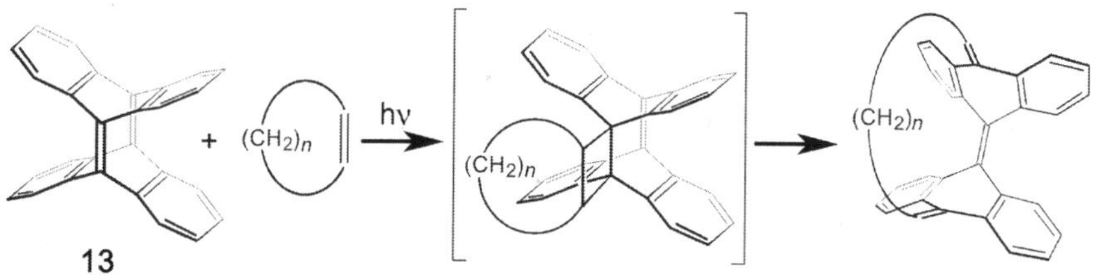

| | | |
|---|---|---|
| cyclopentene | (*n* = 3) | **21**: 11 % |
| cyclohexene | (*n* = 4) | 0 % |
| cycloheptene | (*n* = 5) | **22**: 29 % |
| cyclooctene | (*n* = 6) | **23**: 60 % |
| cyclodecene | (*n* = 8) | **24**: 26 % |
| cyclododecene | (*n* = 10) | **25**: 27 % |
| norbornadiene | | **26**: 61 % |
| norbornene | | **27**: 22 % |

**Scheme 10-8.** REM of TDDA (**13**) with cyclic alkenes.

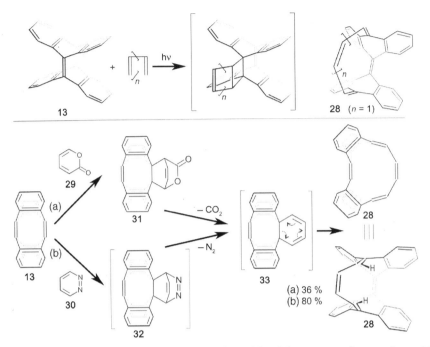

**Scheme 10-9.** REM of TDDA (**13**) with [*n*]annulenes (a) and the corresponding reactions with the cyclobutadiene equivalents α-pyrone (**29**) and 1,2-diazine (**30**) (b). [From R. Herges and H. Hopf (eds.), *Modern Cyclophane Chemistry*, 2004, copyright Wiley-VCH Verlag GmbH & Co. KGaA. Reproduced with permission.]

In contrast to normal arenes with *p* orbitals perpendicular to the plane of the molecule, the *p* orbitals of these structures are perpendicular to the surface of a cylinder. Hence, the inner lobes of the *p* orbitals point toward the axis of the cylinder.

The central double bond in the bridge of **28** is orientated *syn* with respect to the quinoid double bonds, and therefore sterically easily accessible. Hence, the pre-fabricated [*n*]annulene derivative **28** undergoes a further REM with TDDA (**13**), to provide Kammermeierphane (**34**) (Scheme 10-10). Compound **34** includes 20 π electrons in its annulene periphery. Its shape resembles a deformed cylinder with a cavity of 4.8 Å in height and 7.9 Å in width. Kammermeierphane (**34**) is a substructure of an (5,5)-armchair CNT.

Benzene follows cyclobutadiene as the next-higher-order homolog in the [*n*] annulene series. Only few photochemical cycloadditions of benzene are known, and to the best of our knowledge REM reactions have never been observed. However, with TDDA (**13**) compound **35** is formed in 28% yield (Scheme 10-11).[29] Compound **35** is $C_2$-symmetic and has 14 π electrons in its [*n*]annulene periphery. It represents the first rationally synthesized chiral beltlike molecule with a fully conjugated aromatic π system. Molecule **35** can also be viewed as a subunit of a (10,3)-chiral CNT.

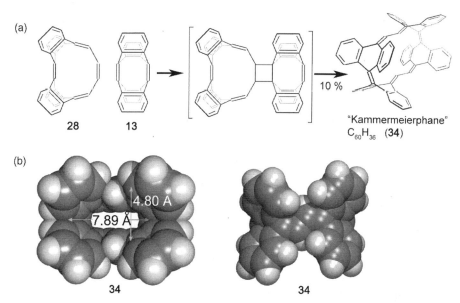

**Scheme 10-10.** (a) REM of **28** with TDDA (**13**) to compound **34**; (b) space-filling model of **34**. [From R. Herges and H. Hopf (eds.), *Modern Cyclophane Chemistry*, 2004. Copyright Wiley-VCH Verlag GmbH & Co. KGaA. Reproduced with permission.]

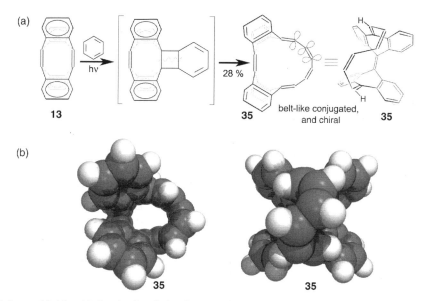

**Scheme 10-11.** (a) Synthesis of the first chiral beltlike conjugated molecule (**35**); (b) space-filling model of **35**. [From R. Herges and H. Hopf (eds.), *Modern Cyclophane Chemistry*, 2004. Copyright Wiley-VCH Verlag GmbH & Co. KGaA. Reproduced with permission.]

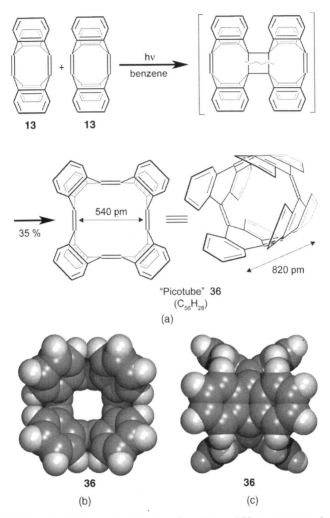

**Scheme 10-12.** Synthesis (a) van der Waals surface (b), and X-ray structure, including one molecule of acetonitrile and carbon disulfide (c) of picotube **36**. [From R. Herges and H. Hopf (eds.), *Modern Cyclophane Chemistry*, 2004. Copyright Wiley-VCH Verlag GmbH & Co. KGaA. Reproduced with permission.]

*Cyclo Dimerization.* TDDA (**13**) undergoes metathetic dimerization on irradiation and forms the tube-shaped structure in **36**. Four anthracenylidene units are connected by quinoid double bonds in a cyclic arrangement (Scheme 10-12).[30] Molecule **36** is a [0$_4$]paracyclophane, as well as a subunit of an (4,4)-armchair CNT, and is consequently termed *picotube* (diameter 540 pm, length 820 pm) in analogy with the larger CNTs.

Picotube **36** contains a fully conjugated but antiaromatic π system with 16 π electrons in its [*n*]annulene subunit. According to calculations at the B3LYP/6-31G$^*$

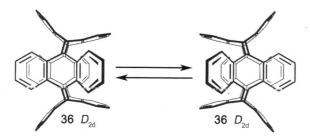

**Figure 10-6.** Time-averaged $D_{4h}$ symmetry of picotube **36**.

level of theory, the most stable conformation of **36** is a $D_{2d}$-symmetric structure, which is formed by a 20° twist of the quinoid double bonds. However, the two possible $D_{2d}$-structures interconvert rapidly at room temperature on the NMR timescale, resulting in a time-averaged $D_{4h}$ symmetry (two signals in the $^1$H- and four signals in the $^{13}$C NMR spectrum) (Figure 10-6). Low-temperature IR spectroscopy[39] revealed the $D_{2d}$ structure of **36**. The alternating bond lengths (quinoid double bonds 136.7 pm, adjacent single bonds 148.9 pm) suggest a quinoid rather than a benzenoid structure of **36**.

Figures 10-7 and 10-8 present an assortment of aromatic rings, belts, and tubular compounds, which represent sections of CNTs, that have been made according to the REM construction scheme.

***Properties, Reactivity, and General Reactions of Picotube.*** Picotube **36** is extraordinarily stable (up to 450°C neither melting nor decomposition is observed), poorly soluble in common solvents, and unreactive toward oxidation (e.g., no reaction occurs with peracids or bromine at room temperature). However, it reacts under Friedel–Crafts alkylation conditions (Scheme 10-13).[40]

Friedel–Crafts alkylation with chloromethane gave a large number of isomers and different degrees of alkylation. Treatment with an excess of $t$-butylchloride/AlCl$_3$ furnished only two major isomers and small amounts of a third isomer. Because of the steric demand of the $t$-butyl groups, the reaction stops after eightfold substitution and the bulky substituents are oriented in such a way that they are placed as far as possible away from each other to avoid steric congestion (Figure 10-9).

This method should also be of interest for derivatizing open CNTs obtained from CVD.

Structure **39** has $D_4$ symmetry and is therefore chiral. Enantioseparation of **39** was achieved by HPLC on a chiral stationary phase.

Guest atoms and molecules inside nanotubes and fullerenes[41] have been used to investigate the electronic and magnetic properties. Because of its similarity to CNT structure, picotube **36** is suitable as a model compound for those studies because it is, in contrast to standard CNTs, readily accessible as a pure compound, moderately soluble, and not endcapped, making the insertion of atoms or cations comparatively easy.

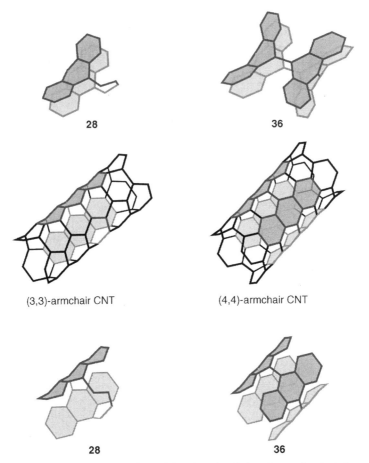

**Figure 10-7.** Beltlike compounds **28** and **36** (reduced to their $\sigma$ frames for clarity) as sections for CNTs.

If treated with Li metal, picotube **36** undergoes a slow chemical reduction to the tetraanion **40** (Figure 10-10).[42] Oxidation of tetraanion **40** leads to an almost quantitative recovery of picotube **36**. No decomposition of tetraanion **40** is observed, even by boiling in THF, which is quite unusual for a pure hydrocarbon.

Two Li$^+$ ions are bound inside, and two outside, the tube. An X-ray structure of the silver complex of the picotube reveals a similar structure.[43] The Ag$^+$ ions inside the tube occupy almost the same positions as do the Li$^+$ ions in **40**.

***Cyclodehydrogenation of Picotube.*** A conventional synthesis of a CNT or a short piece of it is still elusive, even though CNTs are probably the most coveted "nonnatural product" targets in chemistry. Picotube **36** is a promising precursor and a first step toward this target. An eightfold cyclodehydrogenation at the *ortho* positions of the phenyl rings of picotube **36** should lead to a fully conjugated tubular aromatic compound, in this case a subunit of an (4,4)-armchair CNT (**41**).

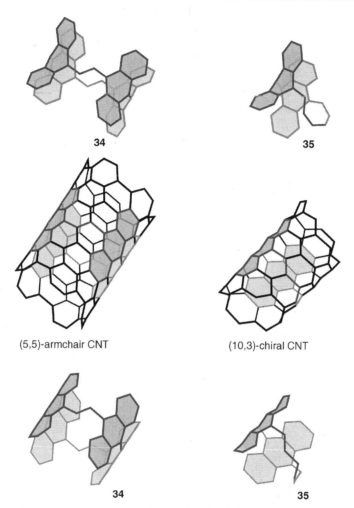

**34**    **35**

(5,5)-armchair CNT    (10,3)-chiral CNT

**34**    **35**

**Figure 10-8.** Belt-like compounds **34** and **35** (reduced to their $\sigma$ frames for clarity) as sections for CNTs.

Similar cyclodehydrogenation reactions leading to planar polyaromatic compounds (fragments of graphene sheets)[44] can be accomplished by oxidation protocols[45] (e.g., under Scholl[46] or Kovacic[47] conditions, irradiation in the presence of iodine[48]). If strained buckybowls are the targets, high-temperature flash vacuum pyrolysis (FVP) (elimination of hydrogen) of suitable precursors is the method of choice. This reaction is known to be able to build up substantial amounts of strain (average strain 47.0 kcal/mol per cyclodehydrogenation step in case of $C_{60}$).[49]

Unfortunately, the picotube **36** is completely unreactive under Kovacic and Scholl conditions. Flash vacuum pyrolysis mainly yields rearrangement products (e.g., **42**) (Scheme 10-14).[50] The structure of cyclic product **42** is confirmed by X-ray analysis,

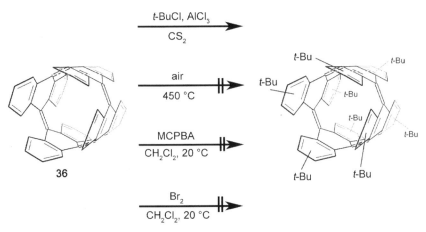

**Scheme 10-13.** Reactivity of picotube **36**.

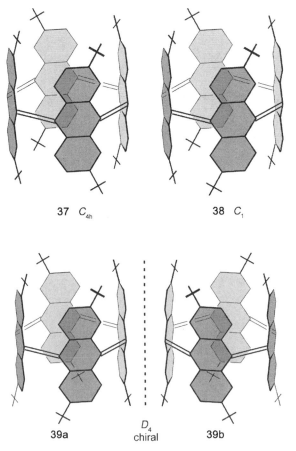

**Figure 10-9.** Three isolated products (**37**, **38**, **39a**, and **39b**) obtained by Friedel–Crafts alkylation of picotube **36** with *t*-butylchloride.

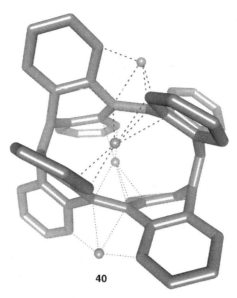

**40**

**Figure 10-10.** Structure of tetraanion **40** optimized at the B3LYP/6-31G* level of theory. The four spheres represent the $Li^+$ ions.

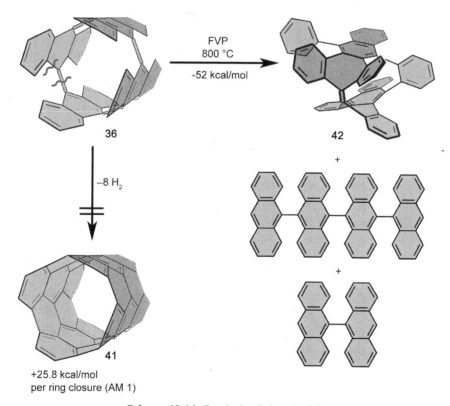

**Scheme 10-14.** Pyrolysis of picotube **36**.

and according to DFT calculations at the B3LYP/6-31G$^*$ level of theory, cyclic product **42** is 52 kcal/mol lower in energy then picotube **36**.

Only small amounts of products were formed by hydrogen elimination that could not be isolated. Semiempirical calculations (AM1) predict that the formation of the (4,4)-armchair nanotube section **41** from picotube **36** by elimination of eight hydrogen molecules is extremely endothermic, by 206 kcal/mol. The ring closure mechanism proceeds stepwise. The reaction enthalpy of the eight cyclodehydrogenation steps were calculated (AM1) to be endothermic by 8.5, 3.2, 3.4, 1.4, 60.2, 48.2, 43.0, and 38.2 kcal/mol. After the fourth step, a $D_{2h}$-symmetric structure is formed that is responsible for the dramatic increase in the reaction enthalpies of the subsequent cyclodehydrogenation steps, especially at the fifth step. Most probably our cyclodehydrogenation failed because the strain energy needed to form the (4,4)-nanotube section **41** is too high. (4,4)-Armchair single-walled nanotubes (SWNTs) indeed have never been observed. However, they are formed as inner tubes inside multiwalled (MWCNTs) and are stabilized by the external layers of the larger CNTs. A straightforward approach to abate the strain problem would be the preparation of larger rings leading to nanotube sections with larger diameters.

**REM of Picotube.** The synthesis of larger rings can be accomplished by REM. Reaction of the pictube **36** with TDDA (**13**) would give a cyclic hexaanthracenylidene,

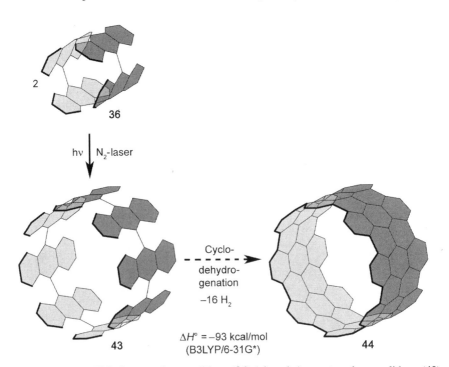

**Scheme 10-15.** REM of tetraanthracenylidene (**36**) (picotube) to octaanthracenylidene (**43**) and the proposed cyclodehydrogenation to a short (8,8)-armchair CNT section (**44**).

which on dehydrogenation, would furnish a section of an (6,6)-armchair nanotube; dimerization of the picotube by REM leads to a cyclic octaanthracenylidene **43** (ring diameter: 11.2 Å), and subsequent cyclodehydrogenation would give an (8,8)-armchair nanotube section (**44**) (Scheme 10-15). According to DFT calculations (B3LYP/6-31G*), the latter cyclodehydrogenation is exothermic (−93 kcal/mol).

Irradiation of picotube **36** with a $N_2$ laser yields the desired octaanthracenylidene **43** in very small amounts, as proved by MALDI mass spectrometry.[51] However, the scaleup to mg amounts, necessary for systematic cyclodehydrogenation experiments, is still under investigation.

## 10-3. CNT GROWING STRATEGIES

So far, all strategies described for a rational synthesis of nanotubes have been based on the preparation of a short template (belt or bowl) that has to be extended to longer

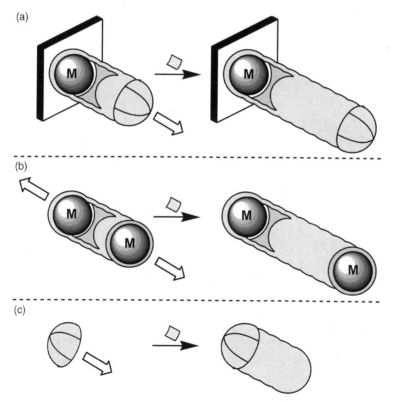

**Figure 10-11.** "Bricking" strategies for growing the templates toward CNTs. Bricking via (a) metal-based chemical vapor deposition (CVD) on a surface, (b) metal-based CVD from both ends, and (c) metal-free cycloaddition of small $C_2$ units to the rim of the nanotube.

tubes in a second independent step. Several approaches have been developed to this end (Figures 10-11 and 10-12):

1. "Bricking" with masked or free $C_2$ units via
   a. Metal-based chemical vapor deposition (CVD) on a surface
   b. Metal-based CVD growth from both ends (amplification method)
   c. Metal-free cycloaddition of small masked $C_2$ units to the rim of a nanotube
2. "Stacking" of
   a. Prefabricated rings or belts
   b. Prefabricated rings or belts in a suitable host

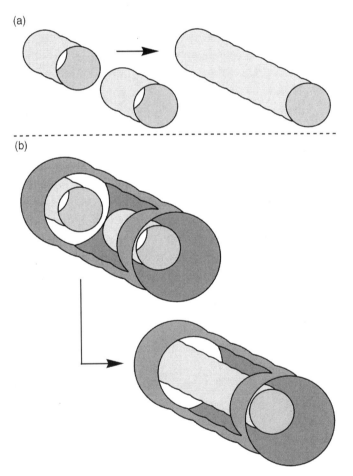

**Figure 10-12.** Stacking strategies for growing the templates toward CNTs. Stacking via (a) prefabricated ring or belts and (b) prefabricated ring or belts in a suitable host.

The proof of principle has been demonstrated for approaches 1a, 1b, and 2b. However, the rational synthesis of bulk amounts of nanotubes remains to be performed.

### 10-3-1. Bricking

***Metal-Based CVD and CVD-Related Processes.*** Currently (as of 2010), most CNTs are produced by the CVD method.[52–55] The tubes grow from metal clusters that usually are deposited on a surface. Carbon from a carbon feedstock gas diffuses into the cluster, forming a solution or metal carbide, until the excess carbon is deposited from the supersaturated solid solution as a fullerene hemisphere around the cluster. Further excretion of carbon pushes the fullerene cap up, which leads to a growing tube (Scheme 10-16).[56] The cluster size determines the diameter of the growing tube.[56] Nevertheless, one has not yet been able to produce monodisperse tubes mainly because it is difficult to generate metal clusters of uniform shape. Even if clusters with identical size could be placed on a surface, different-shape tubes would grow, because there are tubes with the same diameter but different helicities.

Smalley et al.[57] demonstrated an interesting approach to circumvent this problem. They cut a single tube into smaller sections that are used as templates for a CVD-based growth. Each template section is equipped with a metal cluster on both ends and extended in a CVD process. The long tubes are again cut into increasingly small sections[57] (Scheme 10-17). Because of the very large aspect ratio of carbon nanotubes, it should be possible in principle to produce bulk amounts of uniform

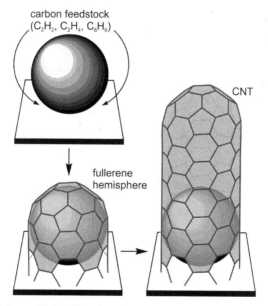

carbon feedstock
($C_2H_2$, $C_2H_4$, $C_6H_6$)

CNT

fullerene
hemisphere

**Scheme 10-16.** Growing of CNTs using the CVD process.

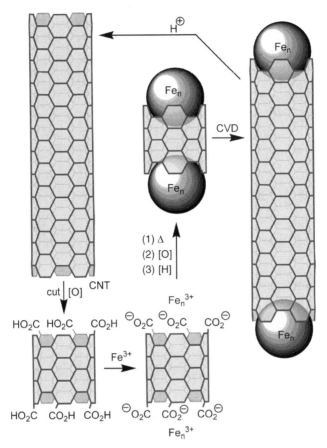

**Scheme 10-17.** Smalley's "amplification" approach for the synthesis of monodisperse nanotubes.

nanotubes from a single nanotube if this "amplification process" is repeated several times.

Whether this approach can be extended to produce bulk amounts of monodisperse nanotubes remains to be seen.[58–61]

***Metal-Free Wet-Chemical Process.*** Scott et al. have proposed a low-temperature, wet-chemical growing strategy that is based on the Clar method to extend planar PAHs such as perylene or bisanthene.[62,63,64] The bay regions of many planar polyaromatics contain a *cis*-butadiene substructure that undergoes Diels–Alder reactions with maleic anhydride. Oxidation of the newly formed cyclohexene ring to benzene and subsequent dicarboxylation forms a new hydrocarbon extended by a benzene ring that exhibits a new bay region. In principle it should be possible to repeat that process at the rim of nanotubes, thereby extending their length (Scheme 10-18).[64]

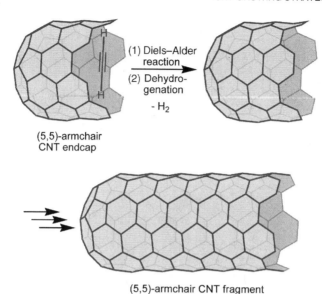

(5,5)-armchair
CNT endcap

(1) Diels–Alder
reaction
(2) Dehydro-
genation
- H$_2$

(5,5)-armchair CNT fragment

**Scheme 10-18.** Scott's approach for growing of CNTs via the Diels–Alder reaction.

Using PAHs such as perylene or 7,14-dimesitylbisanthene and diethylacetylene-dicarboxylate as starting materials, Scott et al. demonstrated that the above mentioned principle can at least be applied to planar aromatic hydrocarbons. A sequence of a Diels–Alder reaction followed by a thermal dehydrogenation step furnishes extended PAHs.

In principle this process can be applied to extend short tubes as well as fullerene hemisphere templates in length using acetylene or an acetylene equivalent as the dienophile. Probably, templates with a relatively large diameter, such as precursors for (8,8)- or even (10,10)-CNTs, are needed to keep the ring strain low enough to obtain sufficiently high yields in the cycloaddition step.

## 10-3-2. General Stacking and Host–Guest Stacking

Stacking of prefabricated rings is an established method used to furnish belt or tubular systems, which has been demonstrated on cyclodextrins,[65,66] cyclopeptides,[67–69] and cyclic hydrocarbons.[70] However, fully conjugated tubes or belts have not been synthesized according to this approach.

Dehydrogenation and stacking of the picotube (**36**) to form nanotubes was not successful either on oxidation in solution or at high temperatures on flash vacuum pyrolysis (FVP).[50] However, it is known whether reactions in well-defined molecular cavities lead to different products than under normal conditions. For example, fullerenes inside carbon nanotubes (peapods) at higher temperatures fuse to give an inner nanotube inside the outer "nano–test tube".[71,72] On FVP the picotube (**36**) rearranged to ring-type products with larger diameters (and lower strain energy).

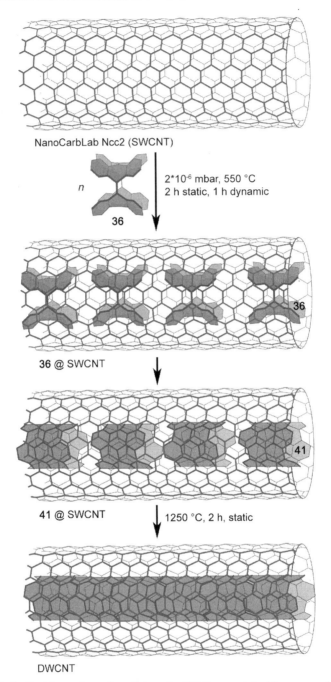

NanoCarbLab Ncc2 (SWCNT)

*n*     36

2*10⁻⁶ mbar, 550 °C
2 h static, 1 h dynamic

**36** @ SWCNT

**41** @ SWCNT     1250 °C, 2 h, static

DWCNT

**Scheme 10-19.** Packing and reaction of picotube (**36**) fragments inside commercial SWCNT.

For a reaction inside a nanotube, this ring expansion would likely be unfavorable. The confinement inside a rigid nanotube should favor reactions that would lead to products of a smaller diameter such as the 4,4-armchair nanotube section (**41**) by dehydrogenation. Also, stacking of the resulting nanotube sections would be facilitated by preorganization of the tube sections in a favorable arrangement.

Following this idea, Kuzmany and Herges and colleagues filled SWCNTs with picotubes **36** (diameter 1.16 nm) into commercially available SWCNTs (mean diameter 1.36 nm) at 550°C and high vacuum (Scheme 10-19).[73]

Under these conditions the **36**@SWCNT spontaneously undergoes dehydrogenation and forms short sections of (4,4)-armchair nanotubes (**41**). Extraction and analysis of the small tubes has not yet been achieved. The structural assignment is based on the comparison of experimentally measured and theoretically (DFT) calculated Raman spectra (Figure 10-13). The ring-breathing mode at 515 nm clearly indicates that a nanotube with a very small diameter must have been formed. The $D$ and $G$ bands are also in very good agreement with theory.

After heating to higher temperatures (1250°C), the Raman bands of the small tubes disappear and double-walled nanotubes form by cyclodehydrogenation (Figure 10-14).

Nanotube sections **13**, **34**, and **36** and a further REM product of **13**[74] were investigated by Raman spectroscopy, because they are well-defined model structures [(2,2)-, (3,3)-, (4,4)-, (5,5)-armchair and (10,3) chiral tubes] for nanotubes that are not available as pure compounds. Some of the vibrational normal modes, particularly the ring-breathing mode, are close in energy to the corresponding modes of the larger nanotubes of similar diameter.

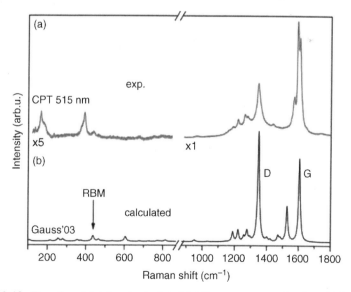

**Figure 10-13.** Experimental (a) and calculated (b) Raman spectra of a short (4,4)-armchair SWCNT.

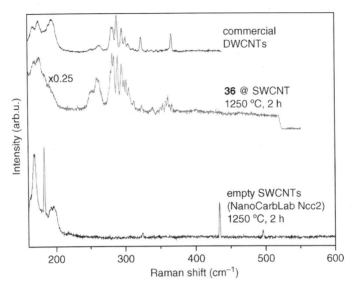

**Figure 10-14.** Experimental Raman spectra of DWCNT, **36**@SWCNT, and empty SWCNT.

Nanotubes are probably the most interesting and the most challenging targets in "nonnatural" chemical synthesis. Promising approaches toward this end have been developed since the mid-1990s. It is extremely exciting to follow the progress in this field, and it will be thrilling to see who will be awarded the coveted trophy.

## REFERENCES

1. M. S. Arnold, A. A. Green, J. F. Hulvat, S. I. Stup, M. C. Hersam, *Nature Nanotech.* **2006**, *1*, 60–65.

2. Y. Kato, Y. Niidome, N. Nakashima, *Angew. Chem.* **2009**, *121*, 5543–5546; *Angew. Chem. Int. Ed. Engl.* **2009**, *48*, 5435–5438.

3. S. R. C. Vivekchand, R. Jayakanth, A. Govindaraj, C. N. R. Rao, *Small* **2005**, *1*, 920–923.

4. L. T. Scott, M. M. Boorum, B. J. McMahon, S. Hagen, J. Mack, J. Blank, H. Wegner, A. de Meijere, *Science* **2002**, *295*, 1500–1503.

5. L. T. Scott, *Angew. Chem.* **2004**, *116*, 5102–5116; *Angew. Chem. Int. Ed. Engl.* **2004**, *43*, 4994–5007.

6. R. Herges, *Nachr. Chem.* **2007**, *55*, 962–963, 966–969.

7. V. M. Tsefrikas, L. T. Scott, *Chem. Rev.* **2006**, *106*, 4868–4884.

8. K. T. Rim, M. Siaj, S. Xiao, M. Myers, V. D. Carpentier, L. Liu, C. Su, M. L. Steigerwald, M. S. Hybertsen, P. H. McBreen, G. W. Flynn, C. Nuckolls, *Angew. Chem.* **2007**, *119*, 8037–8041; *Angew. Chem. Int. Ed. Engl.* **2007**, *46*, 7891–7895.

9. F. H. Kohnke, A. M. Z. Slawin, J. F. Stoddart, D. J. Williams, *Angew. Chem.* **1987**, *99*, 941–943; *Angew. Chem. Int. Ed. Engl.* **1987**, *26*, 892–894.

10. P. R. Ashton, N. S. Isaacs, F. H. Kohnke, J. P. Mathias, J. F. Stoddart, *Angew. Chem.* **1989**, *101*, 1266–1269; *Angew. Chem. Int. Ed. Engl.* **1989**, *28*, 1258–1261.

11. R. M. Cory, C. L. McPhail, A. Dikmans, J. J. Vital, *Tetrahedron Lett.* **1996**, *37*, 1983–1986.

12. M. Stuparu, V. Gramlich, A. Stanger, A. D. Schlüter, *J. Org. Chem.* **2007**, *72*, 424–430.

13. B. Esser, R. Rominger, R. Gleiter, *J. Am. Chem. Soc.* **2008**, *130*, 6716–6717

14. R. Jasti, J. Bhattacharjee, J. B. Necton, C. R. Bertozzi, *J. Am. Chem. Soc.* **2008**, *130*, 17646–17647.

15. H. Takaba, H. Omachi, Y. Yamamoto, J. Bouffard, K. Itami, *Angew. Chem.* **2009**, *121*, 6228–6232; *Angew. Chem. Int. Ed. Engl.* **2009**, *48*, 6112–6116.

16. S. Yamago, Y. Watanabe, T. Iwamoto, *Angew. Chem.* **2010**, *122*, 769–771; *Angew. Chem. Int. Ed. Engl.* **2010**, *49*, 757–759.

17. B. L. Merner, L. N. Dawe, G. J. Bodwell, *Angew. Chem.* **2009**, *121*, 5595–5599; *Angew. Chem. Int. Ed. Engl.* **2009**, *48*, 5487–5491.

18. G. Schröder, W. Martin, *Angew. Chem.* **1966**, *78*, 117; *Angew. Chem. Int. Ed. Engl.* **1966**, *5*, 130.

19. G. Schröder, J. F. M. Oth, *Tetrahedron Lett.* **1966**, *7*, 4083–4088.

20. R. L. Viavattene, F. D. Greene, L. D. Cheung, R. Majeste, L. M. Trefonas, *J. Am. Chem. Soc.* **1974**, *96*, 4342–4343.

21. M. N. Jagadeesh, A. Makur, J. Chandrasekhar, *J. Mol. Model* **2000**, *6*, 226–233.

22. M. Mantina, A. C. Chamberlin, R. Valero, C. J. Cramer, D. G. Truhlar, *J. Phys. Chem. A* **2009**, *113*, 5806–5812.

23. W. T. Borden, *Chem. Rev.* **1989**, *89*, 1095–1109.

24. W. Luef, R. Keese, *Top. Stereochem.* **1991**, *20*, 231–318.

25. R. Herges, H. Neumann, *Liebigs Ann.* **1995**, 1283–1289.

26. R. Herges, H. Neumann, F. Hampel, *Angew. Chem.* **1994**, *106*, 1024–1026; *Angew. Chem. Int. Ed. Engl.* **1994**, *33*, 993–995.

27. R. Herges, S. Kammermeier, H. Neumann, F. Hampel, *Liebigs Ann.* **1996**, 1795–1800.

28. J. Sauer, J. Breu, U. Holland, R. Herges, H. Neumann, S. Kammermeier, *Liebigs Ann./ Receuil* **1997**, 1473–1479.

29. S. Kammermeier, R. Herges, *Angew. Chem.* **1996**, *108*, 470–472; *Angew. Chem. Int. Ed. Engl.* **1996**, *35*, 417–419.

30. S. Kammermeier, P. G. Jones, R. Herges, *Angew. Chem.* **1996**, *108*, 2834–2836; *Angew. Chem. Int. Ed. Engl.* **1996**, *35*, 2669–2671.

31. W. V. Volland, E. R. Davidson, W. T. Borden, *J. Am. Chem. Soc.* **1979**, *101*, 533–537.

32. E. Schaumann, R. Ketcham, *Angew. Chem.* **1982**, *94*, 231–253; *Angew. Chem. Int. Ed. Engl.* **1982**, *21*, 225–247.

33. M. M. Weinshenker, F. D. Greene, *J. Am. Chem. Soc.* **1968**, *90*, 506.

34. S. Kammermeier, P. G. Jones, R. Herges, *Angew. Chem.* **1997**, *109*, 2317–2319; *Angew. Chem. Int. Ed. Engl.* **1997**, *36*, 2200–2202.

35. G. F. Emerson, L. Watts, R. Pettit, *J. Am. Chem. Soc.* **1965**, *87*, 131–133.

36. R. P. Dodge, V. Schomacher, *Nature*, **1960**, *186*, 798–799.

37. L. Watts, J. D. Fitzpatrick, R. Pettit, *J. Am. Chem. Soc.* **1966**, *88*, 623–624.

38. J. C. Barborak, L. Watts, R. Pettit, *J. Am. Chem. Soc.* **1966**, *88*, 1328–1329.

39. R. Herges, M. Deichmann, J. Grunenberg, G. Bucher, *Chem. Phys. Lett.* **2000**, *327*, 149–152.

40. R. Herges, M. Deichmann, T. Wakita, Y. Okamoto, *Angew. Chem.* **2003**, *115*, 1202–1204. *Angew. Chem. Int. Ed. Engl.* **2003**, *42*, 1170–1172.

41. E. Shabtai, A. Weitz, R. C. Haddon, R. E. Hoffman, M. Rabinovitz, A. Khong, R. J. Cross, M. Saunders, P.-C. Cheng, L. T. Scott, *J. Am. Chem. Soc.* **1998**, *120*, 6389–6393.

42. N. Treitel, M. Deichmann, T. Sternfeld, T. Sheradsky, R. Herges, M. Rabinovitz, *Angew. Chem.* **2003**, *115*, 1204–1208; *Angew. Chem. Int. Ed. Engl.* **2003**, *42*, 1172–1176.

43. S. Kammermeier, P. G. Jones, I. Dix, R. Herges, *Acta Crystallogr. Sect. C* **1998**, *54*, 1078–1081.

44. M. Müller, V. S. Iyer, C. Kübel, V. Enkelmann, K. Müllen, *Angew. Chem.* **1997**, *121*, 1679–1682; *Angew. Chem. Int. Ed. Engl.* **1997**, *36*, 1607–1610.

45. S. Hagen, H. Hopf, *Top. Curr. Chem.* **1998**, *196*, 47–82.

46. R. Scholl, K. Meyer, *Chem. Ber.* **1934**, *67*, 1229–1235.

47. P. Kovacic, F. W. Koch, *J. Org. Chem.* **1965**, *30*, 3176–3181.

48. F. B. Mallory, K. E. Butler, A. C. Evans, E. J. Brondyke, C. W. Mallory, C. Yang, A. Ellenstein, *J. Am. Chem. Soc.* **1997**, *119*, 2119–2124.

49. L. T. Scott, M. S. Bratcher, S. Hagen, *J. Am. Chem. Soc.* **1996**, *118*, 8743–8744.

50. M. Deichmann, C. Näther, R. Herges, *Org. Lett.* **2003**, *5*, 1269–1271.

51. Herges et al., unpublished results.

52. S. Iijima, *Nature*, **1991**, *354*, 56–58.

53. S. Iijima, T. Ichihashi, Y. Ando, *Nature*, **1992**, *356*, 776–778.

54. R. S. Ruoff, *Nature*, **1994**, *372*, 731–732.

55. K. Hernadi, A. Fonseca, J. B. Nagy, D. Bernaerts, J. Riga, A. Lucas, *Synth. Met.* **1996**, *77*, 31–43.

56. A. Oberlin, M. Endo, *J. Cryst. Growth.* **1976**, *32*, 335–349.

57. R. E. Smalley, Y. Li, V. C. Moore, B. K. Price, R. Colorado, H. K. Schmidt, R. H. Hauge, A. R. Barron, J. M. Tour, *J. Am. Chem. Soc.* **2006**, *128*, 15824–15829.

58. V. C. Moore, L. A. McJilton, S. T. Pheasant, C. Kittrell, R. E. Anderson, D Ogrin, F. Liang, R. H. Hauge, H. K. Schmidt, J. M. Tour, *Carbon* **2010**, *48*, 561–565.

59. Y. Wang, C. A. Mirkin, S.-J. Park, *ACS Nano* **2009**, *3*, 1049–1056.

60. Y. Yao, C. Feng, J. Zhang, Z. Liu, *Nano Lett.* **2009**, *9*, 1673–1677.

61. D. Ogrin, R. E. Anderson, R. Colorado, B. Maruyama, M. J. Pender, V. C. Moore, S. T. Pheasant, L. McJilton, H. K. Schmidt, R. H. Hauge, *J. Phys. Chem. C* **2007**, *111*, 17804–17806.

62. E. Clar, *Chem. Ber.* **1949**, *82*, 46–60.

63. E. Clar, *Chem. Ber.* **1932**, *65*, 503–519.

64. E. H. Fort, P. M. Donovan, L. T. Scott, *J. Am. Chem. Soc.* **2009**, *131*, 16006–16007.

65. A. Harada, *Farumashia* **1995**, *31*, 1263–1267.

66. P. R. Ashton, C. L. Brown, S. Menzer, S. A. Nepogodiev, J. F. Stoddart, D. J. Williams, *Chem. Eur. J.* **1996**, *2*, 580–591.

67. M. R. Ghadiri, K. Kobayashi, J. R. Granja, R. K. Chada, D. E. McRee, *Angew. Chem.* **1995**, *107*, 76–78; *Angew. Chem. Int. Ed. Engl.* **1995**, *34*, 93–95.

68. M. R. Ghadiri, *Adv. Mater.* **1995**, *7*, 675–677.

69. M. Engels, D. Bushford, M. R. Ghadiri, *J. Am. Chem. Soc.* **1995**, *117*, 9151–9158.

70. H. Meier, K. Müller, *Angew. Chem.* **1995**, *107*, 1598–1600; *Angew. Chem. Int. Ed. Engl.* **1995**, *34*, 1437–1439.

71. D. J. Hornbaker, S.-J. Kahng, S. Misra, B. W. Smith, A. T. Johnson, E. J. Mele, D. E. Luzzi, A. Yazdani, *Science*, **2002**, *295*, 828–831.

72. C. L. Kane, E. J. Mele, A. T. Johnson, D. E. Luzzi, B. W. Smith, D. J. Hornbaker, A. Yazdani, *Phys. Rev. B* **2002**, *66*, 235423-1–235423-15.

73. C. Schaman, R. Pfeiffer, V. Zolyomi, H. Kuzmany, D. Ajami, R. Herges, O. Dubay, J. Sloan, *Phys. Status Solidi B* **2006**, *243*, 3151–3154.

74. N. Rosenkranz, M. Machon, R. Herges, C. Thomsen, *Phys. Status Solidi B* **2008**, *245*, 2145–2148.

# CHAPTER 11

# CYCLOPARAPHENYLENES: THE SHORTEST POSSIBLE SEGMENTS OF ARMCHAIR CARBON NANOTUBES

XIA TIAN and RAMESH JASTI

## 11-1. INTRODUCTION

Cycloparaphenylenes represent the shortest possible units of one type of carbon nanotube (CNT), referred to as an "armchair" carbon nanotube (Figure 11-1). We became interested in the synthesis of the cycloparaphenylenes in an ongoing effort to prepare homogeneous CNTs—a grand challenge in the field of nanoscience. We have envisioned cycloparaphenylenes (and molecules of the like) as "seed" or template molecules to prepare CNTs with predetermined chiralities (see text below). Thus, it became our goal in 2006 to prepare these molecules for the first time. While the prospect of preparing homogeneous CNTs is fascinating and has served as the inspiration for our research, cycloparaphenylenes had garnered interest from the scientific community long before the discovery of CNTs. Their strained and distorted aromatic systems have intrigued synthetic chemists, theoreticians, and physical organic chemists for decades. This chapter begins with a brief summary of carbon nanotube structure and properties as these themes have served as the impetus for our research efforts. The bulk of the chapter presents a historic account of the attempts and recent accomplishments in preparing these hoop-shaped structures. We summarize the early investigations by Vögtle, the first successful synthesis and characterization of the cycloparaphenylenes by Jasti and Bertozzi, and more recent accomplishments by the groups of Itami and Yamago. In addition to the syntheses of these molecules, we also discuss the theoretical investigations and the unique physical properties associated with the cycloparaphenylenes.

*Fragments of Fullerenes and Carbon Nanotubes: Designed Synthesis, Unusual Reactions, and Coordination Chemistry*, First Edition. Edited by Marina A. Petrukhina and Lawrence T. Scott.
© 2012 John Wiley & Sons, Inc. Published 2012 by John Wiley & Sons, Inc.

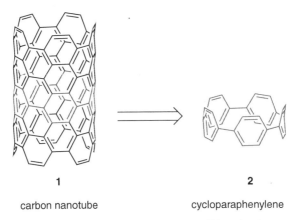

**1**

carbon nanotube

**2**

cycloparaphenylene

**Figure 11-1.** Cycloparaphenylene is the shortest possible unit of an armchair CNT.


## 11-2. CARBON NANOTUBES AS NEW CHALLENGES FOR ORGANIC SYNTHESIS

Sumio Iijima's arch discharge synthesis of CNTs in 1991 is often credited as the first preparation of carbon nanotubes.[1] Since their discovery, carbon nanotubes have received attention from physicists, chemists, and engineers, as well as biologists. In fact, more than 5000 publications related to CNTs were published in 2009 alone.[2] The enormous attention received is a direct result of the fascinating and unique properties of CNTs. Carbon nanotubes are some of the strongest materials known, with a tensile strength and specific strength far exceeding those of even steel.[3] These properties have been exploited in the development of composite materials for lighter and stronger tennis rackets, bicycles, and golf clubs.

While the mechanical properties of nanotubes are quite extraordinary, the electronic and optical properties of CNTs are perhaps the most promising for the nanotechnologies of the future. Certain types of CNTs are metallic and can, in theory, have current densities of $4 \times 10^9$ A/cm$^2$, more than 1000 times that of copper.[4] Other types of CNTs are semiconducting with bandgaps that can vary depending on the diameter and atomic arrangement of the tube.[5] With these unique and tunable electronic properties, CNTs have emerged as one of the most promising materials for next-generation electronics. In addition to their electronic properties, on excitation, carbon nanotubes can emit different wavelengths of light depending on their structure, rendering them as possible components of nanophotonic devices.[6] Carbon nanotubes have the potential to dramatically impact a wide range of fields and therefore have become one of the hottest topics in nanoscience.

The diversity in electronic and optical properties is a direct consequence of the variation in basic cylindrical structure.[7] For example, in the case of an armchair CNT, benzene rings are fused along the long axis of the tube and are linked by single bonds around the circumference [Figure 11-2(a)]. In contrast, in a zigzag tube, the benzene

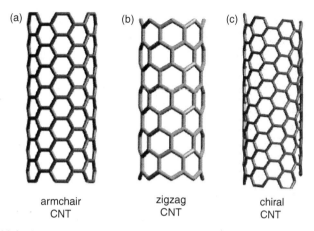

(a)    (b)    (c)

armchair
CNT

zigzag
CNT

chiral
CNT

**Figure 11-2.** CNTs have a wide range of properties depending on their structure.

rings are linked by single bonds along the long axis of the tube and a fused system around the circumference [Figure 11-2(b)]. The differing conjugation systems result in a dramatic change in properties, with armchair CNTs behaving as metals and most zigzag CNTs behaving as semiconductors. The armchair and zigzag tubes can be regarded as sheets of graphene rolled up on axes 90° with respect to each other. The zigzag and armchair CNTs, however, are only two possibilities. In fact, CNTs of all angles in between 0° and 90° are also possible, giving rise to a wide range of electronic band structures [Figure 11-2(c)]. To add even greater diversity, CNTs can be of different lengths and diameters, change chirality along the tube axis, and also incorporate structures other than six-membered rings. The seemingly limitless range of properties of carbon nanotubes is a fascinating consequence of quantum mechanics.

While these tunable electronic and optical properties of CNTs can be advantageous, it is this wide variation in properties that currently impedes the use of CNTs in many applications. The three most common methods for producing CNTs today—laser ablation, arc discharge synthesis, and chemical vapor deposition—all lead to heterogeneous batches of structures with inherently different properties.[8] In order to acquire CNTs of a discrete structure and property, laborious purification techniques are required that are not amenable to large-scale production.[9] Because of the inherent limitations of current synthetic techniques, development of a rational method to synthesize homogeneous batches of carbon nanotubes is paramount to the advancement of these nanomaterials in new technologies.

Our goal to synthesize the cycloparaphenylenes had its genesis in this carbon nanotube heterogeneity problem. Our research group has focused on developing a target-oriented approach to a diverse set of CNTs utilizing well-understood organic transformations. As was shown in Figure 11-1, cycloparaphenylenes are the shortest possible segments of an armchair CNT that retain information on chirality. If cycloparaphenylenes of a given diameter are utilized as a template or monomer unit for CNT synthesis, homogeneous armchair nanotubes potentially could be constructed of a predetermined chirality. This synthetic approach could be extended to other types of carbon nanotubes as well. For instance, the synthesis of cyclacenes might allow for the preparation of homogeneous zigzag CNTs [Figure 11-2(b)]. With this ultimate goal of preparing homogeneous CNTs using organic synthesis, we embarked on a quest in 2006[10] to synthesize the cycloparaphenylenes for the first time.

## 11-3. TIMELINE FOR CYCLOPARAPHENYLENE SYNTHESIS

### 11-3-1. Early Studies Related to the Cycloparaphenylenes

Although no successful syntheses of the cycloparaphenylenes had been reported as of 2006, several articles had been published in regard to both the attempted syntheses of cycloparaphenylenes and theoretical calculations. As these studies were conducted well before the discovery of CNTs, the interest in these molecules at the time was in the unusual geometry and bent $\pi$ system (Figure 11-3).[11] Typical conjugated macrocycles have $\pi$ systems above and below the plane, as illustrated in Figure 11-3(a). The cycloparaphenylenes, however, have $\pi$ systems that are bent and directed into the plane of the macrocycle, as in Figure 11-3(b). As a result of this bending of the $sp^2$ carbons from planarity, the $p$ orbitals share partial $s$ character. This factor results in the differentiation of the $\pi$-electron densities on the inner and outer surfaces of the cycloparaphenylenes. The inner surface is predicted to be electron-rich whereas the outer surface is predicted to be electron-poor.[12] These structural features render the cycloparaphenylenes interesting not only from a theoretical vantage point but also as components of supramolecular host–guest systems.

To our knowledge, the earliest mention regarding the synthesis of cycloparaphenylenes was in 1934 by Parekh and Guha (Figure 11-4).[13] In this work, the authors believed that they had partially desulfurized $p,p$-diphenylenetetrasulfide (**3**) to $p,p$-diphenylenedisulfide (**4**). Complete desulfurization would lead to [2]cycloparaphenylene. The authors speculated on whether this type of cyclic diphenyl molecule could be prepared, noting that earlier work to convert a "*para,para*-diiodo derivative into *para,para*-diphenylene have resulted in failure."

Before the recent flurry of publications, the most extensive research on the synthesis of the cycloparaphenylenes was conducted by Vögtle and coworkers. Several strategies to accomplish this synthetic feat were set forth by Vögtle in 1993 in his work entitled "On the way to macrocyclic paraphenylenes."[14] To Vögtle's credit, many of the strategies described in this pioneering work are very similar to the

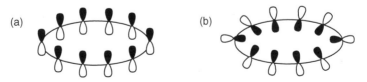

**Figure 11-3.** The $\pi$ system of the cycloparaphenylenes is directed into the ring.

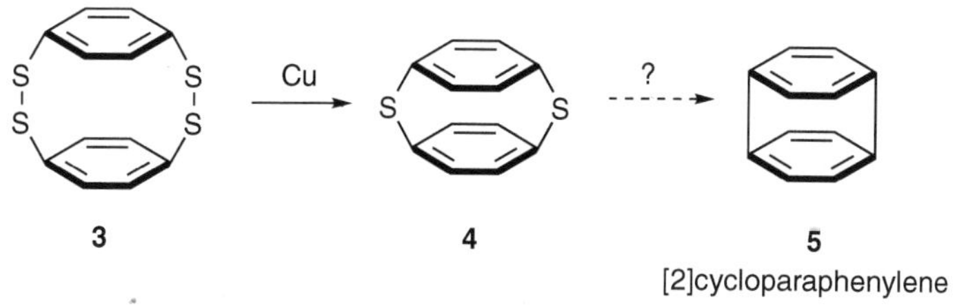

**Figure 11-4.** Early work by Parekh and Guha in 1934.

approaches that proved successful 15 years later—although these strategies were unsuccessful at the time. Figure 11-5 illustrates some of the approaches investigated by Vögtle. One of the first approaches described in this work was a pyrolysis strategy of phenylsulfides [Figure 11-5(a)]—a strategy reminiscent of that suggested by

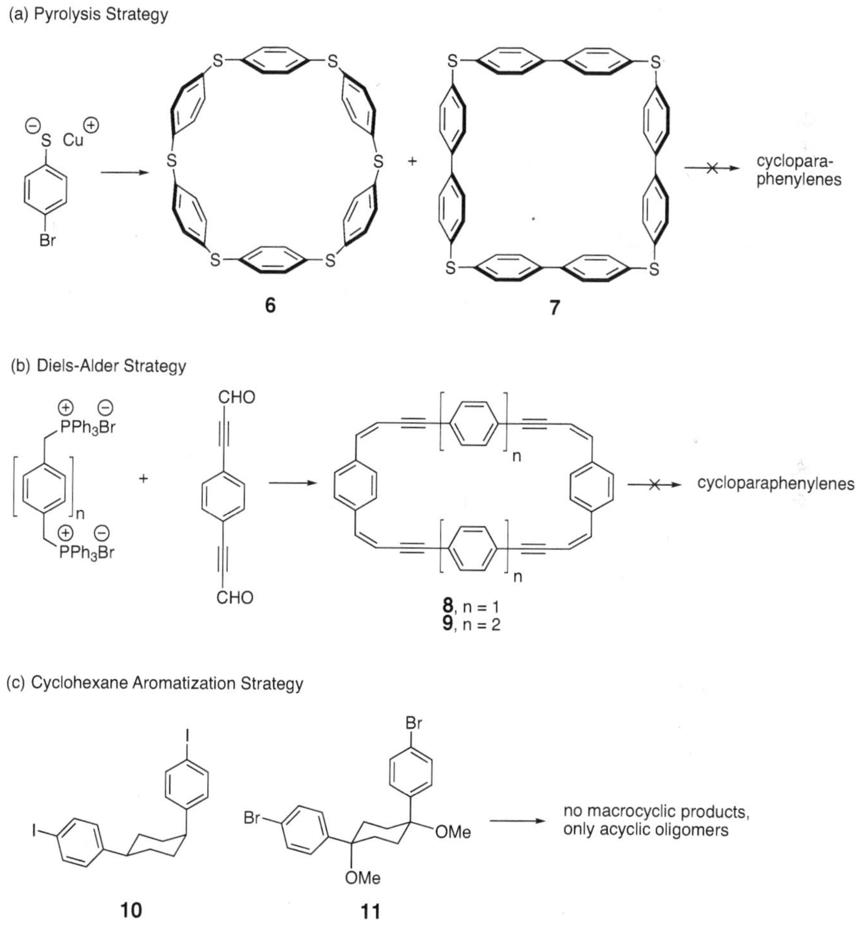

**Figure 11-5.** Vögtle's proposed strategies for synthesizing cycloparaphenylenes; (a) pyrolysis; (b) Diels–Alder approach; (c) cyclohexane aromatization.

Parekh and Guha (Figure 11-4). Macrocycles **6** and **7** were easily prepared from oxidative coupling of 4-bromobenzenethiolates in good yields. Pyrolysis of diphenylsulfides or diphenylsulfones to form aryl—aryl bonds was well documented by the work of Heldt.[15] However, in the case of these macrocyclic components (**6** and **7**), pyrolysis conditions did not lead to the desired cycloparaphenylenes. Vögtle suggested the reason for failure in these systems to be likely due to the increased strain energy. Inspired by Miyahara's synthesis of a strained paracyclophane,[16] Vögtle also investigated a Diels—Alder strategy [Figure 11-5(b)]. Reduction of the alkynes in structures **8** and **9** would lead to a butadiene segment that could potentially undergo Diels—Alder reaction with an appropriate dienophile to generate the required aromatic ring. Again, this strategy proved unsuccessful. A third strategy suggested by Vögtle, which is similar to the strategies executed by both Jasti and Itami, utilized cyclohexane rings as masked aromatic structures [Figure 11-5(c)]. Although structures **10** and **11** were both prepared, only acyclic oligomers were isolated under coupling conditions. In this case, Vögtle surmised that the conformational flexibility of these molecules resulted in the oligomerization of these units as opposed to the desired macrocyclization.

Although the cycloparaphenylenes had not been synthesized until 2008, several related quinoid structures had been prepared earlier. Most notably, Herges had synthesized structure **14**, termed a "picotube" because of its structural relationship to a carbon nanotube (Figure 11-6).[17] Herges very elegantly demonstrated that tetrahydrodianthracene (**12**) undergoes a [2+2] dimerization to generate intermediate **13** under photochemical excitation. A rapid retro-[2+2] then leads to ring-expanded product **14**. Crystal structure analysis reveals structure **14** to have a

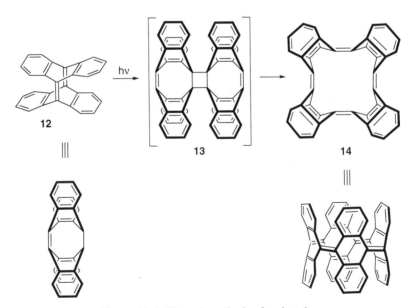

**Figure 11-6.** Herges' synthesis of a picotube.

benzannulated quinoid structure. This structure is most easily described as four benzannulated cyclohexa-1,4-diene units joined by alkenes. In other words, **14** is a π-bond shift isomer of the fully benzannulated [4]cycloparaphenylene. In fact, tetrahydrodianthracene (**12**) is also a π-bond shift isomer of a benzannulated [2] cycloparaphenylene. The work by Herges is more extensively described in Chapter 10 of this book.

In relation to this work, Jagadeesh and coworkers published computational calculations regarding the benzenoid versus quinoid structures of [5]- and [6] cycloparaphenylene.[18] Their findings at the HF/3–21G level of theory support the idea that cyclic arrays of benzene rings can display benzenoid or quinoid character depending on the degree of nonplanarity of the aryl rings. As can be seen by comparison of bond angle **d** in Figure 11-7, [6]cycloparaphenylene has aryl rings that are less askew than those in [5]cycloparaphenylene. This difference in bending translates into the preference for quinoid versus benzenoid character, as seen in bond lengths **a**, **b**, and **c**. For [5]cycloparaphenylene, bonds **a** and **b** are shorter compared to bond **c**, consistent with a quinoid structure [Figure 11-7(a)]. [6] Cycloparaphenylene has bond lengths consistent with a benzene structure— bond **a** is the longest and bonds **b** and **c** are shorter and of virtually the same length [Figure 11-7(b)]. In contrast to these results, the central belt portion of $C_{70}$, a [5] cycloparaphenylene, exhibits significant benzenoid character, as exemplified by bond lengths and reactivity. Presumably, the added connectivity confers aromaticity to the central belt region. This result for $C_{70}$ highlights the importance of the surrounding environment, in addition to bending, for the properties of cycloparaphenylene-type compounds.

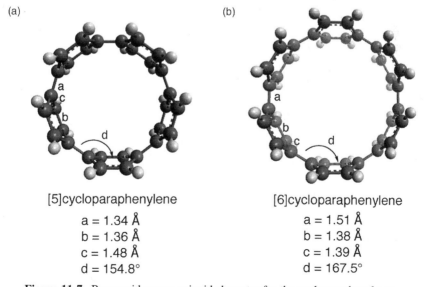

(a)    (b)

[5]cycloparaphenylene

a = 1.34 Å
b = 1.36 Å
c = 1.48 Å
d = 154.8°

[6]cycloparaphenylene

a = 1.51 Å
b = 1.38 Å
c = 1.39 Å
d = 167.5°

**Figure 11-7.** Benzenoid versus quinoid character for the cycloparaphenylenes.

## 11-3-2. The First Synthesis and Characterization of Cycloparaphenylenes

In 2008, more than 70 years after these structures had been theorized, Jasti and co-workers accomplished the syntheses of the first cycloparaphenylenes.[19] As discussed earlier, the heart of the synthetic challenge for these molecules lies in the strain energy of the bent aromatic system. The strategy was to build up strain sequentially during the synthesis, using a 3,6-*syn*-dimethoxycyclohexa-1,4-diene moiety as a masked aromatic ring in a macrocyclic precursor (see structure **16**, Scheme 11-1).[20] This strategy is very similar to Vögtle's early work, except in this case a cyclohexadiene was envisioned instead of the more flexible cyclohexane moiety. It was reasoned that the cyclohexadiene unit would provide both the curvature and rigidity necessary for macrocyclization to afford a marginally strained intermediate. Subsequent aromatization would then provide the driving force necessary to achieve the strain energy of the cycloparaphenylenes.

Synthesis of the macrocyclic precursors to cycloparaphenylenes is depicted in Scheme 11-1. Monolithiation of diiodobenzene and subsequent addition of benzoquinone generated a *syn* diol, which was then alkylated with methyl iodide to produce diiodide **16** in an overall yield of 34%. The stereochemical preference for the *syn* product is attributed to electrostatic repulsion. This charge–charge repulsion favors addition of the second equivalent of aryl lithium to the opposite face of the initially formed lithium alkoxide, delivering the *syn* product.[21] In preparation for a tandem Suzuki coupling/macrocyclization sequence, portions of diiodide (**16**) were converted to diboronate (**17**) in a straightforward manner. Advantageously, both coupling partners could easily be prepared on gram scale without the need for column chromatography. Standard Suzuki reaction conditions afforded macrocycles **18**, **19**, and **20** in 22% combined yield. Formation of macrocycle **18** was unexpected and likely involves homocoupling of diboronate **17** as an intermediate step.[22] Although the yield of macrocyclization products was modest, the synthetic route provided rapid access to three different macrocycles, which were easily separated by

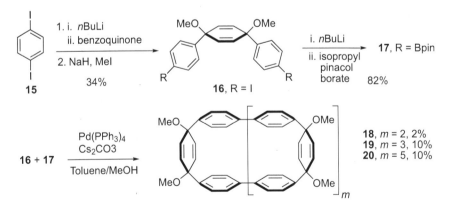

**Scheme 11-1.** Jasti's synthesis of macrocyclic precursors.

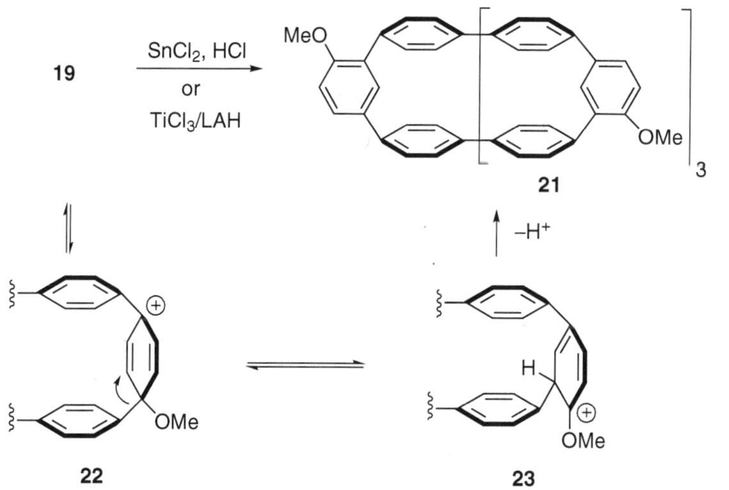

**Figure 11-8.** Macrocyclic precursors are prone to migration under acidic conditions.

column chromatography. It is notable that the construction of macrocycle **20** involved formation of 18 carbon–carbon bonds in only four steps.

With the three macrocycles in hand, the key aromatization step was explored. At the outset, we envisioned using either Stephen's reagent[23] or low-valence titanium[24] to deliver the desired aromatic ring. These reaction conditions had proved successful for the synthesis of a variety of naphthalene and anthracene derivatives. Our case proved to be much more difficult in that rearrangements are extremely facile for these cycloparaphenylene precursors (Figure 11-8). Under these Lewis acidic conditions, **19** rapidly generated cation **22**. A 1,2 carbon shift can then produce stabilized oxonium cation **23**. This structure can then rapidly aromatize by loss of proton to generate rearranged product **21**. This intramolecular migration/aromatization process was presumably faster than an intermolecular reduction process mediated by addition of Sn or Ti. In an effort to capture the initial cation and prevent this rearrangement, we added hydride sources such as $Bu_3SnH$. Even at very high concentrations of reductant, migrations were rapid and delivered significant amounts of rearranged product.

Having discovered the propensity of these molecules to undergo rearrangement processes, we explored other reaction conditions to aromatize model compound **24** (Figure 11-9). One possible strategy to prevent the undesired rearrangements was to remove the problematic alkenes that were leading to allyl cations. A saturated version of **24** (where the central ring is a cyclohexane unit) might undergo elimination to a diene on treatment with acid, which could then be oxidized to the desired compound. Surprisingly, treatment of **24** with standard hydrogenation conditions led to an aromatization process that directly yielded terphenyl (**25**). Although the mechanism of this reaction was not rigorously studied, it likely involves reduction of the C–O bond similar to a benzyl ether deprotection.[25] Interestingly, under the same room temperature conditions, macrocycle **19** did not react. Under more compelling conditions, macrocycle **19** led to an intractable mixture of products.

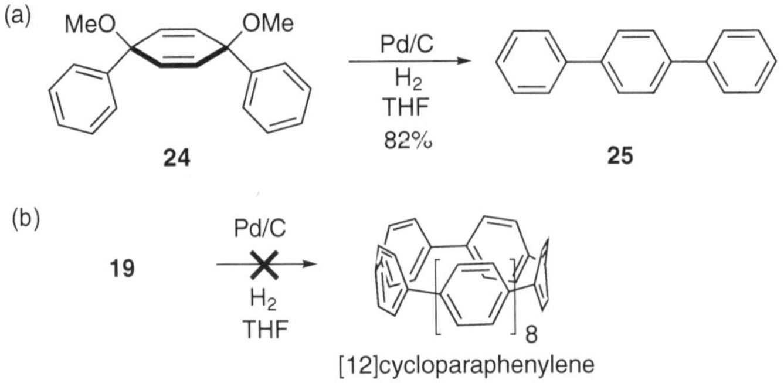

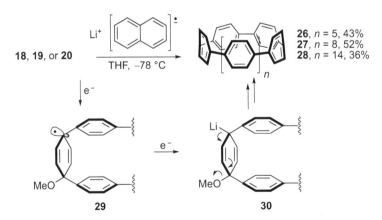

**Figure 11-9.** Unexpected aromatization of model system **24**.

With these results in mind, we investigated other reaction conditions known to reductively cleave benzyl ethers. Lithium naphthalenide, a reagent known to reduce benzylic ethers at low temperature,[26] provided the eventual solution (Figure 11-10). Treatment of each macrocycle (**18**, **19**, and **20**) with lithium naphthalenide at −78°C afforded the corresponding cycloparaphenylenes (**26**, **27**, and **28**) in moderate yields. Mechanistically, the reaction is envisioned to proceed via an initial one-electron transfer, which generates stabilized radical intermediate **29** and an equivalent of lithium methoxide. A second electron transfer then produces the alkyl lithium (**30**), which can aromatize via loss of a second equivalent of lithium methoxide. Especially noteworthy is the formation of cycloparaphenylene (**26**), the most strained hoop structure of the series, using these low-temperature conditions.

With cycloparaphenylenes in hand for the first time, we analyzed their properties using a variety of methods. Interestingly, their UV−visible absorption spectra showed symmetric peaks with a maximum at approximately 340 nm regardless of cycloparaphenylene size [Figure 11-11(a)]. Indeed, the smallest cycloparaphenylene (**26**), which has the poorest geometry for orbital overlap and fewest number of aryl rings, has a slightly smaller optical gap than the larger cycloparaphenylenes. This result

**Figure 11-10.** A novel reductive aromatization reaction.

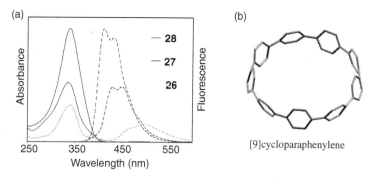

**Figure 11-11.** (a) Absorption (solid line) and fluorescence (dashed line) spectra of cyclopar-aphenylenes; (b) energy-minimized geometry of [9]cycloparaphenylene.

stands in sharp contrast to the acyclic paraphenylenes in which a narrowing of the optical gap occurs with extended conjugation. The fluorescence spectra also provided some unexpected results. As cycloparaphenylene size decreased, the Stokes shift increased. Interestingly, the smallest cycloparaphenylene (**26**) shows an especially broad spectrum with a large Stokes shift of approximately 160 nm.

We further investigated the structural and optical properties of the cyclopar-aphenylene molecules with computational methods based on density functional theory (DFT).[27] Our calculations indicate that the favored geometry for the even-membered macrocycles, [12]- and [18]cycloparaphenylene, is a staggered config-uration in which the dihedral angle between two adjacent phenyl rings alternates between ±33° and ±34°, respectively. This arrangement minimizes steric repul-sion between neighboring aryl rings. For [9]cycloparaphenylene, the situation is slightly more complex. Reduced symmetry due to an odd number of phenyl rings results in a lowest energy conformation in which the dihedral angle varies between ±18°, ±30°, ±31°, and ±33° around the macrocycle [Figure 11-11(b)]. We found that chiral Möbius-strip-like arrangements, where the dihedral angle is succes-sively increased by approximately 30° around the ring, were higher in energy by at least 2 kcal/mol per phenyl ring. The calculated strain energies of cycloparaphe-nylenes **26**, **27**, and **28** were 47, 28, and 5 kcal/mol with diameters of 1.2, 1.7, and 2.4 nm, respectively.

We also performed DFT calculations to understand the counterintuitive trends in the observed optical data. We estimated the average optical absorption energy gap $E_g$ (Figure 11-12) with the following approximate formula: $E_g$ = ionization potential (IP)−electron affinity (EA)−$E_{e-h}$, where $E_{e-h}$ is the electron−hole interaction energy between the highest occupied and lowest unoccupied DFT one-electron wavefunctions.[28] As expected, we observed that IP−EA decreases as the number of phenyl rings increases, in both the cyclic and acyclic cases. Remarkably, however, we found that the magnitude of $E_{e-h}$ grows more dramatically with decreasing number of phenyl rings for cyclic paraphenylenes than for their acyclic counterparts. This computed trend in $E_{e-h}$ more than compensates for the increase in IP−EA and results in an overall decrease in $E_g$ with cycloparaphenylene

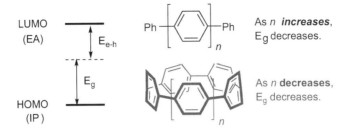

**Figure 11-12.** Opposing trends in optical gap for acyclic versus cyclic paraphenylenes.

diameter, in agreement with the experimentally observed trends. The fluorescence data were also rationalized with DFT methods. The large Stokes shift observed in these spectra is governed by the degree of structural relaxation in the optically excited state. In the smaller cycloparaphenylene, enhanced curvature leads to greater $sp^3$ hybridization and increasingly asymmetric $p$ orbitals, with their smaller lobes oriented inside the nanohoops. This factor reduces steric interactions and facilitates a larger decrease in dihedral angle in the excited state for the smaller nanohoop. Thus smaller hoops have the potential for larger structural relaxation and Stokes shifts. Indeed, constrained DFT calculations confirmed the observed trend that the smallest cycloparaphenylene has the largest relaxation in its excited state.

The trends in optical gap for the cycloparaphenylenes were more extensively investigated by Wong.[29] As shown in Table 11-1, the optical gap narrows from

**TABLE 11-1. Calculated Ionization Potential, Electron Affinity, Electron–Hole Interaction, and Optical Gap Energies for Acyclic and Cyclic Paraphenylenes**

| Number of Benzene Rings | Cyclic | | | | Linear | | | |
|---|---|---|---|---|---|---|---|---|
| | IP (eV) | EA (eV) | $E_{e-h}$ (eV) | $E_g$ (eV) | IP (eV) | EA (eV) | $E_{e-h}$ (eV) | $E_g$ (eV) |
| 5 | 6.34 | 1.17 | 3.09 | 2.08 | 7.05 | 0.51 | 2.62 | 3.91 |
| 6 | 6.34 | 1.22 | 2.82 | 2.29 | 6.90 | 0.66 | 2.46 | 3.78 |
| 7 | 6.35 | 1.25 | 2.60 | 2.50 | 6.80 | 0.76 | 2.32 | 3.71 |
| 8 | 6.50 | 1.10 | 2.44 | 2.96 | 6.71 | 0.84 | 2.21 | 3.66 |
| 9 | 6.45 | 1.16 | 2.29 | 3.00 | 6.64 | 0.92 | 2.10 | 3.61 |
| 10 | 6.39 | 1.25 | 2.16 | 2.98 | 6.58 | 0.98 | 2.02 | 3.58 |
| 11 | 6.43 | 1.20 | 2.07 | 3.16 | 6.52 | 1.04 | 1.93 | 3.54 |
| 12 | 6.43 | 1.20 | 1.99 | 3.24 | 6.48 | 1.08 | 1.87 | 3.54 |
| 13 | 6.37 | 1.27 | 1.90 | 3.20 | 6.45 | 1.12 | 1.81 | 3.52 |
| 14 | 6.39 | 1.25 | 1.85 | 3.30 | 6.41 | 1.16 | 1.75 | 3.51 |
| 15 | 6.36 | 1.28 | 1.79 | 3.29 | 6.38 | 1.19 | 1.70 | 3.50 |
| 16 | 6.34 | 1.31 | 1.73 | 3.30 | 6.35 | 1.22 | 1.65 | 3.49 |
| 17 | 6.31 | 1.33 | 1.68 | 3.30 | 6.33 | 1.24 | 1.61 | 3.48 |
| 18 | 6.32 | 1.33 | 1.65 | 3.35 | 6.31 | 1.26 | 1.57 | 3.48 |

[18]cycloparaphenylene to [5]cycloparaphenylene, with slight deviations for [10]- and [12]cycloparaphenylene. As was predicted from the earlier work by Jasti et al., the HOMO-LUMO gap narrows as the number of phenyl rings increase for both the cyclic and acyclic cases. The electron–hole interaction energy, however, increases much more dramatically in the case of the cyclic systems. Interestingly, [5]cyloparaphenylene is predicted to have an optical gap of only 2.08 eV, corresponding to a maximum absorbance of approximately 600 nm. It should be noted that although the trends are consistent with empirical observations, absolute values do not match. For instance, the optical gap calculated for [9]cycloparaphenylene is 3.00 eV (413 nm). However, the absorbance maximum was empirically determined to be 345 nm.

The first synthesis and characterization of unsubstituted [n]cycloparaphenylene was thus accomplished in five overall steps with only two chromatographic separations required. The straightforward monomer synthesis and cyclization reaction were easily scaled up to provide large quantities of the macrocycle. The novel reductive aromatization allowed for the isolation of a single regioisomer without the complication of aryl migration. Access to three different hoop sizes was also beneficial in that it allowed for a comparison of cycloparaphenylene properties with regard to size.

### 11-3-3. A Selective Synthesis of [12]Cycloparaphenylene

Shortly after our reported synthesis of the cycloparaphenylenes, a second synthesis was reported by Itami and coworkers.[30] The hallmark of this synthesis was the ability to selectively prepare [12]cycloparaphenylene. Their approach is similar to the method originally proposed by Vögtle, whereby a cyclohexyl derivative was used as the connecting lynchpin between fully aromatic rings to provide the curvature necessary to achieve macrocyclization. The final product would be delivered by an oxidative aromatization reaction. Itami calculated that cyclization of a completely aromatic precursor (**31**) to [12]cycloparaphenylene would require a buildup of 55 kcal/mol of strain energy (Figure 11-13).[31] In contrast, cyclization of a partially saturated precursor (**32**) to macrocycle **33** requires only 5 kcal/mol of strain energy.

Synthesis of the coupling partners is illustrated in Scheme 11-2. The mono-anion of 1,4-diiodobenzene (**34**) was added to 1,4-cyclohexadione (**35**) to generate the required *syn* diol **36**. The *syn* product is favored by approximately 4 : 1 over the *anti* product. A portion of the material is treated with methoxymethyl chloride to deliver methoxymethyl ether (**37**), which facilitated chromatographic separation in subsequent steps. A second portion of the material was converted to the bisboronate **38** utilizing palladium-catalyzed borylation conditions. With both diiodide **37** and boronate **38** in hand, the stage was set for the development of a selective macrocyclization.

For the selective synthesis of [12]cycloparaphenylene, the authors explored two approaches. The first approach involved targeting a differentially terminated dimer, which on coupling and cyclization would yield the desired product. The dimer could be produced only in low yields, however, due to competing oligomerization.

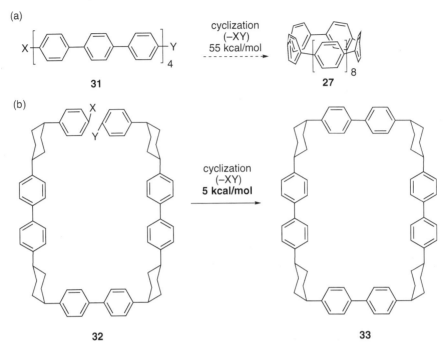

**Figure 11-13.** Strain energies for cyclization of (a) fully aromatic precursors versus (b) partially saturated precursors.

Alternatively, generation of symmetric trimer **39** was much more successful, with the desired product isolated in 81% yield (Scheme 11-3). Trimerization was facilitated by employment of an excess of iodide **37** relative to boronate **38**. With trimer **39** in hand, the final coupling and macrocyclization were investigated. After surveying a variety of Suzuki–Miyaura reaction conditions, the authors found Buchwald's X-Phos ligand to be the most efficient ligand for generating macrocycle **40**.

The final one-pot dehydration/oxidation sequence was performed under microwave irradiation in the presence of stoichiometric $p$-toluenesulfonic acid. Under these

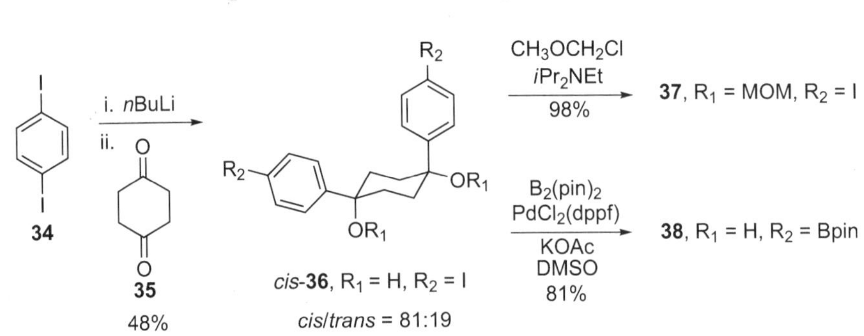

**Scheme 11-2.** Itami's synthesis of substituted cyclohexane coupling partners.

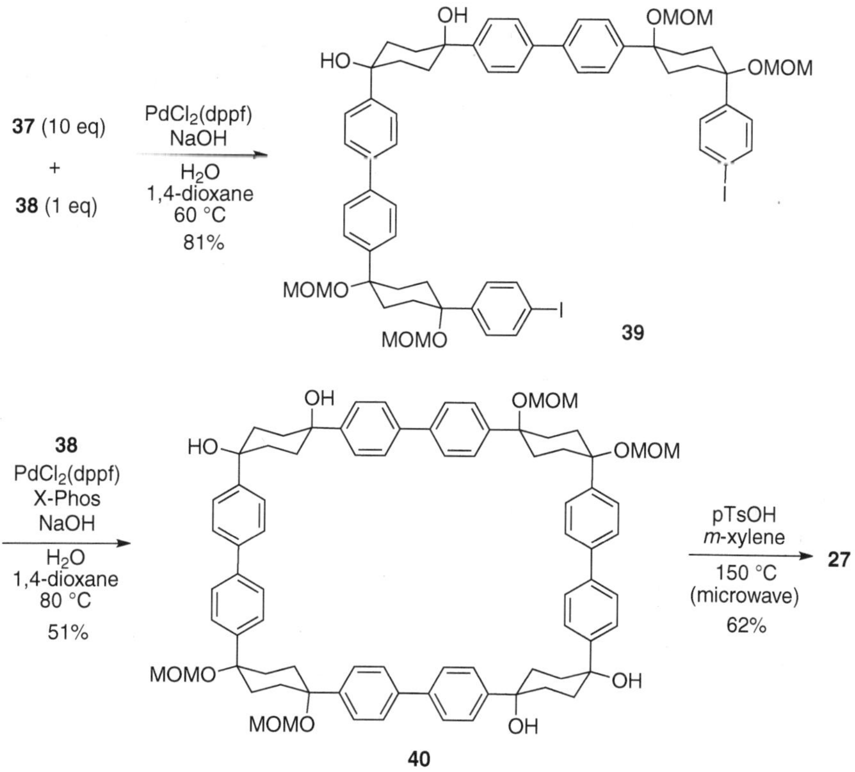

**Scheme 11-3.** Final steps for the selective synthesis of [12]cycloparaphenylene.

forcing conditions, dehydration to the cyclohexadiene followed by oxidation delivered [12]cycloparaphenylene. Interestingly, the final oxidation to the fully aromatic system occurs without an added oxidant. The authors do not speculate on a mechanism for this process, but the liberated formaldehyde from the methoxymethyl (MOM) deprotection or adventitious oxygen could potentially serve as the ultimate electron acceptor. As expected, the synthesized [12]cycloparaphenylene displayed the same spectroscopic properties as those reported by Jasti et al. Overall, Itami's synthesis is noteworthy for the high overall yield and the selective formation of a single product.

## 11-3-4. Synthesis of [8]Cycloparaphenylene by a Reductive Elimination Strategy

The third published synthesis of a cycloparaphenylene utilized a distinctly different approach to the previous syntheses of the cycloparaphenylenes. Yamago and co-workers hypothesized that carbon–carbon bond formation via a reductive elimination process is sufficiently exothermic to generate the inherent strain in [8]cycloparaphenylene.[32] The required box-shaped tetraplatinum complexes such as that shown in

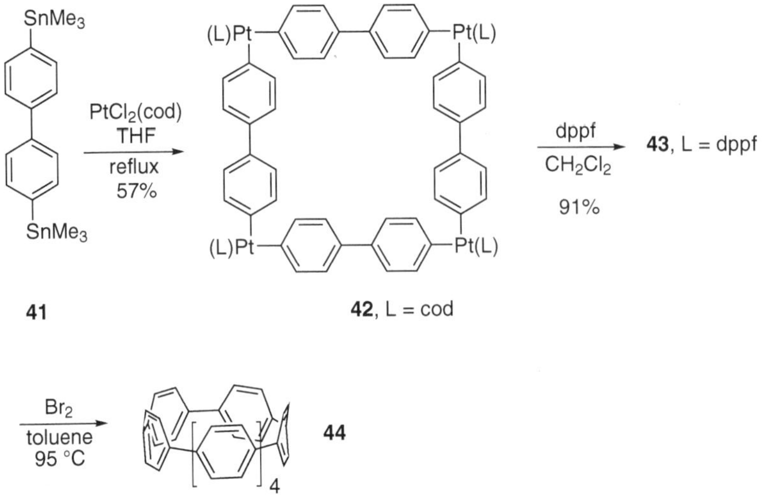

**Scheme 11-4.** Synthesis of [8]cycloparaphenylene via platinum complex.

Scheme 11-4 have been reported,[33] but none with exclusively aryl–metal bonds, as would be required for the synthesis of a cycloparaphenylene. Undeterred by these significant challenges, Yamago set out to synthesize the shortest cycloparaphenylene to date.

The synthesis of [8]cycloparaphenylene is illustrated in Scheme 11-4. Transmetallation of bis(stannylbiphenyl) (**41**) to platinum generated cyclic square planar complex **42**, containing the desired aryl–metal bonds. The choice of 1,5-cyclooctadiene (COD) as an ancillary ligand was presumably necessary for macrocycle formation. With complex **42** in hand, they investigated the crucial reductive elimination.[33] The spectator COD ligand was exchanged for the much more electron-rich 1,1′-bis(diphenylphosphino)ferrocene (dppf) leading to complex **43**. Although the authors do not comment on the ligand exchange, the more donating phosphine ligands were presumably necessary to effect the desired reductive eliminations. Bromine-induced reductive elimination[34] yielded the desired [8]cycloparaphenylene in good yield. The authors note the extensive amount of experimentation that was required to identify appropriate reductive elimination conditions. Interestingly, this synthetic route can be regarded as a modern organometallic version of Vögtle's original sulfide pyrolysis strategy.

Having synthesized a new cycloparaphenylene, Yamago et al. characterized its optical properties. Yamago's results are consistent with those reported for the previously synthesized cycloparaphenylenes. As can be seen in Figure 11-14, [8] cycloparaphenylene has a maximum absorbance of 340 nm, with a fluorescence emission at ~540 nm. As predicted by the computational analysis by Jasti and coworkers, the Stokes shift of [8]cycloparaphenylene is even larger (approximately 200 nm) than that of [9]cycloparaphenylene. This observation can be accounted for by the additional strain energy in [8]cycloparaphenylene.

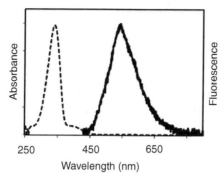

**Figure 11-14.** Optical characterization of [8]cycloparaphenylene.

## 11-4. CONCLUSION AND FUTURE DIRECTIONS

The syntheses of [n]cycloparaphenylenes presented here represent a milestone in the field of strained aromatic systems. These cyclic structures with radially oriented $p$ orbitals have intrigued chemists for decades and have only quite recently yielded to synthesis. The novel methods and strategies described in this chapter can be regarded as general approaches to prepare not only cycloparaphenylenes but also other strained conjugated systems. Furthermore, the preparation of these systems has led to surprising observations in regard to their optical properties. In contrast to acyclic paraphenylenes, the cyclic paraphenylenes absorb lower-energy (higher-wavelength) light as the number of phenyl rings decreases. This finding is counterintuitive, yet was able to be rationalized from first principles.

While the synthesis of cycloparaphenylenes is a monumental achievement in and of itself, accessing the "bare bones" versions is just the starting point of research in this area. Many questions remain to be answered. Can homogeneous armchair carbon nanotubes be prepared by utilizing these molecules as templates? How will the electronic and optical properties of the cycloparaphenylene structures change with functionalization? Can heteroaromatic rings be incorporated into these hoop structures—what effect will this modification have on the properties? Indeed, the future holds many exciting prospects for the cycloparaphenylenes and their derivatives.

## REFERENCES AND NOTES

1. (a) S. Iijima, *Nature* **1991**, *354*, 56–58; (b) N. Hamada, S. Sawada, A. Oshiyama, *Phys. Rev. Lett.* **1992**, *68*, 1579–1581; (c) T. W. Odom, J. Huang, C. M. Lieber, *Ann. NY Acad. Sci.* **2002**, *960*, 203–215; (d) H. Dai, *Acc. Chem. Res.* **2002**, *35*, 1035–1044; (e) X. Liu, C. Lee, S. Han, C. Li, C. Zhou, in *Molecular Nanoelectronics*, M. A. Reed, T. Lee, (eds.), American Scientific Publishers, **2003**, 179–198; (f) M. Terrones, *Int. Mater. Rev.* **2004**, *49*, 325–377; (g) Q. Zeng, Z. Li, Y. Zhou, *J. Nat. Gas Chem.* **2006**, *15*, 235–246.

2. This number was estimated on the basis of a SciFinder Scholar search.

3. M. Yu, O. Lourie, M. J. Dyer, K. Moloni, T. F. Kelly, R. S. Ruoff, *Science* **2000**, *287*, 637–640.

4. Z. K. Tang, L. Zhang, N. Wang, X. X. Zhang, G. H. Wen, G. D. Li, J. N. Wang, C. T. Chan, P. Sheng, *Science* **2001**, *292*, 2462–2465.

5. E. Joselevich, *ChemPhysChem* **2004**, *5*, 619–624.

6. J. A. Misewich, R. Martel, Ph. Avouris, J. C. Tsang, S. Heinze, J. Tersoff, *Science* **2003**, *300*, 783–186.

7. M. S. Dresselhaus, G. Dresselhaus, P. Avouris (eds.), *Carbon Nanotubes: Synthesis, Structure, Properties and Applications*, Springer-Verlag, Heidelberg; *Top. Appl. Phys.* **2001**, 80.

8. (a) E. Joselevich, H. Dai, J. Liu, K. Kenji, A. H. Windle, *Top. Appl. Phy.* **2008**, *111*, 101–164; (b) G. Lolli, L. Zhang, L. Balzano, N. Sakulchaicharoen, Y. Q. Tan, D. E. Resasco, *J. Phys. Chem. B* **2006**, *110*, 2108; (c) T. Nozaki, K. Okazaki, *Plasma Proc. Polym.* **2008**, *5*, 301–321; (d) K. J. MacKenzi, O. M. Dunens, C. H. See, A. T. Harris, *Recent Patents Nanotech*, **2008**, *2*, 25–40; (e) C. H. See, A. T. Harris, *Indust. Eng. Chem. Res.* **2007**, *46*, 997–1012; (f) C. Oencel, Y. Yueruem, *Fullerenes Nanotubes Carbon Nanostruct.* **2005**, *14*, 17–37; (g) C. T. Kingston, B. Simard, *Anal. Lett.* **2003**, *36*, 3119–3145.

9. (a) R. C. Haddon, J. Sippel, A. G. Rinzler, F. Papadimitrakopoulos, *MRS Bull.* **2004**, *29*, 252–259; (b) M. S. Strano, C. A. Dyke, M. L. Usrey, P. W. Barone, M. J. Allen, H. W. Shan, C. Kittrell, R. H. Hauge, J. M. Tour, R. E. Smalley, *Science* **2003**, *301*, 1519; (c) R. Krupke, F. Hennrich, H. v. Lohneysen, M. Kappes, *Science* **2003**, *301*, 344; (d) D. Chattophadhyay, I. Galeska, F. Papadimitrakopoulos, *J. Am. Chem. Soc.* **2003**, *125*, 3370; (e) Z. Chen, X. Du, M. H. Du, C. D. Raneken, H. P. Cheng, A. G. Rinzler, *Nano Lett.* **2003**, *3*, 1245; (f) M. S. Arnold, S. I. Stupp, M. C. Hersam, *Nano Lett.* **2005**, *5*, 713; (g) S. Banerjee, S. S. Wong, *Nano Lett.* **2004**, *4*, 1445; (h) L. An. Q. Fu, C. G. Lu, J. Liu, *J. Am. Chem. Soc.* **2004**, *126*, 10520; (i) C. A. Dyke, M. P. Stewart, J. M. Tour, *J. Am. Chem. Soc.* **2005**, *127*, 4497. (j) W. J. Kim, M. L. Usrey, M. S. Strano, *Chem. Mater.* **2007**, *19*, 1571; (k) M. C. Hersam, *Nature Nanotech.* **2008**, *3*, 387.

10. Professor Jasti began this work as a postdoctoral fellow at the Molecular Foundry, a nanoscience institute at the Lawrence Berkeley National Laboratory.

11. (a) D. J. Cram, J. M. Cram, *Acc. Chem. Res.* **1971**, *4*, 204–213. (b) K. Tahara, Y. Tobe, *Chem. Rev.* **2006**, *106*, 5274–5290 and references cited therein. (c) L. T. Scott, *Angew. Chem. Int. Ed.* **2003**, *42*, 4133.

12. T. Kawase, H. Kurata, *Chem. Rev.* **2006**, *106*, 5250–5273.

13. V. C. Parekh, P. C. Guha, *J. Indian Chem. Soc.* **1934**, *11*, 95–100.

14. R. Friederich, M. Nieger, F. Vögtle, *Chem. Ber.* **1993**, *126*, 1723–1732.

15. W. Z. Heldt, *J. Org. Chem.* **1965**, *30*, 3897–3902.

16. Y. Miyahara, T. Inazu, T. Koshino, *Tetrahedron Lett.* **1983**, *24*, 5277–5280.

17. S. Kammermeier, P. G. Jones, R. Herges, *Angew. Chem. Int. Ed. Engl.* **1996**, *35*, 2669–2671.

18. M. N. Jagadeesh, A. Makur, J. Chandrasekhar, *J. Mol. Model.* **2000**, *6*, 226–233.

19. R. Jasti, J. Bhattacharjee, J. B. Neaton, C. R. Bertozzi, *J. Am. Chem. Soc.* **2008**, *130*, 17646–17647.

20. M. Srinivasan, S. Sankararaman, H. Hopf, B. Varghese, *Eur. J. Org. Chem.* **2003**, 660–665.

21. F. Alonso, M. Yus, *Tetrahedron* **1991**, *47*, 7471–7476.

22. Z. Z. Song, H. N. C. Wong, *J. Org. Chem.* **1994**, *59*, 33–41.

23. M. S. Newman, K. Kanakarajan, *J. Org. Chem.* **1980**, *45*, 2301–2304.

24. H. M. Walborsky, H. II. Wüst, *J. Am. Chem. Soc.* **1982**, *104*, 5807–5808.

25. P. G. M. Green, T. W. Wutts, *Protecting Groups in Organic Synthesis*, 3rd ed., Wiley, New York, **1999**.

26. H. J. Liu, J. Yip, K. S. Shia, *Tetrahedron Lett.* **1997**, *38*, 2253–2256.

27. D. R. Salahub, M. Castro, E. I. Proynov, *NATO ASI Series, Series B: Physics* **1994**, *318*, 411–445.

28. (a) R. O. Jones, O. Gunnarsson, *Rev. Mod. Phys.* **1989**, *61*, 689; (b) A. Hellman, B. Razaznejad, B. I. Lundqvist, *J. Chem. Phys.* **2004**, *120*, 4593.

29. B. M. Wong, *J. Phys. Chem. C* **2009**, *113*, 21921–21950.

30. H. Takaba, H. Omachi, Y. Yamamoto, J. Bouffard, K. Itami, *Angew. Chem. Int. Ed.* **2009**, *48*, 6112–6116.

31. The strain energy reported by Itami is significantly higher than that reported by Jasti et al.[19] The difference arises from variation in the methods utilized to calculate strain energy. See the "Supporting Information" sections of both articles for additional details.

32. (a) G. Fuhrmann, T. Debaerdemaeker, P. Bäuerle, *Chem. Commun.* **2003**, 948; (b) F. Zhang, G. Götz, H. D. F. Winkler, C. A. Schalley, P. Bäuerle, *Angew. Chem.* **2009**, *121*, 6758; *Angew. Chem. Int. Ed.* **2009**, *48*, 6632.

33. S. M. AlQaisi, K. J. Galat, M. Chai, D. G. Ray, P. L. Rinaldi, C. A. Tessier, W. J. Youngs, *J. Am. Chem. Soc* **1998**, *120*, 12149.

34. A. Yahav-Levi, I. Goldberg, A. Vigalok, *J. Am. Chem. Soc.* **2006**, *128*, 8710.

# CHAPTER 12

# CONJUGATED MOLECULAR BELTS BASED ON 3D BENZANNULENE SYSTEMS

MASAHIKO IYODA, YOSHIYUKI KUWATANI, TOHRU NISHINAGA, MASAYOSHI TAKASE, and TOMOHIKO NISHIUCHI

## 12-1. INTRODUCTION

Belt-shaped $\pi$-conjugated molecules such as cycloparaphenylenes, cyclacenes, cyclophenacenes, and related compounds have long been attractive synthetic targets.[1] Especially after the discovery of fullerenes[2] and carbon nanotubes,[3] these types of molecules have drawn considerable interest as partial structures of the new carbon allotropes. In these fascinating $\pi$ systems, the $p$ orbitals are aligned perpendicularly to the belt surface, and the unusual three-dimensional (3D) molecular shapes provide not only theoretical interests as novel aromatic compounds,[4,5] but also various potential applications in supramolecular chemistry and materials science. In this respect, the shape-persistent structure with a central cavity is advantageous as a host molecule in host–guest chemistry, and the characteristic arrangement of $p$ orbitals endows belt-shaped $\pi$-conjugated molecules with distinctive supramolecular properties, namely concave–convex $\pi$–$\pi$ interactions.[1g] As partial structures of fullerenes and carbon nanotubes, interesting electrical conducting and/or optoelectronic properties are also expected.[6]

Among belt-shaped $\pi$-conjugated molecules, cyclacene was first proposed in 1954.[7] Since then, the properties of the curved $\pi$-conjugated systems have been investigated by means of quantum chemical calculations.[4] In particular, the discovery of carbon nanotubes made cyclacenes a more fascinating subject as the partial structure of zigzag-type single-walled carbon nanotubes.[4,8–10] However, various attempts at synthesis have so far been unsuccessful, probably owing to the predicted high reactivity of the products derived from the small singlet–triplet energy gap.[4c,f]

*Fragments of Fullerenes and Carbon Nanotubes: Designed Synthesis, Unusual Reactions, and Coordination Chemistry*, First Edition. Edited by Marina A. Petrukhina and Lawrence T. Scott.
© 2012 John Wiley & Sons, Inc. Published 2012 by John Wiley & Sons, Inc.

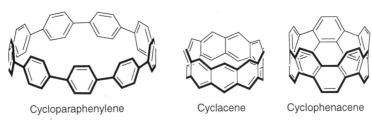

Cycloparaphenylene          Cyclacene          Cyclophenacene

**Chart 12-1.**

As far as the stability of the target molecule is concerned, the synthesis of partial structures of armchair carbon nanotubes is more practical. Thus, cycloparaphenylenes, in which the sidewall segment is benzene, have been synthesized by several groups.[11–13] As a related π-expanded system, cycloparaphenylene ethynylenes (**1**) have also been synthesized.

The interesting host−guest chemistry of these compounds has been investigated.[14] The synthesis of a paracyclophane consisting of four anthracene units, termed *pico-tube* **2**, has been demonstrated.[15] Also, the half-structure of *Vögtle belts* (**3**)[1d,j,16] has more recently been obtained by the use of strained cyclophane structure **4**.[17] We are interested in the synthesis of cyclophenacenes, in which the sidewall segment is phenanthrene. The cyclophenacene structures have already been obtained by chemical modification of fullerenes using regioselective, multiple addition reactions (e.g., **5**).[18] However, the synthesis of cyclophenacenes by a bottom−up approach has not been attained yet. Our approach toward cyclophenacene is the use of all-Z-benzannulenes **6−9** and the related phenanthrene analogue **10** as precursors.[19]

In addition to the abovementioned aspect, all-Z-benzannulenes **6−9** can be regarded as orthocyclophanes bridged by *cis*-olefin units. Cyclophanes have also attracted a considerable interest because of the interesting molecular structure, deformation of the strained arene rings, intramolecular interaction between arene rings, and host−guest chemistry by the use of the macrocyclic cavity.[1a,i,20] Introduction of unsaturated bonds to the bridging carbon chain reduces the flexibility of the macrocyclic skeleton, and extends the conjugated π-electron system. In this chapter, we describe the synthesis and properties of all-Z-benzannulenes **6−9** and observation of the formation of cyclophenacene by laser desorption ionization time-of-flight mass spectroscopy (LD-TOF-MS) from the phenanthrene analog **10**.

Macrocycles composed of rings with sizes other than 6 (i.e., six members) that incorporate cyclooctatetraene (COT) units in the belt together with six- or four-

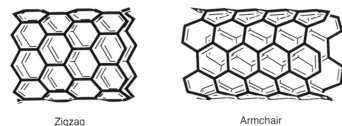

Zigzag          Armchair

**Chart 12-2.**

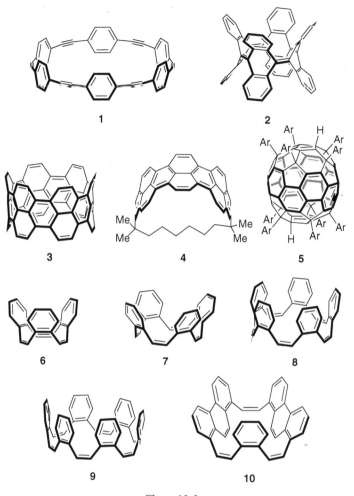

**Chart 12-3.**

membered rings ([4.8]₃cyclacene **11** and [6.8]₃cyclacene **12**) have been synthesized as homolog of cyclacene by Gleiter et al.[1b,c] Because **11** and **12** have tub-shaped COT rings, their internal strains are small and they are stable in both light and air. Although **11** and **12** have unique structures, the cavities of these molecules are too small to include any polyatomic guest molecule, from the viewpoint of host−guest chemistry. In this regard, the cleft-shaped molecule **13**, a partial structure of **12** with two dibenzocyclooctatetraene (DBCOT) units, is expected to work as a molecular tweezer. Because the face-to-face distance of the two terminal benzene rings is calculated to be 6.6 Å in a preliminary model structure at the PM3 level, **13** can incorporate a planar guest molecule between the two benzene rings. The synthesis and recognition properties of π-conjugated molecular tweezers **14−16**, shown here, are described later in the text (Section 12-4).

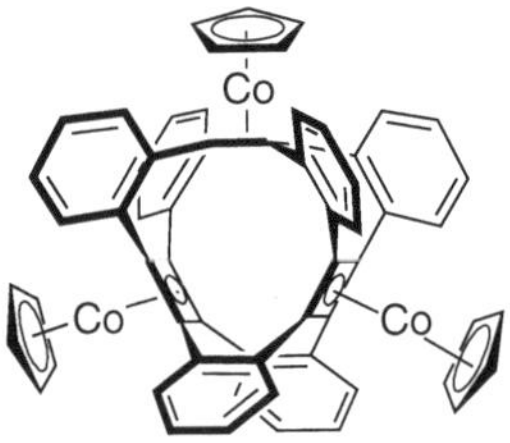

**Chart 12-4.**

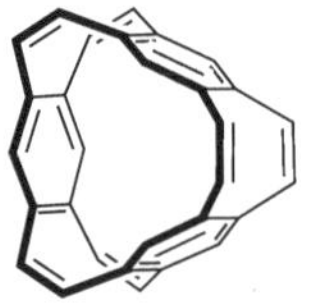

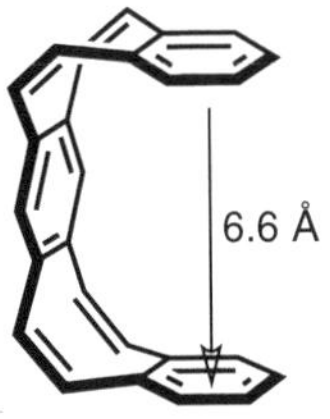

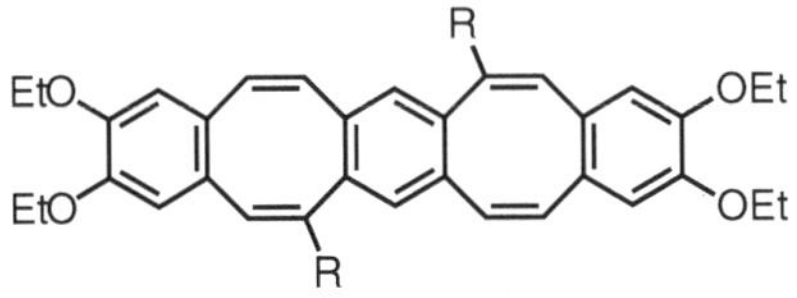

## 12-2.  ALL-Z-BENZANNULENES AND THEIR METAL COMPLEXES

### 12-2-1.  Synthesis of Benzannulenes

As related systems of all-Z-benzannulenes **6–9**, hexabenzo[12]annulene **17** and its atropisomer **18** were synthesized by the reaction of 2,2′-dilithiobiphenyl with various metal salts.[21] Z,Z-Tetrabenzo[12]annulene **19** was also prepared,[22] and its X-ray structure was determined.[23] Furthermore, we[19a] and Vollhardt et al.[25] independently reported the synthesis of all-Z-tribenzo[12]annulene **6** by thermal [2 + 2 + 2] cycloreversion of tris(benzocyclobuteno)cyclohexane (**20**), although it is difficult to apply this method to the synthesis of larger annulenes. Therefore, we decided to develop new synthetic pathways to all-Z-benzannulenes **6–9**.

For the synthesis of all-Z-benzannulenes, formation of the *cis*-stilbene unit and construction of the macrocyclic carbon framework are the key steps. For preparation of the *cis*-stilbene unit, stereoselective reduction of the diphenyl acetylene moiety is considered to be one of the most efficient methods, because the diphenyl acetylene moiety is easily obtained by means of Sonogashira coupling reaction[26] (Scheme 12-1, method 1). Pinacol coupling using low-valence vanadium complex would also be favorable owing to the mild reaction conditions and the high conversion efficiency.[27] The subsequent Corey–Winter procedure[28] of the cyclic pinacol can stereoselectively convert to the desired *cis*-olefin (Scheme 12-1, method 2). Although the requisite stereochemistry of the pinacol for this procedure is *erythro*, the *threo*-isomer can be converted to the desired isomer by the oxidation of the *threo*-pinacol followed by the reduction under chelation-controlled conditions. For the formation of macrocycles, method 2 would also be applicable.

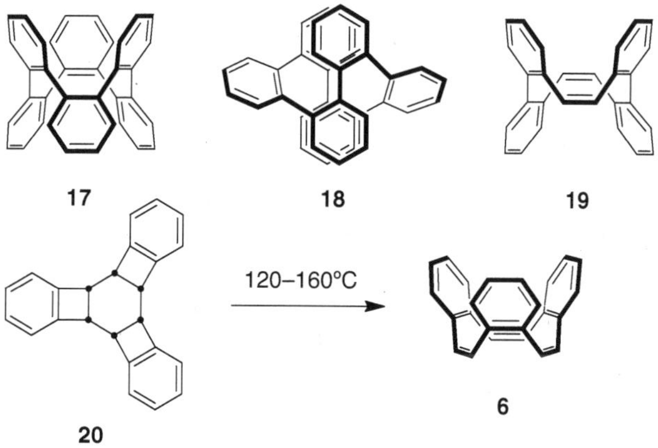
17
18
19
20
120–160°C
6

**Chart 12-5.**

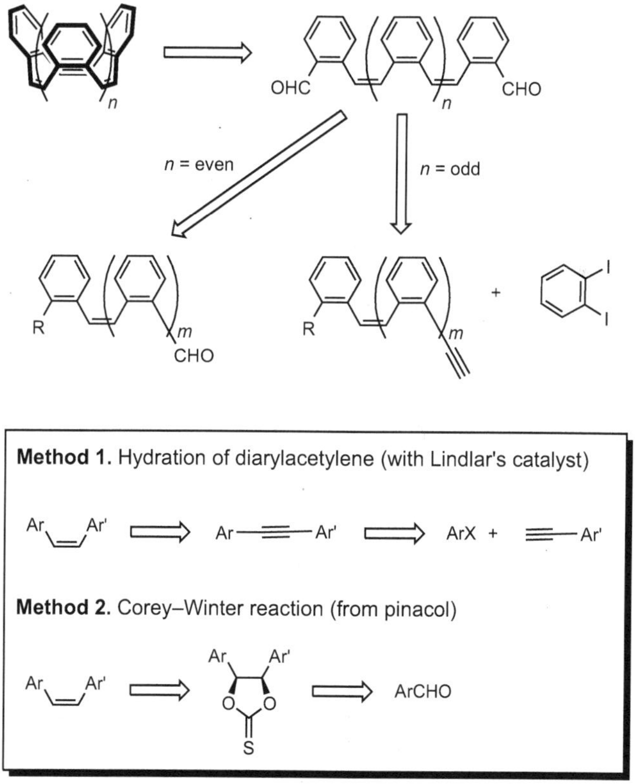
n
OHC
CHO
n
n = even
n = odd
R
m
CHO
R
m
+
I
I
Method 1. Hydration of diarylacetylene (with Lindlar's catalyst)
Ar
Ar'
Ar
Ar'
ArX +
Ar'
Method 2. Corey–Winter reaction (from pinacol)
Ar
Ar'
Ar
Ar'
O
O
S
ArCHO

**Scheme 12-1.** Retrosynthetic analysis for **6–9**.

Using these synthetic tools, the retrosynthetic analysis proceeds as illustrated in Scheme 12-1. The macrocyclic structure can theoretically be constructed from the linear oligomers of *cis*-stilbene possessing two terminal aldehyde functionalities. The linear precursors can be prepared by combinations of the two methods described above. For the linear oligomers with an even number of benzene rings, the central double bond is constructed by method 2. In the case of the compounds with an odd number of benzene rings, two identical components are introduced to the central benzene ring by method 1.

According to the general synthetic strategy mentioned above, compound **6** was synthesized.[19] Sonogashira coupling of *o*-diiodobenzene and 2-ethynylbenzyl alcohol (**21**) gave the diol **22** in a good yield (Scheme 12-2). Reduction of the acetylene units in **22** was achieved by use of the Lindlar's catalyst[29] to afford **23**. After oxidation of the alcohol functionality of **23**, the macrocyclization of dialdehyde **24** by use of a vanadium complex afforded cyclic pinacol **25**, which was proved to be the *threo*-isomer with twisted $C_2$ symmetry. The stereochemistry of pinacol **25** was modified by the sequence of Swern oxidation and reduction with $NaBH_4$ to afford the *erythro*-pinacol **26**. This pinacol was converted to the thionocarbonate **27** by reaction with thiocarbonyldiimidazole (TCDI) in refluxing toluene. The final double-bond formation was completed by reaction of **27** with 1,3-dimethyl-2-phenyl-1,3,2-diazaphospholidine (DMPD) in refluxing benzene (Scheme 12-2).

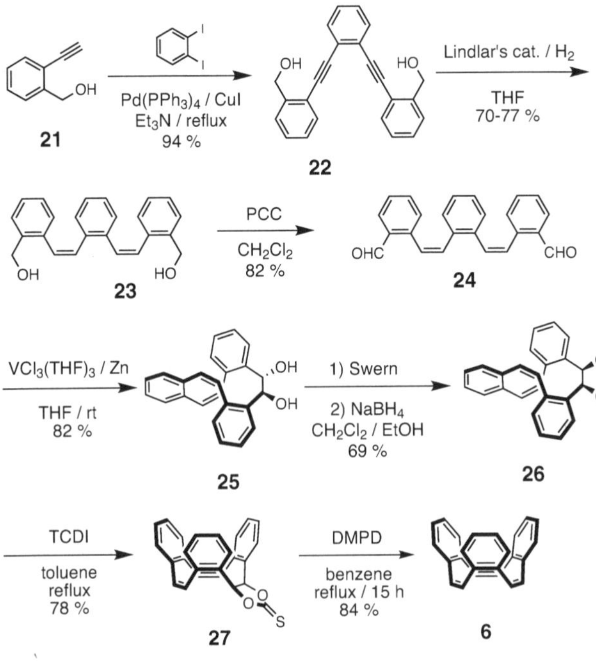

**Scheme 12-2.** Synthesis of benzannulene **6**.

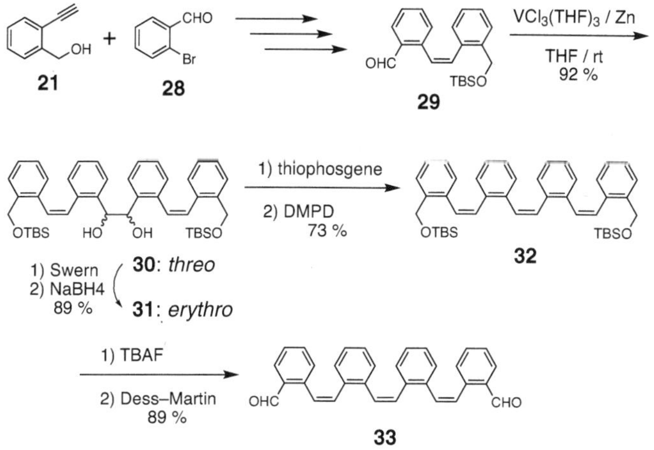

**Scheme 12-3.** Preparation of dialdehyde **33**.

For the synthesis of **7**, which has an even number of benzene rings, linear dialdehyde **33** was prepared by method 2 of Scheme 12-1 (see Scheme 12-3).[19b] A *cis*-stilbene derivative **29** was prepared from **21** and 2-bromobenzaldehyde (**28**) according to method 1. Dimerization of **29** by the pinacol coupling reaction with a vanadium complex selectively afforded *threo*-pinacol **30**, which was converted to the *erythro*-pinacol **31** by the sequence of oxidation and reduction. The corresponding thionocarbonate, which was prepared by the reaction of **31** with thiophosgene,[28] afforded triene **32** by the Corey–Winter reaction with DMPD. Removal of the silyl protecting groups in **32** followed by Dess–Martin oxidation[30] afforded the dialdehyde **33** (Scheme 12-3).

Analogous dialdehydes **36** and **40** were also successfully synthesized by similar pathways (Schemes 12-4 and 12-5). Aldehyde **29** was converted to the terminal alkyne **34** via the corresponding dibromoethylene derivative. Alkyne **34** was introduced to the central benzene ring by the Sonogashira coupling reaction. Oxidation of

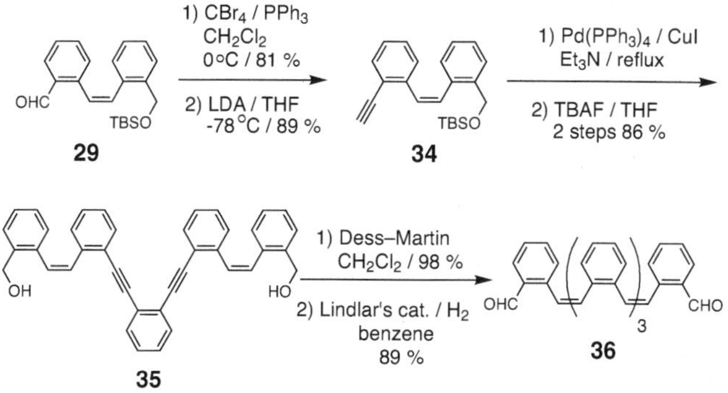

**Scheme 12-4.** Preparation of dialdehyde **36**.

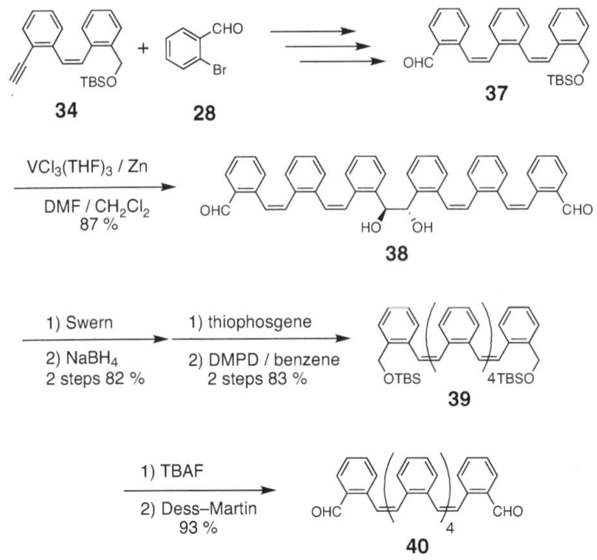

**Scheme 12-5.** Preparation of dialdehyde **40**.

**35** followed by alkyne hydrogenation with the Lindlar catalyst afforded dialdehyde **36**. Dialdehyde **40** was also synthesized by the dimerization strategy of monoaldehyde **37** prepared from **34** and 2-bromobenzaldehyde (**28**).[31] The intermolecular pinacol coupling of **37** afforded *threo*-pinacol **38**, selectively. After the stereochemical transformation of **38** by the oxidation–reduction sequence, dehydroxylation of the resultant *erythro*-pinacol afforded pentaene **39**. Removal of the silyl protecting groups followed by oxidation of the alcohols afforded dialdehyde **40**.

Intramolecular cyclization of **33**, **36**, and **40** by the pinacol coupling reaction with the low-valence vanadium complex afforded the corresponding cyclic pinacols in good yields. However, the diastereoselectivity in the intramolecular pinacol coupling reaction varied with the ring size. The reaction usually shows *threo*-selectivity, which was observed in the intermolecular coupling of **29** and **37** and in the intramolecular cyclization of the aldehyde **24**. The *threo* : *erythro* ratios decreased to about 2 : 1 for the cyclization of **33** and **36**, and inverted to 1 : 5 for the cyclization of **40**.[27b,32] The difference in selectivity may reflect the relative conformational stability in the transition state for the coupling reaction. The following transformation of pinacols **41**–**43** to the corresponding benzannulenes proceeded without any trouble to afford **7**, **8** and **9** as shown in Scheme 12-6.[19b,31]

## 12-2-2. Structures and Dynamic Behaviors

The structure of benzannulenes **6**–**9** were studied by NMR spectroscopy and X-ray crystallography together with the results of theoretical calculations. The smallest analog of this series of benzannulenes is DBCOT **47** which has a tub-shaped structure with dihedral angles of 99° between the two benzene rings in the crystal.[33] The flexibility of this compound was studied by dynamic NMR measurements for the

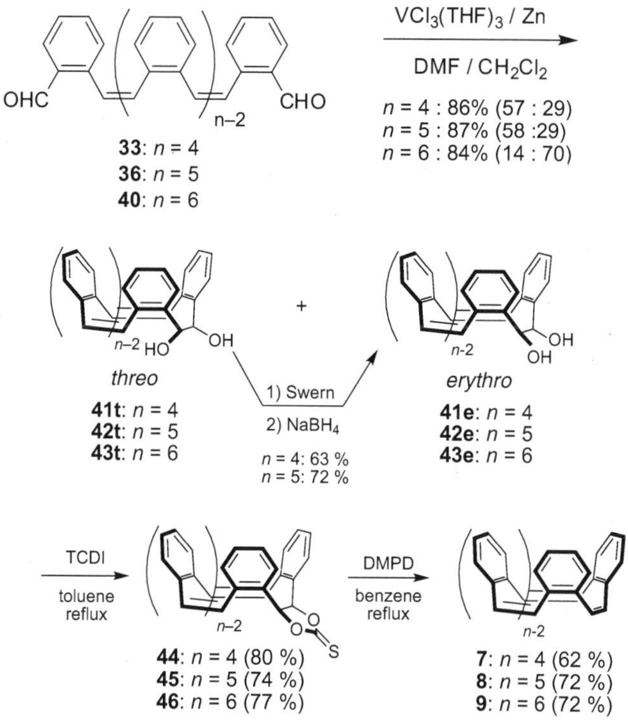

VCl3(THF)3 / Zn
DMF / CH2Cl2
OHC
CHO
n–2
n = 4 : 86% (57 : 29)
n = 5 : 87% (58 :29)
n = 6 : 84% (14 : 70)
33: n = 4
36: n = 5
40: n = 6
+
n–2 HO
OH
threo
41t: n = 4
42t: n = 5
43t: n = 6
1) Swern
2) NaBH4
n = 4: 63 %
n = 5: 72 %
n-2
OH
OH
erythro
41e: n = 4
42e: n = 5
43e: n = 6
TCDI
toluene
reflux
n–2
O
O
S
44: n = 4 (80 %)
45: n = 5 (74 %)
46: n = 6 (77 %)
DMPD
benzene
reflux
n-2
7: n = 4 (62 %)
8: n = 5 (72 %)
9: n = 6 (72 %)

**Scheme 12-6.** Synthesis of benzannulenes **7**, **8**, and **9**.

substituted analog to determine the activation energy of $\Delta G^{\ddagger} = 12.3 \pm 0.2$ kcal/mol at $-5$°C for the ring inversion[34] (Figure 12-1).

Tribenzo[12]annulene **6** is expected to have a $C_{3v}$-symmetric rigid structure. The $^1$H NMR spectrum of **6** showed one olefinic singlet and a set of AA′BB′ aromatic signals, and these signals were found to be temperature-independent. Although the crystal structure of **6** has not been determined, theoretical calculations support the $C_{3v}$ structure [Figure 12-2(a)]. Reflecting the preorganized rigid structure, **6** reacts easily with Ag(I) or Cu(I) salts to form the corresponding complexes. The X-ray structures

**47**

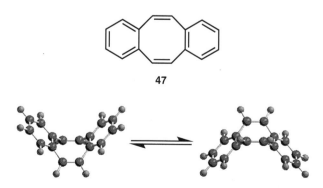

**Figure 12-1.** Ring inversion of DBCOT **47**.

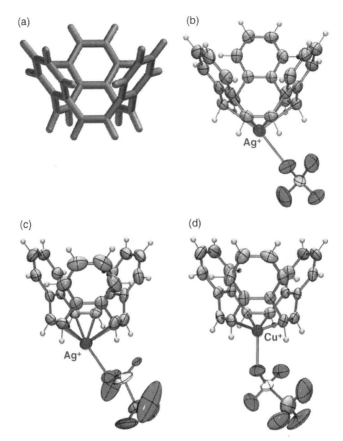

**Figure 12-2.** Molecular structures of **6** and its complexes: (a) optimized structure at the B3LYP/6-31G(d) level; ORTEP (50% probability) drawings of (b) **6**•AgClO₄, (c) **6**•AgOTf, and (d) **6**•CuOTf.

of the complexes clearly showed that the three double bonds effectively coordinate to the metal ions in the central cavity [Figure 12-2(b-d)].[35] The carbon frameworks in all complexes have essentially the same crownlike structure as that obtained by the DFT calculations [B3LYP/6-31G(d)] of **6**. These complexes are also confirmed in solution, judging from the lower-field shifts of the olefinic signals than that of **6** in ¹H NMR spectra.[24,35]

The structure of tetrabenzo[16]annulene **7** is more flexible than that of **6** in solution. The ¹H NMR spectrum of **7** at low temperature shows two sets of AA′BB′ aromatic signals and a pair of AB-type olefinic signals, suggesting a $C_{2v}$-symmetric structure of **7**. The signals broaden with elevation of the temperature and coalesce to the spectrum assigned as a $C_{4v}$-symmetric structure, which is composed of one set of the aromatic AA′BB′ signals and a singlet for the olefin protons. The exchanging process was evaluated by the coalescence temperature method to afford the activation energy of $\Delta G^{\ddagger} = 12.7$ kcal/mol. The X-ray crystallographic analysis of **7** proved the

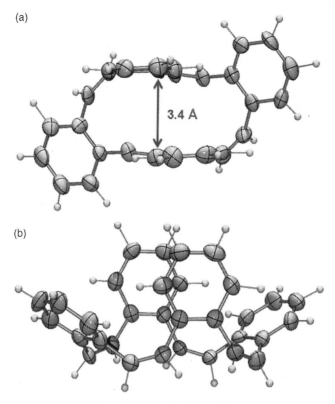

**Figure 12-3.** ORTEP (50% probability) drawings of **7**: (a) top view, (b) side view.

flattened $C_2$-symmetric structure with the two opposite benzene rings assembling at a distance of 3.4 Å (Figure 12-3).[19b] A similar molecular structure was also obtained as the most stable conformer for **7** by DFT calculations [B3LYP/6-31G(d)]. The dynamic process observed by NMR can be regarded as an exchange between a pair of assembling benzene rings and a pair of outside benzene rings. The slipping motion, which is an enantiomerization of the $C_2$-symmetric conformer, is considered to have a barrier that is too low to be observed at the measured temperatures (Scheme 12-7).

Compound **7** also formed silver complexes by reaction with AgClO$_4$ or AgOTf.[36] The $^1$H NMR spectra of both complexes were essentially the same, indicating that both have $C_s$ symmetry in solution. Among the signals for the olefinic protons, a characteristic signal was observed in a high-field region of δ 4.50. The X-ray crystal structure of **7**•AgClO$_4$ demonstrates that the conformation of the annulene ligand is considerably changed on formation of the silver complex [Figure 12-4(a,b)]. The silver ion locates inside the macrocycle, and the coordination geometry of the complex is a distorted trigonal bipyramid, in which perchlorate ion and one olefinic double bond are in the axial positions and three other olefinic double bonds are in the equatorial positions. Thus, the characteristic $^1$H NMR signal mentioned above can be assigned for the inside olefinic protons. In DFT calculations [B3LYP/6-31G(d)], this

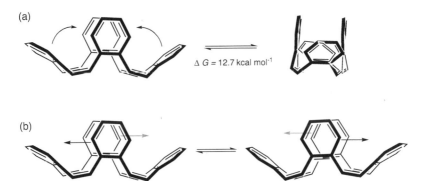

**Scheme 12-7.** Dynamic motions of **7**: (a) exchange of the stacking benzene rings; (b) slipping motion of the stacking benzene rings.

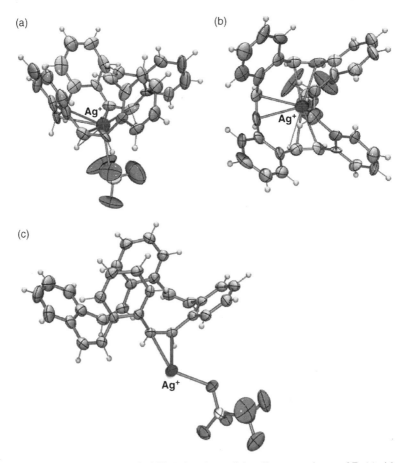

**Figure 12-4.** ORTEP (50% probability) drawings of the silver complexes of **7**: (a) side view of **7**•AgClO$_4$; (b) top view of **7**•AgClO$_4$, (c) molecular structure of **7**•AgOTf.

$C_s$-symmetric conformation of **7** is shown to be a local minimum structure with energy 1.9 kcal/mol higher than that of the $C_2$ conformer. On the other hand, the X-ray crystal structure of **7•AgOTf** shows a different structure, in which the silver ion is coordinated with only one double bond on the outside of a $C_2$ conformer [Figure 12-4(c)]. The silver ion is also coordinated by a triflate anion and an arene part of another molecule to form an infinite coordination polymer.

The larger pentabenzo[20]annulene **8** was proved to have a more flexible structure. The $^1$H NMR spectrum showed an olefinic singlet and a pair of aromatic AA′BB′ multiplets, indicating that **8** has a $C_{5v}$-symmetric structure in solution. The symmetric NMR spectrum does not change even at −90°C, except for slight broadening of the signals. The structure of **8**, determined by X-ray crystallographic analysis [Figure 12-5 (a,b)], shows the nonsymmetric structure in which one double bond comes inside the macrocyclic cavity.[37] The structure observed in the crystal is in good accordance with that obtained by DFT calculations [B3LYP/6-31G(d)]. Therefore, the structure of **8** is very flexible with the low activation energy of < 8 kcal/mol, which is estimated by the coalescence temperature (<−90°C) and the difference between chemical shifts for olefinic protons. Reaction of **8** with AgClO$_4$ afforded the silver complex, which has a

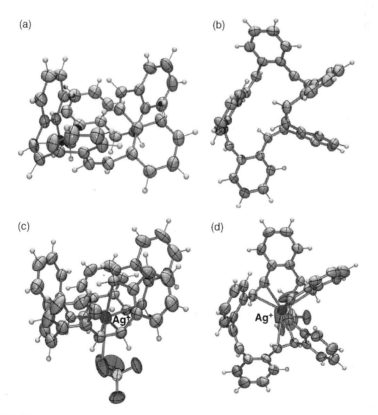

**Figure 12-5.** ORTEP (50% probability) drawings of **8** and **8•AgClO$_4$**: (a) side view of **8**; (b) top view of **8**, (c) side view of **8•AgClO$_4$**; (d) top view of **8•AgClO$_4$**.

silver ion inside the cavity without apparent conformational change in the annulene ligand.[36] As shown in Figure 12-5(c,d), the coordination geometry of the complex is a distorted trigonal bipyramid similar to that of **7**•AgClO$_4$. In this complex, only four of the five olefinic double bonds coordinate to the silver ion to form a nonsymmetric structure. The $^1$H NMR spectrum of the complex at ambient temperature is very broad and turned sharper on cooling. The spectrum at $-50°$C suggests that the complex has a $C_s$-symmetric structure, which can be explained by a rapid exchange of the two coordinating double bonds at the equatorial position. Actually, the signals for these olefinic protons became broad on further cooling. Thus the very flexible molecular skeleton of **8** was rigidified somewhat by the formation of the silver complex.

As shown in Figure 12-6, the molecular structure of hexabenzo[24]annulene **9** was found to be a $C_3$ conformation, in which three benzene rings were projected outward and the others were assembled inside, with dihedral angles $\sim 60°$ to each other.[33] The intramolecular assembly of the inner benzene rings is attributed to the three concurrent CH$-\pi$ interactions among these benzene rings. The average distance between the

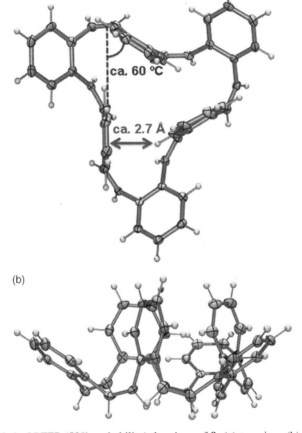

**Figure 12-6.** ORTEP (50% probability) drawings of **9**: (a) top view, (b) side view.

hydrogen atom and the center of the arene is only 2.65 Å (the corresponding C–Ar distance is 3.44 Å).

The $^1$H NMR spectrum shows an olefinic singlet and two broad aromatic signals, which change to sharp AA'BB' multiplets at higher temperatures. The signals broaden on cooling, and clearly separate signals reappear below $-70°$C, which corresponds to a single conformer. The spectrum at $-90°$C is composed of signals from two types of symmetrically disubstituted benzene rings and two sets of *cis* double bonds, which is in good accordance with the $C_3$-symmetric structure observed in the crystal. In the spectrum, the characteristic signal is observed at δ 4.4, which is assigned to the *o*-aryl proton shielded by a neighboring arene ring. Such a large high-field shift suggests a considerable proximity of the proton to the neighboring arene π cloud, as is found in the crystal structure. The distance between the proton and the arene ring can be estimated as 2.50–2.70 Å on the basis of the chemical shift,[38] which is consistent with the X-ray analysis. The activation energy of the conformational exchange process is estimated at the coalescence temperature (approximately $-40°$C) to be $\Delta G^{\ddagger} \sim$ 10 kcal/mol, which suggests a conformational stability of **9**. The thermodynamic and kinetic stability of the $C_3$ conformer in **9** is attributed to the CH–π interactions occurring simultaneously at three sites among the three benzene rings.

### 12-2-3. Conformational Analysis

The conformational analysis of macrocycles **6–9** makes it possible to learn the relationship between stability and ring size. The stabilities of the macrocycles were evaluated by DFT calculations. The optimized geometries of **6–9** and **47** at the B3LYP/6-31G(d) level reproduced the crystallographic structures well, suggesting that the level of theory gives meaningful total energies of the molecules. On the basis of these total energies, the total energy per (Z)-CH=CH-*o*-C$_6$H$_4$– unit structure is calculated, and the results are plotted in Figure 12-7. From this plot, the total energy per unit is found to decrease with increasing ring size in the series of compounds **6–9**.

In order to gain further insight into the relationship, a conformational energy profile of *cis*-stilbene was examined. The energy profile of the optimized structure by a semiempirical method (AM1) under the constraint of the planar ethylene unit

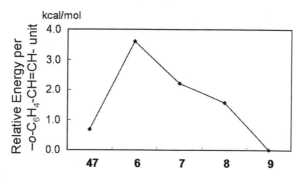

**Figure 12-7.** Relative energies of **6–9** and **47** obtained by DFT calculations (see text).

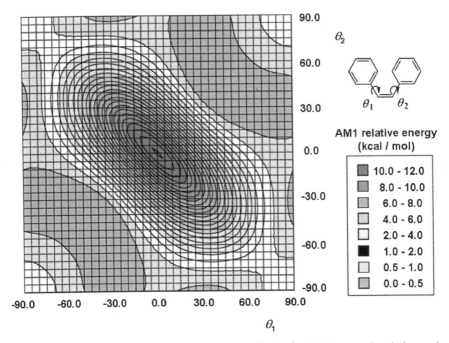

**Figure 12-8.** AM1 contour plot of energy *versus* two dihedral angles between the ethylene unit and phenyl groups.

is shown in Figure 12-8. The global minimum structure of *cis*-stilbene is the *anti*-$C_2$ conformation with torsion angles of 41.2°, which is in good accordance with the reported value.[39] The planar $C_{2v}$ structure is the least stable conformation because of the steric repulsion between two phenyl groups. The region showing the relatively large repulsion in *cis*-stilbene was mainly found in the *syn* conformation ($\theta_1 < 0 < \theta_2$ or $\theta_2 < 0 < \theta_1$), whereas less repulsion is expected in the *anti* conformer ($\theta_1, \theta_2 < 0$ or $\theta_1, \theta_2 > 0$). These results indicate the flexibility in the *anti* conformer for *cis*-stilbene.

Conformation of the *cis*-stilbene moieties in benzannulenes **6–9** and **47** were also analyzed using the pair of torsional angles of the phenyl groups at each ethylene bridge. As shown in Figure 12-9, the smallest cyclic compound **6** has a *syn*-stilbene conformation with a considerable torsional strain. The plotted points for the larger macrocycles appear in the region of *anti*-conformation, and the energy decreases with increasing ring size. This result suggests that the decrease in torsional strain in the larger macrocycles is the principal reason for the thermodynamic stability. In addition to this effect, intramolecular interactions would also play some role. As mentioned above, interesting intramolecular aromatic interactions were observed in **7** and **9** (Figures 12-10 and 12-11). The face-to-face arrangement of two benzene rings in **7** is similar to the stacked benzene dimer, which was reported to have a stabilizing energy of ~2 kcal/mol.[40] The triangular arrangement of three benzene rings in **9** is also consistent with the benzene trimer, the structure of which is

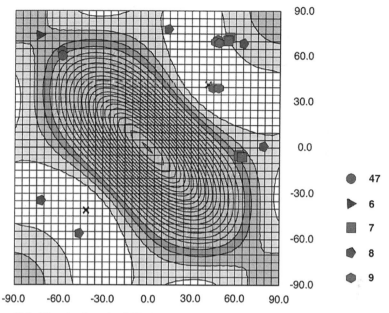

**Figure 12-9.** Plots for the pair of dihedral angles in the structures of **6–9** and **47** on the contour plot for *cis*-stilbene. The parameters are taken from X-ray crystallographic data for **7–9** and **47**, and optimized structure for **6**. The global minimum points for *cis*-stilbene are shown as ×.

theoretically proposed for the gas-phase aggregate of benzene.[41] The binding energy of the benzene trimer determined by a gas phase experiment was reported to be 6.2 kcal/mol.[42] These attractive interactions would be attributable to the conformational stability in **7** and **9**.

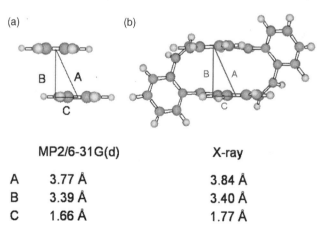

|     | MP2/6-31G(d) | X-ray   |
| --- | ------------ | ------- |
| A   | 3.77 Å       | 3.84 Å  |
| B   | 3.39 Å       | 3.40 Å  |
| C   | 1.66 Å       | 1.77 Å  |

**Figure 12-10.** Comparison of the arrangements of the stacked benzene rings in (a) the theoretically obtained [MP2/6-31G(d)] benzene dimer and (b) the crystal structure of **7**.

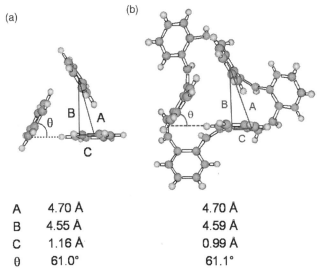

| | (a) | (b) |
|---|---|---|
| A | 4.70 Å | 4.70 Å |
| B | 4.55 Å | 4.59 Å |
| C | 1.16 Å | 0.99 Å |
| θ | 61.0° | 61.1° |

**Figure 12-11.** Comparison of the arrangements of the three benzene rings in (a) the theoretically obtained [MP2/6-31G(d)] benzene trimer and (b) the crystal structure of **9**.

## 12-2-4. Photocyclization Reaction

For the synthesis of phenanthrene, photocyclization of stilbene in the presence of an oxidant such as iodine (Scheme 12-8) is a well-known method. Therefore, we expected that cyclophenacene could be synthesized using a photocyclization reaction of the corresponding unsaturated cyclophanes constructed of stilbene units. Thus,

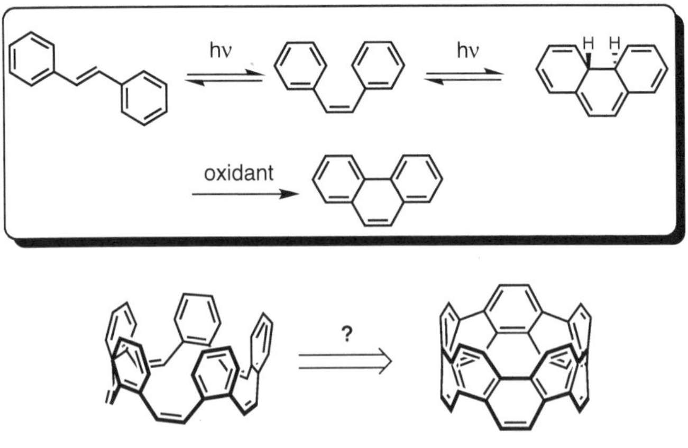

**Scheme 12-8.** Photocyclization reaction of stillbene and approach for the synthesis of cyclophenacene.

(a)                                            (b)

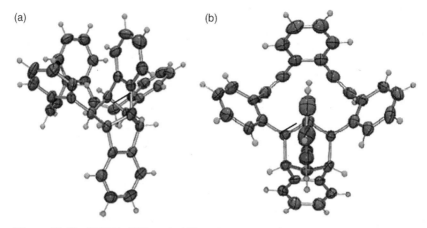

**Figure 12-12.** ORTEP (50% probability) drawings of **8′**: (a) side view, (b) top view.

photoirradiation reactions of the penta- and hexabenzoannulene **8** and **9** were attempted, since cyclo[10]- and [12]phenacenes are considered to be more stable than cyclo[6]- and [8]phenacenes. Although no products could be identified in the photoirradiation of hexabenzoannulene **9**, pentabenzoannulene **8** gave a single product **8′** both with and without an additive such as iodine. From single-crystal X-ray analysis of **8′** (Figure 12-12), we determined that isomerization of one double bond took place by photoirradiation to give the *E,Z,Z,Z,Z*-isomer, which cyclized in a [2 + 2 + 2] manner to produce **8′** (Scheme 12-9).

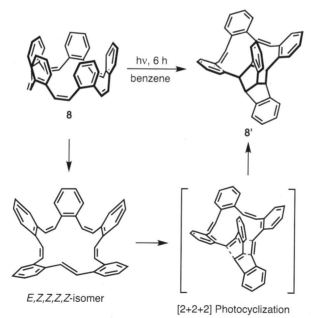

**Scheme 12-9.** Plausible reaction mechanism for the formation of **8′**.

The undesirable isomerization of **8** indicated that a rigid, preorganized structure is required for the synthesis of cyclophenacene; therefore, we planned to prepare the more rigid precursor **10** with two phenanthrene units.

## 12-3. UNSATURATED CYCLOPHANES COMPOSED OF PHENANTHRENE, BENZENE, AND ETHYLENE UNITS

### 12-3-1. Synthesis of All-*Z*-Cyclophanetriene 10

Since triple bonds of tribenzotetradehydro[12]annulene **48** and tribenzohexadehydro [12]annulene **49** can be reduced to double bonds with the low-valence titanium reagent prepared from Ti(O-*i*-Pr)$_4$ and *i*-PrMgBr[43] to produce all-*Z*-tribenzo[12] annulene **6** (Scheme 12-10),[44] this reduction has been applied to the synthesis of macrocycle **10** composed of phenanthrene, benzene, and *Z*-ethylene units.[37]

As shown in Scheme 12-11, 2-bromobenzaldehyde (**28**) was converted into 2,2'-dibromo-(*E*)-stilbene (**51**) by McMurry coupling. Photochemical cyclization of **51** in the presence of I$_2$ in cyclohexane produced 1,8-dibromophenanthrene (**52**). Selective monolithiation of **52**, followed by reaction with DMF, afforded 8-bromophenan-threne-1-carbaldehyde (**53**). Sonogashira coupling of **53** with trimethylsilylacetylene (TMSA), followed by deprotection of trimethylsilyl group with TBAF in THF–methanol, gave **55**. Sonogashira coupling of **55** with *o*-diiodobenzene yielded

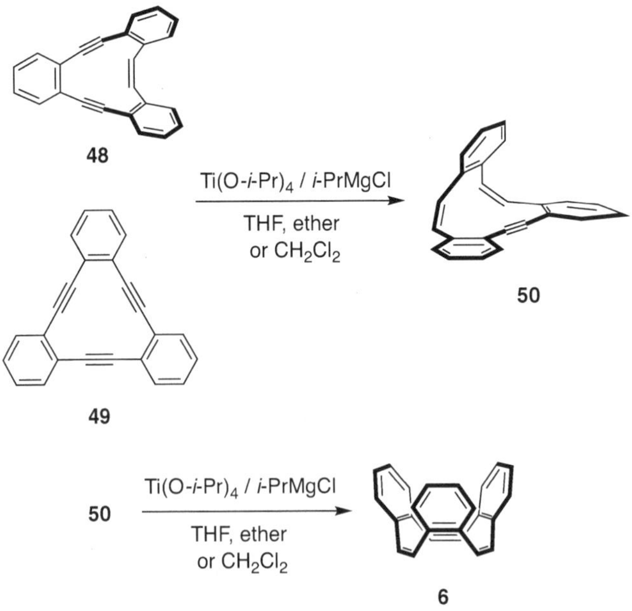

**Scheme 12-10.** Preparation of *all-Z*-tribenzo[12]annulene **6** using the low-valence titanium reagent.

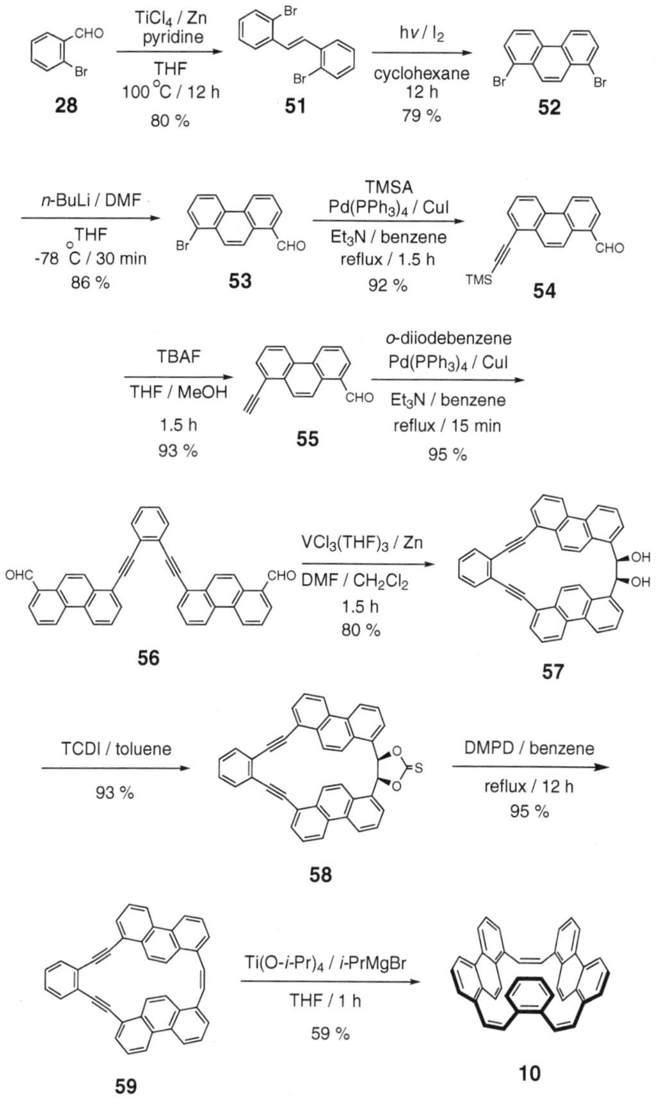

**Scheme 12-11.** Synthesis of *all*-Z-cyclophanetriene **10**.

the insoluble linear dialdehyde **56** (95%). Dialdehyde **56** was intramolecularly cyclized, using the low-valence vanadium reagent prepared from VCl$_3$(THF)$_3$ and activated zinc in DMF and CH$_2$Cl$_2$ at room temperature for 1.5 h to produce *cis*-pinacol **57** selectively in 80% yield. Pinacol **57** was first converted into thiocarbonate **58** with TCDI, followed by treatment with DMPD in refluxing benzene for 12 h, which afforded Z-cyclophadiynene **59** in 88% overall yield (Corey–Winter procedure). Although **59** was rather insoluble in common organic solvents, the reduction of **59**

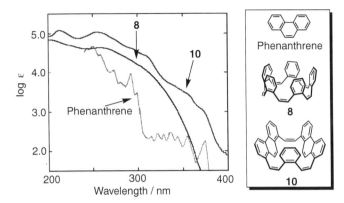

**Figure 12-13.** UV–visible spectra of phenanthrene, **8** and **10**.

with $Ti(OPr^i)_4$ and $i$-PrMgBr[43] in THF at $-78°C$ to room temperature produced all-Z-cyclophanetriene **10** in 59% yield.

## 12-3-2. Structure and Properties of All-Z-Cyclophanetriene 10

All-Z-Cyclophanetriene **10** adopts a cage structure with a large inner cavity. Therefore, the presence of the phenanthrene ring and the 3D structure of **10** result in no cyclic conjugation. Interestingly, however, the UV–visible spectrum of **10** shows a redshift compared with the spectrum of all-Z-pentabenzo[20]annulene **8**, probably owing to the extension of $\pi$ conjugation (Figure 12-13).

As shown in Figure 12-14, X-ray analysis of **10** reveals a rigid cage structure, and the interatomic distances between phenanthrene carbons are 4.25–7.28 Å, whereas the closest interatomic distances between phenanthrene and benzene carbons are 3.45 and 3.51 Å. In the $^1H$ NMR spectrum of **10**, the benzene proton $H_a$ shows an upfield shift at $\delta$ 6.68, reflecting a shielding effect of the closely located phenanthrene ring. Interestingly, the cage molecules stack on each other, using a large cavity to form a columnar structure.

## 12-3-3. Oxidative Cyclization of All-Z-Cyclophanetriene 10

For the synthesis of cyclo[10]phenacene, photochemical cyclization of **10** was attempted. Although the irradiation of **10** in the presence of iodine and propylene oxide in toluene was unsuccessful and resulted in the decomposition of **10**, the LD-TOF-MS of **10** revealed an interesting spectrometric behavior. First, the spectrum was measured under normal conditions to show the molecular ion peak at $m/z$ 506 corresponding to *all-Z*-cyclophanetriene **10** together with a small peak at $m/z$ 500, as shown in Figure 12-15. However, when the laser intensity was increased, the molecular ion peak at $m/z$ 506 decreased, and the molecular ion peak at $m/z$ 500 corresponding to cyclo[10]phenacene increased. Although the attempted isolation of

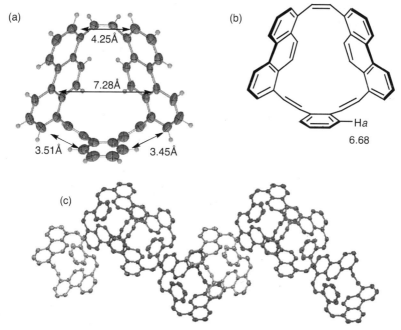

**Figure 12-14.** (a) ORTEP (50% probability) drawing, (b) $H_a$ proton, and (c) packing structure of **10**.

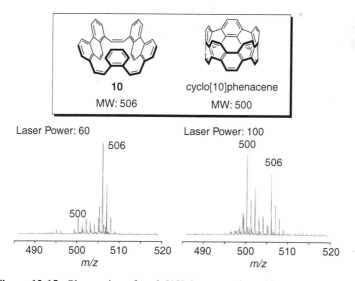

**Figure 12-15.** Observation of cyclo[10]phenacene from **10** by LD-TOF-MS.

cyclo[10]phenacene was unsuccessful, the LD-TOF-MS of **10** showed the formation of cyclo[10]phenacene under the conditions of the LD-TOF-MS measurements.

## 12-4. DYNAMIC MOLECULAR TWEEZERS COMPOSED OF TWO DIBEZOCYCLOOCTATETRAENE UNITS

During the course of the study on the cyclophenacene, we became interested in the tub shape and dynamic motion of DBCOT; the tub shape of DBCOT provides a bent- or beltlike π-conjugated system such as [4.8]₃ cyclacene and [6.8]₃ cyclacene, and the dynamic motion is expected to functionalize the π-conjugated systems from the perspective of supramolecular chemistry. Therefore, we designed a novel bent-π-conjugated system **13** composed of two DBCOT units; molecule **13** is expected to exist as an equilibrium mixture of *syn* and *anti* forms owing to ring inversion of DBCOT, and can work as molecular tweezers[45] as described in the Section 12-1. In this section, the synthesis and dynamic molecular recognition of the π-conjugated dynamic molecular tweezers (DMTs) of **14–16**[46] are presented.

For the synthesis of the title compounds **14–16**, the key step is the formation of COT ring. Although a variety of simple methods for preparation of COT and its derivatives have been reported,[47] we employed a stepwise pathway to construct COT rings of DMTs, as shown in Scheme 12-12. The stepwise reactions enable us to introduce substituent groups at the desired positions. First, **62** was prepared from **60** and two 2-bromo-4,5-diethoxybenzaldehydes (**61**) by Sonogashira reaction. The two acetylene units of **62** were reduced to *cis* olefins by partial hydrogenation using Pd on BaSO₄ to afford **63**. Next, **65** was prepared by deprotection of the *t*-butyldimethylsilyl (TBS) groups of **63**, followed by mesylation of the hydroxyl groups of **64**. When **65** was reacted with NaCN, the formation of cyclized eight-membered ring compounds was observed by ¹H NMR. Therefore, a mixture of linear and cyclized compounds was reacted with DBU without purification to produce molecular tweezer **14** in 58% yield from **65**. Dialdehyde **15** was synthesized in 81% yield by reduction of **14** using DIBAL-H, and the methoxymethyl (MOM) derivative **16** was obtained from **15** in 68% yield.

In host–guest chemistry, ¹H NMR titration is one of the most useful methods not only for estimating the binding constant but also for understanding the binding behavior in solution. EDMT units are equilibrium mixtures of *syn* and *anti* forms, and their ¹H NMR spectra peaks are observed independently, thus giving complicated results. When DDQ was added incrementally to a solution of **14**, no shift of the *anti* protons was observed, while the *syn* protons gradually shift and increase in intensity. These results indicate that only the *syn* form can bind DDQ, and the conformation is fixed toward the *syn* form on complexation (Figure 12-16). Large high-field shifts of the terminal benzene rings (Hₑ and H_f) and low-field shifts of the COTs (H_c and H_d) are attributed to the ring current of the DDQ plane and the interaction (hydrogen bonding) with carbonyl groups of DDQ, respectively.

Detailed information about the complex between 1,2,4,5-tetracyanobenzene (TCNB) and **15** was obtained by X-ray crystallography, as shown in Figure 12-17.

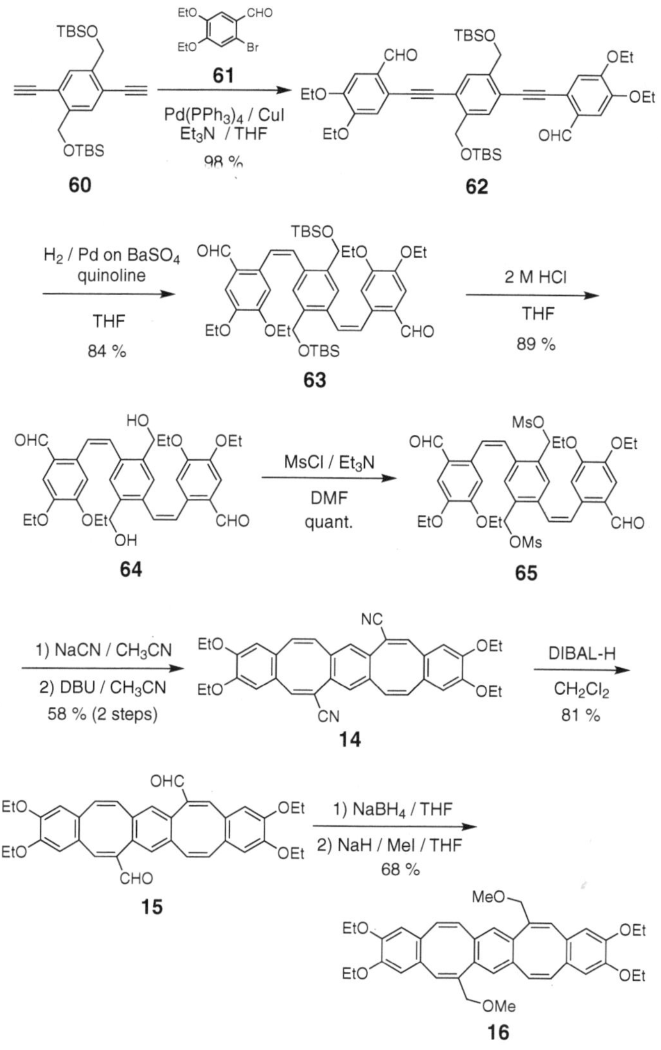

**Scheme 12-12.** Preparation of dynamic molecular tweezers **14–16**.

TCNB and the terminal benzene rings stack through $\pi$–$\pi$ interactions with the distance of 3.33–3.40 Å, and the dihedral angles between terminal and central benzene rings in **15** are approximately 90°. Moreover, the distance between the proton of TCNB and the central benzene ring of **15** is 2.54 Å, indicating the presence of CH–$\pi$ interaction.

Irrespective of slight structural differences among DMTs **14–16**, binding constants were increased by elevating the electron-donating ability of COT substitutions from CN, CHO to $CH_2OMe$ groups. In order to estimate electrostatic effects on the complexation, *syn*-form structures and electrostatic potential surfaces (EPSs) were

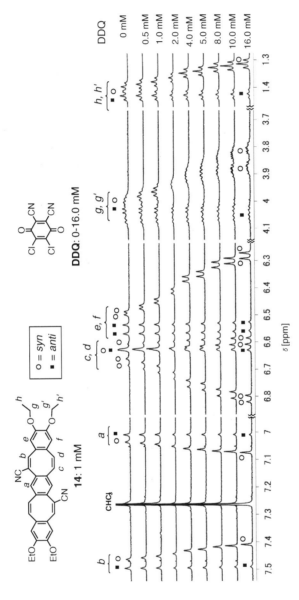

**Figure 12-16.** $^1$H NMR titration of **14** with DDQ (**14**: 1 mM, CDCl$_3$).

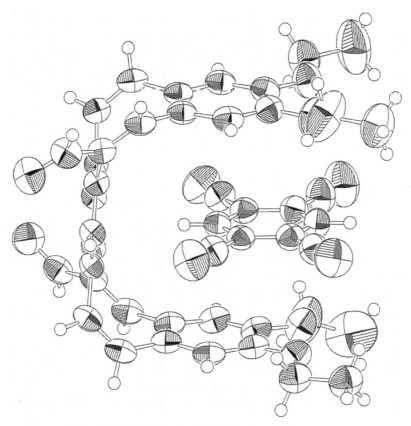

**Figure 12-17.** ORTEP (50% probability) drawing of TCNB@**15** complex.

computed (Figure 12-18). Electron densities of the cavities of **14–16** show a marked differences, and the negative character of the central and terminal benzene rings increases in the order of **14** < **15** < **16**. Therefore, it can be concluded that electrostatic effects of substitutions on COT units efficiently extend into the whole molecule. In

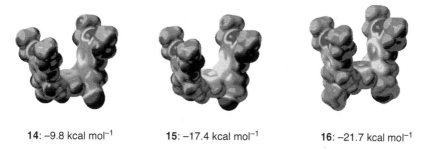

**14**: −9.8 kcal mol$^{-1}$　　　　**15**: −17.4 kcal mol$^{-1}$　　　　**16**: −21.7 kcal mol$^{-1}$

**Figure 12-18.** Electrostatic potential surfaces and the values of molecular electrostatic potentials calculated 2.0 Å above the central benzene ring of **14–16** (B3LYP/6-31G*// B3LYP/6-31G*).

other words, the COT units are effectively conjugated with the central and terminal benzene rings despite the bending π systems.

Interestingly, the solutions of DDQ complexes show thermochromic behavior.[48] Figure 12-19 shows the solutions of the complexes in $CHCl_3$ at various temperatures. The solutions of **14** and **15** with DDQ are green at room temperature. However, the color gradually changes from green to yellow on heating to 60°C and from green to blue on cooling to −40°C. On the other hand, **16** with DDQ shows changes only in the shade of blue color. This thermochromism has good response to the temperature

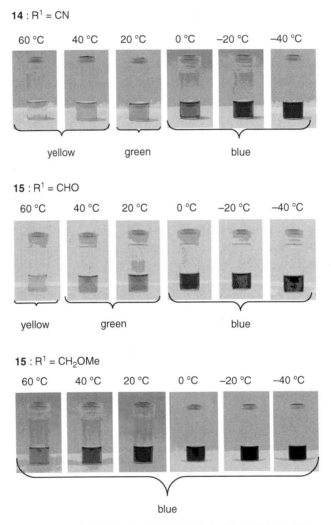

**Figure 12-19.** Solutions of **14–16** with DDQ (DDQ 1 mM, DMTs 1 mM) at various temperatures (60 ∼ −40°C).

changes. Basically, a $CHCl_3$ solution of DDQ is yellow and the DDQ@DMTs complexes are blue. Therefore, the principle pertaining to this thermochromism is related to the mixing ratio of yellow and blue color.

## 12-5. CONCLUDING REMARKS

This chapter reviews our studies on the synthesis and properties of belt-shaped unsaturated cyclophanes and related compounds. Attempted synthesis toward cyclophenacene is still ongoing, and so far we have observed preliminary evidence for the cyclophenacene structure only in the mass spectrum. However, further synthetic efforts could lead to isolation of the desired macrocycles.

On the other hand, during the course of this study, the structure–property relationship of the precursor 3D benzannulenes gave us various insights into the synthetic chemistry of benzenoid hydrocarbons and supramolecular chemistry on the basis of aromatic interactions. Thus, the chemistry of the 3D unsaturated macrocyclic compounds described here as well as other related molecular belts would provide further interesting aspects in novel curved conjugated compounds in the future.

## REFERENCES

1. For reviews, see (a) R. Herges, in *Modern Cyclophane Chemistry*, R. Gleiter, H. Hopf (eds.), Wiley-VCH, Weinheim, **2004**, pp. 337–358; (b) R. Gleiter, B. Esser, S. C. Kornmayer, *Acc. Chem. Res.* **2009**, *42*, 1108–1116; (c) R. Gleiter, B. Hellbach, S. Gath, R. J. Schaller, *Pure Appl. Chem.* **2006**, *78*, 699–706; (d) T. Yao, H. Yu, R. J. Vermeij, G. J. Bodwell, *Pure Appl. Chem.* **2008**, *80*, 533–546; (e) Y. Matsuo, *Bull. Chem. Soc. Jpn.* **2008**, *81*, 320–330; (f) K. Tahara, Y. Tobe, *Chem. Rev.* **2006**, *106*, 5274–5290; (g) T. Kawase, H. Kurata, *Chem. Rev.* **2006**, *106*, 5250–5273; (h) L. T. Scott, *Angew. Chem. Int. Ed.* **2003**, *42*, 4133–4135; (i) B. König, *Top. Curr. Chem.* **1998**, *196*, 91–136; (j) A. Schröder, H.-B. Mekelburger, F. Vögtle, *Top. Curr. Chem.* **1994**, *172*, 179–201.

2. H. W. Kroto, J. R. Heath, S. C. O'Brien, R. F. Curl, R. E. Smalley, *Nature* **1985**, *318*, 162–163.

3. (a) D. S. Bethune, C. H. Kiang, M. S. de Vries, G. Gorman, R. Savoy, J. Vazquez, R. Beyers, *Nature* **1993**, *363*, 605–607; (b) S. Iijima, T. Ichihashi, *Nature* **1993**, *363*, 603–605; (c) S. Iijima, *Nature* **1991**, *354*, 56–58.

4. (a) J. Aihara, *Bull. Chem. Soc. Jpn.* **1975**, *48*, 3637–3640; (b) S. Kivelson, O. L. Chapman, *Phys. Rev. B* **1983**, *28*, 7236–7243; (c) H. S. Choi, K. S. Kim, *Angew. Chem. Int. Ed.* **1999**, *38*, 2256–2258; (d) K. N. Houk, P. S. Lee, M. Nendel, *J. Org. Chem.* **2001**, *66*, 5517–5521; (e) Y. Matsuo, K. Tahara, E. Nakamura, *Org. Lett.* **2003**, *5*, 3181–3184; (f) Z. Chen, D. Jiang, X. Lu, H. F. Bettinger, S. Dai, P. v. R. Schleyer, K. N. Houk, *Org. Lett.* **2007**, *9*, 5449–5452.

5. (a) R. C. Haddon, *Acc. Chem. Soc.* **1988**, *21*, 243–249; (b) R. C. Haddon, L. T. Scott, *Pure Appl. Chem.* **1986**, *58*, 137–142; (c) R. C. Haddon, *Science* **1993**, *261*, 1545–1550.

6. (a) M. S. Dresselhaus, G. Dresselhaus, P. Avouris (eds.), *Carbon Nanotubes: Synthesis, Structure, and Applications, Topics in Applied Physics*, *80*, Springer, Berlin, **2001**; (b) A.

Hebard, M. Rosseinsky, R. Haddon, W. Murphys, S. Glarum, T. Palstra, A. Ramirez, A. Kortan, *Nature* **1991**, *350*, 600–601; (c) W. D. Heer, W. S. Bacsa, A. Chatelain, T. Gerfin, R. H. Baker, L. Forro, *Science* **1995**, *268*, 845–847.

7. E. Heilbronner, *Helv. Chim. Acta* **1954**, *37*, 921–935.

8. (a) U. Girreser, D. Giuffrida, F. H. Kohnke, J. P. Mathias, D. Philp, J. F. Stoddart, *Pure Appl. Chem.* **1993**, *65*, 119–125; (b) F. H. Kohnke, J. P. Mathias, J. F. Stoddart, *Top. Curr. Chem.* **1993**, *165*, 1–69; (c) F. H. Kohnke, A. M. Z. Slawin, J. F. Stoddart, D. J. Williams, *Angew. Chem. Int. Ed. Engl.* **1987**, *26*, 892–894; (d) P. R. Ashton, N. S. Issacs, F. H. Kohnke, A. M. Z. Slawin, C. M. Spencer, J. F. Stoddart, D. J. Williams, *Angew. Chem. Int. Ed. Engl.* **1988**, *27*, 966–969; (e) P. R. Ashton, N. S. Issacs, F. H. Kohnke, J. P. Mathias, J. F. Stoddart, *Angew. Chem. Int. Ed. Engl.* **1989**, *28*, 1258–1261; (f) P. R. Ashton, N. S. Issacs, F. H. Kohnke, G. S. D'Alcontres, J. F. Stoddart, *Angew. Chem. Int. Ed. Engl.* **1989**, *27*, 1261–1263; (g) P. R. Ashton, G. R. Brown, N. S. Issacs, D. Giuffrida, F. H. Kohnke, J. P. Mathias, A. M. Z. Slawin, D. R. Smith, J. F. Stoddart, D. J. Williams, *J. Am. Chem. Soc.* **1992**, *114*, 6330–6353; (h) P. R. Ashton, U. Girreser, D. Giuffrida, F. H. Kohnke, J. P. Mathias, F. M. Raymo, A. M. Z. Slawin, J. F. Stoddart, D. J. Williams, *J. Am. Chem. Soc.* **1993**, *115*, 5422–5429.

9. (a) R. M. Cory, C. L. McPhail, *Adv. Theor. Interesting Mol.* **1998**, *4*, 53–80; (b) R. M. Cory, C. L. McPhail, A. J. Dikmans, *Tetrahedron. Lett.* **1993**, *34*, 7533–7536; (c) R. M. Cory, C. L. McPhail, A. J. Dikmans, J. J. Vittal, *Tetrahedron. Lett.* **1996**, *37*, 1983–1986; (d) R. M. Cory, C. L. McPhail, *Tetrahedron. Lett.* **1996**, *37*, 1987–1990.

10. (a) A. Godt, V. Enkelmann, A. D. Schlüter, *Angew. Chem. Int. Ed. Engl.* **1989**, *28*, 1680–1682; (b) O. Kintzela, P. Lugerb, M. Weberb, A.-D. Schlüer, *Eur. J. Org. Chem.* **1998**, 99–105; (c) W. D. Neudorff, D. Lentz, M. Anibarro, A. D. Schlüter, *Chem. Eur. J.* **2003**, *9*, 2745–2757.

11. R. Jasti, J. Bhattacharjee, J. B. Neaton, C. R. Bertozzi, *J. Am. Chem. Soc.* **2008**, *130*, 17646–17647.

12. H. Takaba, H. Omachi, Y. Yamamoto, J. Bouffard, K. Itami, *Angew. Chem. Int. Ed.* **2009**, *48*, 6112–6116.

13. S. Yamago, Y. Watanabe, T. Iwamoto, *Angew. Chem. Int. Ed.* **2010**, *49*, 757–759.

14. (a) T. Kawase, H. R. Darabi, M. Oda, *Angew. Chem. Int. Ed. Engl.* **1996**, *35*, 2664–2666; (b) T. Kawase, H. R. Darabi, Y. Seirai, M. Oda, Y. Sarakai, K. Tashiro, *Angew. Chem. Int. Ed.* **2003**, *42*, 1621–1624; (c) T. Kawase, H. R. Darabi, K. Tanaka, M. Oda, *Angew. Chem. Int. Ed.* **2003**, *42*, 1624–1628; (d) T. Kawase, K. Tanaka, Y. Seirai, N. Shiono, M. Oda, *Angew. Chem. Int. Ed.* **2003**, *42*, 5597–5600; (e) T. Kawase, K. Tanaka, N. Shiono, Y. Seirai, M. Oda, *Angew. Chem. Int. Ed.* **2004**, *43*, 1722–1724; (f) T. Kawase, N. Fujiwara, M. Tsutumi, M. Oda, Y. Maeda, T. Wakahara, and T. Akasaka, *Angew. Chem. Int. Ed.* **2004**, *43*, 5060–5062.

15. (a) S. Kammermeier, P. G. Jones, R. Herges, *Angew. Chem., Int. Ed. Engl.* **1996**, *35*, 2669–2671; (b) S. Kammermeier, P. G. Jones, R. Herges, *Angew. Chem.* **1997**, *109*, 2317; *Angew. Chem. Int. Ed. Engl.* **1997**, *36*, 2200–2202; (c) D. Ajami, O. Oeckler, A. Simon, R. Herges, *Nature* **2003**, *426*, 819–821; (d) T. Kawase, M. Oda, *Angew. Chem. Int. Ed.* **2004**, *43*, 4396–4398.

16. (a) H. Schwierz, F. Vögtle, *Synthesis* **1999**, 295–305; (b) S. Breidenbach, S. Ohren, F. Vögtle, *Chem. Eur. J.* **1996**, *2*, 832–837; (c) F. Vögtle, *Top. Curr. Chem.* **1983**, *115*, 157–159; (d) G. J. Bodwell, J. N. Bridson, M. K. Cyrañski, J. W. J. Kennedy, T. M. Krygowski, M. R. Mannion, D. O. Miller, *J. Org. Chem.* **2003**, *68*, 2089–2098; (e) G. J.

Bodwell, D. O. Miller, R. J. Vermeij, *Org. Lett.* **2001**, *3*, 2093–2096; (f) G. J. Bodwell, J. J. Fleming, D. O. Miller, *Tetrahedron* **2001**, *57*, 3577–3585; (g) G. J. Bodwell, J. J. Fleming, M. R. Mannion, D. O. Miller, *J. Org. Chem.* **2000**, *65*, 5360–5370; (h) G. J. Bodwell, J. N. Bridson, T. J. Houghton, J. W. J. Kennedy, M. R. Mannion, *Chem. Eur. J.* **1999**, *5*, 1823–1827; (i) G. J. Bodwell, J. N. Birdsong, T. J. Houghton, J. W. J. Kennedy, M. R. Minion, *Angew. Chem., Int. Ed. Engl.* **1996**, *35*, 1320–1321.

17. B. L. Merner, L. N. Dawe, G. J. Bodwell, *Angew. Chem. Int. Ed.* **2009**, *48*, 5487–5491.

18. (a) Y. Matsuo, E. Nakamura, *Chem. Rev.* **2008**, *108*, 3016–3028; (b) X. Y. Zhang, Y. Matsuo, E. Nakamura, *Org. Lett.* **2008**, *10*, 4145–4147; (c) Y. Matsuo, K. Tahara, K. Morita, K. Matsuo, E. Nakamura, *Angew. Chem. Int. Ed.* **2007**, *46*, 2844–2847; (d) Y. Matsuo, K. Tahara, M. Sawamura, E. Nakamura, *J. Am. Chem. Soc.* **2004**, *126*, 8725–8734; (e) E. Nakamura, K. Tahara, Y. Matsuo, M. Sawamura, *J. Am. Chem. Soc.* **2003**, *125*, 2834–2835.

19. (a) Y. Kuwatani, T. Yoshida, A. Kusaka, M. Oda, K. Hara, M. Yoshida, H. Matsuyama, M. Iyoda, *Tetrahedron* **2001**, *57*, 3567–3576; (b) Y. Kuwatani, T. Yoshida, A. Kusaka, M. Iyoda, *Tetrahedron Lett.* **2000**, *41*, 359–363; (c) Y. Kuwatani, A. Kusaka, M. Iyoda, G. Yamamoto, *Tetrahedron. Lett.* **1999**, *40*, 2961–2964.

20. (a) *Top. Curr. Chem.* **1983**, *113;* (b) *Top. Curr. Chem.* **1983**, *115;* (c) G. J. Bodwell, T. Satou, *Angew. Chem. Int. Ed.* **2002**, *41*, 4003–4006; (d) K. P. C. Vollhardt, *Synthesis* **1975**, 765–780.

21. (a) G. Wittig, G. Lehmann, *Chem. Ber.* **1957**, *90*, 875–892; (b) G. Wittig, K.-D. Rümpler, *Liebigs Ann. Chem.* **1971**, *751*, 1–16; (c) H. A. Staab, C. Wünsche, *Chem. Ber.* **1968**, *101*, 887–899.

22. (a) G. Wittig, G. Koening, K. Clauβ *Liebigs Ann. Chem.* **1955**, *593*, 127–156; (b) G. Wittig, G. Skipka, *Liebigs Ann. Chem.* **1973**, 59–74; (c) K. Grohmann, P. D. Howes, R. H. Mitchell, A. Monahan, F. Sondheimer, *J. Org. Chem.* **1973**, *38*, 808–809.

23. H. Irngartinger, *Chem. Ber.* **1973**, *106*, 2786–2795.

24. M. Iyoda, Y. Kuwatani, T. Tamauchi, M. Oda, *J. Chem. Soc., Chem. Commun.* **1988**, 65–66.

25. D. L. Mohler, K. P. C. Vollharrdt, S. Wolff, *Angew. Chem. Int. Ed.* **1990**, *29*, 1151–1154.

26. (a) K. Sonogashira, Y. Thoda, N. Hagihara, *Tetrahedron. Lett,* **1975**, 4467–4470; (b) K. Sonogashira, in *Metal-Catalyzed Cross-Coupling Reactions*, P. J. Stang, F. Diederich (eds.), Wiley-VCH, Weinheim, **1998**, pp. 203–229; (c) E. Negishi, L. Anasytasla, *Chem. Rev.* **2003**, *103*, 1979–2018.

27. (a) J. H. Freudenberger, A. W. Konradi, S. F. Pedersen, *J. Am. Chem. Soc.* **1989**, *111*, 8014–8016; (b) A. S. Raw, S. F. Pedersen, *J. Org. Chem.* **1991**, *56*, 830–833.

28. E. J. Corey, P. B. Hopkins, *Tetrahedron Lett.* **1982**, *23*, 1979–1982.

29. H. Lindlar, R. Dubins, *Org. Syn.* **1973** (*Coll. Vol. 5*), 880–883.

30. (a) D. B. Dess, J. C. Martin, *J. Org. Chem.* **1983**, *48*, 4155–4156; (b) D. B. Dess, J. C. Martin, *J. Am. Chem. Soc.* **1991**, *113*, 7277–7287.

31. Y. Kuwatani, J. Igarashi, M. Iyoda, *Tetrahedron. Lett.* **2004**, *45*, 359–362.

32. In the case of strained-ring formation, a few examples of *erythro*-selective pinacol coupling by the vanadium complex have been reported: (a) M. Iyoda, K. Fuchigami, A. Kusaka, T. Yoshida, M. Yoshida, H. Matsuyama, Y. Kuwatani, *Chem. Lett.* **2000**, 860–861; (b) M. Iyoda, T. Horino, F. Takahashi, M. Hasegawa, M. Yoshida, Y. Kuwatani, *Tetrahedron Lett.* **2001**, *42*, 6883–6886; see also Ref. 27b.

33. H. Irngartingerand W. R. K. Reibel, *Acta Crystallogr.* **1981**, *B37*, 1724–1728.

34. G. H. Senkler, Jr., D. Gust, P. X. Riccobono, K. Mislow, *J. Am. Chem. Soc.* **1972**, *94*, 8626–8627.

35. T. Yoshida, Y. Kuwatani, K. Hara, M. Yoshida, H. Matsuyama, M. Iyoda, S. Nagase, *Tetrahedron Lett.* **2001**, *42*, 53–56.

36. Y. Kuwatani, T. Yoshida, K. Hara, M. Yoshida, H. Matsuyama, M. Iyoda, *Org. Lett.* **2000**, *2*, 4017–4020.

37. M. Iyoda et al. (unpublished results).

38. S. Klod, E. Kleinpeter, *J. Chem. Soc., Perkin Trans. 2*, **2001**, 1893–1898.

39. (a) C. H. Choi, M. Kertesz, *J. Phys. Chem.* **1997**, *101*, 3823–3831; (b) R. R. Monaco, W. C. Gardiner, S. Kirschner, *Int. J. Quant. Chem.* **1999**, *71*, 57–62; (c) M. Traetteberg, E. B. Frantsen, *J. Mol. Struct.* **1975**, *26*, 69–76.

40. (a) S. Tsuzuki, K. Honda, T. Uchimaru, M. Mikami, K. Tanabe, *J. Am. Chem. Soc.* **2002**, *124*, 104–112; (b) R. L. Jaffe, G. D. Smith, *J. Chem. Phys.* **1996**, *105*, 2780–2788.

41. (a) C. Gonzalez, E. C. Lim, *J. Phys. Chem. A*, **2001**, *105*, 1904–1908; (b) O. Engkvist, P. Hobza, H. L. Selzle, E. W. Schlag, *J. Chem. Phys.* **1999**, *110*, 5758–5762.

42. H. Krause, B. Ernstberger, H. J. Neusser, *Chem. Phys. Lett.* **1991**, *184*, 411–417.

43. (a) F. Sato, H. Urabe, S. Okamoto, *Chem. Rev.* **2000**, *100*, 2835–2886; (b) F. Sato, Urabe, S. Okamoto, *Synlett* **2000**, 753–775.

44. S. Sirinintasak, Y. Kuwatani, S.-I. Hoshi, E. Isomura, T. Nishinaga, M. Iyoda, *Tetrahedron Lett.* **2007**, *48*, 3433–3436.

45. (a) C. -W. Chen, H. W. Whitlock, *J. Am. Chem. Soc.* **1978**, *100*, 4921–4922; (b) S. C. Zimmerman, *Top. Curr. Chem.* **1993**, *165*, 71–102; (c) A. E. Rowan, J. A. A. W. Elemans, R. J. M. Nolte, *Acc. Chem. Res.* **1999**, *32*, 995–1006; (d) H. Kurebayashi, T. Haino, S. Usui, Y. Fukazawa, *Tetrahedron* **2001**, *57*, 8667–8674; (e) F.-G. Klärner, B. Kahlert, *Acc. Chem. Res.* **2003**, *36*, 919–932; (f) M. Harmata, *Acc. Chem. Res.* **2004**, *37*, 862–873; (g) X. Peng, N. Komatsu, S. Bhattacharya, T. Shimawaki, S. Aonuma, T. Kimura, A. Osuka, *Nat. Nanotechnol.* **2007**, *2*, 361–365; (h) A. Sygula, F. R. Fronczek, R. Sygula, P. W. Rabideau, M. M. Olmstead, *J. Am. Chem. Soc.* **2007**, *129*, 3842–3843.

46. T. Nishiuchi, Y. Kuwatani, T. Nishinaga, M. Iyoda, *Chem. Eur. J.* **2009**, *15*, 6838–6847.

47. (a) L. F. Fieser, M. M. Pechet, *J. Am. Chem. Soc.* **1946**, *68*, 2577–2580; (b) M. Avram, I. G. Dinulescu, G. Dinu, G. Mateescu, C. D. Nenitzescu, *Tetrahedron* **1963**, *19*, 309–317; (c) M. P. Cava, R. Pohlke, *J. Org. Chem.* **1963**, *28*, 1012–1014; (d) C. E. Griffin, J. A. Peters, *J. Org. Chem.* **1963**, *28*, 1715–1716; (e) J. W. Barton, D. V. Lee, M. K. Shepherd, *J. Chem. Soc., Perkin Trans. 1* **1985**, 1407–1411; (f) H. Ihmels, M. Schneider, M. Waidelich, *Org. Lett.* **2002**, *4*, 3247–3250; (g) P. A. Wender, J. P. Christy, A. B. Tesser, M. T. Gieseler, *Angew. Chem. Int. Ed.* **2009**, *48*, 7687–7690.

48. (a) A. Tsuda, S. Sakamoto, K. Yanaguchi, T. Aida, *J. Am. Chem. Soc.* **2003**, *125*, 15722–15723; (b) S. Nishida, Y. Morita, K. Fukui, K. Sato, D. Shiomi, K. Takui, K. Nakasuji, *Angew. Chem. Int. Ed.* **2005**, *44*, 7277–7280; (c) H. Enozawa, M. Hasegawa, D. Takamatsu, K. Fukui, M. Iyoda, *Org. Lett.* **2006**, *8*, 1917–1920; (d) T. Tsuchiya, K. Sato, H. Kurihara, T. Wakahara, Y. Maeda, T. Akasaka, K. Ohkubo, S. Fukuzumi, T. Kato, S. Nagase, *J. Am. Chem. Soc.* **2006**, *128*, 14418–14419; (e) Y. Morita, S. Suzuki, K. Fukui, S. Nakazawa, H. Kitagawa, H. Kishida, H. Okamoto, A. Naito, A. Sekine, Y. Ohashi, M. Shiro, K. Sasaki, D. Shiomi, K. Sato, T. Takui, K. Nakasuji, *Nat. Mater.* **2008**, *7*, 48–51.

# CHAPTER 13

# TOWARD FULLY UNSATURATED DOUBLE-STRANDED CYCLES

MALTE STANDERA and A. DIETER SCHLÜTER

## 13-1. INTRODUCTION

Fully unsaturated, double-stranded hydrocarbon cycles such as the [*n*]cyclacenes **A** (Figure 13-1), the buckybelts **B**, the cyclo[*n*]phenacenes **C**, and the Vögtle belts **D** have been a dream of chemists for several decades.[1] Besides their aesthetic appeal, they may be important for studying host–guest aspects all the way from hydrogen bonding between weak donors and π systems to the formation of polyrotaxanes, the *endo/exo*-cyclic selectivity of reactions, the dynamics of transition metal fragments complexed to the π system, the redox behavior, including the aspect of maximum chargeability and the like. They also could serve as models for their open-chain linear analogs, the ladder polymers,[2] and some of them could be developed into novel shape-persistent constituents for molecular constructions.[3] Finally, and in view of the topic of the present book, it should be stressed that such compounds could enrich the world of larger all-carbon structures such as fullerenes, carbon nanotubes, and graphene. Although there has been significant progress in the synthesis of these compounds, a breakthrough has not yet been achieved. Several double-stranded cycles with the appropriate carbon skeletons have been synthesized, and even a few chemical modifications with them aiming at an expansion of the conjugated parts could be accomplished.[4] However, attempts to generate any of the fully unsaturated targets mentioned above have historically failed altogether, and only relatively recently has any synthesis come close. Reasons for this considerable complication are target-specific but, by and large, also have something to do with the compounds' curvature, which renders a fully conjugated structure energetically less favorable relative to flat analogues. This energy price makes the final "aromatization" step more difficult to achieve and also the target structure, once attained, more prone to follow-up reactions, which may even render it transient and nonisolable. This chapter tries to provide a

*Fragments of Fullerenes and Carbon Nanotubes: Designed Synthesis, Unusual Reactions, and Coordination Chemistry*, First Edition. Edited by Marina A. Petrukhina and Lawrence T. Scott.

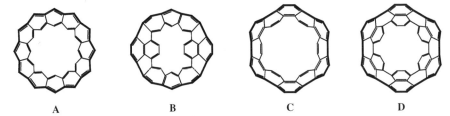

**Figure 13-1.** Fully unsaturated, beltlike compounds such as the [$n$]cyclacenes **A** ($n$=12), the buckybelt **B** (shown with two RUs), and angularly annulated congeners like the cyclo[$n$] phenacenes **C** ($n$=12) and the Vögtle belt **D** (shown with six RUs): fascinating curved molecules that have been on the wish list of chemists for decades.

state-of-the-art picture regarding the synthetic attempts into this field of hydrocarbon research. In Section 13-2 we describe our own laboratory's results starting with our original motivation for using Diels–Alder chemistry to construct the main carbon skeleton. The reader is then guided through some of the ups and downs encountered in a years-long struggle to achieve full aromatization and ultimately to the fact that although there is sufficient evidence for the achievement of buckybelt **B** with two repeating units (RUs), some of the results cannot be interpreted as of now. In Section 13-3 we focus on activities in other laboratories such that the reader finally is left with a rather comprehensive picture of the present situation in the field of double-stranded, fully unsaturated, hydrocarbon belts.

## 13-2. THE DIELS–ALDER APPROACH

In the 1980s and 1990s our group was involved in a major research program supported by the Federal Ministry for Research and Education of Germany and the chemical company BASF, Ludwigshafen, Germany, aimed at the development of double-stranded (ladder-type), conjugated polymers. Whenever pairs of bonds have to be formed repeatedly, there is the danger that the reactions designed to produce the second bond proceed not only intramolecularly but also intermolecularly. The latter lead to crosslinking and in high-molar-mass cases also to the formation of three-dimensional networks. These networks are insoluble and intractable and therefore not normally desired products. In order to prevent the occurrence of crosslinking, it was decided from the very beginning to concentrate on Diels–Alder (DA) chemistry for the critical growth step. This chemistry has the beauty that the two new bonds created in each of these steps are formed in a concerted manner. Unfavorable intermolecular side reactions are thus prevented, and even high-molar-mass, structurally perfect ladder polymers can in principle be obtained.[5] Throughout the project it was observed that bifunctional DA building blocks (monomers) give rise not only to the desired linear growth but occasionally also to the formation of double-stranded cycles consisting of a small number of repeat units (typically two). An early example from 1989 is shown in Figure 13-2. The "masked" DA AB-type monomer (**1**) is forced to

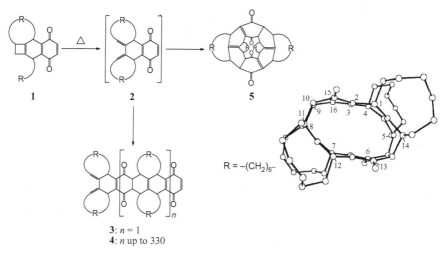

**3**: $n = 1$
**4**: $n$ up to 330

**Figure 13-2.** The first DA route to a high-molar-mass ladder polymer (**4**), which in variable amounts also gave the cyclic dimer **5**. This dimer consists of six linearly annulated six-membered rings and is therefore formally a precursor to a [6]cyclacene derivative. The structure of **5** in the crystal proves its double-stranded nature (right). Note that the cycle carries four hexamethylene substituents for solubility reasons.

undergo an electrocyclic ring-opening reaction to give *in situ* the actual monomer (**2**), which, under the reaction conditions, is a transient species. Its self-cycloaddition furnishes both the linear ladder polymer (**4**) and the cyclic dimer (**5**). As expected, the higher the concentration of **1** the more open-chain material **4** is obtained. The initial products of a DA reaction between two monomers (**2**), the diastereomers of **3**, are then trapped faster by either **1** or **2** instead of yielding **5** by cyclization. Cycle formation could not be totally suppressed, which indicates that, at least for certain diastereomers, the two reactive ends of **3** must be in a spatial arrangement relative to one another that largely favors cyclization.

Although we were aiming at solving the longstanding ladder polymer problem at that time, the potential relevance of this cyclic side product for double-stranded aromatic cycles was recognized. The organic chemistry educational background of the main author of the present chapter had its impact there. Aromatization of **5** was nevertheless not attempted for reasons that will become clear below.

Other exciting developments aiming at double-stranded structures also took place in the 1980s, actually before the abovementioned results were published. This specifically refers to the beautiful work by Stoddart et al., who in 1987 obtained the first (formal) cyclacene precursor (**9**) (Figure 13-3).[6]

Macrocycle **9** was synthesized by two different routes, either by a cyclic dimerization of the AB-type building block **6** (low yield) or by an AA/BB-type DA oligomerization of building blocks **7** and **8** (20% yield). In both cases high pressure was applied to overcome the components' insufficient intrinsic reactivity.[7] Macrocycle **9** is composed of 12 linearly annulated six-membered rings and is thus much larger than **3**, a structural

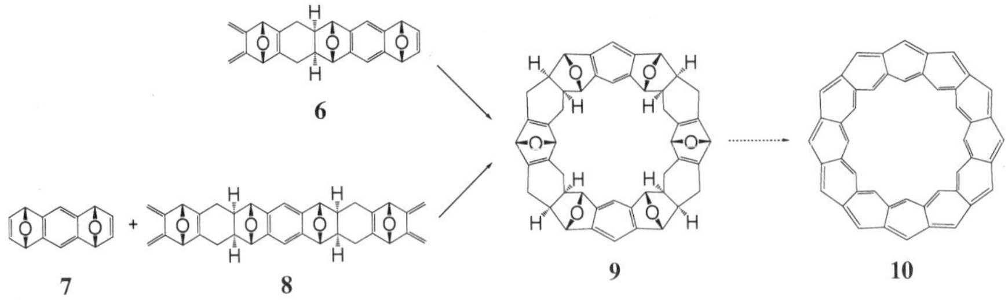

**Figure 13-3.** DA-type synthesis of the first formal cyclacene precursor, the double-stranded **9** by either dimerization of **6** or reaction of **7** with **8** and the chemical structure of the hypothetical [12]cyclacene **10**.

feature that would reduce the overall strain energy on aromatization. Further important developments include related DA homologizations and other DA sequences to longer, open-chain DA oligomers, which had been reported even a few years earlier (in 1983 and 1986) by the laboratories of Johnson[8] and Miller.[9] All these results underline the considerable interest that several researchers had at that time in exploring the low- and later also the high-molar-mass regime of double-stranded compounds and the potential that DA chemistry has in this regard.

The decision not to attempt aromatization of cycle **3** was supported by a study from Stoddart's laboratory following their 1987 publication.[6] In 1988 the results of serious experimentation aiming at the conversion of the beautiful cycle **9** into its fully unsaturated analog, the [12]cyclacene (**10**), were reported.[10] At first glance this starting compound seems almost ideal for aromatization. Its potential for strain is expected to be minimal, and successful performance of the chemical transformations of dehydration, deoxygenation, and dehydrogenation on this carbon skeleton would not appear to be unrealistic. However, it has to be noted that (1) quite a few such reactions would have to be performed to reach the target (**10**), (2) the target will be insoluble because of its conformational inflexibility paired with the lack of solubilizing peripheral substituents, and, most importantly, (3) cycle **10** would be a bent oligoacene and thus would be expected to be a highly sensitive compound. Whereas the synthetic complexity and solubility issues would require careful consideration but could possibly be managed, the sensitivity aspect is a rather serious one. For years, the longest fully proven member of the polyacene series was hexacene. Several attempts to access its longer homologs, not to mention *poly*acene,[11] failed because of the products' high reactivity, for example, toward oxidants and reagents employed in their generation. Acenes have been addressed theoretically, leading to a still unresolved controversy as to their electronic structure and other aspects. Most authors seem to agree that they exhibit increasing reactivity with increasing length.[12] This is corroborated by the fact that only quite recently, thus, approximately 70 years after Clar's pioneering work,[13] Neckers and Bettinger provided unequivocal evidence for heptacene,[14] octacene, and nonacene[15] under the conditions of matrix isolation. Thus [12]cyclacene (**10**) was expected to be a highly sensitive species even though its curvature could possibly increase its instability even further.[16] All this skepticism raises the question of how far the aromatization of cycle **9** could be pushed. The reactions performed were all wet-chemical in nature (and thus not under the perhaps more appropriate conditions of matrix isolation employing a modified structure) [Figure 13-4(a)] and resulted in the double-stranded cycle **11**, in which 8 of the 12 six-membered rings are aromatic. The largest aromatic subunit was (only) anthracene in nature. In a publication on the same topic that appeared 2 years later, this level of unsaturation could not be surpassed, and this situation remains today.[17]

It is not surprising that similar attempts by Cory on the much smaller cycle **12** [Figure 13-4(b)], where, in addition to the product's intrinsic instability, considerably more strain would have to be accommodated, were also unsuccessful. The unsaturation of the elusive [12]cyclacene (**10**) seems to be a destabilizing rather than a stabilizing factor. The other two potential precursors from our own laboratory, compounds **13** and **14** [Figure 13-4(b)] like cycle **3**, were therefore not considered for such transformations.

Some of the DA monomers synthesized in our laboratory provided access to polymers whose carbon skeletons did not resemble polyacene but rather the belt regions of fullerenes. These regions are characterized by vertical naphthalenes sandwiched in between two 5-membered rings, which simultaneously serve as connectors to short segments of linearly annulated six-membered rings. Theoretical predictions of the relative stability of such unsaturated compounds, which could loosely be called "buckybelts," suggested that they would be more stable than the polyacenes.[18] There was, of course, also clear-cut experimental support for this expectation. The very fact that countless fullerenes had been isolated and handled under ambient conditions spoke to their stability. Also, rather extended "buckyribbons" such as **15** were synthesized by conventional methods and found to be stable [Figure 13-5(a)].[19]

These stability considerations made us look into such compounds more closely, of course, and still with the emphasis on high-molar-mass linear polymers. Nevertheless, all polymerization attempts with respective DA monomers were carefully checked for whether cyclic side products had formed during polymerization. Figure 13-5(b) shows important examples, comprising AA/BB-type monomers **16**, **17**,[20] and **19**[21] and the two closely related AB-type monomers, **18a,b**.[21,22] Although it was considered a major success that the corresponding polymers could be synthesized, structurally characterized, and finally even fully aromatized (!), it was disappointing that the gel permeation chromatography (GPC) elution curves of the polymerization raw mixtures did not reveal any indication of cyclic byproducts. Obviously the distances between the two termini of the respective monomers (e.g., **16** and **17**) are too different and the attainment of a curved conformation of **17** (or **19**) is too costly for a cyclic 1:1 adduct to form. The stereoselectivity of the DA reaction between an acenaphthylene-type dienophile and an isobenzofuran-type diene is difficult to predict and may actually be low. The formation of a cyclic 2:2 adduct is therefore also unlikely to occur. The reasons why the AB-type monomers **18a,b** do not cyclize can be manifold and are not known with certainty. A discussion would require too much space for this brief chapter.

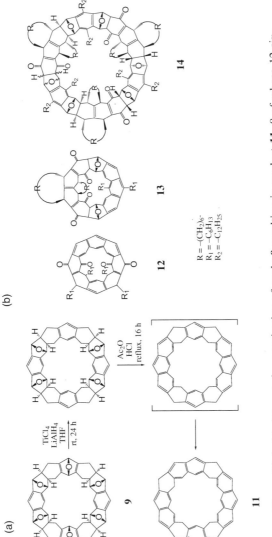

**Figure 13-4.** (a) Attempted aromatization of cycle **9** resulting in product **11**, 8 of whose 12 six-membered rings are aromatic; (b) counterparts to cycle **9**: Cory's cycle **12** and cycles **13** and **14** from the author's laboratory, potential precursors for [8]-, [9]-, and [18]cyclacenes.

348

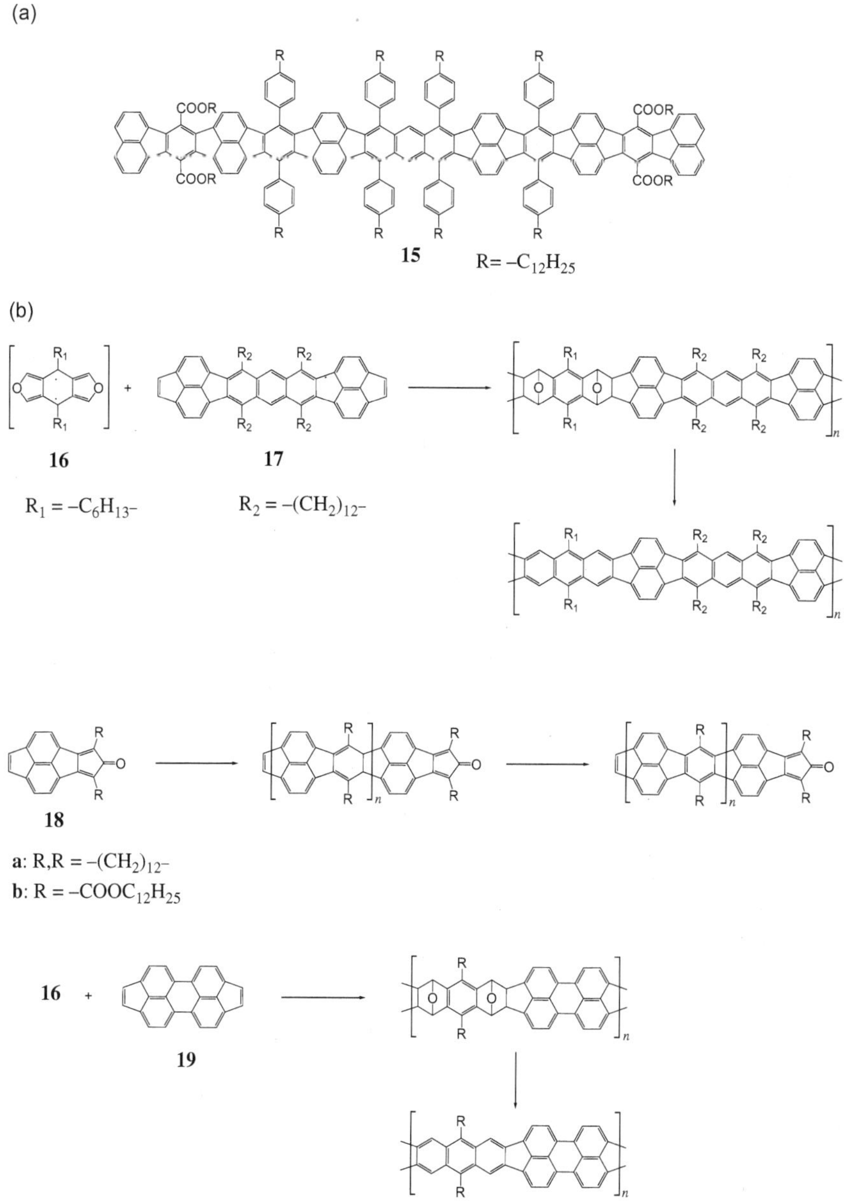

(a)

**15**   R= –C$_{12}$H$_{25}$

(b)

**16**              **17**

R$_1$ = –C$_6$H$_{13}$–      R$_2$ = –(CH$_2$)$_{12}$–

**18**

**a**: R,R = –(CH$_2$)$_{12}$–
**b**: R = –COOC$_{12}$H$_{25}$

**16**   +

**19**

**Figure 13-5.** Chemical structure of the "buckyribbon" **15** (a) and DA polymerization of the AA/BB monomers **16** and **17**, the AB monomers **18a,b**, and the AA/BB monomer **16** and **19** to the corresponding ladder polymers, which were subsequently aromatized.

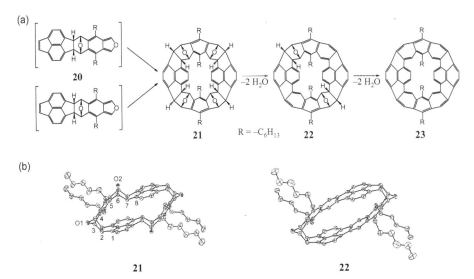

**Figure 13-6.** (a) Synthetic sequences to cycle **21** ("tetrahydrate") from the two diastereomeric monomers **20** generated *in situ* and to cycle **22** ("dihydrate") from the tetrahydrate **21**. The structure of target buckybelt **23** is also shown. (b) single-crystal structures of cycles **21** (left) and **22** (right). Note the almost unstrained nature of **22**.

We therefore developed the route to the key building blocks *exo-* and *endo-***20**, which on dimerization should give cycle **21** (Figure 13-6).[23] This expectation was based on the kinked nature of these monomers by a ball-and-stick consideration of all theoretically possible diastereomers of the DA dimers of *exo-* and *endo-***20**. This clarified the fact that some of them have the correct stereochemistry such that their ends can react with one another to give **21** without much strain to overcome in this step. Thus, at least for those diastereomers, there is no need for strong bending. For a detailed discussion of the stereochemical aspects of this chemistry, the reader is referred to the original publication.[23] An additional advantage of this strategy are the ether bridges in the oxanorbornene units of cycle **21**. Their removal was expected to be relatively easy, specifically, by well-established dehydration chemistry. This is why this cycle (**21**) is sloppily referred to in our laboratory as the "tetrahydrate" of the corresponding buckybelt.

Having reached this point after 17 steps and 3 years of intense laboratory work, we enthusiastically thought that the remaining tetrafold dehydration of **21** was more or less a simple trail to follow on known grounds. One could not have been more wrong with this assessment! It quickly became clear that the dehydration of **21**, a single-crystal X-ray structure of which is shown in Figure 13-6, easily results in a loss of two water molecules already under mild acidic conditions. The reason for this ease is the two kinds of "water" molecules contained in **21**, one at *exo* and the other at *endo* configurated sites. As ball-and-stick models show, only the loss of the *exo* water molecules does not result in an increased strain in the product. It should be noted that the single-crystal X-ray structure of **22** has an ellipsoidal cross section, indicative of a basically unstrained conformation. Not discouraged by this very high dehydration selectivity, dihydrate **22** was subjected to more drastic conditions by stirring it at 150°C in concentrated sulfuric acid. Yet no visible reaction was observed,[24] even though most, if not all, of the conventional dehydrations leading to flat aromatics would, under those conditions, be finished instantaneously. The interpretation that the cycle tries to avoid increasing strain—and, thus, curvature—was supported by another surprising reaction. When the dihydrate was treated with methane sulfonic acid in refluxing toluene,[25a] the oxygen bridges could in fact be removed but were replaced by bridging toluene units (Figure 13-7). Obviously the "desire" of the cycle to avoid the buildup of curvature results in an unprecedented reaction course in which it incorporates solvent molecules rather than doing what all noncyclic congeners would do, namely, aromatize. The incorporation of toluene can be understood in terms of a proton-catalyzed opening of the oxygen bridges with formation of a carbocation, which then attacks a solvent molecule in an electrophilic aromatic substitution. In the end, the bistoluene adduct **24** was obtained as the sole product. Figure 13-7 also shows a single-crystal X-ray structure of **24**. Its cross section, very much like the one of **22**, is rather ellipsoidal and thus only slightly strained.

Removal of the first water molecules from the dihydrate **22** is associated with the buildup of considerable strain, the magnitude of which may even amount, more or less, to the final strain when both remaining waters are removed. Commonly ring strain in hydrocarbons is normalized to the number of carbon atoms that make up the skeleton under consideration. By this measure the strain in the monohydrate or the final buckybelt is very small, just 1–2 kcal/mol per carbon atom.[26] What should be considered here in terms of potential barriers to the reaction, however, is the fact that practically all of the strain is being built up in one step, actually the next dehydration to come (the third of four). This dehydration must therefore be carried out under conditions robust enough to compensate for the cost in terms of strain energy.[27] However, as soon as even more drastic dehydration conditions were applied, the dihydrate (**22**) either survived or was converted into a black, insoluble, possibly

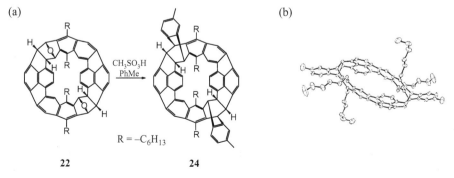

(a)                                                              (b)

CH₃SO₃H / PhMe

$R = -C_6H_{13}$

22                              24

**Figure 13-7.**  (a) Conversion of the dihydrate **22** into the bistoluene adduct **24**; (b) ORTEP plot of the structure of **24** in the single crystal.

networked material that was difficult to analyze. Since several other transformations (shown in Figure 13-8) also failed to lead to any progress in terms of increased unsaturation, attempts to dehydrate **22** ceased.

This failure initiated thoughts about whether it would be possible to enforce the elimination of **22**'s two formal water molecules under mass spectrometric conditions. In the initial publication on compound **22**,[23] its electron ionization (EI) mass spectrum was reported to show a molecular ion peak at *m/z* 968 and also a peak at *m/z* 484,

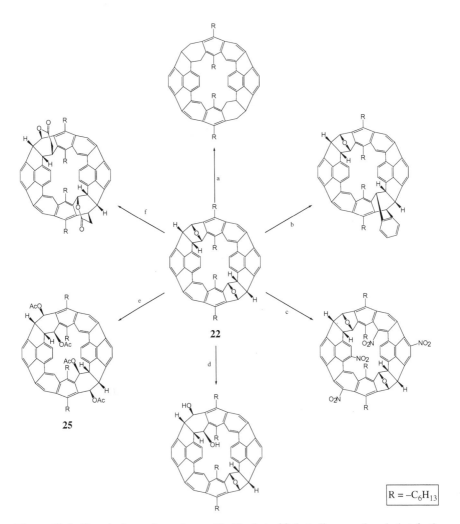

**Figure 13-8.** Chemical transformations with dihydrate **22** that all proceed such that further aromatization is avoided: (a) $(CH_3)_3SiI/CH_2Cl_2$; (b) $CH_3SO_3H/C_6H_6$ at 130°C; (c) $HNO_3/C_6H_4Cl_2$; (d) $BBr_3/CH_2Cl_2$ (aqueous workup); (e) $Ac_2O/ZnCl_2$ at 60°C; (f) $Ac_2O/ZnCl_2$ at 120°C. For details, see Ref. 25b. The tetraacetate **25** serves as an important potential precursor for aromatization experiments.

which had the highest intensity. This latter peak appeared at half the mass of the parent ion, which may indicate that fragmentation under EI conditions passes through a *retro*-Diels–Alder (RDA) channel. Different mass spectrometric conditions were therefore applied to **22** to see whether other fragmentation paths could be activated. Matrix-assisted laser desorption/ionization (MALDI) spectra were obtained in both presence and absence of acidic matrix materials. In either case spectra very similar to the abovementioned EI spectrum were obtained, revealing the presence of both ions. In order to ensure that it was in fact RDA that generated the peak at $m/z$ 484 and not a sequence of fragmentations that may include dehydration, collision-induced dissociation (CID) experiments were performed with cationic compound **22** that was generated in an electrospray ionization (ESI) source connected to an ion cyclotron resonance (ICR) mass spectrometer. This treatment afforded exclusively the product ion of $m/z$ 484 which strongly supports the proposed RDA fragmentation channel and finally made us discontinue all attempts to achieve direct dehydration of **22**.

An overview of the reactions studied for the dihydrate **22** (Figure 13-8) shows that mild Lewis acids such as $ZnCl_2$ in the presence of acetic anhydride lead to an opening of the oxygen bridges with formation of the corresponding tetraacetate (**25**) (Figure 13-9). Since acetate thermolysis is an established method for synthesizing olefins, tetraacetate (**25**), in combination with pyrolytic conditions, was considered an attractive alternative to the unsuccessful dehydration path. The experimentation was rather complicated partially because of the relatively small quantities of **25** (20–40 mg) that were available after approximately 20 synthetic steps. In one of the several pyrolytic experiments that were conducted with **25**, a fraction of a recycling GPC separation was analyzed in somewhat more detail. This led to in-depth insight into the products. Figure 13-10 shows the low- and high-mass regions of a MALDI-TOF spectrum of this fraction, together with calculated isotope distributions of relevant intermediates. From the bottom traces of Figure 13-10 it seems that this fraction consists of three components, which give rise to peaks for the molecular ion minus two, three, and four acetic acid molecules. The signals show an increasing complexity in this order. For an assignment of these peaks, they were compared with calculated isotopic distributions, taking reductions by two hydrogen atoms into account. The experimental signal at $m/z$ 1052 ($[25-2HOAc]^+$) is almost identical to the calculated

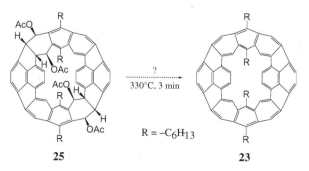

**Figure 13-9.** Thermolysis of **25**: possible access to the target structure, the buckybelt **23**.

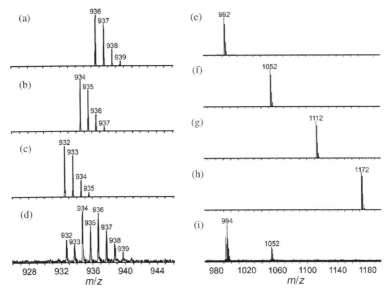

**Figure 13-10.** The low-molar-mass (*d*) and high-molar-mass regions (*i*) of a thermolysis product's MALDI mass spectrum of compound **25** [first fraction (**23**)]. These regions of the spectrum are compared with calculated isotope patterns of the parent molecular ion [**23**]$^+$ (c) and the ion of its first and second reduction products [**23**+2H]$^+$ (b) and [**23**+4H]$^+$ (a), respectively, as well as the molecular ions of **25** (h) and its fragmentation products [**25**-HOAc]$^+$ (g), [**25**-2HOAc]$^+$ (f), and [**25**-3HOAc]$^+$ (e), respectively.

one [Figure 13-10(f)], which indicates that negligible reduction, if any, has taken place. The experimental signal group above $m/z$ 990 consists of two isotopic patterns, one that starts at $m/z$ 992, corresponding to the [**25**-3HOAc]$^+$ ion, and the other at $m/z$ 994, corresponding to [**25**-3HOAc+2H]$^+$ [Figure 13-10(i)]. The latter is thus a reduction product. Finally, the most important signal group appears just above $m/z$ 930 and consists of three superimposed isotopic patterns, as a comparison with the three calculated spectra in Figure 13-10(a–d). First, the pattern of the [**25**-4HOAc]$^+$ ion at $m/z$ 932 can be identified. This is superimposed by two series of reduction products, namely, [**25**-4HOAc+2H]$^+$ starting at $m/z$ 934 and [**25**-4HOAc+4H]$^+$ starting at $m/z$ 936. The fact that the isotopic series starting at $m/z$ 932 can actually be detected shows that one could obtain the true molecular ion of **23**. Compound **23** obviously has a tendency to add two hydrogen atoms in each of the two consecutive reduction steps and not more. It should be noted that this hydrogen addition does not take place in the mass spectrometer. Repeated measurements under different conditions left the proportion between hydrogenated and nonhydrogenated compounds unaltered. This observation supports the presence of two anthracene units in **23**, the reduction of each of which, possibly at the 9 and 10 positions, would be energetically favorable in light of the proven desire of **23** to reduce its strain (see above).

Having these wonderful mass spectra (the elemental composition of the molecular ion was independently confirmed by high-resolution MS), together with the chemical

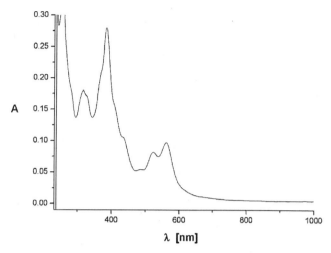

**Figure 13-11.** UV–visible spectrum in chloroform of a chromatography fraction obtained from the thermolysis of the tetraacetate **25**. The same material gave the mass spectra shown in Figure 13-10.

information about the selective addition of hydrogen, could make one lean back and relax. Also, the UV–visible spectrum of this fraction supports the view that we were actually able to obtain the first evidence for the existence of buckybelt **23** (Figure 13-11). It shows a longest wavelength absorption at approximately $\lambda_{max} = 550$ nm, which is where one would intuitively expect it to be in comparison with $C_{60}$ and $C_{70}$[28] as well as with different $C_{84}$ isomers[29] and compound **15** ($\lambda_{max} = 611$ nm, dichloromethane).[19]

However, there were a few observations that made us hesitant to actually claim to have achieved the goal. First, it was not possible to record a $^1H$ NMR spectrum (for a $^{13}C$ NMR spectrum the amount was too small) with reasonable linewidth that would allow for an assignment. All signals were extremely broad (virtually undetectable), suggesting that radicals were present. In fact, EPR measurements confirmed their presence. The spectrum [Figure 13-12(a)] revealed that it was a carbon-centered radical involved in a conjugated system ($g = 2.0026$). A simulated spectrum for a system consisting of 50 conjugated carbon atoms based on reasonable parameters is in good agreement with the recorded spectrum. The resolution of this experiment did not allow for differentiation between neutral and charged radicals. Also, a mixture of positively and negatively charged radicals could be present, and the existence of triplet states could not be excluded. The concentration was roughly estimated by integration against TEMPO as an external standard and found to be small.[30] Another ambiguity resulted from computational studies by Stanger, which point to structural alternatives for **23** with the same mass, namely, isomers[26] that form by 1,5-hydrogen shift.[31] One may argue that such a shift would interrupt the conjugation through the anthracene units of **23**. However, this interruption, and the blueshift of $\lambda_{max}$ that it should result in, could be compensated for by radical species. As mentioned above,

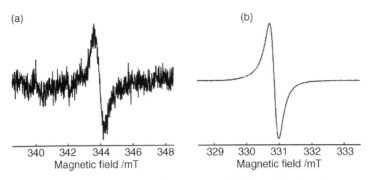

**Figure 13-12.** EPR spectra of the thermolysis product of **25** (a) and the dehydration product of **26** (b). The spectra were recorded under different conditions and on different spectrometers. It was ascertained that in both cases $g = 2.0026$.[30]

the presence of radicals, albeit in low concentration, was confirmed. Various attempts to trap **23** failed or gave inconclusive results.

Given the complications with proving the existence of **23** in its nonrearranged form, it would have been nice to solve all problems by growing single crystals and solving their structure. Obviously such attempts were undertaken, but they were met with insurmountable problems caused by the fact that one always dealt with a practically inseparable mixture of **23** and at least its two hydrogenation derivatives. In addition, the quantities of this mixture did not exceed the 25 mg range. Therefore, we decided to repeat the entire sequence again, this time with substituents that do not carry benzylic hydrogen atoms to eliminate the possibility of hydrogen shifts. The simplest option would be no substituents at all—in other words, to have hydrogen instead of hexyl groups. We did not place hydrogen first on the priority list, though, because bent and unsubstituted anthracenes should easily dimerize across the 9 and 10 positions. This may be an analytical advantage but could cause other serious problems that we did not want to risk encountering. Of the several other options, we selected phenyl. Phenyl would not only render the dimerization less likely but also increase the crystallization tendency, which is important in light of the tiny amounts expected. It has the additional benefit of high thermal and chemical stability. Obviously there was some concern as to whether the unavoidable decrease in solubility by the cycle's decoration with this substituent would cause serious handling and analytical issues. Nevertheless, the entire sequence was carried through with phenyl instead of hexyl substituents. We encountered quite a few differences, such as of yields and diastereoselectivities, but in the end, we obtained the corresponding tetraphenyl tetrahydrate (**26**) as a pure compound on a 40 mg scale.[32]

Somewhat surprisingly, the tetrahydrate **26** showed rather different dehydration reactivity (Figure 13-13). Most importantly, it did not dehydrate at all under conditions where **21** gave the dihydrate **22**, and **26** could be completely recovered! Thus we concentrated on the more forcing dehydration of **26**. For this purpose the methane sulfonic acid used for **21** was replaced by trifluoromethane sulfonic acid. The other conditions, 140°C and nitrobenzene, remained the same. Whereas the hexyl

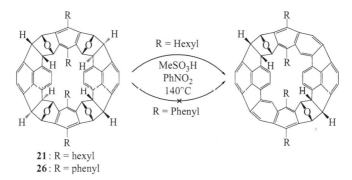

**21** : R = hexyl
**26** : R = phenyl

**Figure 13-13.** The two different tetrahydrates **21** (R = hexyl) and **26** (R = phenyl) exhibiting rather different reactivity toward acids.

analog **21** was fully soluble in nitrobenzene at 140°C without the addition of acid, **26** was only partially soluble and more solvent was therefore used. Still, not all of **26** was dissolved before addition of the acid. On the addition, the solution immediately changed to a dark-red/black color, and some of the product formed a precipitate. This black, fine powder was analyzed by solid-state carbon NMR, and Figure 13-14 compares the spectra of the starting material (**26**) with that of the product, which should be the tetraphenyl analog of **23**. Obviously the dehydration proceeded quantitatively, because the signals of **26** at $\delta = \sim 83$ and 58–60 ppm had completely disappeared. Furthermore, there is no signal to be seen that would indicate a proton-catalyzed retro-DA reaction of **26** with subsequent linear oligomerization. We are, therefore, confident in having solved the problem, although admittedly the solubility of the product is lower than hoped for. Mass spectrometric characterization is presently ongoing but is hampered by some complications. One may be tempted to exclaim at this point: What else could it be! EPR measurements of this material gave a strong signal that had the exact same $g$ value as the previous one, as well as a very similar shape [Figure 13-12(b)]; however, the signal intensity was much higher.

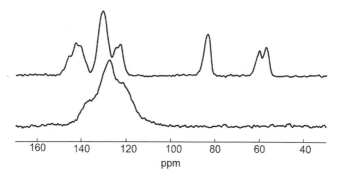

**Figure 13-14.** CPMAS $^{13}$C NMR spectra of starting tetrahydrate **26** (a) and the product after dehydration with trifluoromethanesulphonic acid at 140°C in nitrobenzene (b).

It is interesting to note that both routes, the thermolysis of **25** and the dehydration of **26**, seem to lead to the same radical species.

## 13-3. OTHER APPROACHES AND RELATED STUDIES

Quite a diverse set of strategies and particular chemistries have been attempted by other laboratories to contribute to the burning question of whether access can be provided to fully unsaturated double-stranded cycles.[1c,d] A strategy that has been pursued a number of times aims at creating a single-stranded cycle first. This cycle is designed so as to allow for the second strand to be closed in a separate subsequent step, proceeding preferably with an *in situ* aromatization of the whole compound. For this second step, the protocol used most often is a valence isomerization of *cis*-1,2-diarylvinylenes followed by an *in situ* oxidation (dehydrogenation), abbreviated as VID reaction. Initially, this chemistry was developed for *cis*-stilbene largely by Mallory[33] (Figure 13-15) and later also for pyrene by Boekelheide.[34] Depending on whether this isomerization is induced thermally or photochemically, the two saturated hydrogen atoms formed in the cyclic intermediates are positioned on the same side or opposite sides, respectively. This positioning has an impact on the reagents required for their removal. The other protocol that one encounters utilizes the McMurry reaction,[35] in which two aldehydes are reductively coupled to an olefin. This reaction has been up on the drawing board in a few laboratories for the synthesis of single-

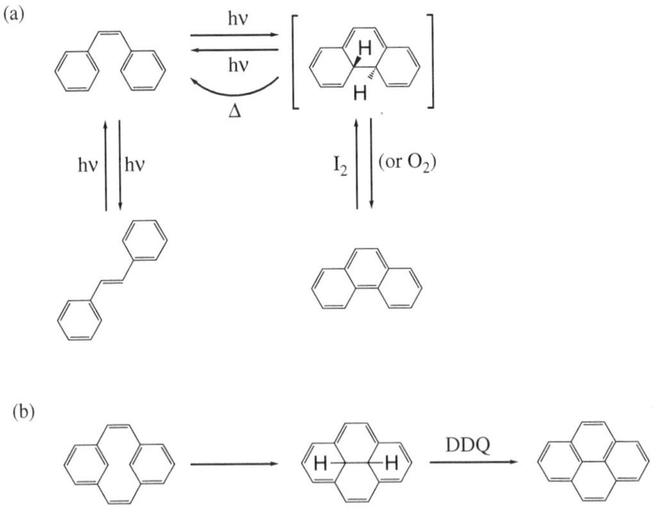

**Figure 13-15.** Two important synthetic tools in the field of double-stranded belts, the photochemically induced *cis*-stilbene condensation, which on oxidative removal of the central hydrogen atoms, affords phenanthrene (a); and the thermally induced analogous valence isomerisation followed by dehydrogenation, affording pyrene, often being carried out with 2,3-dichloro-5,6-dicyano benzoquinone (DDQ) (b).

stranded precursors for quite some time and has even been employed to close the second strand (see the end of this section). The first approaches described here aim at double-stranded cycles of the cyclo[*n*]phenacene type. According to simple Clar sextet considerations and also more sophisticated calculations, they should in principle be stable enough so as to allow for both their synthesis and isolation.

Iyoda was the first to synthesize single-stranded precursors potentially suited for a VID reaction to cyclo[*n*]phenacenes. The cyclic oligo(*ortho*-phenylenevinylene)s **28** and **29**, correctly named as *all-Z*-tribenzo[12]annulene and all-Z-hexabenzo[24] annulene, respectively, are interesting hydrocarbons in themselves and contributed to the field of annulene chemistry (Figure 13-16).[36] Attempts to close the second strand to furnish cyclo[6]- or cyclo[12]phenacenes were undertaken but met with no concrete success.[37] As described by Iyoda in Chapter 12, this group quite recently obtained the first mass spectrometric evidence for the successful cyclization of precursor **30** to the corresponding cyclo[10]phenacene (**31**) under the influence of laser light. Although not all questions seem to have been finally clarified, this is a most interesting development, and one now expects more exciting news from Tokyo in due course.

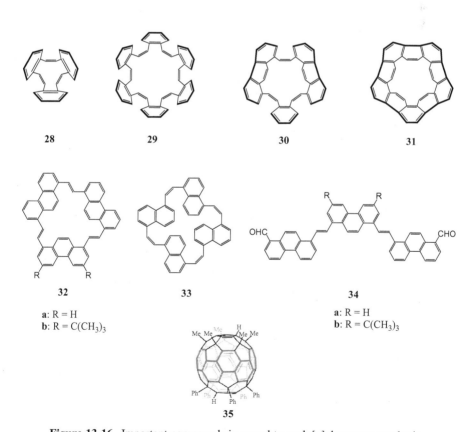

**28**  **29**  **30**  **31**

**32**

a: R = H
b: R = C(CH₃)₃

**33**

**34**

a: R = H
b: R = C(CH₃)₃

**35**

**Figure 13-16.** Important compounds in regard to cyclo[*n*]phenacene synthesis.

Scott's laboratory has also been rather active in a similar direction. Various possible routes into cyclic oligo(*ortho*-arylenevinylene)s were explored, potentially capable of showing VID reactions under the conditions of flash vacuum pyrolysis. Key targets were **32** and **33**, containing phenanthrene and naphthalene subunits. These subunits are more extensive than the ones in the Iyoda annulenes. Despite a considerable effort of Scott's coworkers, testifying to a strong interest in this matter,[38] neither of the targets could yet be prepared. The story behind this is not without some dramatic character, because a potential precursor for **32a**, the dialdehyde **34a**, could actually be obtained. Its anticipated McMurry ring closure, however, could not be made to work, despite considerable effort. The aspects that must have been most frustrating with this are the unknown reason(s) for failure and the following good arguments as to why it should have worked, in principle. Not only does the group have substantial experience in McMurry reactions but also the same coworker who failed with **32a** was capable of synthesizing the bis(*tert*-butyl)-substituted cycle **32b** from the correspondingly substituted **34b** in good yield. Furthermore, there is a substantial body of literature on ring closures of shape-persistent difunctional compounds to the respective cyclics.[39] All this knowledge and know-how does not point toward a factor responsible for the failure to obtain **32a**. It seems that chemistry sometimes guards its secrets very well. We should add here that alkyl-substituted precursors such as **32b** cannot be subjected to pyrolysis directly. The substituents tend to give side reactions.[40]

Although fullerene derivatives obviously can not have a bearing on the rational synthesis of cyclo[*n*]phenacenes, an earlier report by Nakamura on compound **35** should be mentioned.[41] This particular fullerene derivative and many related others were obtained after chemical reactions at both poles of the parent fullerene $C_{60}$, leaving a cyclo[10]phenacene fragment untouched. Compound **35** is thus a cyclo-phenacene derivative, but obviously a very special one without belt topology and an accessible interior. Until the more recent evidence for **31** by Ioyda, fullerene **35** was the closest that chemistry had come to cyclo[*n*]phenacenes. Compound **35** is undoubtedly beautiful and constitutes a major accomplishment for fullerene chemistry.

Besides VID reactions involving *cis*-1,2-diarylvinylenes, those involving pyrene precursors (Figure 13-15) were also used. Bodwell and coworkers very elegantly put this kind of chemistry to work.[42] Since this is covered in a separate chapter of this book (see Chapter 14), we should mention only briefly here that the Bodwell group quite recently achieved the synthesis of the teropyrenophane **38**, which is no less than half of a fully conjugated belt. Its single-crystal X-ray structure is shown in Figure 13-17. This is a truly noteworthy accomplishment, because it proves that (1) such curvature can be generated and (2) the VID sequence is a powerful synthetic tool for exactly such a purpose, robust enough even to work under conditions where considerable strain is being built up. Figure 13-17 shows the key steps of the rather short synthetic sequence. In the first step the precursor for VID (**37**), is synthesized by olefination of the two Z-configurated aldehydes in **36**. Then **37** is heated to 140°C in toluene in the presence of DDQ to drive the valence isomerization and subsequent *in situ* dehydro-genation. This second step renders the isomerization irreversible and generates the

(a)

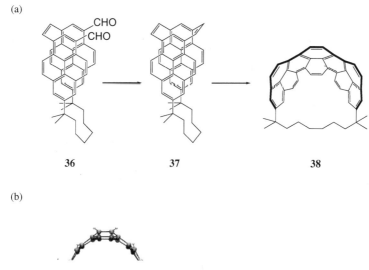

36          37          38

(b)

**Figure 13-17.** (a) The last three steps to the strongly curved half of a long-sought-after fully aromatic belt; (b) single-crystal X-ray structure of the strongly curved half-cycle **38**.

central pyrene unit of **38**. As we are used to multistep sequences with sometimes disappointing yields, it is not without admiration that we point to the almost quantitative yield of the last step. This high yield for a step in which the final strain is being generated underlines what was said above about the usefulness of this chemistry. One can only be curious about the next developments to come from northeastern Canada.

Although slightly outside the scope of the present book on "fragments of fullerenes and carbon nanotubes," the rather exciting more recent developments in Gleiter's laboratory deserve to be mentioned. The Gleiter group has used a strategy different from the ones discussed in this chapter so far. Their method is based on transition-metal-mediated acetylene dimerization. This dimerization leads to cyclobutadienes stabilized with a metal fragment in $\eta^4$ geometry. In this way,[43] as in the DA approach discussed above, two bonds are closed in the same reaction and—when applied to dibenzocyclooctadienediyne **39**—furnish open-chain oligomers and fully unsaturated double-stranded cycles such as **40**.[44] Figure 13-18 illustrates a key sequence. The same group has also successfully hunted for another double-stranded compound, cycle **42**, composed of an alternating sequence of six- and eight-membered rings (Figures 13-18). In their first report, they referred to this beautiful cycle as the "first *fully conjugated,* purely hydrocarbon cyclacene,"[45] while simultaneously assessing the conjugation across the cyclooctatetraene fragments as (only) 31%.[45,46] Given its lack of color, its seems more appropriate to assume a smaller conjugated system. The nomenclature introduced for this and related compounds by Gleiter represents, in our view, a somewhat problematic widening of the generally accepted name, [$n$]cyclacenes. Although it is clearly helpful in terms of covering cycles with different ring sizes (here 6 and 8), this new nomenclature has the disadvantage of including

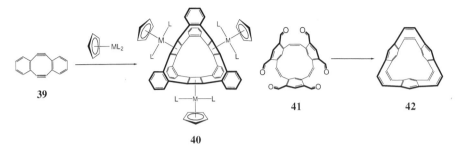

**Figure 13-18.** Synthesis of belts **40** and **42** with their alternating sequences of four- and eight-membered rings and six- and eight membered rings, respectively.

structures that are not at all of the acene type, either by their properties or by the strain introduced in their synthesis. According to the common understanding, acenes are compounds composed exclusively of linearly or angularly annulated six-membered rings. This structural feature forces them to attain a round shape (once they are accessible) and to contain quite some strain energy (depending on size). These aspects are at the heart of the synthetic difficulties encountered with the compounds in this chapter. As soon as eight-membered rings are incorporated, even if they are unsaturated, as in **42**, the 1,2-vinylene units will serve as hinges that reduce the strain but largely interrupt the conjugation. Consequently, synthetic access can be achieved in a more conventional nature. The single-crystal X-ray structure of **42** confirms its kinked structure.

## 13-4. WHERE WILL THE JOURNEY LEAD US FROM HERE?

The endeavors aiming at synthesis of fully unsaturated double-stranded cycles have reached a rather exciting level. Since 2006 or so, different laboratories have advanced much closer to the goal than in all the preceding years together. Especially noteworthy is the achievement of half-cycle **38** with its enormous curvature. This shows that strong bending of conjugated compounds is not per se an insurmountable obstacle and gives great hope to others in the community. Another hot development is the mass spectrometric evidence for cyclo[10]phenacene (**31**). It remains to be established, however, whether this compound can also be made outside the spectrometer. If so, and this is what we all are hoping for, this would constitute the first unequivocal solution to the problem. Both of these findings have in common that VID is the key step in terms of strain build up. This underscores the potential of this particular chemistry in the present context.

We also feel that our own laboratory has a potential winner in the "race." The tetrahydrate **26** has, in the meantime, been doubly dehydrated, and the resulting dihydrate (structure not shown), which is analogous to **22**, has been converted into the corresponding tetraacetate. The mass spectrometric analysis of the acetylation raw product gratifyingly gave the expected molecular ion at $m/z = 900$, which is the mass of the fully unsaturated target. Additionally the entire fragmentation cascade at $m/z = 1140, 1080, 1020$, and $960$ was detected. The matching high-resolution mass spectrum of $M^+$ at $m/z = 900.2812$ (calculated value $900.2811$) removed any last doubts regarding the composition of the molecular ion. In contrast to its tetrahexyl congener **23**, the tetraphenyl belt has no option to escape its strain by rearrangements involving hydrogen shifts. We therefore consider this $M^+$ peak as the first strong evidence for a fully aromatized buckybelt. Attempts to isolate any of the relevant compounds proved rather complicated, and it is presently not clear whether they will ultimately be successful. As mentioned above, one of the problems is the small quantity available. Although computational results point toward 1,5-hydrogen shifts as a stabilizing mechanism of the tetrahexyl belt **23**,[26] the researchers involved in the synthesis never really believed in this hypothesis. In an attempt to shed more light on this important aspect, a trapping experiment was devised in which the tetraacetate (**25**) was thermolized in the presence of excess 2,9-dimethyl-1,10-phenanthroline in order to provoke a reaction between this potential DA dienophile and the newly formed **23**. Mass spectrometric analysis of the raw product mixtures revealed, besides unreacted cycle **23**, two—and *only* two—groups of adduct signals at $m/z = 1140$ and $1348$. The simplest, but not necessarily only, explanation is that the newly formed anthracene units in **23** are trapped once and twice by DA reactions with the phenanthroline derivative.[32] Of course, the appropriate precautions were taken to ensure that the adducts had not formed in the spectrometer.

Since all the effort briefly described in this chapter has been associated with quite a lot of frustration and even more disappointment, one can only hope that there are enough capable young scientists who are willing to devote themselves entirely to this complicated research to give it the last kick needed to turn it into a true success story. *A note of caution*: The true success will not be achieved when having made fully unsaturated, double-stranded compounds without any residual trace of ambiguity left. Undoubtedly this would be nice, very nice, indeed, but nevertheless, it will be only a milestone on a long journey, not marking its end! It will be of utmost importance to not only fill a tiny amount of product into a vial and place it proudly on a shelf but rather to make reasonable quantities and study the property profile in the directions indicated in the introduction (Section 13-1). Only if the scope of these compounds is explored and understood in terms of a comparison to their open-chain analogs will a marked advance for chemistry and (hopefully) materials science be achieved. This is the true goal; let's go for it.

## ACKNOWLEDGMENT

This work was supported by the Swiss National Science Foundation (Grants 200020-113607 and 200020-126451), which is gratefully acknowledged. We are very thankful to Profs. M. Iyoda and L. T. Scott for making results available to us prior to publication. We are indebted to Dr. J. D. van Beek and Prof. G. Jeschke (both at ETHZ) for their invaluable help with solid-state NMR and EPR spectroscopy, respectively. We are especially thankful to the ETH mass spectrometry service unit with R. Häfliger, L. Bertschi, and Dr. X. Zhang for the continuous, engaged, and competent support.

## REFERENCES AND NOTES

1. (a) B. D. Steinberg, L. T. Scott, *Angew. Chem. Int. Ed.* **2009**, *48*, 5400–5402; (b) R. Gleiter, B. Esser, S. C. Kornmeyer, *Acc. Chem. Res.* **2009**, *42*, 1108–1116; (c) T. Yao, H. Yu, R. J. Vermeij, G. J. Bodwell, *Pure Appl. Chem.* **2008**, *80*, 533–548; (d) K. Tahara, Y. Tobe, *Chem. Rev.* **2006**, *106*, 5274–5290; (e) R. Gleiter, B. Hellbach, S. Gath, R. J. Schaller, *Pure Appl. Chem.* **2006**, *78*, 699–706; (f) R. Herges, in *Modern Cyclophane Chemistry*, R. Gleiter, H. Hopf (eds.), Wiley-VCH, Weinheim, **2004**, pp. 337–358; (g) L. T. Scott, *Angew. Chem. Int. Ed.* **2003**, *42*, 4133–4135; (h) A. Schröder, H.-B. Mekelburger, F. Vögtle, *Top. Curr. Chem.* **1994**, *172*, 179–201; (i) F. H. Kohnke, J. P. Mathias, J. F. Stoddart, *Top. Curr. Chem.* **1993**, *165*, 1–69.

2. A. D. Schlüter, *Adv. Mater.* **1991**, *3*, 282–291.

3. J. Sakamoto, J. van Heijst, O. Lukin, A. D. Schlüter, *Angew. Chem. Int. Ed.* **2009**, *48*, 1030–1069.

4. For example, see F. H. Kohnke, J. F. Stoddart, *Pure Appl. Chem.* **1989**, *61*, 1581–1586.

5. The DA reaction is known be reversible. If high-molar-mass products are the target, the monomers have to be designed such that the equilibrium lies much on the side of the growth product.

6. F. H. Kohnke, A. M. Z. Slawin, J. F. Stoddart, D. J. Williams, *Angew. Chem. Int. Ed. Engl.* **1987**, *26*, 892–894.

7. From a polymer perspective, it is unfortunate that the structure of the remaining ≥ 80% of product mass was not described or investigated. It is likely that it was of polymeric nature. In footnote 13 of Ref. 6, compound **6** is described as sensitive and prone to undergo polymerization. The wording suggests that this refers to an ill-defined, radical-type process rather than a DA polymerization to the corresponding ladder polymer. The reaction between the other monomers (**7** and **8**) may have led to a DA oligomer/polymer.

8. R. O. Angus Jr., R. P. Johnson, *J. Org. Chem.* **1983**, *48*, 273–276.

9. L. L. Miller, A. D. Thomas, C. L. Wilkins, D. A. Weil, *J. Chem. Soc., Chem. Commun.* **1986**, 661–663.

10. P. R. Ashton, N. S. Isaacs, F. H. Kohnke, A. M. Z. Slawin, C. M. Spencer, J. F. Stoddart, D. J. Williams, *Angew. Chem. Int. Ed. Engl.* **1988**, *27*, 966–969.

11. T. Vogel, K. Blatter, A. D. Schlüter, *Makromol. Chem., Rapid Commun.* **1989**, *10*, 427–430.

12. M. Pomerantz, R. Cardona, P. Rooney, *Macromolecules* **1989**, *22*, 304–308; C. Raghu, Y. A. Pati, S. Ramasesha, *Phys. Rev. B* **2002**, *65*, 155–204; M. Bendikov, H. M. Duong, K. Starkey, K. N. Houk, E. A. Carter, F. Wudl, *J. Am. Chem. Soc.* **2004**, *126*, 7416–7417; M. Bendikov, F. Wudl, D. F. Perpichka, *Chem. Rev.* **2004**, *104*, 4891.

13. E. Clar, *Polycyclic Hydrocarbons*, Academic Press, London, **1964**, Vols. 1 and 2.

14. R. Mondal, B. K. Shah, D. C. Neckers, *J. Am. Chem. Soc.* **2006**, *128*, 9612-9613. For matrix isolation of heptacene and lower homologs, see also R. Mondal, C. Tönshoff, D. Khon, D. C. Neckers, H. F. Bettinger, *J. Am. Chem. Soc.* **2009**, *131*, 14281–14289.

15. H. F. Bettinger, C. Tönshoff, Pushing the limits in acene chemistry: Synthesis of octacene and nonacene, in *Book of Abstracts*, 13th Int. Symp. Novel Aromatic Compounds, July 19-24, **2010**, Luxembourg: C. Tönshoff, H. F. Bettinger, *Angew. Chem. Int. Ed.* **2010**, *49*, 4125. See also T. Fang, *Heptacene, Octacene, Nonacene, and Related Polymers*, PhD thesis, UCLA, **1986**.

16. For theoretical treatments of cyclacenes, see J.-I. Aihara, *J. Chem. Soc., Perkin Trans.* **1994**, 971–974; H. S. Choi, K. S. Kim, *Angew. Chem. Int. Ed.* **1999**, *38*, 2256;S. Tonmunphean, A. Wijitkosoom, Y. Tantirungrotechai, N. Nuttavut, J. Limtrakul, *Bull. Chem. Soc. Jpn.* **2003**, *76*, 1537; L. Türker, *J. Mol. Struct. (THEOCHEM)* **2000**, *531*, 333; S. W.Yang, H. Zhang, J. M. Soon, C. W. Lim, P. Wu, K. P. Loh, *Diamond Relat. Mater.* **2003**, *12*, 1194; K. P. Loh, S. W. Yang, J. M. Soon, H. Zhang, P. Wu, *J. Phys. Chem. A* **2003**, *107*, 5555.

17. P. R. Ashton, G. R. Brown, N. S. Isaacs, D. Giuffrida, F. H. Kohnke, J. P. Mathias, A. M. Z. Slawin, D. R. Smith, J. F. Stoddart, D. J. Williams, *J. Am. Chem. Soc.* **1992**, *114*, 6330–6353.

18. M. Kertesz, A. Ashertehrani, *Macromolecules* **1996**, *29*, 940–945; M. Kertesz, in *Handbook of Organic Conductive Molecules and Polymers*, H. S. Nalwa (ed.), Wiley, New York, **1997**, Vol. 4, pp. 147–173.

19. B. Schlicke, A. D. Schlüter, P. Hauser, J. Heinze, *Angew. Chem. Int. Ed. Engl.* **1997**, *36*, 1996–1998.

20. M. Löffler, A. D. Schlüter, K. Gessler, W. Saenger, J.-M. Toussaint, J.-L. Brédas, *Angew. Chem. Int. Ed. Engl.* **1994**, *33*, 2209–2212.

21. B. Schlicke, H. Schirmer, A. D. Schlüter, *Adv. Mater.* **1995**, *7*, 544–546.

22. A. D. Schlüter, M. Löffler, V. Enkelmann, *Nature* **1994**, *368*, 831–834.

23. W. D. Neudorff, D. Lentz, M. Anibarro, A. D. Schlüter, *Chem. Eur. J.* **2003**, *9*, 2745–2757.

24. Proton-catalyzed breaking and closing of the oxygen bridges cannot be excluded.

25. (a) M. Stuparu, V. Gramlich, A. Stanger, A. D. Schlüter, *J. Org. Chem.* **2007**, *72*, 424–430; (b) M. Stuparu, D. Lentz, H. Rüegger, A. D. Schlüter, *Eur. J. Org. Chem.* **2007**, 88–100.

26. C. Denekamp, A. Etinger, W. Amrein, A. Stanger, M. Stuparu, A. D. Schlüter, *Chem.Eur. J.* **2008**, *14*, 1628–1637.

27. An example should be noted where strained single-stranded oligo(*p*-phenylene)s were successfully obtained, the synthesis of which in the last step encounters the same kind of problem; see R. Jasti, J. Bhattacharjee, J. B. Neaton, C. R. Bertozzi, *J. Am. Chem. Soc.* **2008**, *130*, 17646.

28. H. Ajie, M. M. Alvarez, S. J. Anz, R. D. Beck, F. Diederich, K. Fostiropoulos, R. Huffman, W. Krätschmer, Y. Rubin, K. E. Schriver, D. Sensharma, R. L. Whetten, *J. Phys. Chem.* **1990**, *206*, 639.

29. J. Crassous, J. Rivera, N. S. Fender, L. Shu, L. Echegoyen, C. Thilgen, A. Herrmann, F. Diederich, *Angew. Chem. Int. Ed.* **1999**, *38*, 1613–1617.

30. M. Standera, G. Jeschke, A. D. Schlüter (unpublished results).

31. E. Clar et al., *Nature* **1949**, *163*, 921; K. N. Houk et al., *Org. Lett.* **2006**, *8*, 4915.

32. M. Standera, R. Häfliger, R. Gershoni-Poranne, A. Stanger, G. Jeschke, J. D. van Beek, L. Bertschi, A. D. Schlüter, *Chem. Eur. J.* **2011**, accepted.

33. F. B. Mallory, C. W. Mallory, *Org. React.* **1984**, *30*, 1–456; F. B. Mallory, K. E. Butler, A. Bérubé, E. D. Luzik Jr., C. W. Mallory, E. J. Brondyke, R. Hiremath, P. Ngo, P. J. Carroll, *Tetrahedron* **2001**, *57*, 3715–3724.

34. R. H. Mitchell, V. Boekelheide, *J. Am. Chem. Soc.* **1970**, *92*, 3510–3512.

35. J. E. McMurry, T. Lectka, J. G. Rico, *J. Org. Chem.* **1989**, *54*, 3748–3749.

36. Y. Kuwatani, T. Yoshida, K. Hara, M. Yoshida, H. Matsuyama, M. Iyoda, *Org. Lett.* **2000**, *2*, 4017–4020; T. Yoshida, Y. Kuwatani, K. Hara, M. Yoshida, H. Matsuyama, M. Iyoda, S. Nagase, *Tetrahedron Lett.* **2001**, *42*, 53–56; Y. Kuwatani, J.-I. Igarashi, M. Iyoda, *Tetrahedron Lett.* **2004**, *45*, 359–362.

37. Private communication with M. Iyoda.

38. M. Brooks, PhD thesis, Boston College, **1998**. H. St. Martin, PhD thesis, Boston College, **2002**.

39. C. Grave, A. D. Schlüter, *Eur. J. Org. Chem.* **2002**, 3075–3098.

40. *Tert*-butyl groups can commonly be removed under electrophilic conditions; however, in the particular case of **32b** attempts in this direction did not work. Skeletal rearrangements were observed instead. It seems that the only possible way out is to find conditions under which pyrolyses are carried out on precursors that have a proton or a metal cation captured on the hydrocarbon skeleton while being pyrolyzed.

41. For example, see E. Nakamura, K. Tahara, Y. Matsuo, M. Sawamura, *J. Am. Chem. Soc.* **2003**, *125*, 2834–2835. More references are available in Ref. 1 (above).

42. B. L. Merner, L. N. Dawe, G. J. Bodwell, *Angew. Chem. Int. Ed.* **2009**, *48*, 5487–5491. See also G. J. Bodwell, J. N. Bridson, M. Cyrański, J. W. J. Kennedy, T. M. Krygowski, M. R. Mannion, D. O. Miller, *J. Org. Chem.* **2003**, *68*, 2089–2098; G. J. Bodwell, D. O. Miller, R. J. Vermeij, *Org. Lett.* **2001**, *3*, 2093–2096; G. J. Bodwell, J. J. Fleming, D. O. Miller, *Tetrahedron* **2001**, *57*, 3577–3585; G. J. Bodwell, J. J. Fleming, M. R. Mannion, D. O. Miller, *J. Org. Chem.* **2000**, *65*, 5360–5370; G. J. Bodwell, J. N. Bridson, T. J. Houghton, J. W. J. Kennedy, M. R. Mannion, *Chem. Eur. J.* **1999**, *5*, 1823–827; G. J. Bodwell, J. N. Bridson, T. J. Houghton, J. W. J. Kennedy, M. R. Mannion, *Angew. Chem. Int. Ed. Engl.* **1996**, *35*, 1320–1321.

43. G. Haberhauer, F. Rominger, R. Gleiter, *Angew. Chem. Int. Ed.* **1998**, *37*, 3376–3377.

44. B. Hellbach, F. Rominger, R. Gleiter, *Angew. Chem. Int. Ed.* **2004**, *43*, 5846–5849; S. C. Kornmayer, B. Hellbach, F. Rominger, R. Gleiter, *Chem. Eur. J.* **2009**, *15*, 3380–3389.

45. B. Esser, F. Rominger, R. Gleiter, *J. Am. Chem. Soc.* **2008**, *130*, 6716–6717.

46. B. Esser, A. Bandyopadhyay, F. Rominger, R. Gleiter, *Chem. Eur. J.* **2009**, *15*, 3368–3379.

# CHAPTER 14

# BENT PYRENES: SPRINGBOARDS TO AROMATIC BELTS?

GRAHAM J. BODWELL, GANDIKOTA VENKATARAMANA,
and UNIKELA KIRAN SAGAR

## 14-1. INTRODUCTION

With sincere apologies to my coauthors (G.V. and U.K.S.), I (G.J.B.) have chosen to write in the first person. Although I often advise students not to do this, I have elected to do so here for two reasons: (1) it is entirely appropriate for a personal account, and (2) it emphasizes the fact that the experiences and opinions expressed herein are purely my own. This also serves to illustrate that there are exceptions to most things we, as academics, tell our students.

Organic chemistry is a multilayered and highly nuanced discipline, which makes it an especially challenging subject to teach at the introductory level. My first efforts to teach introductory organic chemistry focused squarely on ideal situations. Whether it came from mental laziness or the intentional avoidance of long-winded asides, I gave extremely short shrift to the consequences of deviating from the ideal situation. The importance of this issue became clear to me when I later taught the same students at higher levels. To my dismay, many of the sound bites and catch phrases (e.g., "aromatic compounds *are* planar" — more on this later) that I emphasized at the introductory stage had been brought through to intermediate- and advanced-level courses with (in the most extreme cases) blind acceptance and, even worse, staunch resistance to notions that went against them.

Since that time, I have made a real effort to better prepare beginning students for the subsequent layers and nuances without thoroughly confusing them. This has turned out to be a formidable balancing act. In fact, such questions of balance pervade all aspects of my life, and I submit that the value of $K_{eq}$ is not measurable, but rather a matter of personal choice. For example, as a parent, I struggle (perhaps this is too weak a word) to provide my children with nutritious food that they will eat without

*Fragments of Fullerenes and Carbon Nanotubes: Designed Synthesis, Unusual Reactions, and Coordination Chemistry*, First Edition. Edited by Marina A. Petrukhina and Lawrence T. Scott.
© 2012 John Wiley & Sons, Inc. Published 2012 by John Wiley & Sons, Inc.

complaint. Similarly, as a teacher of organic chemistry, I aspire to feed my students the basics in a digestible but nutritious form. Taking this further, as an academic, I work to spend sufficient time on research, teaching, and administrative responsibilities, each of which could easily consume 100% of my working time. As a researcher, I must weigh issues of depth and breadth in my research program in the context of a raft of competing issues. As an organic chemist, I have wrestled with the interplay of strain and aromatic stabilization energy (more on this later), which embodies the entwinement of frustration and reward associated with all of the challenges described above. The happy medium in all of this is a moving target, which, if nothing else, keeps life very interesting.

## 14-2. BACKGROUND

During my graduate career, I had the great fortune to work under the supervision of two important figures in the "novel aromatics" community. Reg Mitchell supervised my MSc research at the University of Victoria (MSc 1986), and Henning Hopf was my "Doktorvater" at the Technical University of Braunschweig (Dr. rer. nat. 1989). In different ways, I was genuinely inspired by my two mentors and, consequently, the set of research proposals that I prepared while applying for academic positions was heavily influenced by the chemistry I had been involved with under their tutelage. For example, one of my proposed projects was entitled "Tethered [2.2]metacyclophanes." The object of this proposal was to synthesize a series of [$n$.2.2](1,3,5)cyclophanes **1** (these are what I dubbed tethered [2.2]metacyclophanes) and determine the value of $n$ at which the innate preference of the [2.2]metacyclophane unit for the *anti* conformation would outweigh the constraints of the long bridge (the tether) (Scheme 14-1).

Academic positions in Canada were scarce during the years following the completion of my doctoral work, and in early 1992 I was beginning to relinquish hope of ever having the opportunity to pursue my own research interests. Thus, when I was offered a position at Memorial University, I accepted it in less than a nanosecond. Perhaps a more circumspect approach would have improved my bargaining position, but this proved to be the best move of my life. The university was in the early stages of developing a strong focus on research, and the city of St. John's, despite the now bygone economic hard times, revealed itself to be one of the most unique, interesting, and endearing places in the world.

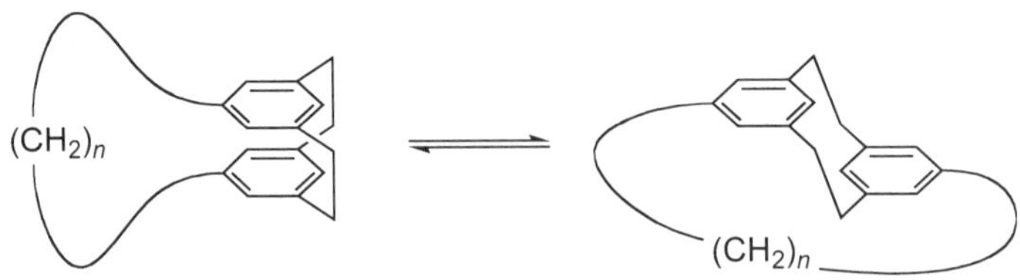

**Scheme 14-1.** *Syn/anti*-isomerism in tethered [2.2]metacyclophanes.

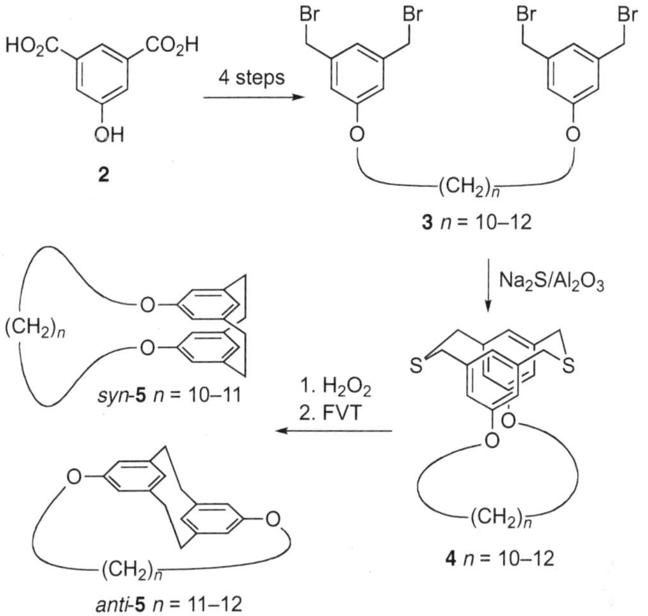

**Scheme 14-2.** Synthetic approach to *syn*-**5** and *anti*-**5**.

Having found my place personally and professionally, I set about the task of establishing my research program, and the tethered [2.2]metacyclophane project was one of several projects (too many projects, in retrospect) that my group began working on. A synthetic route to a series of 1,*n*-dioxa[*n*.2.2](1,3,5)cyclophanes **5** was soon completed (Scheme 14-2).[1] This started with 5-hydroxyisophthalic acid (**2**), which was converted into tetrabromides (**3**) in four steps. Dithiacyclophanes (**4**) were then generated on reaction of **3** with $Na_2S/Al_2O_3$. An oxidation/FVT (flash vacuum thermolysis) sequence then completed the synthesis of **5**. Cyclophane **5** ($n = 10$) was obtained exclusively as the *syn* conformer, and **5** ($n = 12$) was obtained exclusively as the *anti* conformer, but **5** ($n = 11$) was isolated as an easily separable mixture of the two conformers. The equilibration of each conformer of **5** ($n = 11$) could then be observed in the NMR spectrometer. I am indebted to my former graduate student, Tom Houghton, not only for working out this synthetic route but also for fishing the reagent $Na_2S/Al_2O_3$ out of the literature and adapting it for the synthesis of dithiacyclophanes.[2] Our group still uses it on a regular basis.

## 14-3. VALENCE ISOMERIZATION BETWEEN [2.2]METACYCLOPHANE-1,9-DIENES AND 10B,10C-DIHYDROPYRENES

As the tethered metacyclophane project was progressing, the idea of also using dithiacyclophanes (**4**) as precursors to tethered [2.2]metacyclophanedienes (**6**) was born (Scheme 14-3). My first thought was that the unsaturation in the bridges opened

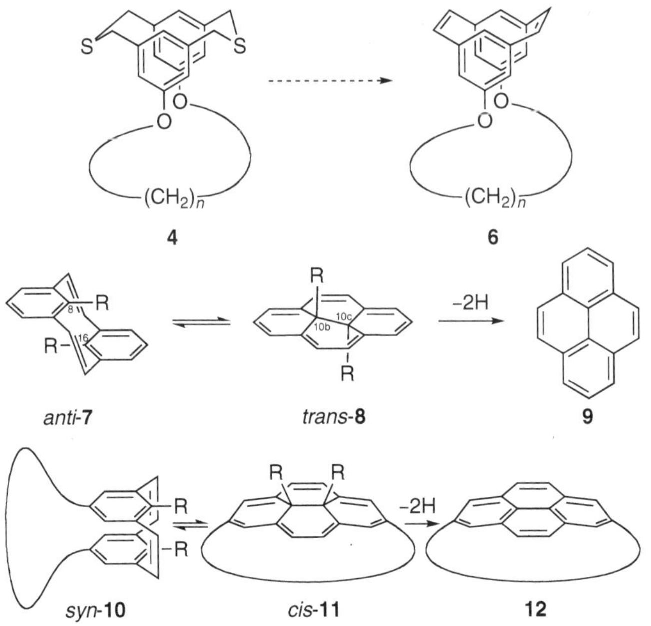

**Scheme 14-3.** Pyrenophanes from tethered [2.2]metacyclophanedienes.

up the possibility of developing new photochromic and/or thermochromic systems. Specifically, the valence isomerization between [2.2]metacyclophanedienes and 10b,10c-dihydropyrenes was already a well-studied photochromic system.[3,4] However, the overwhelming majority of work in this area had been on the switching between *anti*-[2.2]metacyclophanedienes (*anti*-**7**) and *trans*-10b,10c-dihydropyrenes (*trans*-**8**). The twist with my group's systems was that they featured a tether, which we had recently learned would constrain the [2.2]metacyclophane system to the *syn* conformation when $n \leq 12$.[1] This presented us with an opportunity to investigate the interconversion between tethered *syn*-[2.2]metacyclophanedienes (*syn*-**10**) and *cis*-10b,10c-dihydropyrenes (*cis*-**11**) as well as to influence the position of the equilibrium by varying the length of the tether. Although this idea was pursued to some extent by a postdoctoral fellow, Dr. Shu-Lin Chen,[5,6] it was ultimately overshadowed by a closely related idea, which developed into a core research theme in our group.

## 14-4. THE VID REACTION

A common feature of all photochromic metacyclophanediene–dihydropyrene systems with reasonable stability is the presence of "internal" substituents (8 and 16 positions of the metacyclophanediene; 10b and 10c positions of the dihydropyrene). In most cases, methyl groups have been used. These substituents are important because they prevent an irreversible dehydrogenation of the dihydropyrene, which destroys the photochromic system.[7] For the parent system (R = H), the result is a very expensive synthesis of commercially available pyrene (**9**) (Scheme 14-3). The two steps that result in the conversion of *anti*-**7** into **9** are a valence isomerization and a dehydrogenation. As this was quite a mouthful, we quickly got into the habit of calling it the VID reaction. In more general terms, the overall transformation is a cyclodehydrogenation, but we prefer to stick with the more specific term VID because it is descriptive of the chemistry.

In the case of the tethered [2.2]metacyclophanedienes *syn*-**10**, the incorporation of methyl groups at the internal positions proved to be nontrivial, which is a major reason why our interest in the photochromic systems evaporated. On the other hand, the "undesired" dehydrogenation reaction piqued my interest because (at least on paper) it promised to lead to the formation of "tethered pyrenes" (**12**), or using cyclophane nomenclature, [*n*](2,7)pyrenophanes. This system looked extremely attractive because it appeared as though I had the chance to embark on a systematic study of an expanded version of the quintessential cyclophanes, the [*n*]paracyclophanes.[8]

Much as I would like to be able to say that I immediately recognized the various issues associated with this chemistry, my approach was to shoot first and ask questions later. Thus, without investing much thought into the details of the transformation, the VID reaction was quickly put to the test and (much to my delight) was found to be a remarkably powerful reaction for the formation of strained molecules containing nonplanar pyrene systems. As time passed, more careful consideration was given to the VID reaction and, as discussed below, some of the factors that contribute to its effectiveness were revealed.

The first key issue pertains to the valence isomerization step. The interconversion of *anti*-**7** and *trans*-**8** is a conrotatory antarafacial $6\pi$ electrocyclic reaction, which is a thermally forbidden process.[9] Although this reaction does disobey the rules of orbital symmetry conservation and proceed thermally in many cases, it is typically driven by light. In contrast, the valence isomerization between a *syn*-[2.2]metacyclophanediene such as *syn*-**10** and the corresponding *cis*-10b,10c-dihydropyrene (e.g., *cis*-**11**) is a disrotatory suprafacial $6\pi$ electrocyclic reaction, which is a thermally allowed process. As such, we are working with nature instead of against it.

The energetics of the VID reaction are also very important, but this is a complicated issue. Information about all of the thermodynamic parameters for each step of the reaction is needed, as is an understanding of how they all vary with changes in the composition (or more simply stated, the length) of the tether/bridge. In a nutshell, the big picture has yet to be sorted out and, hopefully, progress in this area will be reported soon. Until then, some general comments can be made. Major contributing factors to the change in energy during the valence isomerization step are bonding, aromatic stabilization energy (ASE), and strain, each of which would be expected to vary with changes in the composition of the tether. The same is true for the dehydrogenation step, but there are additional complications arising from the question of what happens to the two hydrogen atoms that are lost (as $H_2$ or as part of an added oxidant).

Some work in this direction has been completed, most of it driven by a question I have been asked numerous times at conferences and academic institutions: How does the aromaticity of the pyrene system vary with the deviation from planarity? Being asked this reasonable question was never surprising, but the expectations of those who posed this question often were. Many people said they expected the aromaticity to fall off sharply as the aromatic unit was distorted from planarity. In fact, it became apparent to me that a major proportion of chemists at all levels had an innate sense of aromatic systems as being rigid entities with little tolerance to bending as far as the aromaticity was concerned. Although I can't remember precisely where my own prejudices lay before I became involved with nonplanar aromatics, I tend to think that I was probably in this camp too. As foreshadowed in the introduction (Section 14-1), my feeling is that this misguided notion comes from being taught in one's introductory course in organic chemistry that planarity is a *criterion* for aromaticity, or that aromatic compounds *are* planar.

So how does the aromaticity of the pyrene system change as it is bent out of planarity? To answer this question, the thorny problem of measuring aromaticity must be dealt with. This has been the subject of debate for many years, and a variety of yardsticks for quantifying aromaticity have been put forward. Fortunately, the answer to the question at hand depends little on whether geometric [harmonic oscillator model of aromaticity (HOMA)],[10–13] magnetic [nucleus-independent chemical shift (NICS)],[14–16] or energetic criteria (ASE)[17–23] are used to quantify the aromaticity of the pyrene system. As pyrene is increasingly bent out of planarity, the aromaticity drops off consistently, but gently. Even in the most severely bent pyrenes that my group has been able to synthesize (see Section 14-5), the majority of the aromaticity (65–80%, depending on the method of quantification) present in planar pyrene is still there.[24,25] In other words, aromaticity (whatever it may be) is quite insensitive to bending out of planarity.

It is also worth noting that the ASE of planar pyrene is large (74.5 kcal/mol).[26] In the context of the VID reaction leading to pyrenophanes, this implies that there is a substantial driving force for the second step (the dehydrogenation), in which a *cis*-10b,10c-dihydropyrenophane is destroyed and a pyrene system is generated. There does not appear to be a published value for the ASE (or other energetic measure of aromaticity) of the [14]annulene system in *cis*-10b,10c-dihydropyrene. A resonance aromatization energy, $E_{\text{aromatization}}$ (not the same as ASE), of 33 kcal/mol has been reported for the [14]annulene system in *trans*-10b,10c-dihydropyrene,[27] which suggests that the value for the *cis* isomer, which has a nonplanar [14]annulene, might be a little lower. Less nebulous comments will require the support of a detailed theoretical treatment of the VID reaction, but it seems for now that the large ASE of pyrene and the insensitivity of its aromaticity to bending out of planarity are key contributors to the ability of the VID reaction to afford strained systems (pyrenophanes) under relatively mild conditions.

## 14-5. PYRENOPHANE SYNTHESIS USING THE VID REACTION

Until my group became involved in this area, the transformation of a [2.2]metacyclophanediene into a pyrene derivative had not been exploited synthetically since Boekelheide and Mitchell's original observation.[7] As an aside, the conversion of [2.2]metacyclophanes into 4,5,9,10-tetrahydropyrenes[28] has seen significant use

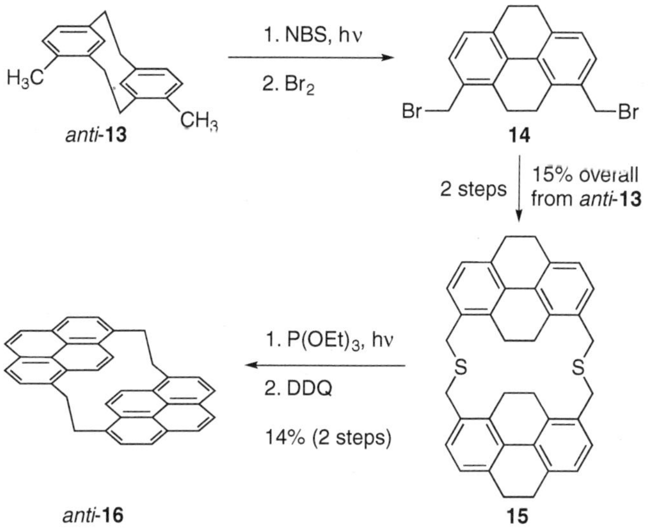

**Scheme 14-4.** Synthesis of *anti*-[2.2](1,8)pyrenophane (*anti*-**16**).

in the synthesis of pyrene derivatives, including pyrenophanes. This reaction is mechanistically different from the VID reaction and requires a separate (and often low-yielding) dehydrogenation step to afford a pyrene system.[29] The synthesis of *anti*-[2.2](1,8)pyrenophane **16** from *anti*-4,14-dimethyl[2.2]metacyclophane **13** (Scheme 14-4) typifies this approach.[30] An investigation of how well this reaction (also a cyclodehydrogenation) works on tethered [2.2]metacyclophanes *syn*-**1** to give tetrahydropyrenophanes would certainly be worth looking into.

The strategy that my group used to synthesize all of our pyrenophanes has three main stages: (1) the tethering of two appropriately substituted benzene units, (2) the formation of a [2.2]metacyclophanediene unit, and (3) VID reaction to create the pyrene system, which can be regarded as a bent board or perhaps a springboard (Scheme 14-5). The placement of the VID reaction at the end of the synthesis is

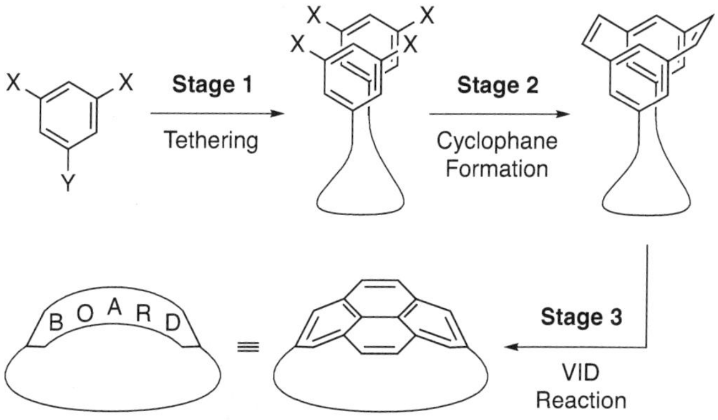

**Scheme 14-5.** General strategy for the synthesis of (2,7)pyrenophanes.

important because the pyrene system is generated in its final nonplanar conformation. As discussed above, the buildup of strain at this point is counteracted to some extent by the substantial gain in ASE associated with the creation of a pyrene system. Earlier construction of a planar pyrene system would necessitate bending at a later stage, and this would require some other means of offsetting the increase in strain energy.

### 14-5-1. [n](2,7)Pyrenophanes

As exemplified by the synthesis of [7](2,7)pyrenophane **26** (Scheme 14-6),[31] several [n](2,7)pyrenophanes have been synthesized.[24, 31–35] The synthesis commenced with 5-hydroxyisophthalic acid (**17**), which was esterified and triflated in preparation for a twofold Sonogashira coupling, which afforded diyne **20** and thus completed stage 1 of the general strategy (Scheme 14-5). Having served their purpose, the triple bonds were catalytically hydrogenated to afford tetraester (**21**), which was then converted into tetrabromide (**22**). This set the stage for cyclophane formation (stage 2), which was accomplished by the reaction of **22** with $Na_2S/Al_2O_3$. Stevens rearrangement of the resulting dithiacyclophane (**23**) followed by Hofmann elimination resulted in the formation of a [2.2]metacyclophanediene system (**25**), which smoothly underwent VID reaction (stage 3) on treatment with DDQ in benzene at reflux to afford **26**. The 26% overall yield for this particular 12-step sequence is atypically high. Overall yields are more likely to be under 10%, mainly because of the often poor (sometimes abysmal) outcome of the Hofmann elimination.

In the synthesis of [n](2,7)pyrenophanes with short bridges ($\leq 8$ atoms), their cyclophanediene precursors can be isolated, purified, and characterized routinely. However, when the tether exceeds 8 atoms in length, the Hofmann elimination of the bis(dimethylsulfonium) salt **27** usually gives mixtures consisting of varying proportions of the cyclophanediene (**28**), the desired pyrenophane (**33**), and the corresponding 4,5-dihydropyrenophane (**32**) (Scheme 14-7). Reaction of the mixture with DDQ in benzene at reflux brings about dehydrogenation of the 4,5-dihydropyrenophane (**32**) and/or VID reaction of the cyclophanediene (**28**) to afford an easily purified sample of the pyrenophane (**33**). The formation of the 4,5-dihydropyrenophane (**32**) can be explained by valence isomerization of the cyclophanediene (**28**) under the conditions of its formation to give a 10b,10c-dihydropyrenophane (**29**), which can then undergo a series of three consecutive [1,5]-H shifts. The accumulation of 4,5-dihydropyrenophane (**32**) is not unreasonable because the ASE of its phenanthrene system (66.6 kcal/mol)[26] is substantially larger than that of the π systems present in **29** ([14]annulene, 33 kcal/mol),[27] **30** (benzene, 29.3 kcal/mol),[26] and **31** (nonaromatic polyene). Dehydrogenation at any point from **29** to **32** affords pyrenophane (**33**). In any future investigation of the VID reaction, it would be worthwhile to consider the rearrangement of **29** to **32** at the same time.

Following the synthesis of my group's first pyrenophanes, one of the first orders of business was to devise a method for quantifying the extent to which the pyrene unit was distorted from planarity. Having already pictured the [n](2,7)pyrenophanes as expanded versions of the [n]paracyclophanes, it seemed logical to extend the metrics that have been used to describe nonplanarity in the benzene ring of the

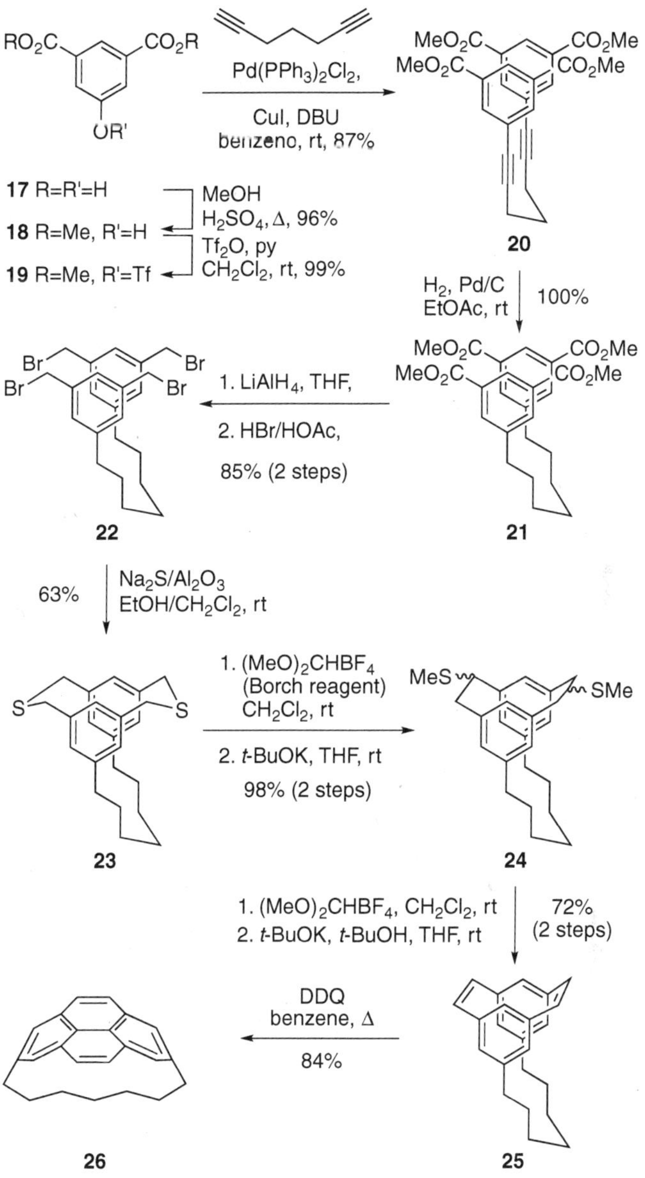

**Scheme 14-6.** Synthesis of [7](2,7)pyrenophane (**26**).

[n]paracyclophanes.[36] In such systems, the angle $\alpha$ is the envelope flap angle in the benzene ring or, in more precise terms, the smallest angle formed by the planes defined by C1−C2−C6 and C2−C3−C5−C6. Similarly, the angle $\beta$ is the smallest angle formed by the plane defined by C1−C2−C6 and the line defined by C1−C1′ (Figure 14-1). Thus, for the [n](2,7)pyrenophanes, the smallest angles formed by

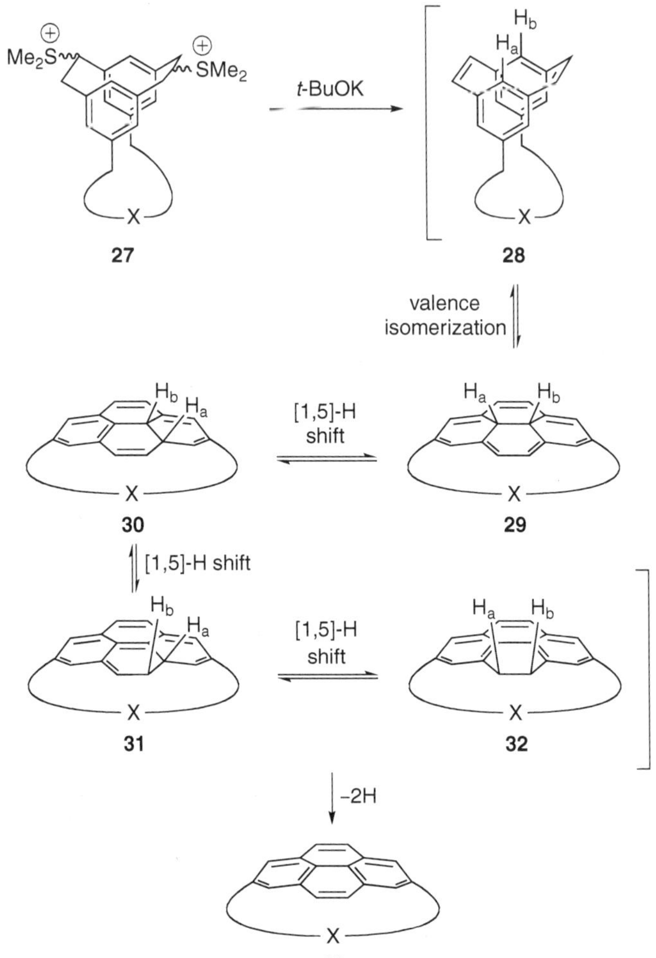

**Scheme 14-7.** Possible mechanism for the formation of dihydropyrenophanes **32** from tethered [2.2]metacyclophanedienes (**28**).

adjacent planes of carbon atoms can provide a measure of local bend, which can be compared directly to the bend in the [*n*]paracyclophanes. At the same time, the smallest angle formed by the two terminal planes in the pyrene system (C1−C2−C3 and C6−C7−C8), can be taken as a measure of the overall bend, which can be used to compare different pyrenophanes. This angle, to which I randomly assigned the Greek character θ, is 0° for planar pyrene and becomes increasingly large as the pyrene system is bent out of planarity.

During the course of collaborative work with Michal Cyrański and Marek Krygowski on the relationship between aromaticity and nonplanarity in the pyrene system, two other parameters were defined to quantify the degree of bend, namely,

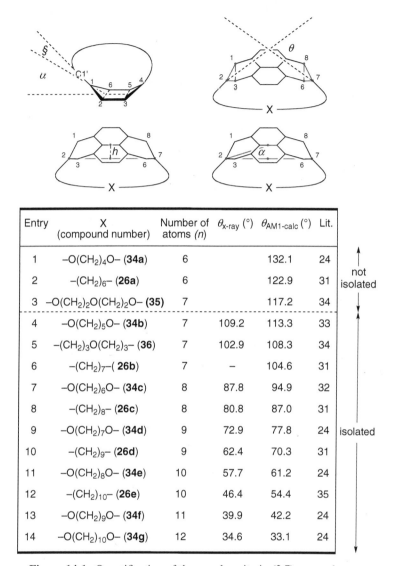

| Entry | X (compound number) | Number of atoms $(n)$ | $\theta_{x\text{-ray}}$ (°) | $\theta_{AM1\text{-calc}}$ (°) | Lit. | |
|---|---|---|---|---|---|---|
| 1 | $-O(CH_2)_4O-$ (**34a**) | 6 | | 132.1 | 24 | not isolated |
| 2 | $-(CH_2)_6-$ (**26a**) | 6 | | 122.9 | 31 | |
| 3 | $-O(CH_2)_2O(CH_2)_2O-$ (**35**) | 7 | | 117.2 | 34 | |
| 4 | $-O(CH_2)_5O-$ (**34b**) | 7 | 109.2 | 113.3 | 33 | |
| 5 | $-(CH_2)_3O(CH_2)_3-$ (**36**) | 7 | 102.9 | 108.3 | 34 | |
| 6 | $-(CH_2)_7-$ (**26b**) | 7 | – | 104.6 | 31 | |
| 7 | $-O(CH_2)_6O-$ (**34c**) | 8 | 87.8 | 94.9 | 32 | |
| 8 | $-(CH_2)_8-$ (**26c**) | 8 | 80.8 | 87.0 | 31 | |
| 9 | $-O(CH_2)_7O-$ (**34d**) | 9 | 72.9 | 77.8 | 24 | isolated |
| 10 | $-(CH_2)_9-$ (**26d**) | 9 | 62.4 | 70.3 | 31 | |
| 11 | $-O(CH_2)_8O-$ (**34e**) | 10 | 57.7 | 61.2 | 24 | |
| 12 | $-(CH_2)_{10}-$ (**26e**) | 10 | 46.4 | 54.4 | 35 | |
| 13 | $-O(CH_2)_9O-$ (**34f**) | 11 | 39.9 | 42.2 | 24 | |
| 14 | $-O(CH_2)_{10}O-$ (**34g**) | 12 | 34.6 | 33.1 | 24 | |

**Figure 14-1.** Quantification of the nonplanarity in (2,7)pyrenophanes.

a bowl depth parameter $(h)^{24}$ and a different bend angle $(\alpha)^{25}$ (Figure 14-1). These parameters proved to be useful and, in the case of $h$, were shown to have a high correlation coefficient with $\theta$.[24] Nevertheless, I prefer to use $\theta$ because of its consonance with how bend has been quantified in the [$n$]paracyclophanes. Bowl depths are seldom used to describe cyclophane structures and, as explained above, there is already an angle $\alpha$ that is used to describe bending in the [$n$]paracyclophanes.

Using synthetic routes similar to those shown in Schemes 14-2 and 14-6, my group was able to synthesize a series of [$n$](2,7)pyrenophanes (Figure 14-1).[24,31–35] The compounds listed below the dotted line were isolated in pure form and characterized, while those compounds appearing above the dotted line were either not formed or not isolated. X-ray crystal structures were determined for all except one of the $[n](2,7)$pyrenophanes that were isolated, and experimental $\theta$ values were obtained. Semiempirical (AM1) calculations were performed on all of the $[n](2,7)$pyreno-phanes listed in Figure 14-1, and the $\theta$ values obtained from the calculated structures were found to be consistently 4–8° higher than the experimental values when there were $\leq 10$ atoms in the bridge. Despite the low level of theory, the quick-and-easy AM1 calculations have been a useful tool in identifying viable targets. As a rule of thumb, I consider any targets with a calculated $\theta$ value below about 115° to have at least a reasonable chance of being synthesized. A target with a $\theta$ value below about 100° is typically expected to be well within the capabilities of the VID reaction.

As expected, the pyrene system becomes more distorted as the length of the bridge becomes shorter. Coarse-tuning of the $\theta$ value can be achieved by adjusting the number of atoms in the bridge $(n)$, while fine-tuning can be accomplished by adjusting the nature of the bridge. Specifically, replacement of a carbon atom with an oxygen atom results in a slight shortening of the tether, mainly because the $C–O$ bond ($\sim 1.41$ Å) is shorter than the $C–C$ bond ($\sim 1.54$ Å). This is seen clearly in the four examples with $n = 7$. As the number of oxygen atoms increases from 0 to 3, the value of $\theta_{calc}$ increases from 104.6° to 117.2° in roughly equal steps. The $\theta$ range is 12.6°, which is about two-thirds of the $\sim 18°$ jump that results from changing the value of $n$ by one.

The current record holder for highest measured $\theta$ value (most distorted pyrene) is 1,7-dioxa[7](2,7)pyrenophane (**34b**) (Figure 14-2, $\theta = 109.2°$),[33] which was actually one of the first pyrenophanes that my group synthesized. Subsequent attempts to synthesize pyrenophanes with even more distorted pyrene units have been generally

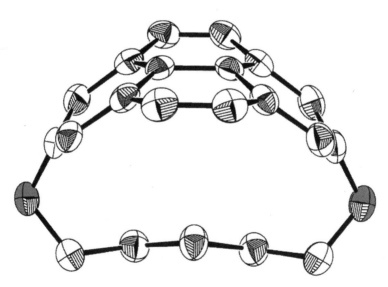

**Figure 14-2.** X-ray crystal structure of 1,7-dioxa[7](2,7)pyrenophane (**34b**).

unsuccessful. For example, the VID reaction leading to 1,4,7-trioxa[7](2,7)pyreno-phane (**35**) (entry 3 in the table in Figure 14-1, $\theta_{calc} = 117.2°$) gave a mixture of products, from which some starting material was recovered by column chromatogra-phy.[34] The $^1$II NMR spectrum of this material indicated that the desired pyrenophane (**35**) had coeluted with the starting material and constituted about 10% of the mixture. Although no pure sample was obtained, this result indicated that 1,4,7-tri-oxa[7](2,7)pyrenophane (**35**) can not only form in refluxing benzene but also survive standard column chromatography.

The calculated $\theta$ value of [6](2,7)pyrenophane (**26a**) (entry 2 in the table in Figure 14-1) is 122.9°. An attempt to synthesize this compound resulted in the consumption of the cyclophanediene and formation of an intractable product.[31] However, an NMR tube experiment showed the appearance of a new compound at low conversion. Traces of this compound were isolated by column chromato-graphy, and its $^1$H NMR spectrum resembled that of a Diels–Alder adduct of [7](2,7)pyrenophane (**26b**) (see Section 14-6). MS analysis of the product showed a strong peak at $m/z = 284$, which corresponds to the [6](2,7)pyrenophane (**26a**). The conclusion from this experiment is that the VID reaction is capable of generating [6](2,7)pyrenophane (**26a**), but that it is not stable under the conditions of its formation. The structure of the new compound that it formed was not determined, but it is probably either a dimer resulting from [4+4] cycloaddition with itself or a Diels–Alder adduct with DDQ (see Section 14-6 for structures).

The most ambitious target my group ever tackled (this was still in the early days) was 1,6-dioxa[6](2,7)pyrenophane (**34a**)[24] (entry 1 in the table in Figure 14-1), which has a $\theta_{calc}$ value of 132.1°. Not surprisingly, in retrospect, the VID reaction leading to this pyrenophane showed absolutely no signs of progress in benzene at reflux. The use of more forceful conditions (reflux in xylenes) resulted in the very slow appearance of a baseline spot [thin-layer chromatography (TLC) analysis]. Thus, somewhere in the $\theta_{calc}$ range of 123–132°, the VID reaction is challenged beyond its capabilities.

Much of the pioneering work on the [n](2,7)pyrenophanes was done by a very talented graduate student, Mike Mannion, who was assisted ably by two equally talented undergraduate students, Jason Kennedy and Jim Fleming.

## 14-5-2. [2.2]Cyclophanes of Pyrene and Benzene

As my group's [n](2,7)pyrenophane count was growing, work aimed at the incorpo-ration of aromatic units in the bridge was initiated. The first success in this area was realized by an excellent graduate student, Rolf Vermeij, who completed the synthesis of [2]paracyclo[2](2,7)pyrenophane (**44**) (Scheme 14-8).[37] The synthesis was similar to that of the parent [n](2,7)pyrenophanes (**26**). Starting with triflate (**19**), Sonogashira reactions were used to install the tether, and a series of functional group interconver-sions then gave tetrabromide (**41**), which was converted into dithiacyclophane (**42**) with some difficulty. Apparently, the rigid tether prevents the [3.3]dithiacyclophane system from adopting its preferred geometry, and this presumably accounts for both the low yield of the reaction and the tendency of **42** to degrade during purification, especially column chromatography. The conversion of **42** into cyclophanediene (**43**)

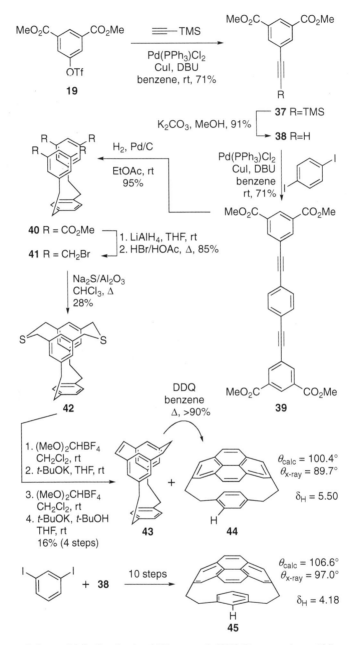

**Scheme 14-8.** Synthesis of [2]paracyclo[2](2,7)pyrenophane (**44**).

was also low-yielding, and it was isolated as a ~1:1 mixture with the desired pyrenophane (**44**). Treatment of this mixture with DDQ (2,3-dichloro-5,6-dicyano-1,4-benzoquinone) proceeded routinely to give uncontaminated **44**. The isomeric [2]metacyclo[2](2,7)pyrenophane (**45**) was synthesized analogously using 1,3-diiodobenzene instead of 1,4-diiodobenzene (Scheme 14-8).[38]

The crystal structures of pyrenophanes **44** and **45** were determined, and the θ values were found to be about 10° less than the values that had been calculated prior to starting the syntheses. Pyrenophane (**44**) also has unusually large bond angles (124°) at the carbon atoms that are benzylic to the benzene ring. In both **44** and **45**, the strong magnetic anisotropic effect of the pyrene system is evidenced by the high-field chemical shifts of the protons attached to the benzene ring (δ 5.50 for all of the benzene ring protons in **44** and δ 4.18 for the internal proton in **45**). There is a third isomer in this series, namely, [2]orthocyclo[2](2,7)pyrenophane. However, this pyrenophane has a calculated θ value of 130.4°, so its synthesis was never attempted.

Having successfully synthesized a couple of pyrene-containing [2.2]cyclophanes, my thoughts turned to the corresponding cyclophanedienes, and this quickly led me to identify the benzannulated cyclophanediene (**54**) as an especially interesting target. This compound attracted my attention because it can also be viewed as a cyclic oligoarylene or a (small) shape-persistent macrocycle.[39] A short synthetic route appeared to be available, and I gave the project to an undergraduate student, Greg Manning, for his Honours project. Gratifyingly (as much for me as for Greg), the synthesis proceeded pretty much according to plan (Scheme 14-9).[40]

A chemoselective Suzuki reaction between boronic acid **47** and 1-bromo-2-iodobenzene (**46**) afforded biaryl **48**, and a second Suzuki reaction, this time with bis(boronic acid) **49**, provided oligophenylene **50**. I had expected that the central phenylene unit would provide some steric hindrance to the methyl groups on the two terminal benzene rings and that this would enable selective monobromination of each methyl group. In practice, free-radical bromination of **50** afforded a mixture of products, the major component (~70% by $^1$H NMR) of which was, indeed, the desired tetrabromide (**51**). Attempts to purify this compound were unsuccessful, so the crude product was reacted with $Na_2S/Al_2O_3$ in the hope that separation could be achieved after the next step. The same situation arose with dithiacyclophane (**52**) (as did the degradation problem that was encountered with **42**), so this crude mixture was then taken through four more steps without purification to afford cyclophanediene (**53**). Considering what Greg had achieved in a short timeframe, the 12% overall yield for the six-step sequence was perfectly acceptable. By chance, the VID reaction of **53** was found to proceed at room temperature. Greg set up a reaction and then remembered that he had to be somewhere almost immediately. Since he wanted to monitor the reaction, he decided to leave it at room temperature and then heat it when he got back to the lab. When he returned to his experiment, its appearance had changed and he ran a TLC plate. A new fluorescent spot had appeared, and this turned out to be pyrenophane (**54**). Owing to time constraints, this reaction was done only once. The yield of this reaction probably could be improved substantially, and other VID reactions, especially those leading to some of the [n](2,7)pyrenophanes, probably could go at room temperature, too.

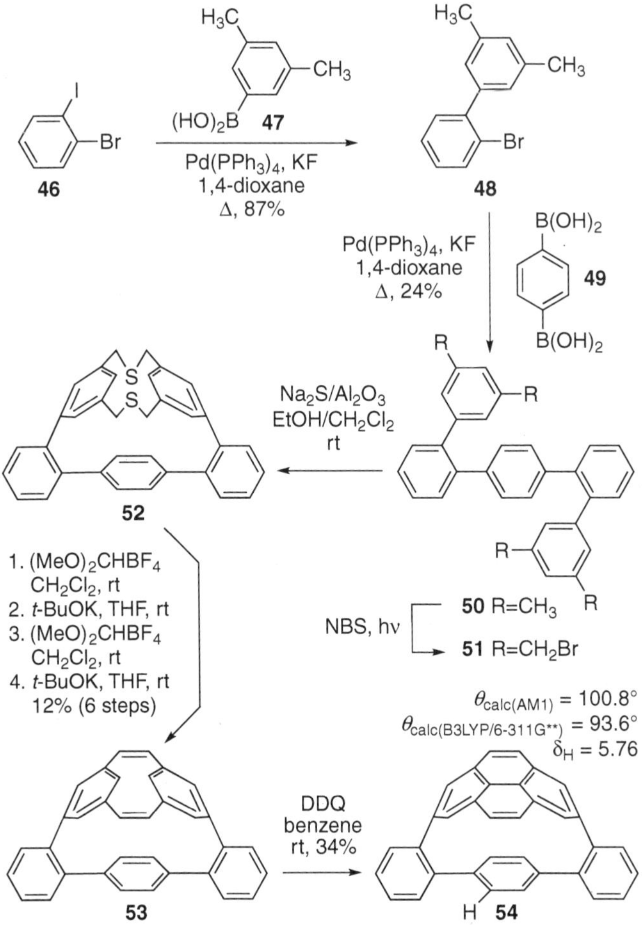

**Scheme 14-9.** Synthesis of 1:2,13:14-dibenzo[2]paracyclo[2](2,7)pyrenophane-1,13-diene (**54**).

Following the completion of Greg's degree, I tried to grow crystals of **54** but failed miserably. The AM1-calculated structure has a θ value of 100.8° (almost the same as **44**), so I expected the actual value to be as much as 10° less than this. Through a very productive collaboration with Michal Cyrański (University of Warsaw), DFT calculations (B3LYP/6-311G$^{**}$) were performed on **54**, and a θ value of 93.6° was predicted. Presumably, the higher-level calculations give more accurate results, but the quick-and-dirty AM1 calculations are still useful first efforts.

A completely different route to the same benzannulated cyclophanediene system was subsequently pursued by a skillful graduate student, Baozhong Zhang.[40] Diynetetraester (**39**), which was a key intermediate in the synthesis of **44** (Scheme 14-9), was reacted with tetraphenylcyclopentadienone (**55**) to afford tetraester **56**, an oligoarylene consisting of 13 benzene units (Scheme 14-10). From this point, the

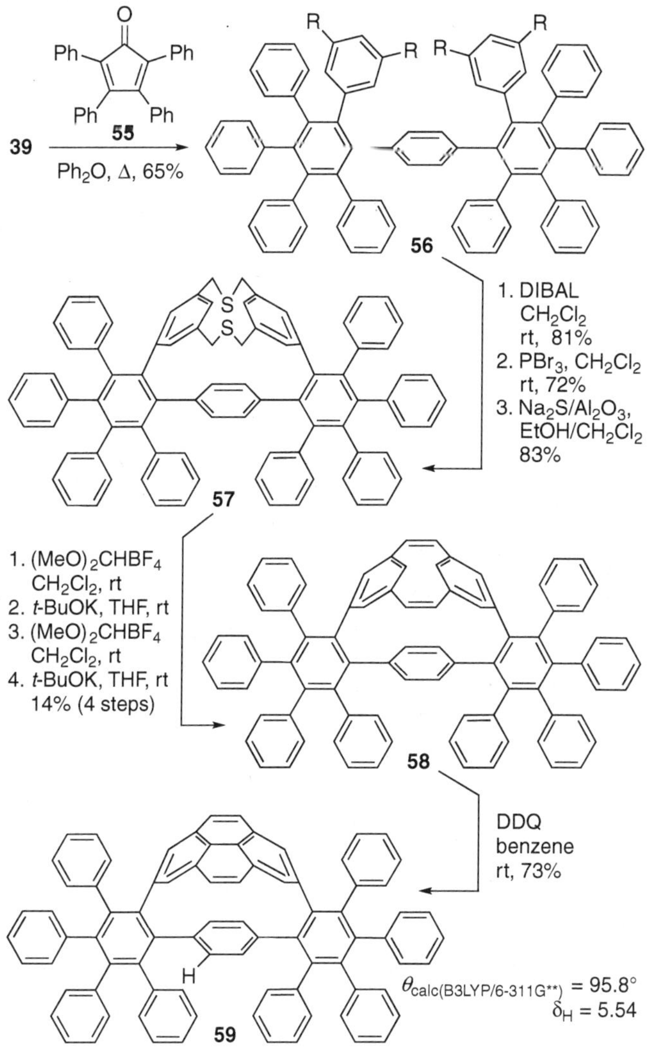

**Scheme 14-10.** Synthesis of an octaphenyl-1:2,13:14-dibenzo[2]paracyclo[2](2,7)pyreno-phane-1,13-diene (**59**).

synthesis followed my group's well-trodden pathway and cyclophanediene (**58**) was obtained in due course. As with several previous pyrenophane syntheses, the Achilles' heel was the Hofmann elimination. A superior method for synthesizing cyclopha-nedienes from dithacyclophanes needs to be discovered.

The VID reaction of **58** was then conducted intentionally at room temperature, and the targeted octaphenylpyrenophane (**59**) was isolated in 73% yield. Again, crystals suitable for an X-ray crystal structure determination could not be grown. Bao tried hard to do this, but I ultimately had to rely on DFT calculations for structural information (thanks again to Michal Cyrański). Although I would have much preferred to have a crystal structure, this did make me feel a little better about my earlier failure to grow crystals of **54**. The calculated (B3LYP/G-311$^{**}$) structure of **59** has a θ value of 95.8°. Similar to **44**, both **54** and **59** have high field resonances in their $^1$H NMR spectra (δ 5.76 and δ 5.54, respectively).

## 14-6. PYRENOPHANE CHEMISTRY

Most, if not all, known pyrenophanes have been the final targets of synthetic studies. Consequently, the chemistry of pyrenophanes has received rather scant attention. Some of the more strained 1,$n$-dioxa[$n$](2,7)pyrenophanes (**34**) and [$n$](2,7)pyreno-phanes (**26**) undergo formal [4 + 2] cycloadditions with reactive dienophiles (Scheme 14-11). In particular, pyrenophanes **26b** and **34b** react with TCNE at room temperature to afford 1 : 1 adducts **60** and **61**, respectively.[31,33] The next-higher homologs **26c** and **34c** are both unreactive toward TCNE at 80°C. 4-Phenyl-1,2,4-triazoline-3,5-dione (PTAD) reacts with **34b** within a few minutes at room temperature to afford 2 : 1 adduct **62b**.[31] In this case, the next-higher homolog **34c** also reacts with PTAD to give a 2 : 1 adduct, but the reaction took 3 h. 1,9-Dioxa[9](2,7)pyrenophane (**34d**) showed only traces of reaction after 5 days. Strain relief is surely a driving force in these cycloadditions. As mentioned in Section 14-5-1, [6](2,7)pyrenophane (**26a**) reacts under the conditions of its formation to give either a [4+2] cycloadduct with DDQ (e.g., **63**), or (less likely, it seems) a self [4+4]-cycloadduct (**64**).[31]

Several attempts to directly brominate 1,10-dioxa[10](2,7)pyrenophane (**34e**) gave mixtures of products from which no pure compound could be isolated. An indirect route to brominated pyrenophanes involving the reaction of cyclophanediene (**6c**) with a Br$_2$ · dioxane complex was also pursued, but this gave low yields of two quite unexpected products, hexabromide (**65**) and bridge-cleaved product (**66**) (Scheme 14-12). Bridge cleavage products (**67b** and **67c**) were also obtained from the reactions of pyrenophanes (**34b** and **34c**) with $t$-BuLi. As with the cycloadditions, strain relief is presumably behind the unusual reactivity.

Work aimed at conducting substitution reactions on the parent [$n$](2,7)pyreno-phanes (**26**) have not yet been performed. Considering that the absence of the oxygen atoms should preclude bridge cleavage, this is something that really needs to be done. More importantly, there are many exciting possibilities that can be derived from the notion of using functionalized pyrenophanes as starting materials in the synthesis of more elaborate systems (see Section 14-9). The downside to this is the amount of effort required to synthesize useful amounts of the [$n$](2,7)pyrenophanes (**26**). As this is the case, the development of shorter and/or more efficient routes to **26** will make work along these lines considerably more attractive.

A very fruitful collaboration with Mordecai Rabinovitz at the Hebrew University of Jerusalem resulted in the discovery of some unusual chemistry on reduction of the [$n$](2,7)pyrenophanes (**26b–e**) with alkali metals (Scheme 14-13).[35,41,42] Treatment of each of the four pyrenophanes with Li metal gives a radical anion, which dimerizes to afford a single diastereoisomer of dianion (**68b–e**). Exactly which diastereoisomer

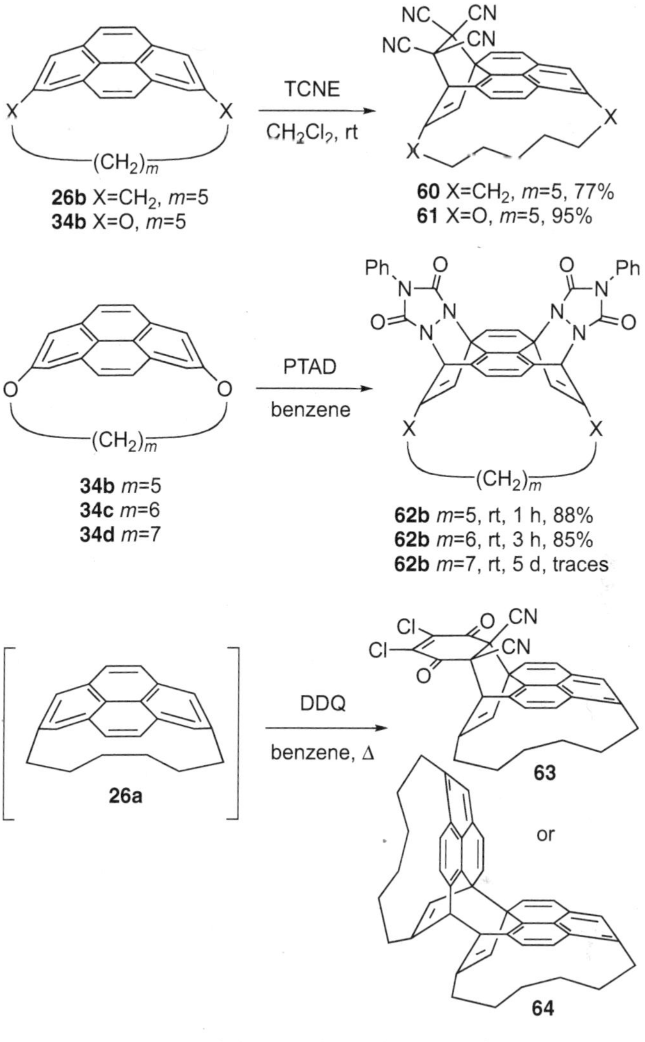

**Scheme 14-11.** Cycloaddition reactions of some [n](2,7)pyrenophanes.

forms was not determined. What happens on further exposure to Li metal depends on the length of the bridge. The least strained dianionic dimers (**68d** and **68e**) behave like pyrene and 2,7-dimethylpyrene and give formally antiaromatic dianions **69d** and **69e**. Dianionic dimer **68c**, which is more strained, does not react further, but the most strained member of the series (**68b**) undergoes a novel rearrangement, in which one of the six-membered rings in the pyrene system is sacrificed for a bicyclo[3.1.0] substructure in dianion **70**. Not only does the bent nature of the bicyclo[3.1.0] system serve to reduce strain; its formation is accompanied by the formation of an aromatic phenalenyl anion. Presumably, **68c** is not strained enough to for allow a sufficient driving force for the

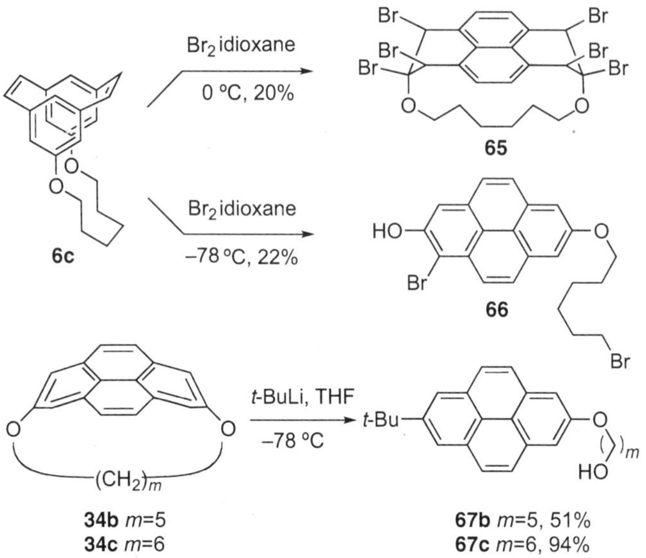

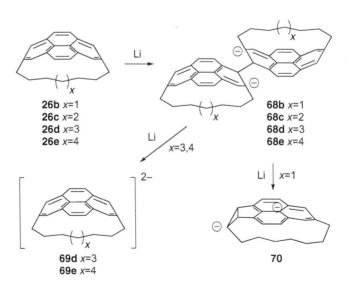

**Scheme 14-12.** Unusual chemistry of some [n](2,7)pyrenophanes.

rearrangement to occur, but is strained enough for the formation of a strained antiaromatic dianion (**69b**) to be unfavorable.

Several interesting or unusual results have come out of the few forays into pyrenophane chemistry. It seems likely that further investigation will produce other surprises that also have their origin in strain relief. However, as mentioned above, the

**Scheme 14-13.** Reduction of [n](2,7)pyrenophanes **26b–e** with Li metal.

development of some unsurprising chemistry is what is really needed if my group is to build grander structures on pyrenophane foundations.

## 14-7. A (2,11)TEROPYRENOPHANE

Having found the VID reaction to be very good at generating nonplanar pyrene systems from two benzene-based units, I naturally wondered whether larger building blocks **71** could be used as starting points for the synthesis of larger polycyclic aromatic systems **73** (Scheme 14-14). For the same general strategy (see Scheme 14-5) to apply, the starting "aromatic boards" (**71**) would require a 1,3-substitution pattern of one functional group (X) at one end of the molecule and a different functional group (Y) at the other end, preferably oriented at 120° to the other two functional groups. Tethering two boards followed by cyclophane formation would afford a cyclophanediene (**72**), subjection of which to a VID reaction should deliver a large [*n*]cyclophane with a long bent arene. Large nonplanar aromatic systems such as these are especially interesting because they resemble segments of single-walled carbon nanotubes. Of course, the same can be said for the nonplanar pyrene systems in the [*n*](2,7)pyrenophanes, or even the benzene rings in the [*n*]paracyclophanes. However, to paraphrase a referee of one of my group's early pyrenophane papers, the link becomes increasingly contrived as the aromatic system becomes smaller.

I considered several aromatic systems for the role of aromatic board and eventually settled on pyrene, mainly because short synthetic routes to appropriately substituted pyrenes appeared to be available. In fact, the routes looked so short (as little as two steps) that I gave the project to an eager undergraduate student, Chad Warford, who got the ball rolling. After Chad's good start, the project sat idle for a couple of years before being taken over by a very enthusiastic graduate student, Brad Merner. It took Brad many months of hard work to navigate his way through a seemingly endless series of dead ends, but he eventually broke through with spectacular results.[43]

Friedel–Crafts alkylation of pyrene (**74**) with dichloride (**75**) accomplished the tethering of two pyrene systems. The use of tertiary alkyl halides was important for

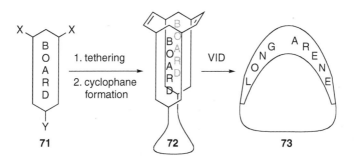

**Scheme 14-14.** Strategy for the synthesis of large [*n*]cyclophanes.

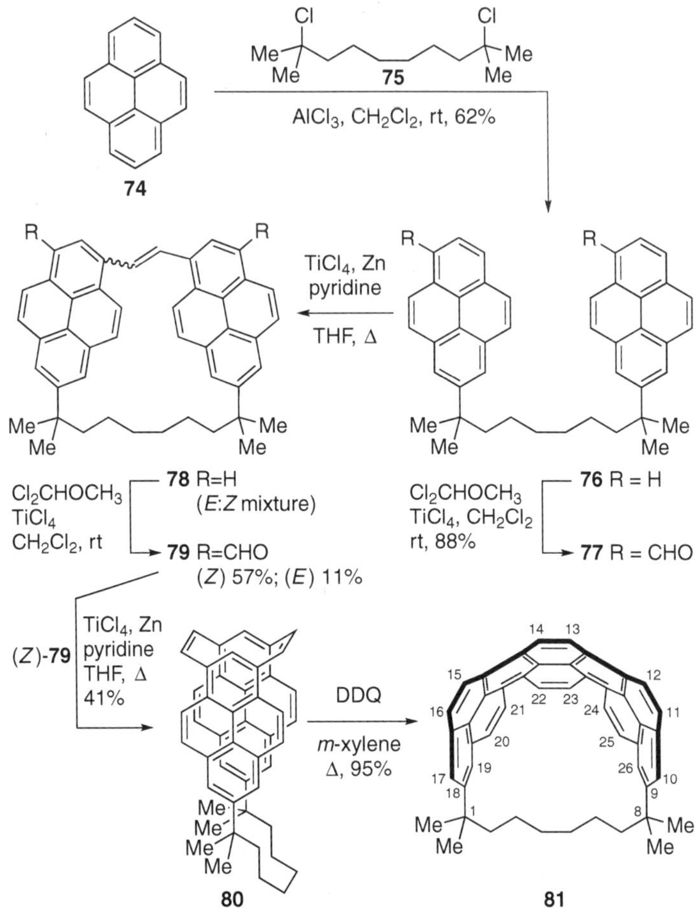

**Scheme 14-15.** Synthesis of 1,1,8,8-tetramethyl[8](2,11)teropyrenophane (**81**).

two reasons: (1) owing to steric hindrance, it caused the alkylation of each pyrene unit to occur at the 2 position instead of the innately more reactive 1 position;[44] and (2) again as a result of steric hindrance, the quaternary benzylic carbon atoms in dipyrenylalkane (**76**) prevented the ensuing Rieche formylation from occurring adjacent to the point of attachment of the tether. The resulting dialdehyde (**77**) underwent an intramolecular McMurry reaction to afford pyrenophane (**78**) as an inseparable mixture of geometric isomers. Another completely regioselective Rieche formylation afforded dialdehyde (**79**), the isomers of which were separable by column chromatography. To my surprise and sheer delight, subjection of (Z)-**79** to McMurry reaction conditions led to the formation of cyclophanediene **80** in 41% yield. This left only the key VID reaction to be attempted, and this proved to be very well behaved. Heating **80** with DDQ in m-xylene induced spot-to-spot conversion into 1,1,8,8-tetramethyl[8](2,11)teropyrenophane **81** (Scheme 14-15).

I have to admit that I rated the chances of success of the second intramolecular McMurry reaction as very low because I knew that the McMurry reaction had a very poor track record in forming small cyclophanes. Brad deserves much credit for attempting this McMurry reaction, anyway. This is a perfect example of why one should be wary about convincing oneself that a reaction is not worth undertaking. Had the conversion of (Z)-**79** to **80** been unsuccessful, Brad would have pressed on through a dithiacyclophane, but the success of the McMurry reaction saved several steps. As a result, the synthesis was just eight steps long (starting from dimethyl suberate) and had an overall yield of 10%. Indeed, the availability of a relatively short synthetic route bodes well for the synthesis of other teropyrenophanes and the likelihood of producing enough material to study their chemistry.

The crystal structure of teropyrenophane **81** was determined using X-ray methods (Figure 14-3). This overall bend ($\theta_{tot}$ = the smallest angle between the two terminal planes of atoms, C9−C10−C26 and C17−C18−C19) of the teropyrene system was found to be 167.6 and 166.4° for the two independent molecules in the unit cell. The individual pyrene units embedded in the teropyrene system had $\theta$ values of 67.5−70.4° for the terminal pyrene units and 92.8−95.9° for the central pyrene unit. The teropyrene system in **81** is comparable to a segment of an (8,8) single-walled carbon nanotube. This is reflected in the virtually parallel arrangement of the

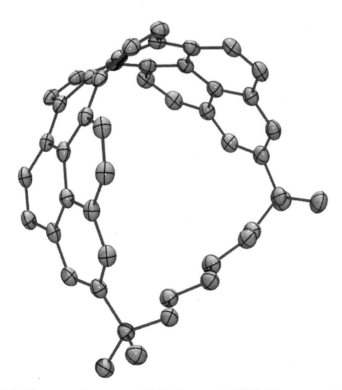

**Figure 14-3.** X-ray crystal structure of 1,1,8,8-tetramethyl[8](2,11)teropyrenophane (**81**).

C(bridgehead)−C(benzylic) bonds (C18−C1 and C9−C8), which correspond to symmetry-related and parallel bonds in an (8,8) single-walled carbon nanotube.

## 14-8. TOWARD AROMATIC BELTS

Having established that the VID reaction can be extended from the generation of nonplanar pyrene systems to the generation of nonplanar teropyrene systems, I was faced with the question as to whether this powerful reaction could be used to construct even larger polycyclic aromatic systems that are bent well beyond 180°. Extending this concept to its limit, I arrived at polycyclic aromatic systems that are wrapped around onto themselves (e.g., **82**) (Figure 14-4). Such systems were first proposed by Vögtle[45–49] prior to the discovery of $C_{60}$ and can be viewed as self-bridged cyclophanes, aromatic belts, or slices of single-walled carbon nanotubes. Whatever the perspective, they are as intriguing as they are synthetically challenging. Every individual benzene ring is bent out of its ideal planar conformation, so any synthetic approach must be capable of dealing with a considerable amount of strain, either incrementally or in one fell swoop. As my group's familiarity with the VID reaction has deepened, so has my conviction that·it will be the key that unlocks the door to aromatic belts, hopefully within a reasonable timeframe (i.e., before I die).

Our strategy for the synthesis of aromatic belts also has three stages (Scheme 14-16):

*Stage 1.* Construction of appropriately substituted aromatic boards from simple starting materials

*Stage 2.* The union of two boards to form a cyclophane, with a [2.2]metacyclo-phanediene unit at both ends

*Stage 3.* Application of a twofold VID reaction to serve as a springboard for the generation of a single polycyclic aromatic hydrocarbon (i.e., an aromatic belt.)

My group's first synthetic efforts were aimed at the $D_{6h}$-symmetric Vögtle belt (**82b**), which has an AM1-calculated θ value of 87.0° for each of the six pyrene substructures (Scheme 14-17). Using chemistry described earlier by Vögtle during his unsuccessful attempts to synthesize the same aromatic belt and its congeners, my

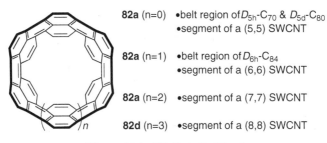

**82a** (n=0)  •belt region of $D_{5h}$-$C_{70}$ & $D_{5d}$-$C_{80}$
•segment of a (5,5) SWCNT

**82a** (n=1)  •belt region of $D_{6h}$-$C_{84}$
•segment of a (6,6) SWCNT

**82a** (n=2)  •segment of a (7,7) SWCNT

**82d** (n=3)  •segment of a (8,8) SWCNT

**Figure 14-4.** Vögtle belts **82a–d**.

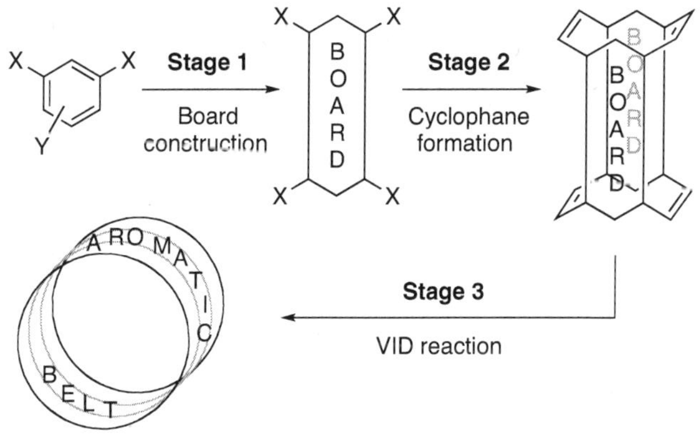

**Scheme 14-16.** Strategy for the synthesis of aromatic belts.

graduate student Rolf Vermeij was able to synthesize tetraester (**83**). My plan was to convert **83** into the partially hydrogenated perropyrene board (**85**) and then bring it forward according to the general strategy (Scheme 14-16). The key step was to be a twofold transannular ring closure of **84** to give **85**. Misumi had reported the synthesis of the parent system (**85** without the esters) using this transformation in the early 1970s, but (much to my consternation) Rolf never had the chance to attempt this conversion

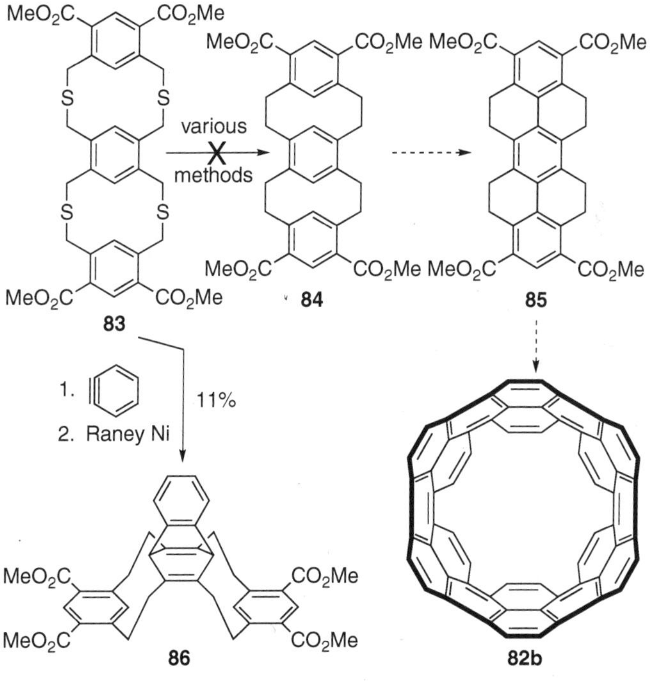

**Scheme 14-17.** Failed approach to Vögtle belt **82b**.

because all attempts to obtain cyclophane (**84**) from tetrathiacyclophane (**83**) went awry. For some reason, the ester groups caused the standard cyclophane chemistry to fail. The best result came from the rarely used benzyne–Stevens rearrangement followed by treatment of the crude product with Raney nickel. This afforded compound **86** in a measly 11% yield. Although the desired ring contraction had been accomplished, the product underwent [4+2] cycloaddition with benzyne. Further work in this area may have been worthwhile, but the project was abandoned at this point.

The lasting lesson from Rolf's initial efforts was that merely adding substituents to a framework of interest can bring with it a variety of synthetic complications, ranging from the failure of standard chemistry to the need to embark on extensive detours. I wrestled with this issue for some time while trying to identify the next *appropriately substituted* aromatic board to set my sights on. Eventually, I opted for the fewest possible carbon atoms at the expense of higher symmetry and decided to whittle off two of the rings in **85**. This left tetrasubstituted dibenzo[*a,h*]anthracene-based board **92** as the next target (Scheme 14-18). Having two of the rings in partially hydrogenated form was part of the plan—it was meant to help maintain solubility throughout the synthesis. Aromatization of these rings

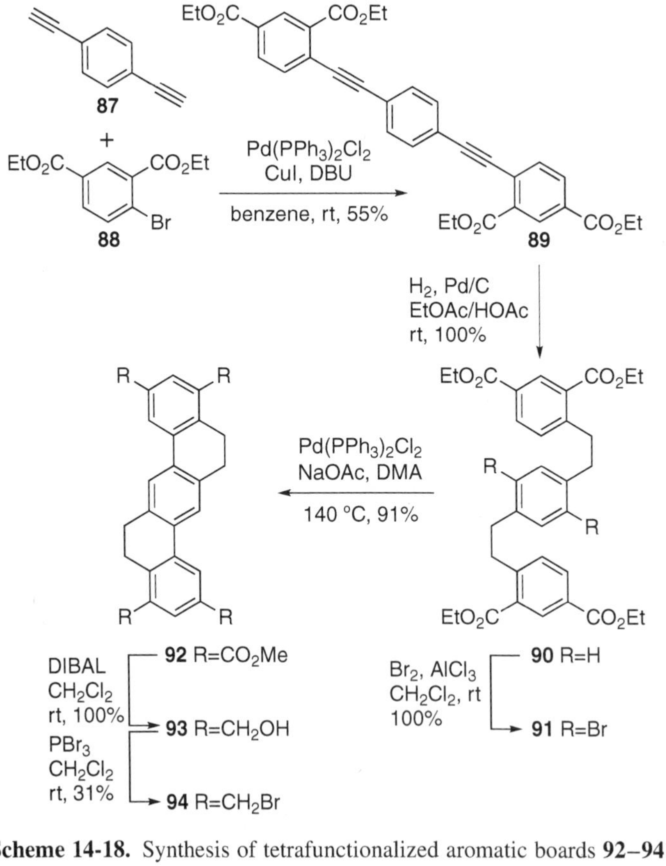

**Scheme 14-18.** Synthesis of tetrafunctionalized aromatic boards **92–94**.

was slated to be accomplished by adding extra DDQ to the VID reaction at the end of the sequence.

Another graduate student, Hao Yu, accepted this project and worked on a synthetic pathway that hinged on a twofold direct arylation reaction (Scheme 14-18).[50] Sonogashira reaction between diyne (**87**) (two steps from 1,4-diiodobenzene) and bromide (**88**) (three steps from *m*-xylene) accomplished the assembly of all of the carbon atoms required for the desired board. Diynetetraester (**89**) was hydrogenated and then dibrominated chemoselectively and regioselectively on the central ring in preparation for the key board-forming reaction. Some experimentation was required, but Hao eventually found conditions that resulted in the efficient formation of the desired board (**92**). However, the celebration was shortlived because solubility problems surfaced at the very next step. Although tetraester (**92**) could be reduced quantitatively to tetraol (**93**), this compound exhibited very low solubility. A small-scale, low-yielding conversion of **93** to tetrabromide (**94**) was achieved, but this compound did not have much better solubility.

Solubilizing groups were clearly needed, and the central ring on the dibenzo[*a,h*]anthracene system was identified as the best place for them. However, this threatened to complicate a synthesis that was already too long for my liking. A major concern was that the twofold direct arylation reaction would now be required to form a hexasubstituted benzene ring. At the time, there was no precedent for this. Although methyl is not a very effective solubilizing group, two of them were included in the next target board (**97**) (Scheme 14-19). The main goal of this exercise was to establish whether the direct arylation reaction could rise to the challenge while minimizing additional synthetic complications.

Tieguang Yao, another graduate student, took over the project from Hao and attempted to synthesize the methylated board (**97**).[51] Bromodiester (**88**) was subjected to Sonogashira reaction with diyne (**95**) (three steps from *p*-xylene), and the resulting diynetetraester was hydrogenated and brominated to afford board precursor **96** in good yield. Fortunately, only a small amount of effort was required before the direct arylation reaction could be conducted in 73% yield. Functional group interconversion then provided tetrabromide (**98**) and tetrathiol (**99**). However, these compounds proved to have insufficient solubility in common organic solvents to be coupled to afford, as I had hoped, tetrathiacyclophanes $C_{2h}$-**100** and $D_2$-**101**. Two isomeric products are expected because the two faces of **98** and **99** are prochiral. The coupling of **98** and **99** involves the addition of a dilute solution of the two components in exactly a 1:1 ratio to a dilute solution of base. Crystallization of one or both of them during the addition occurred, which presumably skewed the ratio. On the bright side, the methyl groups did seem to have a positive effect on the solubility of the boards. Tetraester (**97**) and tetrabromide (**98**) were found to be a little more soluble than their nonmethylated counterparts (**92** and **94**), which augured well for longer alkyl groups.

Tieguang then started to work on the corresponding decyl-substituted boards. As feared, a number of synthetic challenges presented themselves, but he eventually worked out a reliable route to tetrabromide **102** (six steps from **91**) (Scheme 14-20).[52] From this point onward, things got rather messy, and I came to regret having not given higher symmetry the respect it deserves. Reaction of tetrabromide **102** with

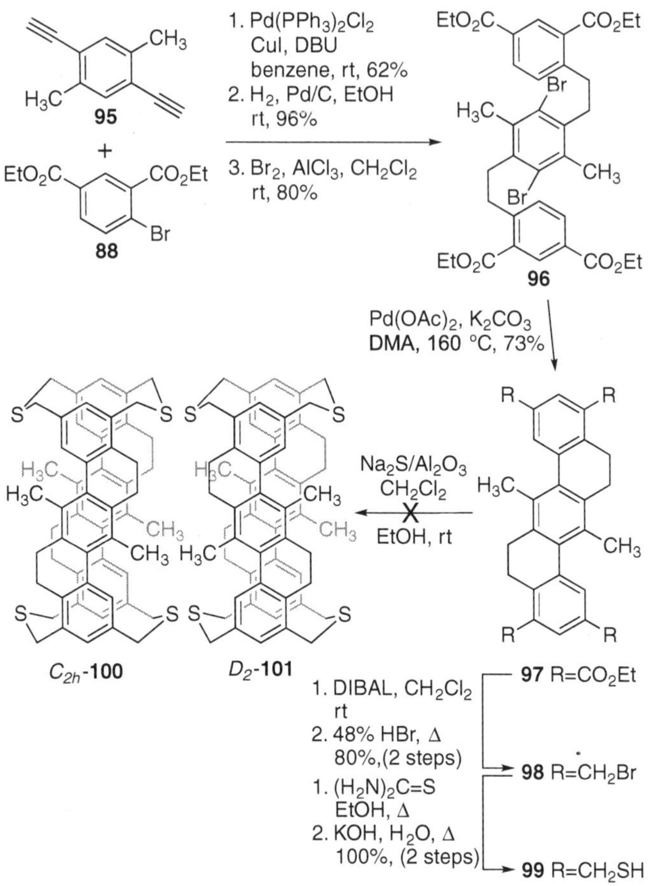

**Scheme 14-19.** Synthesis of tetrafunctionalized aromatic boards **97**–**99**.

Na$_2$S/Al$_2$O$_3$ gave a mixture of what I believe to be tetrathiacyclophanes $C_{2h}$-**103** and $D_2$-**104** in 36% yield. I use cautious language here because the only solid evidence for the formation of these compounds is the mass spectrum. The $^1$H NMR spectrum was not helpful. It was not only complicated; the peaks were also broad. Each of the presumed two isomers, $C_{2h}$-**103** and $D_2$-**104**, would be expected to have a rather complex spectrum on its own. For example, each isomer has 10 unique, highly coupled benzylic protons. Additional complexity may arise from bridge conformational behavior as well as conformational processes (twisting) of the four dihydrophenanthrene units. The situation was essentially the same after an *S*-methylation–Stevens rearrangement–*S*-methylation–Hofmann elimination sequence, which yielded a material that showed the correct mass for cyclophanetetraenes $C_{2h}$-**105** and $D_2$-**106**. With an overall yield of just 4% from **102**, Tieguang did not have much material to work with. Nevertheless, he reacted this material with DDQ a few times on a ~1 mg scale in an attempt to form aromatic belts $C_{2h}$-**107** and $D_2$-**108**. As indicated earlier, excess DDQ was employed to dehydrogenate the boards in addition

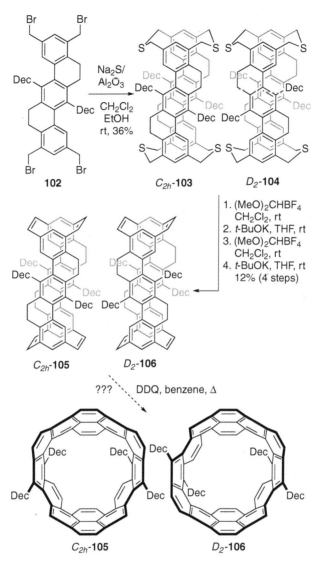

**Scheme 14-20.** Attempted conversion of aromatic board **102** into aromatic belts $C_{2h}$-**105** and $D_2$-**106**.

to participating in the twofold VID reaction. A total of 12 hydrogen atoms should be lost during this transformation, and the mass spectrum of the very small amount of product showed peaks for all stages of dehydrogenation, including a fully dehydrogenated species! The TLC showed the formation of several new, very closely eluting fluorescent compounds. Essentially no separation of this mixture could be achieved.

It is an open question as to whether any of the desired aromatic belts $C_{2h}$-**107** and $D_2$-**108** formed. The mass spectroscopic evidence is tantalizing, but clearly inconclusive. Larger quantities of cyclophanetetraenes $C_{2h}$-**105** and $D_2$-**106** are required to more thoroughly investigate their conversion into aromatic belts, but their synthesis requires 19 steps and suffers very heavy losses during the final few steps. Obviously, more efficient synthetic routes are needed, and my group is working on this topic now. The development of an efficient method for the conversion of dithia[3.3]metacyclophanes into [2.2]metacyclophanedienes would go a long way to addressing this problem, as would the discovery of more concise syntheses of appropriately substituted aromatic boards.

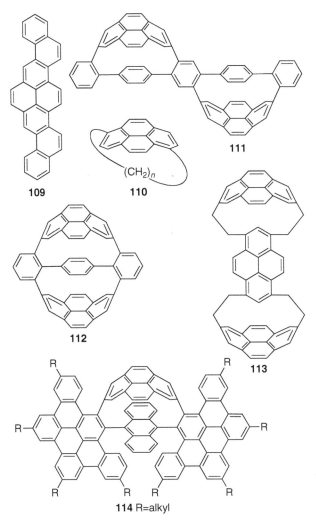

**Figure 14-5.** A selection of targets.

## 14-9. OUTLOOK

Despite the inconclusive end to the synthesis of aromatic belts $C_{2h}$-**107** and $D_2$-**108**, I remain very optimistic that the VID reaction has what it takes to get the job done, if eventually not for these targets, then for other, perhaps larger belts. Accordingly, my group is working on alternative syntheses of appropriately substituted dibenzo[$a,h$]-anthracene-based aromatic boards as well as synthetic approaches to larger boards based on polycyclic aromatic systems such as **109** (Figure 14-5). At the same time, there is still plenty of fun to be had with the original VID reaction. Pyrenophane targets currently under consideration include $C_2$-symmetric [$n$](1,6)pyrenophanes (e.g., **110**), sine-curve-shaped pyrenophanes (e.g., **111**), pyrenophane "burgers" (e.g., **112**), multipyrenophane systems (e.g., **113**), and exotic pyrenophanes coming from the elaboration of smaller pyrenophanes (e.g., **114**). However these projects turn out, I'm sure my group and I will enjoy ourselves, and learn something at the same time. Isn't this why we do chemistry in the first place?

## REFERENCES

1. G. J. Bodwell, T. J. Houghton, J. W. J. Kennedy, M. R. Mannion, *Angew. Chem. Int. Ed. Engl.* **1996**, *35*, 2121–2123.

2. G. J. Bodwell, T. J. Houghton, H. E. Koury, B. Yarlagadda, *Synlett* **1995**, 751–752.

3. R. H. Mitchell, in *Advances in Theoretically Interesting Molecules*, R. P. Thummel (ed.), JAI Press, London, **1989**, Vol. 1, pp. 135–199.

4. R. H. Mitchell, *Eur. J. Org. Chem.* **1999**, 2695–2703.

5. G. J. Bodwell, J. N. Bridson, S.-L. Chen, R. A. Poirier, *J. Am. Chem. Soc.* **2001**, *123*, 4704–4708.

6. G. J. Bodwell, J. N. Bridson, S.-L. Chen, J. Li, *Eur. J. Org. Chem.* **2002**, 243–249.

7. R. H. Mitchell, V. Boekelheide, *J. Am. Chem. Soc.* **1970**, *92*, 3510–3512.

8. S. M Rosenfeld, K. A. Choe, in *Cyclophanes*, P. M. Keehn, S. M. Rosenfeld (eds.), Academic Press, London, **1983**; Vol. 1, pp. 314–325.

9. I. Fleming, *Frontier Orbitals and Organic Chemical Reactions*, Wiley: New York, **1982**.

10. T. M. Krygowski, M. K. Cyrański, *Chem. Rev.* **2001**, *101*, 1385–1419.

11. T. M. Krygowski, *J. Chem. Inf. Comput. Sci.* **1993**, *33*, 70–78.

12. T. M. Krygowski, M. Cyrański, *Tetrahedron* **1996**, *52*, 1713–1722.

13. T. M. Krygowski, M. Cyrański, in *Advances in Molecular Structure Research*, M. Hargittai, I. Hargittai (eds.), JAI Press, London, **1997**.

14. P. v. R. Schleyer, C. Maerker, A. Dransfeld, H. Jiao, N. J. R. van Eikema Hommes, *J. Am. Chem. Soc.* **1996**, *118*, 6317–6318.

15. Z. Chen, C. S. Wannere, C. Corminboeuf, R. Puchta, P. v. R. Schleyer, *Chem. Rev.* **2005**, *105*, 3842–3888.

16. C. Corminboeuf, R. Puchta, P. v. R. Schleyer, *Org. Lett.* **2006**, *8*, 863–866.

17. V. I. Minkin, M. N. Glukhovtsev, B. Ya. Simkin, *Aromaticity and Antiaromaticity. Electronic and Structural Aspects*, Wiley, New York, **1994**.

18. P. v. R. Schleyer, F. Pühlhofer *Org. Lett.* **2002**, *4*, 2873–2876.

19. C. S. Wannere, D. Moran, N. L. Allinger, B. A. Hess, L. J. Schaad, P. v. R. Schleyer, *Org. Lett.* **2003**, *5*, 2983–2986.

20. C. S. Wannere, P. v. R. Schleyer, *Org. Lett.* **2003**, *5*, 865–868.

21. M. K. Cyrański, *Chem. Rev.* **2005**, *105*, 3773–3811.

22. P. v. R. Schleyer, Y. Mo, *Chem. Eur. J.* **2006**, *12*, 2009–2020.

23. M. D. Wodrich, C. S. Wannere, Y. Mo, P. D. Jarowski, K. N. Houk, P. v. R. Schleyer, *Chem. Eur. J.* **2007**, *13*, 7731–7744.

24. G. J. Bodwell, J. N. Bridson, M. K. Cyrański, J. W. J. Kennedy, T. M. Krygowski, M. R. Mannion, D. O. Miller, *J. Org. Chem.* **2003**, *68*, 2089–2098.

25. M. A. Dobrowolski, M. K. Cyrański, B. L. Merner, G. J. Bodwell, J. I. Wu, P. v. R. Schleyer, *J. Org. Chem.* **2008**, *73*, 8001–8009.

26. J. I. Wu, M. A. Dobrowolski, B. L. Merner, G. J. Bodwell, Y. Mo, P. v. R. Schleyer, *Mol. Phys.* **2009**, *107*, 1177–1186.

27. C. H. Suresh, N. Koga, *Chem. Phys. Lett.* **2006**, *419*, 550–556.

28. T. Sato, M. Wakabayashi, Y. Okamura, T. Amada, K. Hata, *Bull. Chem. Soc. Jpn.* **1967**, *40*, 2363–2365.

29. F. Vögtle, A. Ostrowicki, B. Begemann, M. Jansen, M. Nieger, E. Niecke, *Chem. Ber.* **1990**, *123*, 169–176.

30. T. Kawashima, T. Otsubo, Y. Sakata, S. Misumi, *Tetrahedron Lett.* **1978**, 5115–5118.

31. G. J. Bodwell, J. J. Fleming, M. R. Mannion, D. O. Miller, *J. Org. Chem.* **2000**, *65*, 5360–5370.

32. G. J. Bodwell, J. N. Bridson, T. J. Houghton, J. W. J. Kennedy, M. R. Mannion, *Angew. Chem. Int. Ed. Engl.* **1996**, *35*, 1320–1321.

33. G. J. Bodwell, J. N. Bridson, T. J. Houghton, J. W. J. Kennedy, M. R. Mannion, *Chem. Eur. J.* **1999**, *5*, 1823–1827.

34. G. J. Bodwell, J. J. Fleming, D. O. Miller, *Tetrahedron* **2001**, *57*, 3577–3585.

35. I. Aprahamian, G. J. Bodwell, J. J. Fleming, G. P. Manning, M. R. Mannion, B. L. Merner, T. Sheradsky, R. J. Vermeij, M. Rabinovitz, *J. Am. Chem. Soc.* **2004**, *126*, 6765–6775.

36. P. M. Keehn, in *Cyclophanes*, P. M. Keehn, S. M. Rosenfeld, (eds.), Academic Press, London, **1983**; Vol. 1, pp. 69–238.

37. G. J. Bodwell, D. O. Miller, R. J. Vermeij, *Org. Lett.* **2001**, *3*, 2093–2096.

38. G. J. Bodwell, R. J. Vermeij, *Aust. J. Chem.* **2010**, *63*, 1703–1716.

39. C. Grave, A. D. Schlüter, *Eur. J. Org. Chem.* **2002**, 3075–3098.

40. B. Zhang, G. P. Manning, M. A. Dobrowolski, M. K. Cyrański, G. J. Bodwell, *Org. Lett.* **2008**, *10*, 273–276.

41. I. Aprahamian, G. J. Bodwell, J. J. Fleming, G. P. Manning, M. R. Mannion, T. Sheradsky, R. J. Vermeij, M. Rabinovitz, *J. Am. Chem. Soc.* **2003**, *125*, 1720–1721.

42. I. Aprahamian, G. J. Bodwell, J. J. Fleming, G. P. Manning, M. R. Mannion, T. Sheradsky, R. J. Vermeij, M. Rabinovitz, *Angew. Chem. Int. Ed. Engl.* **2003**, *42*, 2547–2550.

43. B. L. Merner, L. N. Dawe, G. J. Bodwell, *Angew. Chem. Int. Ed. Engl.* **2009**, *48*, 5487–5491.

44. Y. Miura, E. Yamano, A. Tanaka, J. Yamauchi, *J. Org. Chem.* **1994**, *59*, 3294–3300.

45. F. Vögtle, *Top. Curr. Chem.* **1983**, *115*, 157–159.

46. F. Vögtle, A. Schröder, D. Karbach, *Angew. Chem. Int. Ed. Engl.* **1991**, *30*, 575–577.

47. A. Schröder, D. Karbach, F. Vögtle, *Chem. Ber.* **1992**, *125*, 1881–1887.

48. W. Josten, D. Karbach, M. Nieger, F. Vögtle, K. Hägele, M. Svoboda, M. Przybylski, *Chem. Ber.* **1994**, *127*, 767–777.

49. W. Josten, S. Neumann, F. Vögtle, K. Hägele, M. Przybylski, F. Beer, K. Müllen, *Chem. Ber.* **1994**, *127*, 2089–2096.

50. T. Yao, H. Yu, R. J. Vermeij, G. J. Bodwell, *Pure Appl. Chem.* **2008**, *80*, 535–548.

51. T. Yao, G. J. Bodwell unpublished results.

52. T. Yao, Y. Yang, G. J. Bodwell unpublished results.

# INDEX